POLYMER
BLENDS
AND ALLOYS

PLASTICS ENGINEERING

Founding Editor

Donald E. Hudgin

Professor
Clemson University
Clemson, South Carolina

POLYMER BLENDS AND ALLOYS

edited by

GABRIEL O. SHONAIKE
Himeji Institute of Technology
Himeji, Hyogo, Japan

GEORGE P. SIMON
Monash University
Clayton, Victoria, Australia

CRC Press
Taylor & Francis Group
Boca Raton London New York

CRC Press is an imprint of the
Taylor & Francis Group, an **informa** business

CRC Press
Taylor & Francis Group
6000 Broken Sound Parkway NW, Suite 300
Boca Raton, FL 33487-2742

First issued in paperback 2019

© 2006 by Taylor & Francis Group, LLC
CRC Press is an imprint of Taylor & Francis Group, an Informa business

No claim to original U.S. Government works

ISBN-13: 978-0-8247-1980-7 (hbk)
ISBN-13: 978-0-367-39974-0 (pbk)

Visit the Taylor & Francis Web site at
http://www.taylorandfrancis.com

and the CRC Press Web site at
http://www.crcpress.com

To our wives and children:

Dayo, Lola, Toyin, Maryrose, and Vincent

Preface

Polymer blends are a key component of current polymer research and technology. This is fed in part by the ease of production of new materials by mixing and the diversity of properties that result. From a scientific standpoint, however, an increasing battery of characterization techniques have also led to an increased understanding of the ways polymers mix, their fundamental interactions, and how these interactions affect their final properties. This link between molecular interactions and physical and engineering properties continues to fascinate because of the scientific insights it produces, and because it recognizes that, in a world of increasingly tight economic circumstances, some outcomes of research are achieved by careful design while others are the result of serendipity.

Because polymer research remains a growing field, we decided to assemble these chapters on polymer blends from laboratories around the world. In this way, many aspects of polymer blends research are represented.

The book is divided into four parts: Compatibilization and Miscibility; Characterization; Morphology; and Recent Developments (although, of course, recent developments are included in all of the parts). The range of topics covered includes synthesis, mechanical properties, computer simulations, new techniques of characterization, reactive blending, and toughening mechanisms, among others. The contributors have made an effort to explain their respective topics to aid readers in "crossing over" from their current areas of expertise into others that may be useful to them.

We hope that this comprehensive book will be a useful reference for academic researchers as well as engineers in polymer and related industries.

Gabriel O. Shonaike
George P. Simon

Contents

PART II   CHARACTERIZATION

PART III   MORPHOLOGY

Contents

Contributors

Stuart R. Andrews, Ph.D.   Department of Chemistry, University of Swansea, Singleton Park, Swansea, Wales

Valeria Arrighi, Ph.D.   Department of Chemistry, Heriot-Watt University, Riccarton, Edinburgh, Scotland

Philippe Béguelin, Ph.D.   Research Associate, Materials Department, Polymer Laboratory, École Polytechnique Fédérale de Lausanne, Lausanne, Switzerland

Witold Brostow, D.Sc., F.R.S.C.   Professor, Department of Materials Science, University of North Texas, Denton, Texas

Chi-Ming Chan, Ph.D.   Professor, Department of Chemical Engineering, and Director, Advanced Engineering Materials Facility, The Hong Kong University of Science and Technology, Clear Water Bay, Kowloon, Hong Kong

Chao-Hsun Chen, Ph.D.   Professor, Department of Applied Mechanics, Institute of Applied Mechanics, National Taiwan University, Taipei, Taiwan, Republic of China

Wen-Yen Chiang, Ph.D.   Professor, Department of Chemical Engineering, Tatung Institute of Technology, Taipei, Taiwan, Republic of China

John M. G. Cowie, B.Sc., Ph.D., D.Sc., C.Chem., F.R.S.C., F.R.S.E.   Professor, Department of Chemistry, Heriot-Watt University, Riccarton, Edinburgh, Scotland

Rudolph D. Deanin, A.B., M.S., Ph.D.   Professor, Department of Plastics Engineering, University of Massachusetts—Lowell, Lowell, Massachusetts

Thomas S. Ellis, B.Sc., Ph.D.*   Staff Research Scientist, Polymers Department, General Motors Research and Development Center, Warren, Michigan

Andy A. Goodwin, B.Sc., Ph.D.†   Senior Lecturer, Department of Materials Engineering, Monash University, Clayton, Victoria, Australia

Roberto Greco, Dr.   Senior Scientist, Institute of Research and Technology of Plastic Materials, National Research Council of Italy, Arco Felice, Naples, Italy

Qipeng Guo   Professor, Department of Polymer Science and Engineering, University of Science and Technology of China, Hefei, People's Republic of China

Kuo-Huang Hsieh, Ph.D.   Professor, Department of Chemical Engineering, National Taiwan University, Taipei, Taiwan, Republic of China

Tsung-Tang Hsieh, M.Sc.   Department of Chemical Engineering, Monash University, Clayton, Victoria, Australia

Chi-Yuan Huang, Ph.D.   Associate Professor, Department of Materials Engineering, Tatung Institute of Technology, Taipei, Taiwan, Republic of China

Takashi Inoue, Dr.Eng.   Professor, Department of Organic and Polymeric Materials, Tokyo Institute of Technology, Tokyo, Japan

Umaru Semo Ishiaku, B.Ed., M.Sc., Ph.D.   Polymer Technology Section, School of Industrial Technology, Universiti Sains Malaysia, Penang, Malaysia

Hanafi Ismail, Ph.D.   Polymer Technology Section, School of Industrial Technology, Universiti Sains Malaysia, Penang, Malaysia

---

Current affiliations:
* Delphi Automotive Systems Research and Development Center, Warren, Michigan.
† Boral Plasterboard, Port Melbourne, Victoria, Australia.

H. H. Kausch, Ph.D.   Professor, Polymer Laboratory, Federal Institute of Technology, École Polytechnique Fédérale de Lausanne, Lausanne, Switzerland

H. Kihara, Dr.Eng.   Petrochemical Research Laboratory, Sumitomo Chemical Company Ltd., Sodegaura-shi, Chiba, Japan

József Karger-Kocsis, Ph.D., D.Sc.   Professor, Institute for Composite Materials Ltd., University of Kaiserslautern, Kaiserslautern, Germany

Yiu-Wing Mai, Ph.D.   Professor, Centre for Advanced Materials Technology (CAMT), Department of Mechanical and Mechatronic Engineering, University of Sydney, Sydney, New South Wales, Australia

Margaret A. Manion, B.S., M.L.S.   Librarian, Lydon Library, University of Massachusetts—Lowell, Lowell, Massachusetts

Takaaki Matsuoka, Dr.Eng.   Principal Researcher and Manager, Computational Materials Engineering Laboratory, Toyota Central Research and Development Laboratories, Inc., Nagakute, Aichi, Japan

S. Mitsui, M.Eng.   Petrochemical Research Laboratory, Sumitomo Chemical Company Ltd., Sodegaura-shi, Chiba, Japan

H. Miyagi, M.Eng.   Petrochemical Research Laboratory, Sumitomo Chemical Company Ltd., Sodegaura-shi, Chiba, Japan

Zainal Arifin Mohd Ishak, Ph.D.   Associate Professor, Polymer Technology Section, School of Industrial Technology, Universiti Sains Malaysia, Penang, Malaysia

René Muller, Ph.D.   Professor, Polymeric Materials and Processes Laboratory, European Engineering School for Chemistry, Polymers and Materials (ECPM), Strasbourg, France

Y. Okamoto, Dr.Eng.   Manager, Petrochemical Research Laboratory, Sumitomo Chemical Company Ltd., Sodegaura-shi, Chiba, Japan

Toshiaki Ougizawa, Dr.Eng.   Associate Professor, Department of Organic and Polymeric Materials, Tokyo Institute of Technology, Tokyo, Japan

Christopher J. G. Plummer, Ph.D.   Research Associate, Materials Department, École Polytechnique Fédérale de Lausanne, Lausanne, Switzerland

H. J. Din Rozman, B.App.Sc., M.Sc., Ph.D.   Lecturer, School of Industrial Technology, Universiti Sains Malaysia, Penang, Malaysia

Gabriel O. Shonaike, M.Sc., Ph.D.   Associate Professor, Department of Chemical Engineering, Himeji Institute of Technology, Himeji, Hyogo, Japan

George P. Simon, Ph.D.   Reader, Department of Materials Engineering, Monash University, Clayton, Victoria, Australia

Ph. Teyssié, Ph.D.   Professor Emeritus, Center for Educational and Research on Macromolecules (CERM), University of Liège, Sart-Tilman, Liège, Belgium

Carlos Tiu, Ph.D.   Reader, Department of Chemical Engineering, Monash University, Clayton, Victoria, Australia

Yasuhisa Tsukahara, Dr.Eng.   Associate Professor, Department of Chemistry and Materials Technology, School of Engineering and Design, Kyoto Institute of Technology, Kyoto, Japan

Wan Rosli Wan Daud, B.Sc., M.Sc., Ph.D.   Associate Professor, School of Industrial Technology, Universiti Sains Malaysia, Penang, Malaysia

Martin Weber, Ph.D.   Senior Scientist, Polymer Research Laboratory, BASF AG, Ludwigshafen, Germany

Graham Williams, Ph.D.   Professor, Department of Chemistry, University of Swansea, Singleton Park, Swansea, Wales

Barbara A. Wood, Ph.D.   Research Associate, Central Research and Development, DuPont Co., Wilmington, Delaware

Jingshen Wu, Ph.D.   Assistant Professor, Department of Mechanical Engineering, The Hong Kong University of Science and Technology, Clear Water Bay, Kowloon, Hong Kong

Marcus D. Zipper, Ph.D.*   Research Fellow, Department of Materials Engineering, Cooperative Research Centre for Polymers, Monash University, Clayton, Victoria, Australia

----

* Current affiliation: Product Development Engineer and Senior Chemist, Hunstman Chemical Company Australia Pty Ltd., West Footscray, Victoria, Australia.

POLYMER
BLENDS
AND ALLOYS

1

Compatibilization of Polymer Blends*

Rudolph D. Deanin and Margaret A. Manion
University of Massachusetts—Lowell, Lowell, Massachusetts

I. INTRODUCTION

A. History and Commercial Importance of Polymer Blends

Polymer blending has a long commercial history in the rubber, coatings, and adhesives industries, and it entered the plastics industry a half century ago. Since then it has been growing so rapidly that it is becoming an increasingly important portion of the entire plastics industry, 20–40% of the total plastics market by various estimates. With growing understanding and command of the science and engineering involved, it may well continue to offer increasing contributions to plastics and the other polymer industries.

B. Miscibility vs. Compatibility

When plastics processors first tried to blend polymers with each other, they were shocked to find that most pairs of plastics were immiscible and had very poor properties. Plastics chemists reasoned that they must develop miscible homogeneous

---

* This chapter is abstracted from the book *Compatibilization of Polymer Blends* to be published by Technomic Publishing Company.

systems to achieve useful properties. They tended to use the terms *miscibility* and *compatibility* interchangeably. They did discover a number of polymer pairs that were completely miscible to give a homogeneous single phase, with properties proportional to the ratio of the two polymers in the blend, and several of these were of commercial importance. They also illuminated the thermodynamics of polymer miscibility and immiscibility and delineated the factors that favored miscibility or immiscibility. But meanwhile, practical plastics technologists were developing a much larger number of polymer blends that were immiscible but very useful, combining some of the best practical properties of each polymer in the blend; they tended to use the term *compatible* for these. Unfortunately most of the authors in the field continued to use the terms *miscible* and *compatible* rather casually and indiscriminately, so that much of the literature is ambiguous or confusing.

In the present review, the term *miscible* will be used to describe polymer blends that have theoretical thermodynamic miscibility down to the segmental level; the term *compatible* will be used to describe polymer blends that have useful practical properties, regardless of whether they are theoretically miscible or immiscible.

C. Factors in Miscibility and Immiscibility

A number of specific features may contribute to miscibility/immiscibility of polymer blends (1). These may be listed in order of commercial importance:

1. Polarity

Polymers that are similar in structure or, more generally, similar in polarity are less likely to repel each other and more likely to form miscible blends (2,3). Diverging polarities generally produce immiscibility.

2. Specific Group Attraction

Polymers that are drawn to each other by hydrogen bonding, acid–base, charge-transfer, ion-dipole, donor–acceptor adducts, or transition metal complexes are less common, but when such attractions occur they are very likely to produce miscibility (1,4–6).

3. Molecular Weight

Lower molecular weight permits greater randomization on mixing and therefore greater gain of entropy, which favors miscibility (7). More surprisingly, polymers of similar molecular weights are more miscible, while polymers of very different molecular weights may be immiscible, even if they both have the same composition.

4. Ratio

Even though two polymers appear immiscible at a fairly equal ratio, it is quite possible that a small amount of one polymer may be soluble in a large amount of the other polymer, as understood in conventional phase rule. This consideration is extremely important in natural compatibility, as will be explained later.

5. Crystallinity

When a polymer crystallizes, it already forms a two-phase system, with important consequences for practical compatibility. In a polymer blend, when a polymer crystallizes, this adds another phase to the system. If both polymers in a blend crystallize, they will usually form two separate crystalline phases; it is quite rare for the two polymers to cocrystallize in a single crystalline phase (12).

D. Practical Compatibility in Polymer Blends

Polymer blends have commercial importance because they offer properties, or a balance of properties, not available in a single polymer. These properties depend very much on the microstructure of the blend.

1. Homogeneous Blends

When two polymers are completely miscible down to the segmental level, they form a single homogeneous phase, and properties are generally proportional to the ratio of the two polymers in the blend. This gives the compounder quick and economical control over the balance of properties for different applications. Major examples are in the coatings, adhesives, and rubber industries. In the plastics industry, the major example is blends of polyphenylene ether with polystyrene (8,9).

2. Phase Rule, Morphology, and Phase Stability

When two polymers are immiscible, phase rule explains quantitatively the extent to which they separate and the extent to which each phase is actually—not pure polymer A and pure polymer B, but rather—a solution of B in A and a solution of A in B. It is extremely important to remember this distinction in trying to understand the practical properties of such two-phase systems. Generally the major phase will form the continuous matrix and control most properties, while the minor phase will form dispersed microdomains and contribute certain specialized properties to the blend. Another factor is rheology: the less viscous phase tends to form the continuous matrix (even if it is present in rather minor amount!), while the more viscous phase tends to form the dispersed domains.

The structure of the dispersed domains is referred to as the morphology. The simplest shape of a dispersed domain, which is trying to minimize its surface energy, is spherical, and most dispersed domains appear in this form. Generally, increasing attraction between phases tends to decrease the size of the spheres and increase practical compatibility. With increasing concentration of the minor phase, the dispersed domains may tend to become rodlike; and at fairly equal concentrations, the two phases may become lamellar. Another factor in morphology is shear flow during melt processing, which tends to elongate spherical domains into plate-like or fibrillar form. Such nonspherical morphology is generally believed to have important effects on practical properties.

Kinetics of phase separation and morphology formation are slowed by the entanglement of large polymer molecules and the high viscosity of such large entangled molecules. While mechanical shear, temperature, and chemical environment may change the size and shape of the dispersed domains, the system tends to revert to thermodynamic equilibrium over a period of time. This creates problems with the stability of polyblend morphology and properties.

3. Interface and Interphase

If two polymers repel each other so much that they separate into phases, the interface could be very weak and fail under any kind of mechanical or thermal stress (3,8). It is true that most polymer pairs form blends that fail under stress, have generally poor properties, and are therefore defined as incompatible. Compatible commercial polyblends must therefore have a strong interface that permits successful stress transfer across it. Many theorists believe that a thin sharp *interface* creates steep property gradients, which fail under stress; whereas compatible polyblends must have a broad gradual transitional region, which modulates property gradients and has the ability to resist stress (3,7). They refer to this transitional region as the *interphase*. There is a growing belief that this distinction between a sharp weak interface and a broad strong interphase explains the difference between incompatible and compatible polymer blends.

E. Natural Compatibility in Two-Phase Blends

There are some polymer pairs that naturally form compatible two-phase blends even without human intervention. When this occurs, it may be possible to apply existing thermodynamics, morphology, and interface theory to explain their compatibility; and conversely, analysis of such systems helps to increase our understanding of the bases of compatibility and guides our work in producing new man-made compatible polyblends. Major features that contribute to natural compatibility include covalent bonding, partial miscibility, and interfacial tension and adhesion.

1. Covalent Bonding

When the two phases are joined by permanent primary covalent bonds across the interface, this gives the greatest interfacial stability and ability to transmit stress and resist failure. Covalent bonds across the interface can occur in several ways.

Crystallinity

When a polymer crystallizes, it forms two phases, crystalline and amorphous. Many or all molecules meander from one phase to the other, tying the two phases together very strongly. This produces strong interfaces and good stress transfer across the interface, and it accounts for their ability to combine some of the best properties of each phase. While crystalline polymers are not normally considered to be polymer blends, they have much to teach us about two-phase systems, the nature of the interface, and its effect on practical properties.

Block and Graft Copolymers

Most polymer structures are immiscible with each other and form two-phase systems. When the two structures are joined together in a block or graft copolymer, they still separate into two phases; but since each molecule extends across the interface, the interface is composed of a great number of primary covalent bonds. Thus it is much easier to produce compatibility in a block or graft copolymer than it is in a mixture of two separate polymers. While such copolymers are not strictly polymer blends, their morphology and properties all appear to obey the same general laws, and their use is intimately involved in the commercial development of compatible multiphase polymer systems in general. The development of thermoplastic elastomers has been based primarily on block copolymers, and most commercial polymer blends depend on a fraction of block or graft copolymer to strengthen the interface and produce practical compatibility.

2. Partial Miscibility

Most polymer pairs are immiscible and separate into two-phase systems. If they are very immiscible, the domain size is coarse, irregular, and unstable, and the interface is sharp and weak, giving poor properties and practical incompatibility. When a two-phase blend has naturally good properties and practical compatibility, the reason is usually to be sought in partial miscibility. As we learn from phase rule, separation into two phases does not produce pure A and pure B, but rather a solution of B in A and a solution of A in B. Thinking more locally about the *interface* between the two phases, it is probable that, rather than a sharp transition from pure A to pure B, there is a broader gradual change in concentration from the high-A phase to the high-B phase, an interphase, producing a gradual modulation in properties (4). Such an *interphase* is better able to resist stress and deliver useful properties. In many naturally compatible polymer blends, such partial miscibility is the most likely explanation for their good properties.

3. Interfacial Tension and Adhesion

When two polymers repel each other and form a two-phase blend, the interfacial tension is high, giving a coarse and unstable structure, and adhesion between the two phases is low, giving poor stress transfer across the interface (1,10). Whereas covalent bonding and partial miscibility are clear microstructural mechanisms to explain practical compatibility, interfacial energy and adhesion are basic micro-properties that may help to explain practical macroproperties. They may simply be different theoretical ways of analyzing and understanding practical compati-bility. It is also possible that other factors are involved in reducing interfacial ten-sion and increasing interfacial adhesion; and some theoretical studies do allude to possible mechanisms that may be involved.

4. Optimum Degree for Different Properties

Between the extremes of complete miscibility and absolute immiscibility, there is a continuous spectrum of partial miscibility and intermediate attractions between polymers in their blends. Individual practical properties respond differently to their position in this spectrum. Some are optimum at complete miscibility, some at high miscibility, some at intermediate miscibility, some at low miscibility, and some at complete immiscibility (2). For a product that requires only one critical property, the choice is obvious. Most products require a balance of important properties, which respond differently to different degrees of miscibility and interfacial attrac-tion. Thus the choice is usually a compromise, and each individual product requires a different level of compromise to optimize its balance of properties. In most cases, this requires human intervention to produce the optimum level of phase separation, morphology, and interfacial attraction (8). This we call *compatibilization*.

F. Compatibilization

Polymer chemists and engineers can produce and control compatibility by physi-cal processes, physical additives, and reactive processes. For two-stage control, they can also premodify polymers to facilitate either physical processes or reac-tive processes.

1. Physical Processes

When polymers crystallize, processors can usually control crystal size and orien-tation by nucleation, thermal history, and mechanical stretching. When polymer blends separate into phases, thermal history, mixing energy, and shear flow can have major effects on domain size and structure. If we then freeze in morphology at room temperature, it may be stable enough to have major effects on practical properties.

2. Physical Additives

Compatibility is often improved by adding one or more ingredients, without in-volving any chemical reaction. The added ingredient is usually a third polymer, but it may also be monomeric or particulate in nature.

3. Polymer Modification for Physical Compatibilization

In many cases, the polymer chemist must first modify one or both polymers to prepare them for later physical blending. Such chemical modification may be performed during polymerization or may be applied to the base polymer as a postpolymerization reaction.

4. Reactive Blending

Since primary covalent bonds may be the most effective way to strengthen the interface, most current research is concerned with chemical reactions that the processor can accomplish during the blending process. These may be reactions directly between the two polymers in the blend, or they may be produced by adding a third polymer that can react with one or both of the primary polymers. In some cases, the reaction may be produced by adding a monomeric ingredient, which can serve as catalyst, initiator, or coreactant in the reaction between the two polymers.

5. Polymer Modification for Reactive Blending

In an increasing number of systems, the commercial polymers that will be used in a blend do not have the reactivity necessary to compatibilize them directly. In such cases, the polymer chemist can first modify the standard polymers to give them the needed reactivity. He can either modify them during the initial polymerization reaction, or he can accept the standard polymer as a starting material and then run postpolymerization reactions on it to give it the reactivity needed for reactive blending.

These five methods of compatibilization may appear very clear in theory, but classification of individual practical polymer blends may be much more problematical. Thus this classification system is primarily useful for theoretical conceptualization, and any particular polymer blend might be classified in two or more places, depending upon the point of view of the chemist or engineer carrying out the development.

II. PHYSICAL PROCESSES

Most polymer blending is carried out by mechanical mixing of the molten polymers. Disentanglement and viscous flow are difficult, slow processes, so temperature, shear, and time are major factors, and final quenching may freeze-in a nonequilibrium morphology (4,7). Consequently it is not surprising that different workers may report different degrees of success or failure when using such physical processes as the only force for compatibilization.

Generally, increasing temperature may increase or decrease thermodynamic miscibility. It will almost always lower viscosity, though the quantitative effects on the two phases may be very different. And it will certainly accelerate the approach to equilibrium; but conversely, return to room temperature may produce a serious shift in equilibrium.

Generally, increasing shear will decrease domain size, within the limits permitted by melt viscosity. Directional shear vectors can convert spherical domains into fibrillar morphology, which might contribute mechanical reinforcement to the matrix polymer, or into lamellar morphology, which has been used to reduce permeability; and of course the same vectors can produce orientation, which would enhance these effects.

Extreme shear forces, particularly at low temperatures and very high viscosities, can actually break otherwise stable polymer molecules into macroradicals. When A—A and B—B polymer molecules are sheared into macroradicals A··A and B··B, cross-combination of these radicals can then produce A—B block or graft copolymers (6), which would properly be classified later as reactive compatibilization (Sect. V.A).

When either polymer crystallizes during cooling of the melt, the crystals of course form a separate phase. When both polymers crystallize, this can add two phases to the system. In rare cases, when both polymers have a similar crystal structure, they may be able to cocrystallize; when this happens, it actually contributes to intermolecular attraction and compatibilization (6).

III. PHYSICAL ADDITIVES

Compatibility of polymer blends can often be produced or enhanced by simple physical addition of monomeric or polymeric materials, without the need for any chemical reaction to produce the desired improvements.

A. Monomeric Additives

Monomeric ingredients that have occasionally been reported to increase compatibility include solvents, plasticizers, surfactants, fibers, and fillers.

1. Solvents and Plasticizers

When two polymers are immiscible with each other, it may be possible to dissolve each of them in the same solvent, or at least in miscible solvents. Then mixing these solutions, at high dilution, may produce a homogeneous mixed solution of the two polymers. This technique is most often used commercially in the coatings and adhesives industries, and it has been used in basic polymer research as well. Of course, as the solvent is evaporated and the polymer concentrations increase, they will reach the point at which phase separation will occur. Thus this requires careful control and understanding of the consequences and seriously limits the usefulness of solution blending.

Of more practical consequence, compatibility of an immiscible polymer blend may sometimes be enhanced by addition of a plasticizer that is miscible with each of the two polymers. This can be rationalized in simple technological terms, or more elegantly by the fundamental thermodynamics of entropy. Even aside from simple considerations of miscibility, plasticization may solve practical problems such as stiffness, inextensibility, or brittleness and thus earn the label of compatibilization.

2. Surfactants

Theoretical analysis of compatibilizers often explains their action by analogy with the effects of monomeric surfactants on monomeric oil-and-water mixtures. While polyblends, and most of their compatibilizers, are much larger molecules, a number of theorists have attempted to use the analogy to explain their effects. It is not clear whether simple monomeric surfactants could be used successfully to enhance compatibility in polymer blends. While they could not penetrate as deeply into the polymeric phases, they should still be able to exert their powerful effects on interfacial tension.

3. Fibers and Fillers

When the polymers in a blend are immiscible, repel each other, and form weak interfaces of poor mechanical properties, it has occasionally been reported by plastics technologists that the addition of reinforcing fibers can successfully bridge across the weak interfaces, connect the strong individual polymer phases with each other, and thus enhance mechanical properties and practical compatibility. This is probably analogous with the way fiber reinforcement of thermoset plastics successfully bridges across the weak interfaces, between the strong micelles that have been observed in most thermoset plastics.

Considerably more questionable is the occasional report that the addition of simple particulate fillers improved compatibility in a polymer blend. While such fillers might occasionally improve a few properties, it is doubtful that there would be very many practical overall benefits.

B. Polymeric Additives

Compatibilization theory is based primarily on the addition of block or graft copolymers at the interface between two immiscible polymers (2,3,5,7,8,11). If such a copolymer A—B is simply added to the system during melt mixing of polyA with polyB, it should tend to migrate to the interface. Block A should be miscible in polyA and block B should be miscible in polyB. This should lower interfacial energy and tension, decrease domain size and stabilize it, increase interfacial attraction and adhesion, and strengthen mechanical properties. Sophisticated modeling starts with the use of monomeric surfactants to emulsify oil-and-water mixtures and extends the treatment to two-phase molten polymer blends. This requires some difficult assumptions and simplifications, but they appear to work fairly well. Alternatively, some theorists even refer to a surface adhesive effect regardless of penetration into the separate phases, but this concept is not as clearly defined.

In attempts to refine the basic model, theorists generally agree that linear block copolymers should fit across the interface more efficiently than graft copolymers, because the latter might suffer steric hindrance and inaccessibility of the backbone. They also agree that A—B diblock copolymers should fit more efficiently than A—B—A triblock copolymers, because the latter would have to loop back into the interface again. And they generally agree that, while identical A and B blocks are ideal, it is quite possible to use C and/or D blocks, provided that these are miscible or at least compatible with polyA and/or polyB.

One suggestion concerns a tapered block copolymer (7). In random copolymerization of A + B, if monomer A enters the copolymer faster than monomer B, the first part of the polymer molecule will be almost polyA. As the concentration of monomer A decreases, monomer B will enter the growing molecule more frequently. Toward the end of the process, the last part of the molecule will be almost polyB. It has been suggested that such a structure would be a more efficient compatibilizer than a simple pure A—B block copolymer.

Use of modeling to quantify molecular weight of the blocks has been interesting but somewhat problematical. Theoretically, low-molecular-weight blocks should flow and mix most easily, whereas very long blocks would suffer from entanglement and inaccessibility. On the other hand, equilibrium penetration and attraction of block A into polyA and of block B into polyB would increase with the molecular weight of the blocks. Various theoretical guesstimates for the optimum length of the blocks have varied from degree of polymerization of 10–15 up to molecular weights of 100,000–150,000 or even higher, and some have even argued that the molecular weight of the block must be greater than the molecular weight of the homopolymer phase.

According to conventional monomeric surfactant theory, a thin interfacial layer should only require 0.1–5.0% of block copolymer to provide efficient compatibilization, and some experimental work actually confirms this. When practical development studies require much larger amounts than this, then theoretical explanation shifts to other mechanisms such as solvation, partial miscibility, or impact modification.

Specific examples can be classified according to whether the added compatibilizer has the same structure as the major polymers in the blend, or whether it is another structure whose blocks are nevertheless miscible, similar, or at least compatible with the major polymers. It may also be of interest to note whether the compatibilizer was a block or a graft copolymer, or simply a polymer that was miscible or compatible with both of the major polymers.

1. Same Structure

Low-density polyethylene + polystyrene + block or graft copolymer (3,12)
EPDM rubber + SBR + graft copolymer (3)
Polybutadiene + polystyrene + block copolymer (3)
Polyisoprene + polystyrene + block copolymer (3)
Polystyrene + ABS + S-B block copolymer (3)
Polystyrene + polyester + block copolymer (12)
Polyethyl acrylate + cellulose + graft copolymer (2)
Polyacrylonitrile + cellulose acetate + graft copolymer (3)

2. Miscible or Compatible Structure

Polyethylene + polystyrene or high-impact polystyrene + SEBS (4,5,12)
Low-density polyethylene + polypropylene + EPDM (1)
Low-density polyethylene + polystyrene + SEB or SEBS (3,9)
Low-density polyethylene + polyvinyl chloride + chlorinated polyethylene (3)
High-density polyethylene + polypropylene + EPR (1,9)
High-density polyethylene + nylon + ethylene ionomer (5)
Ethylene-propylene rubber + nitrile rubber + chlorinated polyethylene (12)
Polypropylene + polystyrene + SEBS (4,9)
Natural rubber + polyvinyl chloride + NR-g-MMA (3)
Polyvinyl chloride + cellulose or starch + cellulose-g-ethyl acrylate (5)
Polyphenylene ether + high-impact polystyrene (graft copolymer) (9)
Polyphenylene ether + polyvinylidene fluoride + PS-b-PMMA (5)

3. Other Polymeric Compatibilizers

Occasionally it is reported that addition of a third polymer improved compatibility of the two major polymers, but it is not clear that the added polymer was really miscible or even compatible with the major polymers. This suggests that other mechanisms may be involved in compatibilization.

Most often, the added polymer may be a rubbery impact modifier and may introduce some ductility into an otherwise weakly interfaced polyblend. Thus SEBS has been reported to improve compatibility of polyethylene + polycarbonate, high-density polyethylene + poly(ethylene terephthalate), and polyesters + engineering thermoplastics in general (1,4,5). Similarly, styrene-butadiene diblock copolymer has been reported to compatibilize polystyrene + ABS (8), and EPDM has been reported to compatibilize high-density polyethylene + poly(ethylene terephthalate) (1).

In another case, compatibility of polycarbonate with nylon 6 was improved by addition of low-molecular-weight polycaprolactone (7). The researchers theorized that PCL dissolved in PC and lowered its melt viscosity, making it closer to the low viscosity of nylon, thus facilitating uniform melt mixing.

IV. POLYMER MODIFICATION FOR PHYSICAL COMPATIBILIZATION

Polymer blends are usually prepared by mixing molten polymers. This is a convenient one-step mechanical process and does not involve any chemical reactions. In some cases the polymers are basic commercial materials and are readily available. In most cases, however, the polymer chemist must first modify the basic polymers to make them suitable for physical compatibilization. He may do this either during the original polymerization reaction or else by postpolymerization reactions on the basic polymers.

A. Modification During Polymerization

There are four ways in which the polymer can be modified during the original polymerization reaction in order to prepare it for physical compatibilization later: block copolymerization, random copolymerization, attachment of terminal functional groups, and control of molecular weight.

1. Block Copolymerization

Theoretical analysis generally predicts that properly designed block copolymers will be the most efficient for producing compatibility, and experimental results frequently confirm this prediction (3). Thermoplastic elastomers are almost always based on 100% block copolymers to give maximum high performance. Many of the compatibilizers described above (Sect. III) were specifically prepared for use in making polyblends; major commercial examples are styrene-b-butadiene-b-styrene, styrene-b-ethylene-butylene-b-styrene, and chlorinated polyethylene. Careful synthesis gives maximum control of structure and most efficient compatibilization. On the other hand, synthesis is difficult and expensive, so it is more popular in research than in commercial production.

2. Random Copolymerization

Whenever a homopolymer does not have the proper polarity, hydrogen bonding, or ionic groups for intermolecular attraction, these can usually be incorporated by copolymerization, to the precise degree desired, and thus produce effective compatibilization with a wide range of different polymers. Major commercial examples are polyethylene ionomers, ethylene-vinyl acetate copolymers, butadiene-acrylonitrile rubbers, and styrene-acrylonitrile and styrene-maleic anhydride copolymers. Other comonomers mentioned in the literature include carbon monoxide, vinyl alcohol (by hydrolysis of vinyl acetate units), acrylic and methacrylic acids and esters, hydroxyethyl methacrylate, maleic and fumaric acids, and vinyl pyridine for addition polymers; and a variety of polyols for polyesters, polyurethanes, and polyamides (5,9).

An unusual phenomenon has been observed when the comonomer units within a copolymer are very incompatible and repel each other. In such cases, blending with a second polymer may mediate between them and produce an unexpected compatibility. A classic example is blends of polyphenylene ether with chlorinated polystyrenes. While PPE is incompatible with either *o*-Cl or *p*-Cl styrene polymers, it is compatible with *o-p* copolymers because it reduces the repulsion between them (7).

3. Terminal Groups

In vinyl polymerizations it is possible to use heterofunctional initiators or chain-transfer agents that attach a desired functional group to the ends of the polymer molecule. These are most often used for reactive compatibilization, which will be discussed later. In one case they were used as intermediates for the preparation of a modified polymer to be used as an additive for physical compatibilization. Butadiene and acrylonitrile were copolymerized to a low-molecular-weight telomer that was terminated by amine end groups, hence the name "amine-terminated butadiene-nitrile" or ATBN. Polypropylene was maleated by grafting maleic anhydride onto it. When these two polymers were reacted with each other, maleic anhydride groups and amine groups reacted to form amide groups, producing a graft copolymer. This was then used as a physical compatibilizer for polyblends of polypropylene + nitrile rubber, to produce an oil-resistant thermoplastic elastomer (5).

4. Control of Molecular Weight

According to thermodynamic theory, miscibility depends on a high degree of randomization to produce a large gain of entropy. Thus decreasing the molecular weight of the polymer should increase miscibility. While the effect is not large, in critical borderline cases it might provide the added advantage needed to produce practical compatibility.

B. Modification After Polymerization

While modification during polymerization is more precise and therefore preferred in theory, the production of small batches of such specialized polymers is expensive and not often favored in commercial practice. Here it is generally easier and less expensive to take a major commercial polymer and run postpolymerization reactions on it in order to modify it into the form most suitable for physical compatibilization. These may be classified as addition reactions and substitution reactions.

1. Addition Reactions

Graft copolymerization is by far the most popular method of producing compatibility in polymer blends. When it is accomplished by the processor during melt blending, it is commonly classified as reactive compatibilization. On the other hand, when it is performed by the polymer chemist before melt blending and is used to improve the physical blending process, it may better be classified as postpolymerization modification, in preparation for later physical blending.

Major commercial examples are high-impact polystyrene, ABS, and impact modifiers for rigid polyvinyl chloride. High-impact polystyrene is produced by dissolving polybutadiene rubber in styrene monomer and then polymerizing the styrene to polystyrene (3,9). During the polymerization reaction, some of the polystyrene chains are grafted onto some of the polybutadiene rubber backbones. In the early days of polyblending, theoretical polymer chemists assumed that their ultimate goal was production of a 100% graft copolymer; but the closer they approached to this goal, the more they realized that the best high-impact polystyrene should be mostly polystyrene homopolymer matrix and polybutadiene homopolymer domains, with just enough graft copolymer to go to the interface, stabilize the rubber domains, and produce strong interfacial adhesion and stress transfer across the interface. Thus the optimum product results from graft copolymer acting as physical compatibilizer for the polymer blend.

ABS is made similarly by dissolving polybutadiene rubber in a mixture of styrene and acrylonitrile monomers and then copolymerizing them to produce primarily styrene-acrylonitrile copolymer matrix and polybutadiene homopolymer domains, with just enough graft terpolymer to go to the interface, stabilize the rubber domains, and produce strong interfacial adhesion and stress transfer across the interface (9). In commercial production, it is common practice to add more styrene-acrylonitrile copolymer and/or polybutadiene rubber to modify the balance of properties for different commercial grades. Here again the optimum product results from graft terpolymer acting as physical compatibilizer for the polymer blend.

Rigid polyvinyl chloride is almost always blended with graft copolymers, small amounts to improve processability, and larger amounts to increase impact strength (3,9). The most common graft copolymers are methyl methacrylate-butadiene-styrene, methyl methacrylate-acrylic rubber-styrene, and acrylonitrile-butadiene-styrene, prepared as described above, but optimized for use in polyvinyl chloride. Here the physical compatibilization is produced by the polymethyl methacrylate or styrene-acrylonitrile chains grafted onto the rubber backbones.

The resulting synergistic increase in impact strength can easily be 10–20 times greater than that of rigid polyvinyl chloride alone.

Grafting addition of maleic anhydride onto polymers is commonly called maleation. While it is mostly used to introduce functional groups for later reactive compatibilization, it may also be used simply to introduce polarity and hydrogen bonding for physical compatibilization of SBS or SEBS with polar polymers such as ionomers, ethylene-vinyl acetate, ethylene-vinyl alcohol, acrylonitrile, polyacetal, polyphenylene ether, polysulfone, polyphenylene sulfide, polycarbonate, polyesters, and polyurethanes (5). Similarly, grafting of diethyl maleate onto polyolefins compatibilizes them with polyvinyl chloride.

A more specialized addition reaction is the epoxidation of the double bonds in natural rubber, either to increase polarity for direct physical compatibilization or for further compatibilization reactions (7). Thus natural rubber can be converted into a thermoplastic elastomer by blending with sulfonated EPDM or by further reaction with maleated polyethylene to form a physical compatibilizer for polyblends of polyethylene + natural rubber.

2. Substitution Reactions

Commercial polymers can also be modified by substitution reactions to introduce polar, hydrogen-bonding, or ionic groups for use as physical compatibilizers. Chlorination of polyethylene can be carried out either in solution to produce random substitution or in suspension to produce block copolymers of polyethylene-b-chlorinated polyethylene structure. In random copolymers, degree of chlorination controls polarity and hydrogen bonding and thus miscibility; while in block copolymers, the nonpolar polyethylene block and the polar chlorinated polyethylene block offer an attractive structure for interfacial bonding, for example to compatibilize blends of polyethylene with polyvinyl chloride (3).

Another substitution reaction of considerable interest is sulfonation, introducing the sulfonic acid group into polymers. This is used in commercial production of EPDM seamless roofing and in experimental compatibilization of polystyrene with more polar polymers (6).

V. REACTIVE COMPATIBILIZATION

Current research on compatibilization of polymer blends is focused primarily on reactive processes to produce the block or graft copolymers directly during the blending process (1,6,8,11). While the reactions, structures, and products are not as well controlled nor as clearly understood as in physical blending, the single-step process is much more economical and therefore is becoming more popular in commercial practice. Particularly when carried out in a twin-screw extruder, the process is fast, irreversible, low-exotherm, and very economical (6).

If polyA and polyB contain functional groups that are capable of reacting directly with each other, they can form block or graft copolymers during melt blending. Since only a small amount of compatibilizer is generally required for interfacial action and adhesion, a small percent of reaction is usually sufficient or even optimal. Actually, some highly reactive systems may continue on until they approach random copolymer structures and homogeneous single-phase systems, which may no longer give the benefits of a good two-phase system. When grafting reactions become cross-linking processes, they produce thermoset polymers, which may or may not be desirable, depending on the specific product.

If polyA and polyB are not capable of reacting directly with each other, they may usually be compatibilized by adding a third polymer that is capable of reacting with one or both of them. For example, the added polymer may be A-D, where block A is compatible with polyA, and functional groups in D are reactive with polyB. Or the added polymer may be C-D, where functional groups in C are reactive with polyA and functional groups in D are reactive with polyB. This general concept greatly increases the ability of the versatile organic polymer chemist to design polyblend systems that will achieve reactive compatibilization during melt blending.

A. Functional Groups and Their Reactions (8)

1. *Carbon-carbon single bonds* are usually stable enough for most purposes, but mechanical shear can sometimes break them into radicals that can couple indiscriminately with each other, thus producing block copolymers. This might appear to be simple physical blending, but it is very definitely reactive compatibilization.

2. *Carbon-carbon double bonds* are moderately to highly reactive, depending on their position in the polymer molecule. They can react with each other or with a variety of free radicals and active hydrogen compounds to produce grafting and even cross-linking.

3. *Carbon-hydrogen bonds* are often activated by adjacent groups, making them available for free radical, oxidation, and substitution reactions. On the benzene ring in phenols and anilines, the ortho- and para-hydrogens are extremely activated and available for compatibilization and cross-linking reactions.

4. *Hydroxyl groups* of alcohols and phenols react readily with epoxy, acid, ester, isocyanate, and oxazoline groups. As methylol groups they are extremely reactive.

5. *Epoxy groups* react readily with hydroxyl and very readily with acid and amine groups.

6. *Carboxylic acid and anhydride groups* react readily with hydroxyl, epoxy, ester, amine, and isocyanate groups and ionomers.

7. *Ester groups* exchange readily with hydroxyl, acid, ester, amine, and amide groups.

8. *Amine groups* react readily with epoxy, acid, ester, amide, isocyanate, and organohalogen groups and ionomers.

9. *Amide groups* exchange readily with ester, amine, and amide groups.

10. *Isocyanate groups* react readily with hydroxyl, acid, and amine groups and less readily with amide and urethane groups.

11. *Carbodiimides* react readily with acids.

12. *Oxazolines* react readily with hydroxyl and acid groups.

13. *Organobromine* reacts fairly readily with amines.

14. *Sulfonic acid groups* react very readily with amines and ionomers.

15. *Phosphonic acid groups* react very readily with amines and ionomers.

B.   Reaction Between Primary Polymers

Direct reaction between the primary polymers in a blend can be illustrated by several major commercial examples.

1. *High-impact polystyrene and ABS* were discussed earlier (Sect. IV.B.1) as physical blends of two major polymers with a small amount of graft copolymer to strengthen the interface. This is how they would appear to the plastics processor. As the polymer producer describes the grafting process, however, it might well be called reactive processing (9). Whatever it is called, it represents two of the earliest and largest-volume commodity compatibilized polymer blend systems.

2. *Urea-formaldehyde and phenol-formaldehyde* resins are most commonly reinforced by cellulosic fibers. When these resins are cured by reaction between methylol groups, these groups also react directly with the hydroxyl groups of cellulose, producing graft copolymers that contribute greatly to the reinforcement (2).

3. *Coating enamels* on autos and appliances are mostly made by blending reactive hydroxyl oligomers of acrylic esters and polyesters, respectively, with reactive methylol oligomers of melamine-formaldehyde resins. These two species coreact during cure to produce high-performance coatings of excellent appearance and durability.

4. *Epoxy resins* cross-link so highly during cure that they become excessively brittle. A popular way of preventing this is to use carboxy-terminated butadiene-nitrile (CTBN) oligomers as the curing agent. These separate as rubbery domains during cure, contributing ductility and toughness to the cured epoxy resin (9).

5. *Polycarbonate* may suffer environmental stress cracking in solvents, which can be remedied by blending with crystalline polyesters such as poly(ethylene or butylene terephthalate). These blends are not naturally compatible, but during melt processing there is sufficient ester interchange to produce some block copolymer, which strengthens the interface and produces good compatibility (5,6,9,11).

C. Addition of Reactive Polymer

If the two major polymers in an incompatible blend are not reactive with each other, it is often possible to design a third polymer that can react with one or both of the major polymers to produce compatibilization. Organic polymer chemists have suggested and demonstrated many such possibilities in basic research (1,3,6,7,8,11). Any of the 15 reactive groups described above (Sect. A) may be a candidate for such an approach.

In commercial practice, the major example is impact modification of nylon by adding elastomers, particularly ethylene copolymers (4,5,6,9,11). These are not naturally compatible with nylon. They are most often compatibilized by grafting maleic anhydride onto the elastomer. When this maleated elastomer is added to the polymer blend, the polyolefin backbone is compatible with the elastomer, and the maleic anhydride reacts with the nylon to produce amide or imide groups that bind them into a graft copolymer. Presumably most high-impact nylons are made in this way.

D. Addition of Monomeric Reactants

When two polymers form an incompatible blend, and are not spontaneously reactive with each other, it is sometimes possible to add a small amount of monomeric material to react with them, or catalyze a reaction between them, and thus produce the block or graft copolymer structure required for compatibilization (6). Normally, such additives are effective in very low concentrations, typically 0.1–3.0%. Some examples from the literature illustrate the possibilities.

1. *Peroxides* decompose during melt processing, producing radicals that can attack polymer molecules, adding to C=C bonds, or cleaving C-H and even C-C bonds, to produce polymer radicals (6,7). When polyA·radical and polyB· radical meet, they can easily combine to produce block or graft copolymers.

2. *Covulcanization* of elastomer blends should ideally produce cross-links, not only between A-A and B-B but also between A-B (6,11). Many rubber technologists actually demand this for practical performance.

3. *Maleic anhydride* is the leading monomeric candidate for reactive compatibilization. Particularly in the production of high-impact nylons, many processors simultaneously blend nylon, elastomer, maleic anhydride, and peroxide initiator to produce the polyamide or polyimide graft copolymer that is the effective compatibilizer (4,5,6,9,11). In exploratory research, maleic anhydride has been used similarly in a variety of other polymer blends.

4. *Catalysts* are particularly useful for transfer reactions between polyester and/or polyamide homopolymers to produce block copolymers (6,11). Commercial thermoplastic polyesters sometimes still contain enough of the catalyst from the original polyesterification reaction to be useful again in later transfer reactions. Other catalysts often added are typically *p*-toluene sulfonic acid and organic phosphites.

5. *Dimethylaminoethanol* was used as a unique reagent to graft styrene-maleic anhydride copolymer to bromobutyl rubber (6,8).

6. *Organosilanes* XR-Si-OR can be used in a two-step process to graft two polymers to each other. In the first step a functional group XR- is grafted onto both polyA and polyB. So long as these are kept clean and dry, they form a thermoplastic polyblend. After melt processing into the final product, exposure to water permits the -OR groups to hydrolyze to -OH groups, which immediately condense to siloxane cross-links (6).

7. *Ionomer* groups have much higher fluidity and can easily exchange with each other to form ionic intermolecular bonds and produce compatibilization. In one example, two carboxylic acid polymers were mixed with zinc acetate, liberating acetic acid and forming zinc carboxylate ionic cross-links between the two polymers (2).

E.   Interpenetrating Polymer Networks

One unique process for reactive compatibilization is the synthesis of interpenetrating polymer networks (IPN). In its most precise form, it is a five-step process: (1) Polymerize monomer A to polyA. (2) Lightly cross-link polyA. (3) Swell it with monomer B. (4) Polymerize monomer B to polyB. (5) Cross-link polyB. This can give much more control to the blending process, produce a finer scale of dispersion, and stabilize it against a later change of morphology. It has given some unusual examples of synergistic improvement of properties and is of particular interest in reduction of vibration and noise.

VI.   POLYMER MODIFICATION FOR REACTIVE COMPATIBILIZATION

Practical reactive compatibilization requires two polymers that can react with each other or a third polymer that can react with one or both of them, or at least a monomeric additive that can react with them or catalyze a reaction between them. Whenever the basic polymers do not have such reactivity, they can be modified before blending in order to introduce the desired reactivity into them. All of the functional groups discussed above (Sect. V.A) can be considered if there is a practical way to introduce them into the polymers as needed (6,8). They may be introduced either during the original polymerization or (generally more economically) by postpolymerization reactions.

A.   Modification During Polymerization

Most functional groups can be introduced by use of the appropriate comonomer. This is particularly neat for theoretical work. On the other hand, the manufacture of small quantities of such specialty polymers is much more expensive than conventional commodity polymerization. In some cases it may be justifiable; in other cases it is simply uneconomical. Practical examples include acrylic and methacrylic acid, their hydroxy and amino esters, and vinyl pyridine in vinyl polymers; and branched polyols, acids, isocyanates, and amines in condensation polymers. These are mostly useful for compatibilization through graft copolymers.

Functional end groups can sometimes be produced directly during the polymerization reaction. In vinyl addition polymerization this is done by the choice of initiators and/or chain-transfer agents. In condensation polymerization it is done by the use of a slight excess of the proper monomer. In either case, the original end group can also be converted into a more desirable one by a postpolymerization reaction. Popular end groups include hydroxyl, carboxylic acid, amine, and isocyanate. These can be used for compatibilization through either block or graft copolymers.

B. Modification After Polymerization

In most cases, reactivity can be added less precisely but more economically by postpolymerization reactions. The most popular in current research and development is the peroxide-initiated grafting of maleic anhydride onto nonpolar polymers, mainly polyolefins and elastomers, to increase their polarity, hydrogen bonding, or reactivity with more polar polymers, particularly nylons. Another major example is the use of organosilane coupling agents on glass fibers to reinforce polymeric binders, actually compatibilizing the inorganic glass polymer with the organic polymer (2). Some other experimental examples illustrate a few of the possibilities:

   Hydrolysis of vinyl acetate units to vinyl alcohol units
   Epoxidation of the $C=C$ bonds in rubber
   Maleation of polystyrene or polyphenylene ether (5,8)
   Sulfonation of polystyrene or polyphenylene ether (5,8)
   Bromination of polystyrene or polyphenylene ether (5,8)

For practical commercialization, these reactions are best carried out in twin-screw extrusion. Ideally, for optimum economy, the postpolymerization modification reaction should be carried out in the back of the extruder, followed by reactive blending further toward the front of the extruder, in what may be called a one-step sequential functionalization and polymer blending (4,6).

REFERENCES

1. L. A. Utracki. Polymer Alloys and Blends. New York: Hanser, 1990.
2. N. G. Gaylord. Role of compatibilizers in polymer utilization. Chemtech 392–395, June 1976.

3. D. R. Paul. Interfacial agents ("compatibilizers") for polymer blends. In: D. R. Paul, S. Newman, eds. Polymer Blends, Vol. 2. New York: Academic Press, 1978, pp 35–62.
4. J. W. Barlow, G. Shaver, D. R. Paul. Compalloy '89:221–244, 1989.
5. N. G. Gaylord. Compatibilizing agents: structure and function in polyblends. J Macromol Sci Chem A26(8):1211–1229, 1989.
6. M. Xanthos, S. S. Dagli. Compatibilization of polymer blends by reactive processing. Polym Eng Sci 31(13):929–935, 1991.
7. R. L. Markham. Introduction to compatibilization of polymer blends. Adv Polym Tech 10(3):231–236, 1991.
8. N. C. Liu, W. E. Baker. Reactive polymers for blend compatibilization. Adv Polym Tech 11(4):249–262, 1992.
9. D. R. Paul, J. W. Barlow, H. Keskkula. Encyc Polym Sci Eng 12:399–461, 1988.
10. T. Tang, B. Huang. Interfacial behaviour of compatibilizers in polymer blends. Polym 35(2):281–285, 1994.
11. L. Mascia, M. Xanthos. An overview of additives and modifiers for polymer blends: facts, deductions, and uncertainties. Adv Polym Tech 11(4):237–248, 1992.
12. C. C. Chen, J. L. White. Compatibilizing agents in polymer blends: interfacial tension, phase morphology, and mechanical properties. Polym Eng Sci 33(14):923–930, 1993.

2

Compounding and Compatibilization of High-Performance Polymer Alloys and Blends

Wen-Yen Chiang and Chi-Yuan Huang
Tatung Institute of Technology, Taipei, Taiwan, Republic of China

I. INTRODUCTION

Engineering plastics (EP) possess several excellent properties such as high specific strength, dimensional stability, low creep, and good electrical and chemical properties. Therefore they are commonly used in electrical and electronic applications, automobiles, buildings and construction, industrial machinery, and consumer products. However, issues of toughness, processibility, and cost often hinder these high-performing plastics from becoming commercially useful. Blending is a simple, fast, effective, and economical method of resolving the above-mentioned problems and also provides a high performance-to-cost ratio.

In the past decade much attention has been paid to the development of polymeric blends. Polyblends offer the possibility of combining the unique properties of available materials and thus of producing materials with tailor-made properties, which often have advantages over the development of a completely new polymeric material.

Mixtures of polymers are classified as polymer alloys (A), polymer blends (B), and polymer composites (C). This chapter is sectioned by this classification. Utracki (1) divided polymer blends into two categories: the "immiscible" polymer system, which is a polymer alloy, and the miscible polymer system, which is a polymer blend. On the other hand, a polymer mixed with a filler forms a polymer composite.

The properties of polymer alloys, polymer blends, and polymer composites depend on the quality of the dispersed phase, the microstructure of all phases, the adhesion and cohesion between phases, and the morphology of the system. Compatibility and molecular miscibility are two important properties of polymeric components.

There are three situations that occur when fillers are blended into polymers: the surface of the filler is wetted by the polymer, the surface of both polymer and filler are in close contact, or the filler and polymer are chemically bonded. If a good affinity exists between the polymer and the filler, the mechanical properties of the blend will be significantly increased. The affinity can be improved by activating the filler's surface or modifying the polymer and is also determined by whether the polymer or the filler bears polar functionalities.

Compatibility of the materials is also a problem. This problem has been solved in many cases by adding a small fraction of an additive known as a compatibilizer (2–4). There are general block and graft copolymers or chemically reactive species that are concentrated at the interface and act as emulsifiers to wet the interface. The function of the in-situ (or in-vivo) formed compatibilizer is reported to reduce interfacial tension and to decrease the size of dispersed phases (5–9) to improve the adhesion between two immiscible polymers. The compatibilizing reaction should be fast and irreversible, and compatibilizers must be able to stand high processing temperatures. Compatibilization can also be done via the addition of low-molecular-weight components to promote copolymer formation or cross-linking reactions (10,11).

A phase compatibilizer (block, graft, or random copolymer) tends to reside along the interface but not exclusively; some will dissolve in both components of the blend. Even without a compatibilizer, a certain degree of mutual solubility still exists in any immiscible polymer pair. A compatibilizer will certainly increase such mutual solubility. Distribution of the compatibilizer molecules in any immiscible polymer blend depends on blend components, processing conditions, molecular weight and distribution, and type of copolymer. The compatibilizer, distributed within the blend components to induce further mutual solubility between components, will certainly alter the inherent toughness of these components. However, the toughness of a compatibilized blend may increase or decrease relative to a noncompatibilized blend, depending on the competition between the advantages of better phase adhesion and dispersity and the disadvantage of loss of the inherent toughness of the blend components. A good adhesion and fine domain size in a compatibilized polymer does not guarantee its mechanical toughness.

This article deals with polymers such as acrylonitrile-butadiene-styrene (ABS), high-impact polystyrene (HIPS), nylon, polycarbonate (PC), polyethylene (PE), polyacetal(polyoxymethylene) (POM), polyphenylenesulfide (PPS), polypropylene (PP), and polystyrene (PS), which can be blended with fillers in the presence of compatibilizers to form functional polymer blends with excellent electrical, magnetic, physical, and thermal properties.

II. POLYMER ALLOYS

A. PC/ABS Alloy

Acrylonitrile-butadiene-styrene (ABS) and polycarbonate (PC) are widely used thermoplastics with good physical properties. However, the general mechanical properties of ABS are not so good as those of other engineering plastics and have limited application in many fields. One approach to upgrade the properties of ABS is to blend ABS with other high-performance engineering plastics such as PC (12–14). Blending with PC may be viewed as increasing the performance of ABS and increasing the processibility of PC (15–30).

Thermoplastics are more or less easily combustible. For example, the limited oxygen indices (LOI) of ABS and PC are about 17 and 22–24, respectively. Consequently, efforts to develop flame-retarding plastics have increased with the increasing use of thermoplastics. As a result, flame-retarding formulations are available for all thermoplastics to reduce the probability of their burning in the initial phase of a fire. Flame-retardant plastics secure the scope of utilization for thermoplastics and in fact increase their range of application (31). However, addition of a large amount of flame retardant decreases the properties of thermoplastics and causes some problems with regards to processibility (14). Flame-retardant plasticizing polymers generally decrease thermomechanical properties. Moreover, nonsoluble solid-state flame retardants in polymers significantly decrease their impact strength. For this reason, special pretreatments of most flame retardants are required. Antimony oxide (Sb_2O_3) is frequently used as a synergistic additive with various reactive flame retardants. As indicated above, however, addition of a flame retardant and antimony oxide will decrease the mechanical properties of the PC/ABS alloy, they significantly reduce, for example, its impact strength. As has been reported in the literature, PC/ABS alloy is a partially miscible system, and the alloy has two glass transition temperatures (T_g). To improve this phenomenon, the compatibilizers MBS (methacrylate-butadiene-styrene), SMA (styrene-maleic-anhydride copolymer), and EVA (ethylene-vinyl-acetate copolymer) have been added to the alloy to improve interface adhesion and reduce the interfacial tension of the alloy (14). When more than 3 phr (parts per hundred resins) of MBS was added, the tensile strength was increased significantly for the PC/ABS (80/20) alloy containing 10 phr and 15 phr flame retardant (14). Adding 1 phr and 3 phr EVA into the alloy with 10 phr flame retardant markedly increased the impact strength, whereas adding SMA did not affect the impact strength of the flame retardant PC/ABS blend (14). The addition of SMA will increase the tensile strength and modulus. The glass transition temperatures of alloys were also influenced by the addition of these three kinds of compatibilizers. Without adding the compatibilizers, the flame retardant PC/ABS alloy had two values of T_g. After compatibilization, the modified alloy had only one value of T_g. This implies that the alloy system became compatible (14).

Table 1 Mechanical Properties of Nylon6/
ABS Blends

Nylon6/ABS (wt%)	TS[a] (MPa)	M[b] (GPa)	E[c] (%)
100/0	64	2.39	125
90/10	57	2.43	38
70/30	44	1.84	25
50/50	40	1.84	25
30/70	34	1.69	23
10/90	35	1.76	20
0/100	43	2.27	20

[a] TS: tensile strength.
[b] M: modulus.
[c] E: elongation.

B. Nylon6/ABS Alloy

Nylon6 and ABS (32) are widely used plastics, but these two are incompatible.
Therefore the mechanical properties of an ABS/nylon6 blend are lower than those
of nylon6 (Table 1). To increase the compatibility of nylon6 and ABS, styrene-
maleic-anhydride (SMA) has been found to be an effective compatibilizer (Fig. 1).
It is known that the compatibilizer concentrates at the interface, and just a thin
molecular layer will be effective. As a result, two situations may take place. First,
when the content of the compatibilizer is inadequate, fracture occurs between
phases, because the interfacial adhesion is not formed perfectly yet. Second, when
the content of the compatibilizer is excessive, there are fractures from layer to
layer, because there are no special interactions between layers. The Izod impact
strength, tensile strength, and Young's modulus of nylon6/ABS blends were enhanced by adding about 1 wt% of compatibilizer (SMA). SMA is like two hands:
one hand (MA) scratches nylon6, the other scratches ABS. With mechanical interlocking thus formed, phase separation will not occur. A molecular layer is
formed when 1 wt% of SMA is added, and the properties of blends change from
an antagonistic effect to a synergistic one.

C. Polymer/PLC Alloy

The modification of a polymer's properties through a physical blending process
(33) has been of increasing importance. Because the cost of physical blending to
improve properties of polymers is much lower than that of synthesizing new polymers or chemically changing an existing polymer's structure by block and/or graft

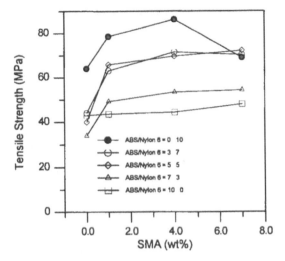

Figure 1 The tensile strength of ABS/nylon6 filled with SMA compatibilizer.

copolymerization, the role of blending is important in polymer science and engineering.

The development of reinforced plastics has traditionally involved the addition of a particulate filler or fiber (34). These fibers and fillers usually tend to increase the stiffness of the composite material. The extent of the reinforcement depends upon many factors, such as the shape and amount of the filler, the interfacial adhesion between the filler and the polymer matrix, the distribution of fiber in the matrix, the surface treatment of the fiber, and the properties of the fiber (35).

Although the addition of fibers to a polymer can improve the mechanical properties, it can also have a contrary effect on the processibility of the polymer. Because the presence of fibers in a polymer melt increases the viscosity, the energy consumption also goes up during the processing of a high viscosity melt (35). Since a glass fiber remains solid during processing and is extremely hard, equipment such as screws and barrels will be worn and need periodic replacement (36). The size distribution of the fiber can be greatly affected by breakage during the compounding. In addition, the control of the filler concentration and distribution is difficult in the compounding step. If nothing else, the compounding step is an extra expense because of the cost of the machinery, the time required to use it, and the cost of the inventory of different grades with different filler levels.

Thus there is an incentive to find an approach by which the reinforcing species does not exist before the processing of the resin but comes into existence during the processing. The term in-situ composite is used to describe materials of this type.

Three approaches have been reported to form an in-situ composite, that is,

1. In-situ crystallization
2. In-situ polymerization
3. In-situ composites from blends with thermotropic polymeric liquid crystalline (PLC)

It has been reported (37–47) that blending thermotropic polymeric liquid crystalline (PLC) with conventional plastics results in a more cost-effective polymer that has unique mechanical and chemical properties. Kimura and Porter (35) discussed the phase behavior and some mechanical properties of blends of poly(butylene terephthalate) (PBT) and a liquid crystalline copolyester of ethylene terephthalate and p-oxybenzoate.

Siegmann, et al. (39) investigated the properties of blends made from Celanese Vectra-A900 (a PLC, on 6-hydroxy-2-naphthoic acid <HNA> and p-hydroxybenzoic acid <HBA>) and an amorphous polyamide. They found that the blend viscosity was always much lower than that of the parent polymers. The polymers were immiscible, and the PLC phase in injection-molded samples ranged from ellipsoidal particles to a fibrillar structure with the increase of the PLC content. These two phase morphologies resulted in reinforced compositions, though the properties were not homogeneous throughout the specimen. A skin-core morphology was obtained, in which the orientation of PLC and the mechanical properties in the skin were higher than those in the core. With the same PLC, Chung (40) investigated the physical properties and morphology of the blends of nylon 12. It was stated that nylon 12 exists preferentially in the skin rather than in the core, although this is inconsistent with previous reports (40).

Blizard and his coworkers (41–43) have studied the blends of polycarbonate and nylon 6,6 with a PLC based on 60 mol% p-hydroxybenzoic acid/40 mol% poly(ethylene terephthalate). It is found that only a small weight fraction of PLC was required to reduce the viscosity of the thermoplastic to that of the polymeric liquid crystal. The mechanical properties were improved by the formation of the PLC fibrous domain in the flow direction.

The properties of polycarbonate (PC)/acrylonitrile-butadiene-styrene (ABS) blends were studied by Chiang and Hwung (12), who found that by blending with ABS, the processability of PC was improved and the T_g of PC reduced. Apicella et al. (44) showed that a significant improvement of dimensional stability of a drawn thermoplastic material can be obtained by adding a small percentage of noncompatible polymeric liquid crystals, while no improvement was observed for compatible polymeric liquid crystal and polystyrene blends. Weiss et al. (45) studied the thermal rheological and mechanical properties of blends of polystyrene with a PLC and also found that the immiscible PLC behaved like short reinforcing fibers, improving the mechanical properties of the blend, but that the short fibers resulted in a lower melt viscosity. The properties of blends of polystyrene and two miscible liquid-crystalline low-molecular-weight additives were studied by Huh et al. (46), who found that the additives act plasticizers, as evidenced by the decreases in the glass transition temperature and the melt viscosity. Paci et al. (47) studied the thermal properties of solution-prepared blends of PBT and liquid crystalline poly(4,4 - diphenyl sebacate) (PB8) and pointed out that the melting and crystallization temperature of PBT decreases with the increase in the content of the liquid crystalline component. Kiss (36) studied PLC blends with various engineering resins such as polyethersulfone, polyarylate, polyacetal, polyester, and others and found that PLCs play an important role in reinforcing the resultant polymers.

The blends of a PLC based on HBA and HNA with polycarbonate (PC) and polyphenylenesulfide (PPS) were investigated by Chiang and Young (33). The relation of mechanical and thermal properties with composition was discussed as follows. PPS exhibits rigid and brittle properties with a high modulus and an extremely small elongation, whereas PLC has greater elongation. The tensile strength of blends was lower than that of their parent polymers. It was found that the Young's modulus of PPS/PLC decreases with increasing weight percentage of PLC. The minimum of tensile strength and notched Izod impact strength occurred at 50% PLC. The blend of PPS with PLC does not show any improvement of the mechanical properties compared with those of the parent materials. It is suggested that the high crystallization temperature and high crystallinity of PPS inhibits the formation of the extended, fibrous domains of PLC. On the other hand, the presence of PLC interfered with the crystallization of PPS, resulting in the poor mechanical properties of the blends.

However, the tensile strength and the Young's modulus of the PC/PLC blend increase with the increasing weight percentage of PLC (Fig. 2). The tensile modulus for injection-molded specimens increased nearly 60% with the addition of 25 wt% PLC. Poor notched Izod impact strength was observed in the PC-rich composition with the minimum impact strength of blends appearing at 25% PLC. It should be noted that the elongation to break of the polymers was drastically reduced when the PLC was added. A fibrous structure in the injection-molded specimen was observed. Fracturing such a specimen gives a failure surface with numerous fibrils visible to the naked eye. It was proposed that the thermotropic PLC would form fibrous domains in the matrix polymer when the melt blended. With the inherent strength and stiffness of the thermotropic polymers, these fibrils would then act as a reinforcement much like chopped glass.

The densities of PPS/PLC and PC/PLC blends were also studied (33). The density of a PC/PLC blend increases with increasing weight percentage of PLC and follows the additive rule. While the density of a PPS/PLC blend decreased with increasing content of PLC, it was lower than the value calculated by the additive rule, indicating that the crystallinity of PPS in the blends was reduced by the interference of the PLC.

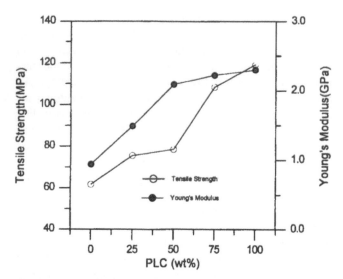

Figure 2 Tensile strength and Young's modulus vs. PLC content for PC/PLC composites.

III. POLYMER BLENDS

A. POM/EPDM Blends

POM/EPDM (48) blends have been prepared that contain up to 50% EPDM by weight (48), and it is observed that their tensile strength and tensile modulus show negative deviations from the rule of mixtures (so-called antagonism) (48). The EPDM was added to the POM matrix to increase the toughness and elongation to failure of the POM. The major differences between the behaviors of the two neat polymers are the higher value of Young's modulus and tensile strength for POM and the greater values of elongation at break point and higher impact strength for EPDM. The tensile strengths of POM/EPDM blends decrease with increasing amount of EPDM. Also, the elongation of the blends reaches a maximum at 7.5 wt% EPDM, because the crystallinity of the blend decreases with increasing concentration of EPDM and the Young's modulus of the blend decreases (48). The various properties of POM/EPDM blends are shown in Table 2.

Addition of a small amount of one polymer to another can often result in a change of structure of the polymer and in an increase in its strength (49–51). Strength varies with small amounts of additives because an additive acts as a damper and redistributes the internal stresses or fills defects within the microstructure of the bulk polymer. However, a detailed study of the changes in the

Table 2 Mechanical Properties of POM/EPDM Blends

POM/EPDM (wt%)	TS[a] (MPa)	M[b] (GPa)	E[c] (%)	IS[d] (kg cm/cm)
100/0	61.8	1.08	45	6.30
97.5/2.5	56.9	0.73	48	6.34
92.5/7.5	52.8	0.70	69	6.86
95.0/5.0	49.0	0.69	77	7.01
90/10	46.1	0.63	60	6.32
85/15	37.6	0.52	56	4.41
80/20	29.8	0.46	61	4.25
70/30	20.0	0.32	45	3.00
50/50	4.5	0.07	25	1.55

[a] TS: tensile strength.
[b] M: modulus.
[c] E: elongation.
[d] IS: notched Izod impact strength.

properties of polymers due to small additions of other polymers has not been made so far, and the mechanism of their effects remains unclear (52). Lipatov (53) thought that mechanical characteristics were connected with interpenetration of the additive component into the surface defects of the bulk component. It is known that the tensile strength of a toughened plastic decreases with the rubber content, while elongation at break increases, within a certain range, with rubber content. The POM/EPDM blends studied in this work follow this rule. For POM/EPDM blends where EPDM is the dispersed phase and POM the matrix, we believe that EPDM in the soft rubbery phase has improved the impact resistance by inducing crazing of POM or absorbing more energy during impact structure. It is well known that the effectiveness of toughening plastics with rubber depends on the shape, size, and distribution of rubber particles (34,54,55). The impact strength of POM/EPDM polyblends reaches a maximum and increases slightly at 7.5 wt% EPDM. It is believed that the EPDM particles are too small to resist the cracks that developed (48). It is understood that in notched specimens there are macrocracks or macroflaws present. Such specimens require large and more numerous rubber particles to stop microcrack propagation (56). Kaplan (57) used the mechanical damping peak (T_g) to discuss the miscibility of two polymers. The peak maximum of the damping curve in dynamic mechanical analysis is due to the glass transition temperature. If two materials are miscible, there is a single T_g. If they are immiscible there are two glass transition temperatures, and if partial miscibility occurs, the temperature locations of the peaks move closer together. Therefore, if two polymers are immiscible, an increase in content of one component will change the height of the polymers that are miscible and shift the transitional temperature. Dynamic mechanical analysis shows that when the EPDM content increases, only the height of the lower damping peak will change, but there is no transition temperature shift. It indicates that the miscibility of POM and EPDM is poor (48). However, the T_gs of ABS and PC moved close to each other for PC/ABS blends (12).

Scanning electron microscopy (SEM), in most cases, is a powerful tool for direct visualization of distinct phases in heterogeneous blends. In POM/EPDM blends, the particle sizes of the dispersed EPDM are in the range of 0.25–1.0 μm in diameter (48). Kaplan (57) used an N_c value ($N_c \equiv 150$ Å/domain size of dispersed polymer) to define the compatibility of two-component polymer blends. When $N_c = \infty$, the polymer blend is compatible; $N_c = 1$, semicompatible, and $N_c = 0$, incompatible. The N_c values vary from 0.015 (150 Å/1.0 μm) to 0.06 (150 Å/0.25 μm) in this study (48). Thus the POM/EPDM blends are referred to as incompatible systems. It is known that in order to obtain improvement in certain properties of a blend, the two components to be blended should be compatible to some degree (58). Because the POM/EPDM system is an incompatible system, the impact strength of blends will not improve markedly.

B. POM/PU Blends

POM/PU blends have been prepared that contain up to 50% by weight of polyurethane (PU) (59). The tensile strength and the modulus of blends both decreased with increasing PU content. On the other hand, the percent elongation at break deviates markedly from additivity connecting the values for pure POM and PU. The elongation of blends reaches a maximum at 20 wt% PU content (59).

The rubbery PU was added to the rigid POM matrix in order to increase the toughness and elongation of the POM break point. It is known that plastics' toughness decreases with rubber content, while elongation at break point increases with rubber content (56).

There are three principal ways of preparing blends with high toughness or impact resistance (34). The first method involves blending by mechanical techniques. The second method is the solution-graft copolymer technique, while the third involves emulsion polymerization. This study used the first method to prepare toughened plastic. The effectiveness of toughening with rubber depends on a number of factors including (60)

1. The concentration of the rubber
2. The size and dispersion of the rubber particles
3. The level and interfacial adhesion between matrix and rubber
4. The inherent ductility of the polymer matrix
5. The shear modulus and glass transition temperature (T_g) of the rubber
6. The craze initiation stress and shear yield stress of the matrix

The various theories thus proposed for explanation of rubber toughening may be categorized as follows (58):

1. Energy absorption by the rubber particles
2. Energy absorption by the yielding of the continuous phase: ductility enhanced by strain-induced dilatation near the rubber occlusion
3. Craze formation involving cavitation and polymer deformation with the craze
4. Shear yielding as a source of energy absorption and crack termination
5. Stress distribution and relief
6. Rubber particles acting as craze termination points and obstacles to crack propagation

The notched Izod impact strength of blends reaches a maximum at 10 wt% of PU for various POM/PU blending systems (59). For the same composition, the effectiveness of POM/PU blends is POM/S80A (ester base, shore A80, Elastollan) > POM/1190A (ether base, shore A90, Elastollan) > POM/S90A (ester base, shore A90, Elastollan) > POM/CVP760AW (ester base, shore A70, Elastollan). While the hardness of PU is shore A90, the ether-based PU is much more effective in increasing the impact strength of POM/PU blends than that of ester-based PU (59). While the hardness of PU is larger than shore A80, soft PU has much more effectiveness in increasing the impact strength of POM/PU blends than that of hard PU (59). The composition of POM/PU blends is shown in Table 3. The various properties of POM/PU blends are shown in Table 4.

The dynamic mechanical behavior of polymers is of great interest and is very important. Damping is often the most sensitive indicator of the different kinds of molecular motions occurring in a material in the solid state. For dynamic mechanical analysis [59], as the PU content increases, the glass transition temperatures of POM and PU do not move closer together, but the height of the damping peak changes. This result is similar to that of POM/EPDM blend and indicates that the miscibility of POM and PU is poor. The damping peak is associated with the partial loosening of the polymer structure so that groups and small chain segments can move. For dynamic mechanical measurement of POM/PU (90/10) blends, the best adhesion is achieved by the POM/1190A system because it has a small damping height at glass transition. For POM/ester-based PU blends, soft PU has higher damping peak values than hard PU (59).

SEM was used to examine the phase morphology of POM/PU blends of composition 90/10, 80/20, 70/30, and PU islands were observed (59). The cross section of the discrete islands has dimensions of 1–10 μm, depending on the precise composition. The spherical sizes of the dispersed PU, either ether-based PU or ester-based PU, are in the neighborhood of 1–3 μm in diameter at the composition of 10 wt% PU. For compositions containing 20 wt% PU, blends exhibit larger spherical domains of PU from 3 to 5 μm. In the other POM/PU blends, it was found that the size of PU particles increased with increase of the PU content in the blend, and that PU and POM both existed in the continuous phase (i.e., phase inversion occurred) at a 50/50 composition. The PU particles (below 30 wt% content) are spherical in

Table 3 Description of the Composition of
POM/PU Blends

No.	POM[a]	PU[c]	POM/PU
900	M90[b]	—	100/0
971	M90	CVP706AW[d]	90/10
972	M90	CVP706AW	80/20
973	M90	CVP706AW	70/30
975	M90	CVP706AW	50/50
981	M90	S80A[e]	90/10
982	M90	S80A	80/20
983	M90	S80A	70/30
985	M90	S80A	50/50
991	M90	S90A[f]	90/10
992	M90	S90A	80/20
993	M90	S90A	70/30
995	M90	S90A	50/50
911	M90	1190A[g]	90/10
912	M90	1190A	80/20
913	M90	1190A	70/30
915	M90	1190A	50/50

[a] POM: Copolymer-type polyactal, polyplastics,
  Japan.
[b] M90: Duracon M90-2.
[c] PU: Polyurethae, Elastollan, Elastogran
  Polyurethane-Elastomere GmbH, Germany.
[d] CVP706AW: Ester-type polyurethane, shore A70.
[e] S80A: Ester-type polyurethane, shore A80.
[f] S90A: Ester-type polyurethane, shore A90.
[g] S1190A: Ether-type polyurethane, shore A90.

shape with spherical inclusions of the POM matrix, since spherical particles of dispersed PU have the lowest energy in most compositions (61).

In two-component blends, the domain size of the dispersion polymer will be changed by the compatibility of these two polymers. Kaplan (57) has suggested that 150 Å was a universal constant. Findings indicate that a domain size of 0.015 μm for $N_c = 1$ is not a universal constant (62); it is strongly dependent upon the nature of the polymers involved in the blends.

C. POM/PTFE Blends

Polyacetal exhibits high crystallinity, and this dense crystalline structure accounts for many of the resin's properties, which include stiffness, fatigue endurance,

Table 4 Various Properties of POM/PU Blends

No.	TS[a] (MPa)	M[b] (GPa)	E[c] (%)	IS[d] (kg cm/cm)
900	61.8	1.08	45	6.3
971	49.5	0.70	54	6.2
972	38.4	0.58	63	5.3
973	27.5	0.41	57	3.7
975	17.0	0.20	50	2.1
981	50.2	0.70	71	8.2
982	40.9	0.61	76	7.2
983	29.4	0.47	72	5.6
985	13.2	0.15	62	4.0
991	50.9	0.72	58	6.8
992	38.9	0.59	70	6.3
993	26.8	0.44	63	5.0
995	20.9	0.27	56	3.5
911	52.5	0.71	63	7.6
912	37.5	0.57	74	6.6
913	33.7	0.45	65	5.3
915	20.6	0.22	60	3.6

[a] TS: tensile strength.
[b] M: modulus.
[c] E: elongation.
[d] IS: notched Izod impact strength.

durability, high temperature resistance, good solvent resistance, low coefficient of friction and stick-slip, good appearance, and high creep resistance. Therefore polyacetals are used in sliding applications, where their frictional and wear characteristics are most important. These inherent properties of polyacetals are coupled with low wear resistance and a low coefficient of friction imparted by PTFE (63–74).

It is known that fluoropolymers such as PTFE have low wettability and bondability due to low surface energy, incompatibility, chemical inertness, the presence of contaminants, and weak boundary layers. For these reasons, the adhesion between POM and PTFE is poor; therefore, the tensile strength, modulus, elongation, and notched Izod impact strength of POM/PTFE blends decrease with increase of PTFE. In POM/PTFE blends, the agglomerative particle size of PTFE powder increases with an increase in PTFE powder content. As the PTFE powder content is raised to 15 wt%, the PTFE powder aggregation is obvious (63). For this reason, the tensile strength of POM/PTFE (pure powder) blends decreases markedly when the PTFE powder content is increased above 15 wt%.

Due to the incompatibility of PTFE and POM, the surface modification of PTFE was attemped. The PTFE powder was treated with sodium-naphthalene complex in tetrahydrofuran to improve its surface property. IR and ESCA spectra showed that the surface of treated PTFE powder formed a —CF=CF— double bond structure (64,65). After being etched, most of the NaF salt formed was adsorbed onto the surface of the treated PTFE; only a small amount of the NaF doped into treated PTFE. The NaF salt adsorbed surface modified PTFE (CPTFE) was added to POM to improve the compatibility of POM and PTFE. The NaF acted as a coupling agent and resulted in very strong adhesion between POM and CPTFE (64,65). The tensile strength and the Young's modulus of POM/CPTFE blends were more than twice those of POM/PTFE blends (64,65). When the NaF salts adsorbed on the surface of CPTFE are rinsed off, the resultant PTFE is indicated as WPTFE. The tensile strength and the Young's modulus of POM/WPTFE blends (66) are higher than those of POM/PTFE blends. In this blend, the WPTFE disperses in POM homogeneously, and the particle size of the agglomerated WPTFE is smaller than that of agglomerated PTFE with less entrapped air and weak points in WPTFE/POM blend (66). On the other hand, the surface tension, polarity, wettability, and bondability of fluoropolymer are increased by the surface treatment (75–77). However, the friction coefficient of POM/WPTFE was higher than those of POM/PTFE blends because the smooth surface of the fluoropolymer was destroyed (66).

Another strategy is to add a small amount of WPTFE into PTFE and then add the WPTFE + PTFE into the POM (63). The WPTFE acts as a compatibilizer for POM and PTFE because the total attractive energy between POM and (WPTFE + PTFE) is higher than that between POM and PTFE. Therefore the tensile strength of POM(PTFE + WPTFE) is higher than that of POM/PTFE. On the other hand, the friction coefficient and wear resistance of POM/(PTFE + WPTFE) blends are higher than those of POM/PTFE blends.

D. POM/PTFE Fiber Blend

Polytetrafluoroethylene (PTFE) exhibits a very low coefficient of friction and easy formation of a thin film transferred on the counter surface in sliding wear. (Fig. 3). However, the compatibility of POM and PTFE fibers is poor. Therefore the improvement of compatibility of POM and PTFE fiber is also important for POM/PTFE fiber blends. The compatibility of POM and PTFE can be improved by using a chemical PTFE fiber that is surface modified or by adding surfactant in PTFE during manufacturing (SPTFE) (78). Using plasma to modify the surface chemical structure of PTFE fiber (PPTFE) is another method that increases the compatibility between POM and PTFE. After plasma surface modification, the PPTFE could be grafted by using UV radiation to enhance the interface bonding (78).

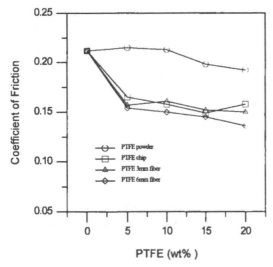

Figure 3 Coefficient of friction vs. various *L/D* ratios of PTFE content for POM/PTFE composites.

IV. POLYMER COMPOSITES

A. POM/Carbon Black Composite

The absorption (79) of UV light in the presence of oxygen causes a set of complex reactions leading to the degradation of polyacetal. Wavelengths shorter than 300 nm are filtered out by the atmosphere, and those longer than 400 nm are ineffective in the bond-cleavage process, and thus the 300–400 nm range is the cause of deleterious effects of sunlight. Polyacetal under the irradiation of short wavelength UV light at 254 nm and 360 nm undergoes depolymerization because of the cleaving of the main chain backbone (80,81). Once the degradation of polyacetal begins, its crystallinity increases and the size of the crystals will also change. Polyacetal degrades when exposed to an outdoor environment, and this leads to a sharp decrease not only in molecular weight but also in impact strength and elongation at break (82,83).

Carbon black in the polyacetal substrate imparts heat and light resistance, which is vital to ensure that the desired optimum end properties and characteristics are maintained. Polyacetal, being sensitive to elements of the environment including sunlight and rain, has to be tested after long periods of outdoor exposure, which in some cases involves several years. Accelerated weathering, however, can give reliable data in a matter of days or weeks, provided the correct conditions and appropriate apparatus are used.

Temperature, water, and ultraviolet light interact to increase the rate of degradation of polyacetal. Rain or dew condensed from an air–water vapor mixture is in effect in electrolyte bearing dissolved oxygen. The pH of rain or dew is slightly acidic, which, when combined with hot weather, can have a strong degradative effect. For this reason, it is necessary to use an apparatus that combines the effects of temperature, water, and ultraviolet light into a close analogy to the accelerated effects of the natural environment. The type of apparatus used for this purpose is the carbon arc.

The influence of ultraviolet irradiation on polyacetal was that it made the surface of polyacetal degrade to form a white powder; thus carbon black was added to enhance the weather stability. However, since compatibility between these two materials was very poor, oligomeric polyethylene glycol (PEG), MW 400, was used to modify the compatibility between them (79), since otherwise carbon black cannot disperse in POM homogeneously. The determination of the effectiveness of carbon black on the useful service life of polyacetal in outdoor applications necessitates testing the mechanical properties of the polymer before and after weathering.

Sufficiently long irradiation brings about remarkable deterioration in ultimate mechanical properties. This is attributed to the heterogeneous nature of the photo-oxidative degradation process (84,85), which is concentrated in a finite number of sites, thus forming crack precursors rather than changing the material properties in bulk. This is remarkably similar to the effect of introducing artificial flaws into the specimens.

Polyacetal under the irradiation of UV light undergoes depolymerization as is illustrated in Fig. 4 (79). The impact strength has low values, due to the higher concentration of —CO— (carbonyl). These factors increase the rigidity of the chains and hence lower the values of impact strength (Fig. 5).

Photochemical degradation leads to deterioration of mechanical properties, causes cracking, and eventually results in complete disintegration of the material.

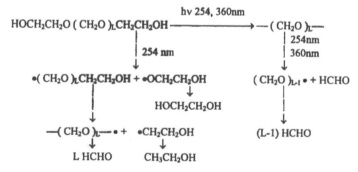

Figure 4 Polyacetal depolymerization mechanism under UV irradiation.

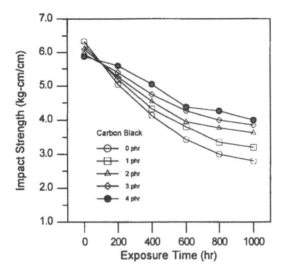

Figure 5 Notched Izod impact strength vs. exposure time for POM/carbon black composites with different carbon black content.

Ultimate mechanical properties (especially elongation at break) of polyacetal are more sensitive to irradiation than to chemical changes (86). The deterioration of mechanical properties is a final consequence of the series of processes involved in photochemical degradation. The obvious cause of such a peculiar behavior is localization of a destructive process in a finite number of sites, mainly on the sample surface. The ultimate behavior responds to localized structure irregularities and defects in a much more pronounced way than to the average properties of the material in bulk.

There are two reasons for the sensitivity of the ultimate properties to photooxidative degradation. The first is the key role of defects or cracks in the fracture behavior of materials expressed theoretically in the classical Griffith concept and in the statistical weakest-link hypothesis. The other reason is the heterogeneous character of the degradation process itself. One should bear in mind that photodegradation proceeds preferentially in a comparatively localized region near the initiation site (such as a photosensitive impurity, a structural element in the polymer or additive). Such a localization of the process is a consequence of the low molecular mobility both of the polymer and of the photoactive site in the solid state. When the body is subjected to mechanical stress, the degraded weak sites act as stress concentrators and crack nuclei.

Carbon black is an effective inhibitor of oxidation in many systems, and its structure is best described as agglomerates of complex, condensed polycyclic aromatic rings with hydrogen and reactive functional groups around the edges of the essentially planar molecules. The typical functional groups (H, COOH, O, OH) are believed to be present in many carbon blacks (87). The phenolic groups could account for part of the activity as an antioxidant, but the presence of both quinone and polynuclear aromatic structures suggests the ability of carbon blacks to trap the free radical to form stabilized radicals and thus to terminate the kinetic chain.

Its effectiveness in protecting polymers against outdoor weathering, particularly at a high temperature, can be attributed in part to its ability to inhibit thermal oxidation, which often occurs concurrently with photodegradation. In addition, carbon black is a very effective light screen and functions as an ultraviolet absorber through energy level transitions in its polynuclear aromatic structure. Its effectiveness, coupled with low cost, accounts for the wide use of this pigment in protecting polymers against photodegradation. However, it is obvious that its use is limited to applications in which black compositions are acceptable.

The effectiveness of carbon black as a light screen depends on several additional factors, for example concentration, particle size, and dispersion. The relation between these variables and their protective effect has been described by Wallder et al. (88) and reviewed in several recent publications. Both particle size and concentration influence the dispersion of the pigment in the polymer matrix, and this ultimately determines its effectiveness. This can be interpreted as an indication of the relative importance of light screening in comparison to the other mechanisms by which carbon black can function as a stabilizer.

To ensure sufficient UV resistance of POM, it is necessary to add the carbon black into POM, and the dispersion of carbon black to POM homogeneously is very important. PEG, mw 400, is useful to modify the dispersion of carbon black and the compatibility between carbon black and POM (79).

B. POM/Glass Fiber Composite

In a number of studies (89) of dynamic dispersion of glass fiber reinforced polymer composites, it was reported that some motion-restricted microphase boundaries that were formed around the surface of glass fiber and characteristics of a secondary dispersion appeared at a temperature above the primary transition of major dispersion. In recent years, the performance of glass fiber reinforced polymer composites has been enhanced because the properties of the glass fiber itself have been greatly improved. In order to take the most advantage of glass fiber reinforced polymer composites, it is essential to improve the adhesion between the glass fiber and the substrate. There are several methods for enhancing the adhesion between the glass fiber and the substrate, such as oxidation, corrosion, and coating polymer on the surface of the glass fiber. Through this surface treatment of the glass fiber, the strength of adhesion to the substrate is evidently enhanced. In the past, there was no effective method for testing the strength of adhesion and there was also little research on surface treatment of fibers in terms of the adhesivity between the glass fiber and the substrate. Polymers such as PU, ethylene-vinyl acetate (EVA), acrylonitrile-butadiene-styrene (ABS), and high impact polystyrene (HIPS) can be coated on the glass fiber and then serve as reinforcement for polyacetal composites (89). This modification leads to good mechanical properties, good thermal and dimensional stabilities, and good color appearance when compared to composite using glass fiber without modification. These effects are attributed to the improvement in interfacial adhesion between the glass fiber and the polyacetal substrate.

For the composites with glass fiber treated with polymer coatings, it was found that the ultimate strength at break and tensile strength increased drastically compared with pure POM or POM reinforced with untreated glass fiber (Fig. 6).

In spite of superior mechanical properties of glass fiber itself, the composite of polyacetal and glass fiber does not show improved mechanical properties because of weak adhesion at the interfacial boundaries between glass fiber and polyacetal. It is found that the surface properties of the fiber such as wettability and adhesive properties influence the mechanical properties of composite more effectively than the mechanical properties of the reinforcing fiber (89). The strength of the adhesion between the surface of modified glass fiber and the matrix is considered to be stronger than that of the untreated glass fiber and the matrix. In the case of composite with glass fiber of which the surface was treated by PU, EVA, ABS, HIPS, the fiber was easily dispersed into the matrix because of the great affinity between them, whereas it was difficult to disperse untreated glass fiber into the matrix. On the other hand, the ultimate strength and the Young's modulus for the composite of glass fiber modified by PU with the poly-

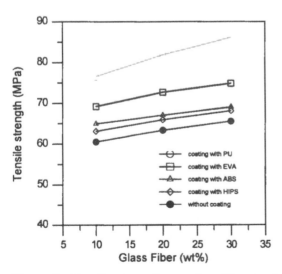

Figure 6 Tensile strength vs. glass fiber content for POM/glass fiber composites.

acetal matrix was greatly increased in extensibility, compared with those of the composite reinforced with the glass fiber only (89). This fact means that the adhesion between the surface of the glass fiber and polyacetal is enhanced and forms a more uniform interface. As the glass fiber was coated by polymer, the growth of cracks initiated by the microvoids at the ends or the interfacial boundaries of the glass fiber might become ineffective due to energy dissipation by the coated polymer.

The impact strength of composites is even more complicated than that of unfilled polymers because of the added reinforced fiber. As with tensile properties, the interface between reinforcement and matrix also plays an important role. For a material to be very tough with high impact strength, there must be some mechanism for dissipating the stored forced energy throughout the bulk of the sample. If the energy is concentrated on only a small area, the material turns brittle and the impact strength will be low. Reinforcements have the following two mechanisms for dissipation of forced energy (90,91). The first is that fibers may pull out of the matrix and dissipate energy by mechanical friction. In addition, the pulling out of the fiber prevents localization of stress in one area along the fiber. The second is controlled dewetting of the fiber, which dissipates energy in the dewetting process, spreading the region of stress concentration throughout a large region, which tends to stop the propagation of the crack. Fibers also tend to reduce the impact strength because (a) they generally drastically reduce the elongation to break and thus may reduce the area under the stress–strain curve or (b) stress concentrates occur at the regions around the fiber end, areas of poor adhesion, and regions where fibers contact one another. Thus, depending upon the nature of the composite and the type of impact test, fibers can cause the apparent impact strength either to increase or to decrease.

The impact strengths are greatly improved in the reinforced composite systems with the glass fiber coated by polymers in comparison with the untreated glass fiber system (Fig. 7). Among polymers used, PU shows the best improvement and is followed by EVA, ABS, and HIPS. Impact strengths decrease with the increase of the glass fiber content.

The rule of additivity of properties of components for the composites was applied to the data to evaluate the fiber efficiency factors for strength (K_σ) and modulus (K_e). The expressions for the rule of additivity for tensile strength and tensile modulus of discontinuous fiber reinforced composites made with random-in-plane fiber orientation are (93)

$$\sigma_c = K_\sigma \sigma_f V_f + \sigma_m (1 - V_f)$$
$$E_c = K_e E_f V_f + E_m (1 - V_f)$$

where σ_c is the ultimate strength of the composite, K_σ is the fiber efficiency factor for strength, σ_f is the ultimate strength of the reinforcing fiber, V_f is the fiber volume fraction, σ_m is the matrix stress at the fracture strain of the composite, E_c

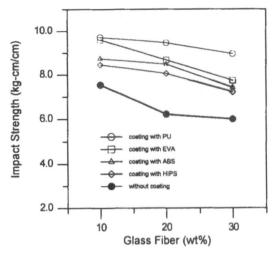

Figure 7 Notched Izod impact strength vs. glass fiber content for POM/glass fiber composites.

is the Young's modulus of the composite in the plane of the fibers, K_e is the fiber efficiency factor for the modulus, E_f is the modulus of the reinforcing fiber, and E_m is the matrix modulus. K_e and K_σ values for the experimental composites are given in Table 5. The values imply that the surface modification of the glass fiber by polymers results in a greater value that plays an important role in the formation of a stronger interfacial layer between the glass fiber and the matrix. There is therefore a great improvement in mechanical properties of the composite of the random-in-plane oriented discontinuous glass fiber.

In the case of a glass fiber reinforced composite, the fiber efficiency factor at room temperature for modulus and strength indicates that these values are far less than those obtained in polymer-treated glass fiber reinforced composites. In other

Table 5 Fiber Efficiency Factors for Modulus (K_σ) and Strength (K_e)

	K_σ	K_e
M90/Glass fiber	0.0070	0.0030
M90/Glass fiber coated with PU	0.0083	0.0064
M90/Glass fiber coated with EVA	0.0075	0.0056
M90/Glass fiber coated with ABS	0.0072	0.0050
M90/Glass fiber coated with HIPS	0.0068	0.0043
M90: Polyacetal		

words, all the glass fiber efficiency factors are very low. This suggests that the adhesive strength between glass fibers and polyacetal copolymer is very weak.

To enhance the adhesion between the reinforcement (glass fiber) and the matrix (polyacetal), polymers such as ester type polyurethane (PU), EVA, ABS, and HIPS were used as compatibilizing agents in coating treatment to form a thin layer on the glass fiber (89). Composites made from polymer-pretreated glass fiber have better mechanical and thermal properties as a result of the better adhesion between them. The calculated fiber efficiency factors for modulus (K_σ) and strength (K_e) were larger than those of the composite reinforced by the glass fiber but not treated with polymers. These results reveal that the adhesion between copolymer type polyacetal and glass fiber is enhanced by the coating treatment of polyurethane (89).

C. LDPE/Magnetic Powder Composite

Magnetic properties of plastic magnets (94–96) composed of polymer matrices and magnetic powder are inferior to general casting or sintered magnets; nevertheless, they have some advantages, e.g., high producibility, possible production of complicated small and thin shapes with precision, etc. Therefore they have been widely used in various fields and have played important roles in the rapid development of electronic and communications instruments, household utensils, audio equipment, and so on (97). Low-density polyethylene (LDPE) is a widely used, general purpose plastic for commodity products. Its properties such as extreme toughness, chemical inertness, low-temperature brittleness, low-temperature flexibility, high environmental stress crack resistance, etc. show excellent processibility and highlight their usefulness as the matrix of plastic magnets. However, a variety of inorganic particles dispersed in LDPE yields substantial incompatibility in many processes and leads to performance problems exemplified by poor dispersion, agglomeration, elevated viscosity, cosmetic defects, and adhesion failure. On the other hand, titanate coupling agents (CA) are molecular bridges at the interface between two substrates. They will generally improve the dispersion of the fillers and the interface of the blend. Therefore CA-treated magnet powders for plastic magnets have many physical properties better than those of the untreated blends that are traditionally used.

For the magnetic powder (HM-170), $SrFe_{12}O_{19}$ was pretreated with CA, such as neoalkoxy tri(dioctylpyrophosphate) titanate (Lica 38) or neopentyl (dially)oxytri-(N-ethylene diamino)ethyl titanate (Lica 44), in various concentrations from 0.5 to 2.5 phf (parts per hundred fillers) on the weight percent of fillers and then blended with LDPE (NA-289) pellets (94). The magnetic powder fillers (40 to 60 vol% of the total volume) are added to LDPE with or without pretreatment by the coupling agent and were compressed and molded to form samples for determination of magnetic and physical properties (94). The magnetic properties, such as the residual magnetic flux density, the coercive force, and the maximum energy product, of the plastic magnets usually increase with an increase in the concentration of the magnetic powder. CA proved to be effective in improving magnetic properties for concentrations of greater than 1.0 phf. The most suitable concentration for pretreated magnetic powder was found at the 1.5 phf of Lica 44 and the 2.0 phf of Lica 38 (94). The effectiveness is likely because the CA improves the magnetic power due to a more uniform dispersion of powders in the LDPE matrix, decreasing the gaps between filler particles. The results of our study are that the residual flux density (Br) is 1.6 K gauss, the cohesive force (Hc) is 1.2 K Oe, and the maximum energy product (BH)max is 0.5 G gauss-Oe (94).

The impact strength of the plastic magnet decreases with the increase of the volume percentage of the magnetic powder. However, the CA-pretreated magnetic powder showed better results than the magnetic powder without pretreatment (Fig. 8). The most suitable amount was found at 1.5 phf of Lica 44 and 2.0 phf of Lica 38, with Lica 44 showing better interaction than Lica 38. These results reveal that the CA is capable of preventing the crack from forming in the interface between the resin and the magnetic powder. Because the titanate CA has two components (one is a hydrolyzable group, and the other is organic functional groups), it can act as a compatibilizer between the filler and the resin, in fact, resulting in a hydrogen bridge in the interface between the magnetic powder and the resin. In

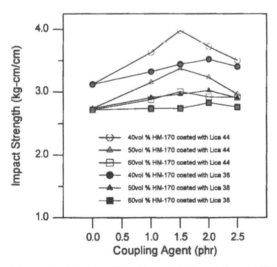

Figure 8 Notched Izod impact strength vs. HM-170 vol% coated with various concentrations of coupling agent for LDPE.

other words, CA enhances the binding force and improves the possibility of preventing crack growth.

From the SEM micrographs of the fracture surface, it is apparent that the surface of the magnetic powder with a pretreated titanate CA has a more uniform dispersion in the plastic magnet than that without such pretreatment, and there are
fewer gaps between the magnetic powder and the resin for pretreated powder (94).
This result reveals that the density of the plastic magnet increases with an increase
in the amount of the CA and the improvement of the magnetic properties of the
plastic magnets.

D. PP/mica Composite

The tensile strength of PP/mica composite increases with the addition of mica
(98). However, the tensile strength of PP/mica composite could be increased by
grafting acrylic acid (AAc) on to PP (PP-g-AAc) (98). With a filler of 30~60
phr, the maximum value of tensile strength is obtained, and this value will increase with increasing grafting ratio. The -COOH functional group contained on
PP side chains will be increased as the grafting ratio is increased and the chance
for the -COOH to bond with mica is increased (98). On the other hand, if the mica
is treated by the silane coupling agent (Z6020, N-(2-aminoethyl)-γ-amino-
propyl-trimethoxy silane), the synergistic effect of the coupling agent and AAc
grafting ratios on the improvement of the tensile strength of the composites is
significant when compared with unmodified PP composites (Fig. 9). The en-

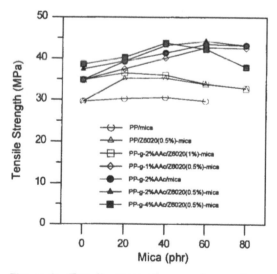

Figure 9 Tensile strength vs. mica content with various kinds of surface
modified method for PP/mica composites.

hancement occurs both due to the amount of mica treatment with coupling agent and to the amount of AAc added to the graft copolymerization. It was found that the addition of 1% AAc is preferable to an addition of 2% or 4% AAc when the mica content is lower than 40 phr. The PP-g-AAc with a polar group of -COOH can easily produce hydrogen bonding in PP matrixes and could form ionic bonding with mica, even though only 1% of AAc was added (98). In terms of impact properties, grafting will make PP brittle because a partial ionic bond cross-linking will be formed when mica is used as a reinforcement. For this reason, the impact strength of PP-g-AAc/mica composites decreased with increasing amount of mica not treated with coupling agent. For these composites, the impact strength was largely determined by dewetting. However, the mica with coupling agent treatment not only showed no deterioration but also led to an increase in impact strength with an increase in mica content.

E. PP/Mg(OH)$_2$ Composite

In these blends (99), the filler Mg(OH)$_2$ acts as a flame retardant and smoke suppressant and makes the matrix less flammable. However, a large amount of filler will seriously degrade the mechanical properties. This phenomenon can be improved by using PP-g-AAc. AAc was grafted to PP as a side chain to make PP molecules difficult to crystallize. Therefore the amorphous phase of the matrix increased, and the increase of impact strength was due to the decrease of PP crystallinity. The effect of AAc is to enhance the mechanical properties, since PP is a nonpolar polymer and dispersion of the polar filler Mg(OH)$_2$ in the polymer may not be ideal. However, the grafting reaction of PP by AAc improves the dispersion of the filler and the adhesion of the interphase between the filler and the polymer matrix. The impact strength, tensile strength, and Young's modulus are all increased.

V. CONCLUSIONS

To achieve special physical properties of polymeric materials, the general method is to make the functional materials by mixing two or more polymers (or a polymer and a filler) to produce polyblends, alloys, or composites. The compounding and compatibilization of these high-performance polymer alloys, blends, and composites is very important. In this chapter, some methods were offered to increase the compatibility between the polymers or between the polymer and the filler, in addition to enhancing the properties of polymeric materials. These methods include interface modification, adding compatibilizers, and choosing miscible polymers for blending.

REFERENCES

1. L. A. Utracki. Polymer Alloys and Blends. New York: Hanser, 1989, pp 1–4.
2. M. E. Fowler, H. Keskkula, and D. R. Paul. Distribution of MBS emulsion particles in immiscible polystyrene/SAN blends. J. Appl. Polym. Sci. 37:225, (1989).
3. D. Debier, J. Devaux, and R. Legras. Influence of a core/shell rubber phase on the morphology and the impact resistance of a PC/SAN blend (75/25). Polym. Eng. Sci. 34:613 (1994).
4. D. J. Ihm and J. L. White. Interfacial tension of polyethylene/polyethylene terephthalate with various compatibilizing agents. J Appl. Polym. Sci. 60:1(1996).
5. F. C. Chang and Y. C. Hwu. Styrene-maleic anhydride and styrene-glycidyl methacrylate copolymers as in situ reactive compatibilizers of polystyrene/nylon 6,6 blends. Polym. Mater. Sci. Eng. 64: 155 (1991).
6. P. C. Lee, W. F. Kuo, and F. C. Chang. In situ compatibilization of PBT/ABS blends through reactive copolymers. Polymer 35:5641 (1994).
7. I. Park, J. W. Barlow, and D. R. Paul. The in situ reactive compatibilization of nylon 6 polystyrene blends using anhydride functionalized. J. Polym. Sci. Part B: 30: 1021 (1992).
8. D. G. Peiffer and M. Rabeony. Physical properties of model graft copolymers and their use as blend compatibilizers. J. Appl. Polym. Sci. 51: 1283 (1994).
9. R. Holsti-Miettinen and J. Seppala. Effects of compatibilizers on the properties of polyamide/polypropylene blends. Polym. Eng. Sci. 32:13 (1992).
10. F. Ide and A. Hasegawa. Studies on polymer blend of nylon 6 and polypropylene or nylon 6 and polystyrene using the reaction of polymer. J. Appl. Polym. Sci. 18 (1974).
11. N. R. Choudhuny and A. K. Bhowmick. Compatibilization of natural rubber-polyolefin themoplastic elastomeric blends by phase modification. J. Appl. Polym. Sci. 38:1091 (1989).
12. W. Y. Chiang and D. S. Hwung. Properties of polycarbonate/acrylonitrile-butadiene-styrene blends. Polym. Eng. Sci. 27(9): 632 (1987).
13. C. C. Wu. The Investigation of EMI Shielding of PC/ABS Conductive Composites, M. S. thesis, Tatung Institute of Technology, June, 1994.
14. W. Y. Chiang and G. L. Tzeng. The effect of the compatibilizers on flame retardant polycarbonate/acrylontrile-butadiene-styrene alloy. J. Appl. Polym. Sci., 65:795 (1997).
15. J. Kuczynski, R. W. Snyder, and P. P. Podolak. Physical property retention of PC/ABS blends. Polym. Degrad. Stab. 43: 285 (1994).
16. H. Suarez, J. W. Barlow, and D. R. Paul. Mechanical properties of ABS/polycarbonate blends. J. Appl. Polym. Sci. 29: 3253 (1984).
17. W. N. Kim and C. M. Burns. Thermal behavior, morphology, and some melt properties of blends of polycarbonate with poly(styrene-co-acrylonitrile) and poly(acrylonitrile-butadiene-styrene). Polym. Eng. Sci. 28: 1115 (1988).
18. M. P. Lee, A. Hiltner, and E. Baer. Fractography of injection molded polycarbonate acrylonitrile-butadiene-styrene terpolymer blends. Polym. Eng. Sci. 32: 909 (1992).
19. J. J. Herpeles and L. Mascia, Effects of styrene-acrylonitrile/butadiene ratio on the toughness of polycarbonate/ABS blends. Eur. Polym. J. 26: 997 (1990).
20. J. H. Chun, K. S. Maeng, and K. S. Suh. Miscibility and synergistic effect of impact strength in polycarbonate/ABS blends. J. Mater. Sci. 26: 5374 (1991).

21. M. P. Lee, A. Hiltner, and E. Baer. Phase morphology of injection-moulded polycar-
     bonate/acrylonitrile-butadiene-styrene blends. Polymer 33: 685 (1992).
22. J. C. Huang and M. S. Wang. Recent advances in ABS/PC blends. Adv. Polym. Tech-
     nol. 9: 293 (1989).
23. F. J. Balta Calleja, T. A. Ezquerra, D. R. Rueda, and J. J. Alonso Lopez. Conductive
     polycarbonate-carbon composites. J. Mater. Sci. Lett. 3: 165 (1984).
24. A. S. Wood. Big buildup in polycarbonate supply; big buildup in polycarbonate val-
     ues. Modern Plastics International, Dec, 39 (1989).
25. J. D. Keitz, J. W. Barlow, and D. R. Paul. Polycarbonate blends with styrene/acry-
     lonitrile copolymers. J. Appl. Polym. Sci. 29:3131 (1984).
26. S. K. Sikdar. The world of polycarbonate. Chemtech, Feb., 112 (1987).
27. M. P. Lee, A. Hiltner, and E. Bear. Formation and break-up of a bead-and-string
     structure during injection moulding of a polycarbonate/acrylonitrile-butadiene-
     styrene blend. Polymer 33: 675 (1992).
28. D. Quintens and G. Groeninckx. Visco-elastic properties related to the phase mor-
     phology of 60/40 PC/SAN blend. Polym. Eng. Sci. 31: 1207 (1991).
29. D. Quintens and G. Groeninckx. Phase morphology characterization and ultimate me-
     chanical properties of 60/40 PC/SAN blend: influence of the acrylonitrile content of
     SAN. Polym. Eng. Sci. 31: 1215 (1991).
30. D. R. Paul and J. W. Barlow. A binary interaction model for miscibility of copoly-
     mers in blends. Polymer 25: 487 (1984).
31. J. Edenbaum, ed. Plastics Additives and Modifiers Handbook. New York: Van Nos-
     trand Reinhold, 1992, p. 1019.
32. J. S. Lee. The functions of compatibilizers in nylon-6/ABS blends and its EMI shield-
     ing effects. M. S. thesis, Tatung Institute of Technology, June, 1994.
33. W. Y. Chiang and Y. T. Young. Thermal and mechanical properties of blends con-
     taining a liquid crystalline polymer. Tatung J. 19: 105 (1989).
34. J. A. Manson and L. H. Sperling. Polymer Blends and Composite. New York:
     Plenum, 1976.
35. L. E. Nielsen. Mechanical Properties of Polymers and Composites, Vol. 2. New York:
     Marcel Dekker, 1974.
36. G. Kiss. In situ composites: blends of isotropic polymers and thermotropic liquid
     crystalline polymers. Polym. Eng. Sci. 27(6): 410 (1987).
37. M. R. Nobile, E. Amendola, L. Nicolais, D. Acierno, and C. Carfagne. Physical prop-
     erties of blend of PC and LC copolyesters. Polym. Eng. Sci. 29:244 (1989).
38. M. Kimura and R. S. Porter. Compatibility of poly(butylene terephthalate) with a liq-
     uid-crystalline copolyester. J. Polym. Sci., Polym. Phys. Ed. 22: 1697 (1984).
39. A. Siegmann, A. Dagan, and S. Kenig. Polyblends containing a liquid crystalline
     polymer. Polymer 26:1325 (1985).
40. T. S. Chung. Thermotropic liquid-crystal polyester/nylon 12 blends. ANTEC SPE:
     1404 (1987).
41. K. G. Blizard and D. G. Baird. Blending of liquid crystalline polymers with engi-
     neering thermoplastics. ANTEC SPE: 311 (1986).
42. R. Ramanathan, K. G. Blizard, and D. G. Baird. The processing of thermotropic liq-
     uid crystalline polymers with engineering thermoplastics. ANTEC SPE: 1399 (1987).
43. K. G. Blizard and D. G. Baird. The morphology and rheology of polyblends contain-
     ing a liquid crystalline copolymer. Polym. Eng. Sci. 27(9): 653 (1987).

44. A. Apicella, P. Iannelli, L. Nicodemo, L. Nicolais, A. Roviello, and A. Sirigu. Dimensional stability of polystyrene/polymeric liquid crystal blends. Polym. Eng. Sci. 26(9): 600 (1986).

45. R. A. Weiss, W. Huh, and L. Nicolais. Novel reinforced polymers based on blends of polystyrene and thermotropic liquid crystalline polymer. Polym. Eng. Sci. 27(9): 684 (1987).

46. W. Huh, R. A. Weiss, and L. Nicolais. Thermal and rheological properties of blends of polystyrene and thermotropic liquid crystals. Polym. Eng. Sci. 23(14): 779 (1983).

47. M. Paci, C. Barone, and P. L. Magagnini. Calorimeter study of blends of poly(butadiene terephthalate) and a liquid crystalline polyester. J. Polym. Sci., Polym. Phy. Ed. 25: 1595 (1987).

48. W. Y. Chiang and C. Y. Huang. Properties of copolymer-type polyacetal/ethylene-propylene-diene terpolymer Blends. J. Appl. Polym. Sci. 47: 105 (1993).

49. R. J. M. Borggreve, R. J. Gaymans, and J. Schuijer. Impact behaviour of (Nylon-rubber blends: 5 influence of the mechanical properties of the elastomer. Polymer 30: 71 (1989).

50. E. Lebedev, Yu. S. Lipatov, and V. Privaiko. Morphological evaluation of the interaction of polyethylene with poly(oxymethylene) in a mixture obtained by extrusion. Vysokomol. Soedyn. A17: 148 (1975).

51. B. Z. Jang, D. R. Uhlmann, and J. B. Vander Sande. Crystalline morphology of polypropylene and rubber modified polypropylene. J. Appl. Polym. Sci. 29: 4377 (1984).

52. Y. Lipatov. Structure, mechanical, and rheological properties of polyethylene-poly(oxymethylene) blends. J. Appl. Polym. Sci. 22: 1895 (1978).

53. Y. Lipatov. Pure Appl. Chem. 43: 273 (1975).

54. C. B. Bucknall, Toughened Plastics. London: Applied Science Publishers, 1977.

55. J. Silberger and C. D. Han. The effect of rubber particle size on the mechanical properties of high-impact polystyrene. J. Appl. Polym. Sci. 22: 599 (1978).

56. D. Yang, B. Zhang, Y. Yang, Z. Fang, G. Sun, and Z. Feng. Morphology and properties of blends of polypropylene with ethylene-propylene rubber. Polym. Eng. Sci. 24: 612 (1984).

57. D. S. Kaplan. Structure-property relationships in copolymers to composites: molecular interpretation of the glass transition phenomenon. J. Appl. Polym Sci. 20: 2615 (1976).

58. D. R. Paul and S. Newman, eds. Polymer Blends, Vol. 1. New York: Academic Press, 1978.

59. W. Y. Chiang and C. Y. Huang. The effect of the soft segment of polyurethane on copolymer-type polyacetal/polyurethane Blends. J. Appl. Polym. Sci. 38: 951 (1989).

60. S. Y. Hobbs, R. C. Bopp, and V. H. Watkins, Toughened nylon resins. Polym. Eng. Sci. 23: 380 (1983).

61. S. Bywater. Block polymers. Characterization and use in polymer blends. Polym. Eng. Sci. 24: 104 (1984).

62. O. Olabisi, L. M. Robeson, and M. T. Shaw. Polymer-Polymer Miscibility. New York: Academic Press, 1979, p. 121.

63. W. Y. Chiang and C. Y. Huang. Synergistic effect of PTFE and sodium etched PTFE on polyacetal ternary blends. Eur. Polym J. 29(6): 843 (1993).

64. W. Y. Chiang and C. Y. Huang. Polyacetal/poly(terefluoroethylene) blends I. The effect of Na-treated poly(terefluoroethylene) on polyacetal. Angew. Makromol. Chem. 196: 21 (1992).

65. W. Y. Chiang and C. Y. Huang. Polyacetal/polytetrafluoroethylene blends II. The effect of chemical surface treatment of polytetrafluoroethylene. Eur. Polym. J. 28(6): 583 (1992).

66. W. Y. Chiang and C. Y. Huang. The Effect of surface modified polytetrafluoroethylene on polyacetal/polytetrafluoroethylene blends. J. Appl. Polym. Sci. 47: 577 (1993).

67. R. E. Estell. Acetal-PTFE resin blend improves properties and processing. Plastic Design and Processing 26: 35 (1971).

68. Y. Yamaguchi, I. Sekiguchi, K. Sugiyama, and N. Suzuki. Effects of carbon fiber and PTFE [poly(tetrafluoroethylene)] powder on the friction, wear, and PV [pressure sliding speed] value of polyacetal. Kogakuin Daigaku Kenkyu Hokoku 31: 26 (1972).

69. Y. Okubo, Y. Yamaguchi, and I. Sekiguchi. Effect of polyblend upon the friction and wear properties of plastics. III. Effects of blending polybutadiene and polytetrafluoroethylene on the properties of polyoxymethylene and polyamide. Kogakuin Daigaku Kenkyu Hokoku 53: 22 (1982).

70. Taiho Kogyo Co. Antifriction Materials. Jpn., Kokai Tokyo Koho, JP 60 72, 952.

71. K. Ziemianski. Analysis of the usability of selected thermoplastic materials in sliding Pairs. Trybologia 19(6): 4 (1988).

72. I. Sekiguchi, Y. Yamaguchi, Y. Okubo, H. Kondo, and S. Onda. Improvement in tribological of plastics by using polymer alloy. Kogakuin Daigaku Kenkyu Hokoku 64: 49 (1988).

73. K. Tanaka and Y. Yamada. Influence of counterface roughness on the friction and wear of poly(tetrafluoroethylene)-and-polyacetal-based Composites. J. Synth. Lubr. 5(2): 115 (1988).

74. H. Endo and M. Umeda. Jpn. railroad sliding parts. Kokai Tokyo Koho, JP 62 54, 753.

75. D. T. Clark and D. R. Hutton. Surface modification by plasma techniques. I. The interactions of a hydrogen plasma with fluoropolymer surfaces. J. Polym. Sci. Polym. Chem. 25: 2643 (1987).

76. D. M. Brewis, R. H. Dahm, and M. B. Konieczko. Reactions of polytetrafluoroethylene with electrochemically generated intermediates. Angew. Mikromole. Chem. 43: 191 (1975).

77. S. Wu. Polymer Interface and Adhesion. New York: Marcel Dekker, 1982, pp. 280–283.

78. C. C. Cheng. The research and applications of lubricant fiber composited polyacetal engineering plastics. M. S. thesis, Tatung Institute of Technology, June, 1996.

79. W. Y. Chiang and J. J. Luor. The effect of oligometric polyethylene glycol modofoed carbon black on the aging of polyacetals. Tatung J. 19: 91 (1989).

80. R. D. Deanin, S. A. Orroth, R. W. Eliasen, and T. N. Greer. Ultraviolet degradation of plastics. Polym. Eng. Sci. 10(4): 228 (1970).

81. J. W. S. Hearle and B. Lomas. Controlled degradation of polybutylene. J. Appl. Polym. Sci. 21: 1103 (1977).

82. J. Pabiot and J. Verdu. Fillers and reinforcements for plastics. Polym. Eng. Sci. 21: 32 (1981).

83. B. Ranby and J. F. Rabek. Photophysical processes of plastics. J. Appl. Polym. Sci. Appl. Polym. Symp. 35: 243 (1979).
84. P. Vink. Crystalline olefin polymers. J. Appl. Polym. Sci. Appl. Polym. Symp. 35: 265 (1979).
85. M. Y. A. Younan, M. A. Elrifai, R. Mohsen, and I. M. El-Hennawi. Polymer stability. J. Appl. Polym. Sci. 28: 3247 (1983).
86. M. Y. A. Younan, R. Mohsen, I. M. El-Hennawi, and M. A. El-Rifai. Additives for plastics. J. Appl. Polym. Sci. 28: 3349 (1983).
87. M. J. Astle and J. R. Shelton. Organic Chemistry, 2d ed. New York: Harper, 1949, pp. 731.
88. V. T. Wallder, W. J. Clarke, J. B. DeCoste, and J. B. Howard. Fillers and reinforcements for plastics. Ind. Eng. Chem. 42: 2320 (1950).
89. W. Y. Chiang and J. J. Luor. The properties of polymer-coated glass fiber reinforced copolymer-type polyacetal composite. Tatung J. 21: 123 (1991).
90. M. M. Coleman and J. Zarian. Mechanical properties of polymers and composites. J. Polym. Sci. Polym. Phy. 17: 837 (1979).
91. T. Kodama and Y. Furuhashi. Fiberglass and advanced plastics composite. J. Soc. Fiber Tech. (Japan) 34: T-72 (1982).
92. H. S. Katz and J. V. Milewski. Handbook of Fillers and Reinforcements for Plastics. Section III. New York: Van Nostrand Reinhold, 1978.
93. B. F. Blumentritt, B. T. Vu, and S. L. Cooper. Reinforced plastics/composites inst. Polym. Eng. Sci. 15: 428 (1975).
94. W. Y. Chang and G. M. Yang. Properties of LDPE/magnetic powder/coupling agent blends for isotropic plastic magnet. Tatung J. 22: 169 (1992).
95. Y. C. Hung, W. S. Ko, M. J. Tung, L. K. Chen, and W. C. Chang. Effects of additives on the orientation and strength of plastic ferrite magnet. IEEE Trans. Magn. 25(5): 3287 (1989).
96. H. Yoneno and Y. Hayashi. Magnetic properties of plastic magnets molded in a magnetic field. Kobunshi Ronbunshu 40(4): 181 (1983).
97. Z. Sawa and K. Kawauchi. Effect of polymer matrices on magnetic properties of plastic magnets. J. Mater. Sci. 23(7):2637 (1988).
98. W. Y. Chiang and W. D. Yang. Polypropylene composires. III. Chemical modification of the interphase and its influence on the properties of PP/mica composites. Polym. Eng. Sci. 34(6): 485 (1994).
99. W. Y. Chiang and C. H. Hu. The improvements in flame retardance and mechanical properties of polypropylene/FR blends by acrylic acid graft copolymerization. Eur. Polym. J. 32: 385 (1996).

3

Miscibility and Interfacial Behavior in Polymer–Polymer Mixtures

Toshiaki Ougizawa and Takashi Inoue
Tokyo Institute of Technology, Tokyo, Japan

I. INTRODUCTION

To estimate polymer–polymer miscibility, measurements of the Flory–Huggins interaction parameter χ have been carried out. Some negative values of χ were sometimes experimentally measured for miscible polymer–polymer mixtures. On the other hand, there are only a few values of experimentally determined positive χ because of difficulties in measuring. Even when data having positive χ are available, they are usually taken at a fixed temperature. However, the data are very important for controlling the morphology of immiscible polymer mixtures.

Interfacial properties between immiscible polymers, e.g., interfacial thickness and interfacial tension, are related to polymer–polymer miscibility. There are many theoretical (1–8) and experimental (9–13) studies on the interface between immiscible polymers. Among them, the most interesting is the recent theory by Broseta et al. (8). It is a modified version of the Helfand–Tagami theory (2,3) and deals with the effect of molecular weights and polydispersity of the component polymers on the interfacial thickness and interfacial tension. The theory indicated the relationship between the equilibrium interfacial thickness and the interaction parameter χ (positive) in immiscible polymer–polymer mixtures and stimulated experimental studies on the polymer interface to reveal the relationship between the interfacial properties and the molecular and thermodynamic parameters of the component polymers. Ellipsometry is a powerful tool for measuring the interfacial thickness in polymer–polymer mixtures. These measurements can be carried out for immiscible polymer blends as well as miscible blends. For miscible blends the investigation of changes of the interfacial thickness with time at a fixed temperature makes it possible to calculate mutual diffusion coefficients (14). For immiscible blends, on the other hand, the interaction parameter χ can be deduced from measurements of the interfacial thickness in an equilibrium state, by use of the theory of Broseta et al. (8). Thus it provides one of the rare opportunities to study the positive χ parameter for immiscible polymer–polymer mixtures.

For the last decade, the study of the miscibility window for mixtures containing random copolymer(s) has been carried out extensively. The miscibility window behavior has been reported for many blends, e.g., poly(methyl methacrylate) (PMMA) and styrene-acrylonitrile random copolymer (SAN) (15). A homopolymer forms a miscible blend with random copolymer in a certain copolymer composition and temperature range, when there is a strong unfavorable interaction between comonomer units of copolymer, though all segmental interactions are positive. Therefore the ellipsometric studies are suitable for evaluating the miscibility of copolymer blends showing miscibility window behavior.

In this chapter polymer–polymer miscibility is investigated on the basis of the interaction parameter χ_{AB} obtained by measuring the interfacial thickness. From the copolymer composition and temperature dependences of χ, we will discuss the phase behavior of polymer blends from the viewpoint of thermodynamics.

II. THEORETICAL BACKGROUND

A. Thermodynamics of Polymer–Polymer Miscibility

We start with a discussion of phase stability. The phase behavior of any mixture at constant pressure P and temperature T is directed by the Gibbs free energy of mixing ΔG^M, which is given by

$$\Delta G^M = \Delta H^M - T \Delta S^M \tag{1}$$

where ΔH^M and ΔS^M are the enthalpy and the entropy of mixing, respectively. According to the second law of thermodynamics, two components will only mix if the ΔG^M is negative ($\Delta G^M < 0$). Furthermore, the condition for phase stability in a binary mixture of composition ϕ (volume fraction) at fixed temperature and pressure is

$$\left(\frac{\partial^2 \Delta G^M}{\partial \phi^2} \right)_{P,T} > 0 \tag{2}$$

To discuss the phase stability in polymer mixtures, one needs an accurate expression of ΔG^M. Some basic models to describe a mixture were proposed and used to explain the phase behavior.

The most popular expression of ΔG^M for polymer–polymer mixtures is the classical Flory–Huggins equation (16). For a mixture consisting of polymer A and polymer B, it is given by

$$\frac{\Delta G^M}{RTV} = \frac{\phi_A}{V_A} \ln \phi_A + \frac{\phi_B}{V_B} \ln \phi_B + \phi_A \phi_B \frac{\chi_{AB}}{V_r} \tag{3}$$

where V is the total volume, V_r the molar volume of the segment, and R the gas constant. The first two terms of the right-hand side of Eq. (3) represent the combinatorial entropy of mixing, where ϕ_i is the volume fraction and V_i ($=r_i V_r$, r_i; number of segments per chain) is the molar volume of a polymer chain of component i. The third term contains the Flory–Huggins interaction parameter χ_{AB}. Since the combinatorial entropy terms are very small for polymer–polymer mixtures, this χ_{AB} is the crucial quantity for the thermodynamic description of phase behavior. Therefore, for most polymer–polymer mixtures, the miscibility or phase behavior is discussed on the basis of the value of χ_{AB}.

The critical values can be found from Eq. (3) by the condition that the second and the third derivatives of ΔG^M with respect to the composition ϕ are zero and that χ_{AB} is not a function of ϕ.

$$\phi_{cr,B} = \frac{1}{1 + \sqrt{r_B/r_A}} = \frac{1}{1 + \sqrt{V_B/V_A}} \tag{4}$$

$$\frac{\chi_{cr}}{V_r} = \frac{1}{2V_r}\left(\frac{1}{\sqrt{r_A}} + \frac{1}{\sqrt{r_B}}\right)^2 = \frac{1}{2}\left(\frac{1}{\sqrt{V_A}} + \frac{1}{\sqrt{V_B}}\right)^2 \tag{5}$$

Here $\phi_{cr,B}$ and χ_{cr} are the composition and the interaction parameter at the critical point. The value of χ_{cr} leads to the criterion of phase stability. For mixtures of high-molecular-weight polymers, χ_{cr} is nearly zero. So miscibility results from negative χ_{AB}, which is due to specific interaction such as hydrogen bonding.

As another criterion the χ parameter at the spinodal (χ_s) is given by

$$\frac{\chi_s}{V_r} = \frac{1}{2V_r}\left(\frac{1}{\phi_A r_A} + \frac{1}{\phi_B r_B}\right) = \frac{1}{2}\left(\frac{1}{\phi_A V_A} + \frac{1}{\phi_B V_B}\right) \tag{6}$$

At this point it should be noticed that it seems better to take the stability limit of the homogeneous region of the blend systems under investigation as $\chi_{AB} < \chi_s$ rather than as $\chi_{AB} < \chi_{cr}$. This difference, even if it is very small, should be considered, because the miscibility windows under later discussion are designed with dependence on copolymer composition and temperature for a fixed blend ratio, not a critical composition. Since this blend ratio might be far away from the critical point, Eq. (6) is sometimes used as the criterion of phase stability.

B. χ_{AB} of Polymer Mixtures Including Random Copolymers

In random copolymers a different mechanism causing miscibility becomes effective without specific interaction. The interaction parameter χ_{AB} between homopolymer A with segments of type 1 and a random copolymer B of type $2_\beta 3_{(1-\beta)}$ is given in the framework of a mean-field theory by (17,18)

$$\chi_{AB} = \beta\chi_{1/2} + (1-\beta)\chi_{1/3} - \beta(1-\beta)\chi_{2/3} \tag{7}$$

where β is the volume fraction of segment 2 in a random copolymer and $\chi_{i/j}$ are the segmental interaction parameters. As can be seen, it is possible that χ_{AB} becomes negative, even if all segmental interaction parameters are positive. The condition for miscibility is that $\chi_{2/3}$ must be greater than both $\chi_{1/2}$ and $\chi_{1/3}$. The strong repulsive intramolecular interaction between the comonomer segments in a random copolymer, causing the miscibility, is usually called the repulsion effect.

The χ_{AB} is dependent on the copolymer composition calculated by using the three sets of segmental interaction parameters. Figure 1 shows one example for the copolymer composition dependence of χ_{AB} in the PMMA/SAN for different temperatures (19). All three curves have in common a β range where χ_{AB} is negative. This copolymer composition range for a fixed blend ratio and in dependence on temperature is usually called the miscibility window. It is common that lower critical solution temperature (LCST) behavior, which is miscibility at lower temperatures and immiscibility at higher temperatures, can be observed at the edges of the miscibility window. Two border values of β in the miscibility window occur where χ_{AB} equals χ_s. Replacing χ_{AB} by χ_s in Eq. (7) gives just two equations but three unknown parameters.

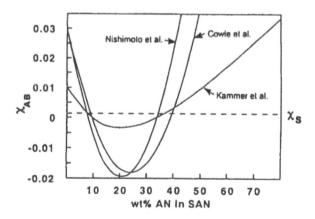

Figure 1 Copolymer composition dependence of χ_{AB} between PMMA and SAN, calculated by using Eq. (7) on the basis of the values of $\chi_{i/j}$ in Refs. 20–22. The values of Nishimoto et al. are at 130°C; Kammer et al. at 180°C; Cowie et al. gave no temperature. (From Ref. 19.)

If the values of $\chi_{i/j}$ and their temperature dependence are obtained by using Eq. (7) from measurements of the interfacial thickness, we can discuss the miscibility of polymer–polymer mixtures including random copolymer(s).

C. Relationship Between Interaction Parameter and Interfacial Thickness

There are various theories that relate the interfacial properties of two immiscible polymers to the interaction parameter χ_{AB}. Helfand and Tagami constructed a self-consistent field theory of polymer–polymer interfaces (2,3). They found, in the limit of infinite molecular weight, a very simple expression relating the interfacial thickness λ at the equilibrium state to the χ_{AB}:

$$\lambda_\infty = \frac{2b}{(6\chi_{AB})^{1/2}} \tag{8}$$

where b is the Kuhn segment length. Equation (8) has been found to describe the essential features of high-molecular-weight polymer–polymer interfaces. It was reported that this equation was in qualitative agreement with experiments (11).

However, the limitation of infinite chain length is not available for real polymer samples. Recently, the Helfand–Tagami theory was modified by Broseta et al. (8) to involve the molecular weight dependence of λ:

$$\lambda = \frac{2b}{\sqrt{6}\chi_{AB}} \left[1 + \frac{\ln 2}{\chi_{AB}} \left(\frac{1}{r_A} + \frac{1}{r_B} \right) \right] \tag{9}$$

where r_i is the number of segments per chain in polymer i. If A and B polymers have infinite molecular weights ($r_A, r_B \to \infty$), Eq. (9) leads to Eq. (8). Thus it is possible to convert the λ into the χ_{AB} parameter.

III. MEASUREMENTS OF INTERFACIAL THICKNESS

A. Ellipsometry

Study of the interface between dissimilar polymers is very important and leads to a better understanding of the adhesion between the two phases and the bulk properties of polymer blends. Various techniques have been developed for investigating the 'thick interface', but there have been few studies on the 'thin interface', of the order of 1–50 nm, which is expected to appear in immiscible polymer pairs. A promising approach by ellipsometry has been put forward by Kawaguchi et al. (10). Recently, Yukioka and Inoue revised the experimental procedure to prepare a bilayer specimen for ellipsometry (23). Ellipsometry is an easy method to handle. Its main limitations are that the refractive index difference between polymers should be larger than 0.02 and that a flat surface is required. The error of the method becomes tremendous when the system has a small refractive index difference and a thin interface which is equivalent to a large positive polymer–polymer interaction parameter χ_{AB}. Therefore copolymer mixtures that show broader interfaces than homopolymer pairs (such as mixtures of PMMA/SAN) are suitable for ellipsometric studies.

In order to measure the interfacial thickness by ellipsometry, bilayer specimens were prepared containing a thick substrate (about 1 mm) and a thin film on top (in the range from 15 to 500 nm). The substrates were melt-pressed between two silicon wafers to get an optical flat surface. The thin films were prepared by spin-coating the polymer solutions onto a silicon wafer and floating off the resulting films onto a water surface. Then the floating films were picked up with the substrate polymer. The specimens dried at elevated temperatures in a vacuum oven are used for ellipsometry.

The measurements are carried out using an ellipsometer. Incident light, for example He-Ne laser ($\lambda' = 632.8$ nm), was applied to the bilayer specimen at an incident angle of 70°. The retardation Δ and reflection ration tan Ψ of reflected light were determined from the ellipsometric readings. For data analysis, the four-layer model was used as shown in Fig. 2a (14). Since the values of the refractive indices n_1, n_2, and n_4 and the thickness d_2 are known, one can estimate n_3 and d_3 by selecting the best set of these values to fit the observed values of Δ and tan Ψ for the four-layer model:

$$\rho = \frac{R_m^p}{R_m^s} = \frac{|R_m^p|\exp(i\Delta_p)}{|R_m^s|\exp(i\Delta_s)} = \frac{|R_m^p|}{|R_m^s|}\exp\{i(\Delta_p - \Delta_s)\} = \tan\psi\,\exp(i\Delta) \qquad (10)$$

$$R_m^v = \frac{r_m^v + R_{m+1}^v\exp(-iD_{m+1})}{1 + r_m^v R_{m+1}^v\exp(-iD_{m+1})} \qquad (v = p,s) \qquad (11)$$

$$D_m = 4\pi n_m d_m\,(\cos\theta_m)/\lambda' \qquad (12)$$

where ρ is the relative amplitude of the parallel (R_m^p) to the perpendicular (R_m^s) reflection coefficient in the incident plane, n_m and d_m represent the refractive index and thickness of the mth layer, respectively, and r_m is the Fresnel reflection coefficient at the boundary between the mth and (m + 1)th layers:

$$r_m^p = \frac{n_{m+1}\cos\theta_m - n_m\cos\theta_{m+1}}{n_{m+1}\cos\theta_m + n_m\cos\theta_{m+1}}$$

$$r_m^s = \frac{n_m\cos\theta_m - n_{m+1}\cos\theta_{m+1}}{n_m\cos\theta_m + n_{m+1}\cos\theta_{m+1}} \qquad (13)$$

$$n_1\sin\theta_1 = n_2\sin\theta_2 = n_3\sin\theta_3 = n_4\sin\theta_4 \qquad (14)$$

Numerical calculation for the best fit was carried out. The four-layer model in Fig. 2a implies that the reflective index at the interface is approximated to being uni-

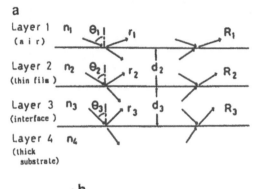

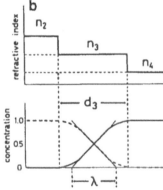

Figure 2 (a) Four-layer model for ellipsometric analysis. (b) Concentration profile at polymer–polymer interface and a stepwise approximation of the refractive index profile for ellipsometry. (From Ref. 14.)

form and equal to $n_3 = (n_1 + n_2)/2$ (Fig. 2b) (14). Taking account of the composition profile at the interface layer (3,7,8), the interfacial thickness λ was determined as $\lambda = d_3/1.7$ (Fig. 2b).

B. Other Methods

A shortcoming of ellipsometry is that concentration profiles usually cannot be obtained, i.e., correction factors are necessary to convert the step interface into a real one. Thus, confirmation of ellipsometric data by independent and direct methods is highly recommended.

For this purpose transmission electron microscope (TEM) measurements are available. Fig. 3 shows a TEM micrograph of PMMA/SAN-38.7 annealed for 24 h at 130°C (24). The bilayer specimens, which were ultramicrotomed in a direction normal to the film surface and had a thickness of approximately 70 nm, were stained with RuO_4 in the gas phase. RuO_4 stained only the benzene ring in PS or SAN and did not stain PMMA. So the white bottom layer represents the PMMA substrate and the dark layer is the SAN film. The sample exhibits an excellent flat surface and a very uniform interface. To compare this interfacial thickness with one of PS/PMMA, concentration profiles were obtained by densitometric measurements of TEM images across the interface using the image processing system (24). The smoothed and normalized best fit curves thus obtained can be seen in Fig. 4. Normalized means that the gray value of the PMMA bulk phase was set to 0 and the value of SAN-38.7 (PS) was set to 1. As can be seen, there is a good resolution in the concentration profiles, especially for the random copolymer system. The inset shows the definition of the interface used. The tangent was drawn at the inflection point creating a so-called wedge shape. Thus the sample containing SAN-38.7 and annealed for 24 h at 130°C gives an interfacial thickness of about 32 ± 3 nm, which is in excellent agreement with the results of ellipsometry as described later. By using the same procedure, the interfacial thickness in the system PS/PMMA is 5 ± 3 nm. This is in fairly good agreement with a value of 3 nm measured by ellipsometry at 140°C.

As another method, heavy ion elastic recoil detection (ERD) is also able to provide concentration profiles at the polymer–polymer interface. By using

Figure 3 TEM micrograph in a bilayer specimen of PMMA/SAN-38.7 annealed for 24 h at 130°C. The benzene ring of SAN was selectively stained by RuO_4. (From Ref. 24.)

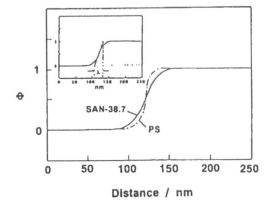

Figure 4 Normalized best fit curves of the concentration profiles of PMMA/SAN-38.7 and PS/PMMA samples. Φ is the concentration of SAN-38.7 or PS. The insert shows schematically the definition of the interfacial thickness λ. (From Ref. 24.)

$^{35}Cl^{6+}$ as primary ions, it is possible to obtain the oxygen and nitrogen concentrations as a function of depth, which can easily be converted in PMMA and SAN concentration profiles. The monoenergetic ion beam was directed at an angle of 11° to the surface of the bilayer specimen, and the recoiled atoms were also detected under an angle of 11°. The number of atoms belonging to one isotope can be measured as a function of their energy (25). To calculate a concentration profile of the different isotopes, the sample is divided into virtual slabs, and each slab corresponds to a certain energy interval of the different recoils. Figure 5 shows the atomic concentration profiles of nitrogen (SAN-38.7) and oxygen (PMMA) for a specimen, a thin film of SAN-38.7 is placed on top of a PMMA substrate (24). It can be seen that the Gaussian type fit is in good agreement with experimental data. Also here oxygen traces in the SAN-38.7 film yield an apparent PMMA content in the range smaller than 50 nm. The interfacial thickness obtained from the oxygen profile is 68 nm and from the nitrogen profile 49 nm. The value obtained by using the nitrogen profile is in fairly good agreement with TEM and ellipsometric data if one takes into consideration the relatively large error. The larger deviation in the interfacial thickness obtained by using the oxygen profile might be caused by a number of reasons. First there is the problem that all samples contain a certain amount of oxygen. Furthermore, a small but significant amount of oxygen enrichment could be observed at the surface of the polymer specimen.

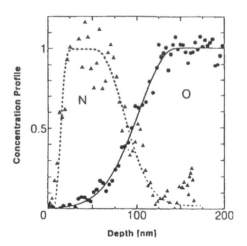

Figure 5 Concentration profile of SAN-38.7(····) and of PMMA(—). The experimental points were obtained from the nitrogen (▲) and oxygen (●) profile, respectively. The apparent PMMA content in the range smaller than 50 nm is not significant. (From Ref. 24.)

After having confirmed the reliability of ellipsometric data, it is possible to deduce thermodynamic data from the temperature dependence of the interfacial thickness.

IV. PS/PMMA

The mixture of PS and PMMA is a typical immiscible system and the χ_{AB} is positive. Figure 6 shows the equilibrium interfacial thickness λ measured by ellipsometry for the mixtures PS/PMMA, PMMA/SAN-5.7, and PMMA/SAN-38.7 (19). Mixtures containing random copolymers show a relatively thick interface. This is caused by their small polymer–polymer interaction parameter χ_{AB} and will be discussed in the next section. The interfacial thickness in PS/PMMA increases slightly with temperature. The value at 120°C was obtained by neutron reflectivity (13).

It is possible to convert the interfacial thickness, given in Fig. 6, into χ_{AB} parameters as shown in Fig. 7 (19), using Eq. (9). The interaction parameter $\chi_{PS-PMMA}$ between PS and PMMA decreases with increasing temperature and is very small at high temperature. This should result in the possibility of upper critical solution temperature (UCST) behavior in low-molecular-weight mixtures such as oligomer mixtures. Figure 8 shows the result of the cloud point measurement for a mixture of PS-1.25K (Mw = 1250) and PMMA-6.35K (Mw = 6350)

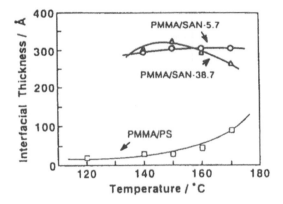

Figure 6 Temperature dependence of the equilibrium interfacial thickness in PMMA/PS and PMMA/SAN mixtures. (From Ref. 19.)

(26). At higher temperatures a miscible region appeared, and the UCST behavior was observed as predicted. The dashed and full lines are the binodals calculated by using the Flory–Huggins theory of Eq. (3) and Flory's equation-of-state theory, which is discussed later, respectively.

For comparison, the temperature dependence of $\chi_{PS\text{-}PMMA}$ was obtained from the relationship between the critical temperatures of some phase diagrams and the molecular weights. From the Flory–Huggins theory, the value of χ_{cr} at critical temperature is calculated for molecular weight pairs by Eq. (5). If one can mea-

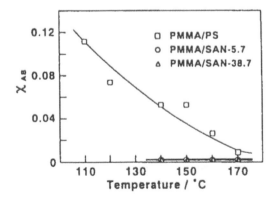

Figure 7 Temperature dependence of χ_{AB} in PMMA/PS and PMMA/SAN mixtures calculated by using Eq. (9) on the basis of the values of Fig. 6. (From Ref. 19.)

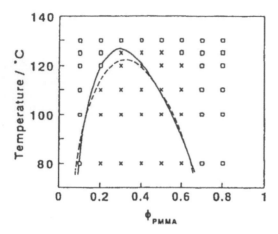

Figure 8 Phase diagram of PS-1.25K/PMMA-6.35K mixture: (x) cloud samples, (□) transparent samples. The dashed and full lines are the binodals calculated by the Flory–Huggins theory, Eq. (3), and Flory's equation-of-state theory, Eq. (11), respectively. (From Ref. 26.)

sure the phase diagrams for some molecular weight pairs, some values of χ are obtained at some temperatures because the critical temperatures are different in molecular weight pairs. So, by obtaining a number of critical data by measuring some phase diagrams it is possible to obtain the temperature dependence of $\chi_{PS\text{-}PMMA}$. Figure 9 shows the temperature dependence of $\chi_{PS\text{-}PMMA}$ obtained by phase

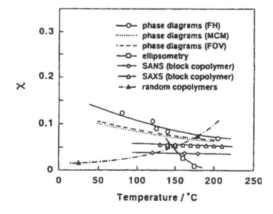

Figure 9 Temperature dependence of χ obtained by different methods in PS/PMMA mixtures. The temperature dependence from ellipsometric results is somewhat strong. (From Ref. 26.)

diagrams and applying the Flory–Huggins theory (open circles) (26). Further-more, a number of literature data and the temperature dependence of $\chi_{\text{PS-PMMA}}$ calculated with Flory's equation-of-state theory can be seen. Using Flory's equa-tion-of-state theory (27), the interaction parameter χ_{AB} is given by (28)

$$\chi_{\text{AB}} = \frac{V_r^*}{\phi_A \phi_B RT} \left[\phi_A P_A^* \left(\frac{1}{\tilde{V}_A} - \frac{1}{\tilde{V}} \right) + \phi_B P_B^* \left(\frac{1}{\tilde{V}_B} - \frac{1}{\tilde{V}} \right) \right.$$

$$\left. + 3\phi_A P_A^* \tilde{T}_A \ln \frac{\tilde{V}_A^{1/3} - 1}{\tilde{V}^{1/3} - 1} + 3\phi_B P_B^* \tilde{T}_B \ln \frac{\tilde{V}_B^{1/3} - 1}{\tilde{V}^{1/3} - 1} + \frac{\phi_A \phi_B X_{\text{AB}}}{\tilde{V}} \right] \qquad (15)$$

where the starred quantities are the characteristic parameters and obtained from pressure-volume-temperature measurements (28). θ_i is the site fraction of the component i, X_{AB} is the exchange energy parameter, k is the Boltzmann constant, and the reduced quantities $\tilde{V}$ and $\tilde{T}$ were obtained from Flory's equation-of-state theory. The parameter χ_{AB} outlined here is different from the exchange energy pa-rameter defined in the Flory–Huggins theory, because it also takes into account the free volume contributions. Therefore also the temperature dependence of this χ_{AB} parameter might be different from the prediction using the classical Flory–Huggins theory. But, as can be seen in Fig. 7, the temperature dependence of $\chi_{\text{PS-PMMA}}$ using the Flory–Huggins theory is not so different from that using Flory's equation-of-state theory. This is because the free volume contributions in blends of PS and PMMA are very small (29). Furthermore, it turns out that it is better to compare the temperature dependence of $\chi_{\text{PS-PMMA}}$ rather than absolute values. As can be seen, the result measured by neutron scattering (30) gives a somewhat smaller temperature dependence of $\chi_{\text{PS-PMMA}}$ than that calculated by the Flory–Huggins theory and Flory's equation-of-state theory, and the tempera-ture dependence from the ellipsometric data is somewhat stronger (19). Absolute values are often influenced by model parameters like segment volume or segment length. This might explain some of the differences of $\chi_{\text{PS-PMMA}}$ in Fig. 9.

V. PMMA/SAN

The equilibrium interfacial thickness dependence on temperature for blends of PMMA/SAN-5.7 and PMMA/SAN-38.7 is shown in Fig. 6, as well as one of PS/PMMA (19). The interfacial thickness of PMMA/SAN-5.7 increases slightly, whereas the interfacial thickness of the system PMMA/SAN-38.7 decreases slightly in the temperature range from 140 to 170°C. According to Eq. (9) the tem-perature dependence of χ_{AB} in PMMA/SAN-5.7 and PMMA/SAN-38.7 are shown in Fig. 7. The χ_{AB} between PMMA and SAN-5.7 or SAN-38.7 is very small in comparison with that between PS and PMMA, and the temperature dependence is also very weak.

In Eq. (7) the interaction parameter χ_{AB} and the segmental interaction parameter $\chi_{S/MMA}$ which equals to $\chi_{PS\text{-}PMMA}$ in PS/PMMA are known for various temperatures. Furthermore, there are for every temperature two different copolymer compositions β (5.7 and 38.7 wt% acrylonitrile in SAN) in some temperatures. Thus there are two unknown parameters and two equations for every temperature under investigation. Taking these data, it is possible to calculate the two unknown segmental interaction parameters. Figure 10 shows the calculated segmental interaction parameters $\chi_{AN/MMA}$ and $\chi_{S/AN}$, as well as the $\chi_{S/MMA}$ parameter obtained from interfacial thickness data with temperature dependence (19). Each segmental interaction parameter becomes smaller with increasing temperature. The values of $\chi_{AN/MMA}$ and $\chi_{S/AN}$ are well above χ_s at 170°C. That means that the blends of PAN and PMMA, as well as PAN and PS, are strongly repulsive and always in the two-phase region at the temperatures under discussion.

At measured temperature, the value of $\chi_{S/AN}$ is always the largest. This large positive $\chi_{S/AN}$, which corresponds to $\chi_{2/3}$ in Eq. (7), causes a negative χ_{AB}. The miscibility window behavior of PMMA/SAN mixtures is due to this large positive $\chi_{S/AN}$ between comonomer segments in SAN. However, the decrease of all three segmental interaction parameters with increasing temperature does not seem to be the free volume contribution that increases χ. Since the free volume contribution produces LCST behavior, this does not agree with the fact that the PMMA/SAN mixtures display LCST behavior. Figure 11 shows the copolymer composition dependence of χ_{AB} at some temperatures calculated by Eq. (7) based on the values of Fig. 10 (31). Figure 11 is almost same as Fig. 1, and there is clearly a region of negative χ_{AB}. This is also in good agreement with the experimentally miscible region. The LCST behavior occurs when the χ_{AB} parameter increases further with temperature and becomes larger than the χ_{cr} value. Figure 12 shows the calculated miscibility window behavior for some χ_{cr} based on the results of Fig. 11 (31). For the infinite or high-molecular-weight components the calculated miscibility window shows the well-known shape (solid curve), and in some ranges of copolymer composition there appears the LCST behavior. This is in fairly good agreement with the experimentally determined miscibility window. However, for lower molecular weights of the components, i.e., for larger χ_{cr} systems, the calculated miscibility window becomes at first temperature-independent (dash-dot curve). There are many cases of reported miscibility window that show hardly any temperature dependence, or the transition from miscibility to immiscibility occurs within an extremely small copolymer composition range. For mixtures where at least one component has a low molecular weight, the behavior changes dramatically, and only UCST can be observed (dash curve). Thus the miscibility window behavior in PMMA/SAN mixtures can be explained by the values of the χ parameter obtained from the interfacial thickness data.

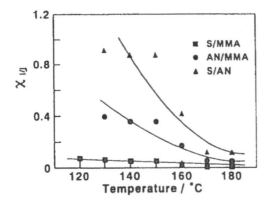

Figure 10 Temperature dependence of three segmental interaction parameters $\chi_{i/j}$ included in PMMA/SAN mixtures. The curves are least-squares fits. (From Ref. 19.)

Generally, the LCST behavior is explained by the free volume effect based on the equation of state theories of polymer mixtures. In the above discussion the free volume effect for the LCST behavior of PMMA/SAN mixtures is not clear. Here the miscibility of PMMA/SAN mixtures is considered by separating χ into two contributions; an interaction term χ_{inter}, which corresponds to the exchange energy term of the van Laar type, and a free volume term χ_{free}, which reflects the differ-

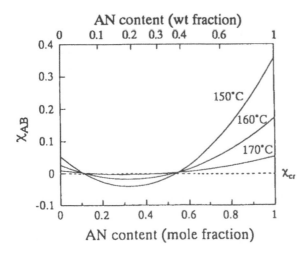

Figure 11 Copolymer composition dependence of χ_{AB} in PMMA/SAN mixtures at different temperatures calculated by Eq. (7) on the basis of the values of Fig. 10. (From Ref. 31.)

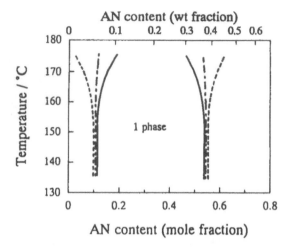

Figure 12 Miscibility window of PMMA/SAN mixtures calculated on the basis of the results of Fig. 11 for some χ_{cr} values: (—) 0; (- —-) 0.002; (----) 0.005. (From Ref. 31.)

ence of the free volumes. The χ parameters in Eq. (7) may contain both interaction and free volume terms. Theoretically, the possibility exists that the different temperature dependences of the contributions of interactions to the segmental interaction parameters $\chi_{i/j}$ may lead to LCST behavior without any obvious influence of the free volume changes. When suitable sets of the temperature-dependent χ parameters are chosen, the explanation for the simultaneous occurrence of the LCST and UCST behavior can easily be given, such as for this PMMA/SAN mixture.

In order to investigate the free volume effect for PMMA/SAN mixture, χ was separated into the two contributions on the basis of Flory's equation-of-state theory. Shiomi and Imai (32) and Jo and Lee (33) suggested that X_{AB} in Eq. (15) can be calculated from segmental exchange energy parameters $X_{i/j}$ in the case of homopolymer **A** and random copolymer **B**, where homopolymer **A** is composed of monomer 1 and copolymer **B** of monomers 2 and 3:

$$X_{AB} = \theta_B^{(2)}X_{1/2} + \theta_B^{(3)}X_{1/3} - \frac{S_A}{S_B}\phi_B^{(2)}\theta_B^{(3)}X_{2/3} \tag{16}$$

where $\phi^{(i)}{}_B$ is the copolymer composition of monomer i ($\phi^{(2)}{}_B + \phi^{(3)}{}_B = 1$) and $\theta^{(i)}{}_B$ is the site fraction of monomer i ($\theta^{(2)}{}_B + \theta^{(3)}{}_B = 1$), and S_j is the number of contact sites of component j. They also suggested that the characteristic parameters of copolymer **B** are given by

$$P_B^* = \phi_B^{(2)}P_2^* + \phi_B^{(3)}P_3^* - \phi_B^{(2)}\theta_B^{(3)}X_{2/3} \tag{17}$$

$$V_{SP_B}^* = m_B^{(2)}V_{SP_2}^* + m_B^{(3)}V_{SP_3}^* \tag{18}$$

$$T_B^* = \frac{P_B^*}{\phi_B^{(2)}(P_2^*/T_2^*) + \phi_B^{(3)}(P_3^*/T_3^*)} \tag{19}$$

where m_1 is the mass fraction ($m_B^{(2)} + m_B^{(3)} = 1$). It is important to note that Eqs. (17)–(19) for copolymers are completely identical with the equations for binary homopolymer blends consisting of components 2 and 3.

For a comprehensive discussion on the miscibility of PMMA/SAN blends, one has to come back to the χ parameter given by Eq. (15). The last term in the bracket of Eq. (15) represents the effect of the exchange energy, and we call term χ_{inter}. The other terms, which vanish at $\tilde{V} = \tilde{V}_A = \tilde{V}_B = 1$, correspond to the free volume effect and are called χ_{free}. In the calculations the values of the characteristics parameters for each SAN were determined by fitting to pressure-volume-temperature data.

Figure 13 shows the copolymer composition dependence of $X_{PMMA\text{-}SAN}$ calculated from Eq. (16) by using each segmental X_{ij} (34). This curve does not depend on temperature according to Flory's equation-of-state theory. Although all X_{ij}s are positive, there appears a region in which $X_{PMMA\text{-}SAN}$ is negative. The negative $X_{PMMA\text{-}SAN}$ is caused by repulsion between S and AN in the SAN copolymer. Thus the miscibility window appears in the $T\text{-}\phi_2$ phase diagram because of the segmental exchange energy effect. Figure 14 shows the calculated χ_{inter} (a), χ_{free} (b), and total of $\chi(= \chi_{inter} + \chi_{free})$ (c) as a function of copolymer composition (core volume fraction) of AN in SAN at three temperatures (34). Figure 14a was obtained by using the copolymer composition dependence of $X_{PMMA\text{-}SAN}$ shown in Fig. 13. The blend should be miscible in the range where χ is smaller than χ_{cr}. The copolymer composition dependence of χ_{inter} is parabolic due to the strong unfavorable interaction between S and AN segments in SAN copolymers. For the particular composition for which X_{AB} vanishes (17 and 37% of AN in SAN), χ_{inter} also vanishes and does not depend on temperature. In the copolymer composition range from 17 to 37%, χ_{inter} is negative and increases with increas-

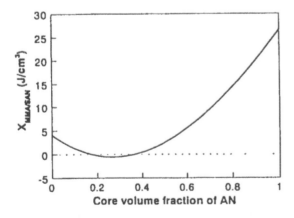

Figure 13 Copolymer composition dependence of $X_{PMMA\text{-}SAN}$ calculated by Eq. (12). (From Ref. 34.)

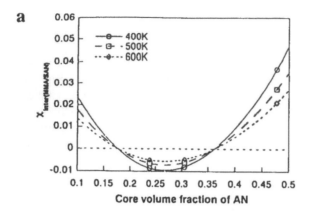

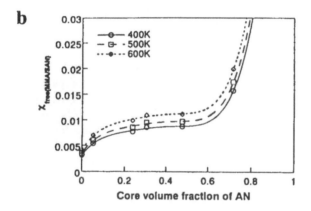

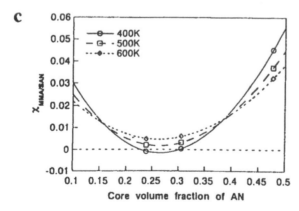

Figure 14 Copolymer composition dependence of χ parameter in PMMA/SAN mixtures calculated by Flory's equation-of-state theory at 400K, 500K, and 600K: (a) χ_{inter}; (b) χ_{free}; (c) total χ ($=\chi_{inter} + \chi_{free}$). (From Ref. 34.)

ing temperature, but the sign never changes from negative to positive. This means that χ_{inter} never becomes greater than χ_{cr} when χ_{inter} increases with increasing temperature. The miscibility window can be explained, but the LCST behavior observed in some PMMA/SAN blends cannot. On the other hand, outside the negative χ_{inter} region, χ_{inter} is positive and decreases with increasing temperature. This leads to the UCST behavior if the value of χ_{inter} crosses χ_{cr}. This agrees with the experimental finding that the low-molecular-weight mixtures of PMMA with SAN having low AN content (and PS) show the UCST behavior (Sect. IV).

From Fig. 14b, for all copolymer compositions, χ_{free} is positive and increases monotonically with increasing temperature. Finally, the copolymer composition dependence of total χ ($= \chi_{inter} + \chi_{free}$) at three temperatures is shown in Fig. 14c. There is a copolymer composition range in which χ increases with increasing temperature and its sign changes from negative to positive. In the region where χ increases, χ crosses the value of χ_{cr} and the LCST behavior can appear. The LCST behavior can be explained only by considering both χ_{inter} and χ_{free}.

So, from this analysis by the equation-of-state theory, it was shown that the miscibility in the limited copolymer composition range is explained by the copolymer composition dependence of only χ_{inter}, but χ_{free} is necessary to explain the LCST behavior of PMMA/SAN mixtures.

VI. PC/PS AND PC/SAN

Blends of bisphenol A polycarbonate (PC)/ABS (acrylonitrile-butadiene-styrene) have been widely used as commercial products for many years. One of the reasons that useful properties appear in these blends is that there is good compatibility between PC and SAN, which is the matrix of ABS. However, the mixtures of PC and SAN are immiscible. Callaghan et al. calculated the interfacial thickness by the theory of Helfand [Eq. (1)] and predicted that it had a maximum in PC/SAN-25 (35). This seems to relate to the observation that PC/SAN blends often show optimum physical properties at around 25 wt% AN content. One can evaluate the copolymer composition and temperature dependences of χ_{AB} by ellipsometry and discuss the phase behavior of PC/SAN blends from the viewpoint of thermodynamics.

Figures 15a and 15b show the equilibrium interfacial thickness λ measured by ellipsometry as functions of AN content in SAN and temperature, respectively (36). The interfacial thickness of PC/PS was about 12–16 nm, while the interfacial thickness of PC/SAN was thicker than that of PC/PS at all the measured temperatures, and the thickest interface of PC/SAN was about 45 nm. From Figure 15a, there is an AN content that gives the maximum interfacial thickness at each temperature. The AN content having thickest interface is 25 wt% at 180°C but 40–50 wt% at higher temperature. At higher temperatures, this is not always consistent with the AN content (around 25 wt%), which exhibits good mechanical properties (35). In Fig. 15b it was shown that with the increase of temperature, the

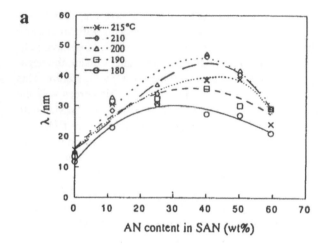

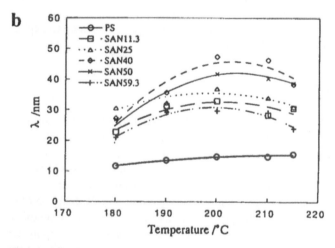

Figure 15 Equilibrium interfacial thickness measured by ellipsometry as functions of (a) AN content in SAN and (b) temperature in PC/PS and PC/SAN mixtures. (From Ref. 36.)

interface of PC/PS increased slightly and monotonically. However, PC/SAN blends exhibited a maximum interfacial thickness at about 200°C.

Using Eq. (2) (here $b = 8$ Å is used), if we estimate r_i on the basis of monomer units, it is easy to calculate the interaction parameter χ_{AB} between two polymers from the interfacial thickness measured by ellipsometry. Figures 16a and 16b show χ_{AB} as functions of AN content in copolymer and temperature, respectively. From Fig. 16a, all the values of χ_{AB} are positive, and the AN content

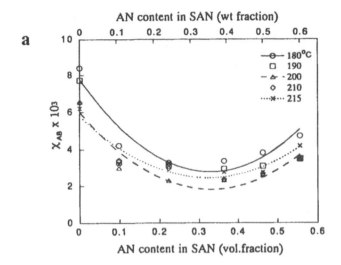

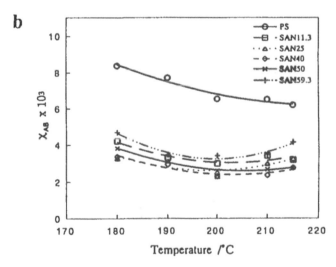

Figure 16 χ_{AB} calculated from Eq. (9) based on the results of Fig. 15 as functions of (a) AN content in SAN and (b) temperature in PC/PS and PC/SAN mixtures. (From Ref. 36.)

dependence of χ_{AB} shows a concave curve at each temperature. If one rewrites Eq. (7) for a PC/SAN mixture, the interaction parameter between PC(A) and SAN(B) is given by

$$\chi_{AB} = (1 - \beta)\,\chi_{S/C} + \beta\,\chi_{AN/C} - \beta\,(1 - \beta)\,\chi_{S/AN} \qquad (20)$$

where β means volume fraction of AN in SAN. It is understood that the concave curve in Fig. 16a is caused by repulsion between S and AN in the SAN copolymer (large positive $\chi_{S/AN}$). This behavior is similar to PMMA/SAN blends, though χ_{AB} of PC/SAN does not become negative. From Fig. 16b, χ_{AB} of PC/PS decreased with the increase of temperature monotonically, but the temperature dependence of χ_{AB} shows a concave curve in all the measured SAN. The minimums of χ_{AB} appeared at about 200°C.

From Fig. 16a (or b) one can draw the miscibility diagram for PC/SAN by using the value of χ_{AB} at the critical point χ_{cr}, which is given by Eq. (5). In the case of smaller χ_{AB} than χ_{cr}, one phase is stable but unstable in the opposite case. Figure 17 shows the miscible–immiscible boundary estimated from χ_{AB}-AN content curves in Fig. 16a for two values of χ_{cr} (36). There appears a closed loop type miscibility region (a "miscibility egg") in the PC/SAN system, instead of the usual miscibility window. As χ_{cr} decreases, i.e., the molecular weight of the constituent polymers increases, the miscibility egg is reduced. For high-molecular-weight blends of PC/SAN, the miscibility egg disappears, and one cannot find a one-phase region. However, for suitable molecular weight specimens, one can observe the miscibility egg. In a PC/SAN-25 blend having suitable molecular weight specimens, for example, the system should undergo a two-phase, one-phase, two-phase change with increasing temperature. This im-

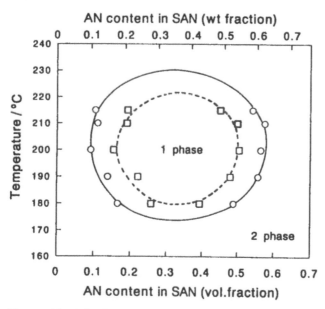

Figure 17 Miscibility diagram of PC/SAN system for two χ_{cr} values: (----) 0.003, which corresponds to $r_A = r_B = 667$, and (—) 0.004, which corresponds to $r_A = r_B = 500$. (From Ref. 36.)

plies that the PC/SAN-25 blend displays the coexistent behavior of UCST and LCST in phase diagrams of suitable molecular weight pairs. This is also understood from Fig. 16b in which PC/SAN blends show the U-shape temperature dependence of χ_{AB}. From Fig. 16a and 16b, it is also predicted that PC/PS blend exhibits UCST phase behavior in the low-molecular-weight specimens, since χ_{PC-PS} decreases monotonically.

Since high-molecular-weight mixtures of PC/SAN are immiscible, UCST or LCST behavior has not been observed. However, from the above results, UCST or LCST behavior is predicted in PC/SAN blends with specimens of suitable molecular weight. To examine this phase behavior of PC/SAN blends, cloud point measurements were carried out using low-molecular-weight PC (Mw = 6000). Figures 18a, 18b and 18c show the phase diagrams of the blend of PC/PS (Mw = 2700), PC/SAN-11.3, and PC/SAN-25, respectively. There appears the UCST type phase behavior in PC/PS (Fig. 18a), as predicted. This is consistent with the result of Callaghan and Paul (37). In PC/SAN-11.3 blend (Fig. 18b) there appears an hourglass type phase behavior, though one phase region cannot be observed in PC-rich blends. In PC/SAN-25 blends (Figure 18c), there appears both UCST and LCST behavior, also as predicted. The coexistent behavior of UCST and LCST has appeared in PC/SAN blends. However, the temperature region in which the miscibility gap in the phase diagrams appears is different from the one predicted from Fig. 17.

By using at least three sets of χ_{AB} from Fig. 16b with different β values in Eq. (20) all the $\chi_{i/j}$ s were calculated. Here the experimental value of $\chi_{S/C}$ is used as χ_{AB} between PS and PC, and one can obtain the values of $\chi_{AN/C}$ and $\chi_{S/AN}$ from two sets of PC/SAN-x. The temperature dependence of $\chi_{i/j}$s is shown in Fig. 19 (36). Since some combinations of two sets of PC/SAN-x exist, the values of $\chi_{i/j}$ are scattered, but it is clear that each $\chi_{i/j}$ decreases monotonically with increasing temperature. In particular, the values of $\chi_{AN/C}$ and $\chi_{S/AN}$ decrease remarkably.

The χ_{AB} can be separated into two contributions, χ_{inter} and χ_{free}, based on the equation-of-state theories for mixtures. As described in the previous section, for PMMA/SAN mixtures the miscibility window behavior is explained by both χ_{inter} and χ_{free}, i.e., the contribution of χ_{inter} is the main reason for the miscibility window behavior, but the two-phase behavior at higher temperature (LCST behavior) is driven by χ_{free}. χ_{free} always shows an unfavorable contribution for mixing and increases with increasing temperature. Since all of the $\chi_{i/j}$s in the PC/SAN system decrease monotonically in Fig. 19, it is understood that the contribution of χ_{free} is remarkably small in each $\chi_{i/j}$. So there is the possibility that the LCST behavior in PC/SAN blends is not driven by χ_{free}.

In Fig. 16a the curves of the AN content dependence of χ_{AB} at three temperatures are shown. Though the χ_{AB} in the PC/PS system decreases with increasing temperature nonotonically, the χ_{AB} in PC/SAN systems decreases (180 $\rightarrow$ 200°C) and then increases (200 $\rightarrow$ 215°C). This behavior can be interpreted as follows.

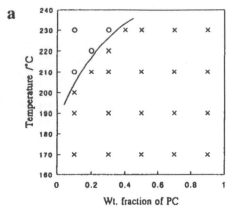

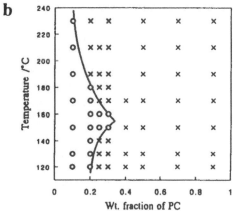

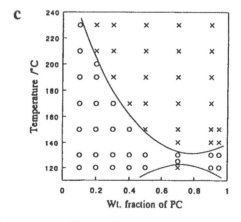

Figure 18 Phase diagrams of oligomeric PC(Mw=6000)/SAN mixtures measured by the cloud point method: (x) cloud samples; (O) transparent samples. (a) PC/PS; (b) PC/SAN-11.3; (c) PC/SAN-25. (From Ref. 36.)

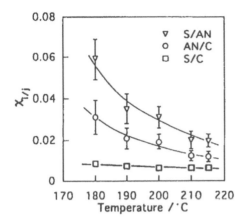

Figure 19 Temperature dependence of three segmental interaction parameters $\chi_{i/j}$ included in PC/SAN mixtures. (From Ref. 36.)

From Fig. 16b, $\chi_{S/AN}$ and $\chi_{AN/C}$ (especially $\chi_{S/AN}$), decrease remarkably in comparison with $\chi_{S/C}$. From Eq. (20) the concave curve of the AN content dependence of χ_{AB} is due to the large $\chi_{S/AN}$, i.e., repulsive interaction between S and AN in SAN copolymer. If $\chi_{S/AN}$ becomes very small with increasing temperature, χ_{AB} can increase at high temperature. This can be considered a plausible reason for LCST behavior in PC/SAN blends without the free volume contribution. This may also explain the LCST behavior observed in other blends including SAN, such as PMMA/SAN and poly(ϵ-caprolactone)/SAN (38). Moreover, this can be one of the reasons for the coexistent behavior of UCST and LCST in high-molecular-weight polymer blends.

VII. CONCLUDING REMARKS

The polymer–polymer interactions of immiscible mixtures on the basis of interfacial thickness measured by ellipsometry were estimated. It was shown that the data of interfacial thickness and its temperature dependence from ellipsometric measurements were useful for estimating values of the interaction parameter χ and its temperature dependence for immiscible systems. It was also shown that the phase behavior of some mixtures could be explained by χ obtained from the ellipsometric results. However, as shown in Fig. 9 for PS/PMMA, the temperature dependence of χ_{AB} obtained from ellipsometric data is somewhat different from those obtained by other methods. When χ_{AB} is a very small positive value (weak segregation condition), the relationship between λ and χ_{AB} in Eq. (9) may not be satisfied. Further investigations are necessary.

Equation (7) is useful for explaining the miscibility in polymer–polymer mixtures including random copolymer. It was reported that Eq. (7) was applicable in many cases. However, the free volume contribution in $\chi_{2/3}$ between segments in a random copolymer was not obviously observed in the case that $\chi_{2/3}$ was calculated from Eq. (7). So the polymer–polymer interaction χ_{23} between two and three homopolymers may not be simply applicable to the segmental interaction $\chi_{2/3}$ between two and three segments in a random copolymer. On the contrary, the equation-of-state theory for random copolymer mixtures suggested by Shiomi and Imai (32) and Jo and Lee (33) introduced only the free volume contribution between homopolymer A and random copolymer B, without considering one between comonomer segments. Eq. (7) corresponds only to an exchange energy term expressed by Eq. (16) without the free volume term. It is necessary to examine the free volume contribution in $\chi_{2/3}$, not only theoretically but also experimentally.

Another method of measuring interfacial thickness is neutron reflectivity (39). Though this method can measure thinner interfaces, it is not easy, because deuterated samples and large-scale apparatus are necessary. In order to measure the thin interfacial thickness in most immiscible polymer mixtures, simple measuring methods are required.

There are clearly difficulties and unclear points in measurements of polymer–polymer interactions of immiscible mixtures. Systematic investigations between the miscibility (interactions) and interfacial behavior in polymer–polymer mixtures are necessary.

REFERENCES

1. A. Vrij. Equation for the interfacial tension between demixed polymer solution. J Polym Sci A-2 6:1919–1932, 1968.
2. E. Helfand, Y. Tagami. Theory of the interface between immiscible polymers. J Polym Sci Polym Lett 9:741–746, 1971.
3. E. Helfand, Y. Tagami. Theory of the interface between immiscible polymers. II. J Chem Phys 56:3592–3601, 1971.
4. R.-J. Roe. Multilayer theory of adsorption from a polymer solution. J Chem Phys 60:4192–4207, 1974.
5. R.-J. Roe. Theory of the interface between polymers or polymer solutions. I. Two components system. J Chem Phys 62:490–499, 1975.
6. D. J. Meier, T. Inoue. Polymer blends: theory of interface. Polym Prep Jpn 24:293–296, 1975.
7. T. Nose. Theory of liquid–liquid interface of polymer systems. Polym J 8:96–113, 1976.
8. D. Broseta, G. H. Fredrickson, E Helfand, L Leibler. Molecular weight and polydispersity effects at polymer–polymer interfaces. Macromolecules 23:132–139, 1990.
9. S. Wu. Interfacial and surface tensions of polymers. J Macromol Sci Rev Macromol Chem C10:1–73, 1974.

10. M. Kawaguchi, E. Miyake, T. Kato, A. Takahashi. Ellipsometric measurements of thickness of interfacial layer between coating polymer layer and bulk polymer. Koubunshi Ronbunshu 38:349–353, 1981.

11. S. H. Anastasiadis, J. K. Chen, J. T. Koberstein, J. E. Sohn, J. A. Emerson. The determination of polymer interfacial tension by drop image processing: comparison of theory and experiment for the pair, poly(dimethyl siloxane)/polybutadiene. Polym Eng Sci 26:1410–1418, 1986.

12. S. H. Anastasiadis, I. Gancarz, J. T. Koberstein. Interfacial tension of immiscible polymer blends: temperature and molecular weight dependence. Macromolecules 21:2980–2987, 1988.

13. M. L. Fernandez, J. S. Higgins, J Penfold, R. C. Ward, C Shackleton, D. J. Walsh. Neutron reflection investigation of the interface between an immiscible polymer pair. Polymer 29:1923–1928, 1988.

14. S. Yukioka, K. Nagato, T. Inoue. Ellipsometric studies on mutual diffusion and adhesion development at polymer–polymer interfaces. Polymer 33:1171–1176, 1992.

15. D. J. Stein, RH Jung, K.-H. Illers, H. Hendus. Phänomenologische untersuchungen zur mischbarkeit von polymeren. Angew Makromol Chem 36:89–100, 1974.

16. P. J. Flory. Principles of Polymer Chemistry. Ithaca: Cornell Univ. Press, 1953.

17. D. R. Paul, J. W. Barlow. A binary interaction model for miscibility of copolymers in blends. Polymer 25:487–494, 1984.

18. G. ten Brinke, F. E. Karasz, W. J. MacKnight. Phase behavior in copolymer blends: poly(2,6-dimethyl-1,4-phenylene oxide) and halogen-substituted styrene copolymers. Macromolecules 16:1827–1832, 1983.

19. N. Higashida, J. Kressler, S. Yukioka, T. Inoue. Ellipsometric measurements of positive χ parameters between dissimilar polymers and their temperature dependence. Macromolecules 25:5259–5262, 1992.

20. N. Nishimoto, H. Keskkula, D. R. Paul. Miscibility of blends of polymers based on styrene, acrylonitrile and methyl methacrylate. Polymer 30:1279–1286, 1989.

21. J. M. G. Cowie, V. M. C. Reid, I. J. McEwen. Prediction of the miscibility range in blends of poly(styrene-co-acrylonitrile) and poly(N-phenyl itaconimide-co-methyl methacrylate): a six-interaction-parameter system. Polymer 31:486–489, 1990.

22. M. Suess, J. Kressler, H. W. Kammer. The miscibility window of poly(methylmethacrylate)/poly(styrene-co-acrylonitrile) blends. Polymer 28:957–960, 1987.

23. S. Yukioka, T. Inoue. Ellipsometric analysis of polymer–polymer interface. Polym Comm. 32:17–19, 1991.

24. J. Kressler, T. Inoue. Thermodynamics and interfaces of polymer blends. In: P. N. Prasad, ed. Frontiers of Polymers and Advanced Materials. New York: Plenum, 1994, pp 561–574.

25. Q. Qiu, K. Kurimoto, M. Nakajima, M. Ogawa, E. Arai. ERDA of hydrogen and helium using an 8 MeV^{16}O beam. Nucl Instr and Meth B45:186–189, 1990.

26. J. Kressler, N. Higashida, K. Shimomai, T. Inoue, T. Ougizawa. Temperature dependence of the interaction parameter between polystyrene and poly(methyl methacrylate). Macromolecules 27:2488–2453, 1994.

27. P. J. Flory, R. A. Orwoll, A. Vrij. Statistical thermodynamics of chain molecule liquids. I. An equation of state for normal paraffin hydrocarbons. J Am Chem Soc 86:3507–3514, 1964.

28. T. Ougizawa, T. Inoue. Characterization of phase behavior in polymer blends. In: NP Cheremisinoff, ed. Elastomer Technology Handbook. Boca Raton: CRC Press, 1993, pp 701–729.

29. D. Patterson, A. Robard. Thermodynamics of polymer compatibility. Macromolecules 11:690–695, 1978.

30. T. P. Russell, R. P. Hjelm, P. A. Seeger. Temperature dependence of the interaction parameter of polystyrene and poly(methyl methacrylate). Macromolecules 23:890–893, 1990.

31. N. Higashida, J. Kressler, T. Inoue. Lower critical solution temperature and upper critical solution temperature phase behavior in random copolymer blends: poly(styrene-co-acrylonitrile)/poly(methyl methacrylate) and poly(styrene-co-acrylonitrile)/poly(ε-caprolactone). Polymer 36:2761–2764, 1995.

32. T. Shiomi, K. Imai. A consideration on miscibility behaviour in random copolymer blends based on the equation-of-state theory. Polymer 32:73–78, 1991.

33. W. -H. Jo, M. -S. Lee. Thermodynamics analysis on the phase behavior of copolymer blends: an equation of state approach. Macromolecules 25:842–848, 1992.

34. K. Shimomai, N. Higashida, T. Ougizawa, T. Inoue, B. Rudolf, J. Kressler. Studies on miscibility in homopolymer/random copolymer blends by equation of state theory. Polymer 37:5877–5882, 1996.

35. T. A. Callaghan, K. Takakuwa, D. Paul, A. R. Padwa. Polycarbonate-SAN copolymer interaction. Polymer 34:3796–3808, 1993.

36. H. Lee, R. Fujitsuka, Y. Yang, T. Ougizawa, T. Inoue. Studies on miscibility in polycarbonate/poly(styrene-co-acrylonitrile) blends by ellipsometry. Polymer 40: 927–933, 1999.

37. T. A. Callaghan, D. Paul. Estimation of interaction energies by the critical molecular weight method: 1. Blends with polycarbonates. J Polym Sci Polym Phys 32:1813–1845, 1994.

38. S. -C. Chiu, T. G. Smith. Compatibility of poly(ε-caprolactone)(PCL) and poly(styrene-co-acrylonitrile) (SAN) blends. II. The influence of the AN content in SAN copolymer upon blend compatibility. J Appl Polym Sci 29:1797–1814, 1984.

39. B. B. Sauer, D. J. Walsh. Use of neutron reflection and spectroscopic ellipsometry for the study of the interface between miscible polymer films. Macromolecules 24:5948–5955, 1991.

4

Miscibility and Relaxation Processes in Blends

John M. G. Cowie and Valeria Arrighi
Heriot-Watt University, Riccarton, Edinburgh, Scotland

I. INTRODUCTION

The design, selection, and performance of polymer blends crucially depends on our ability to predict and control the phase behavior of selected polymer–polymer pairs. For this reason a large number of studies have been devoted to the understanding of the parameters that govern miscibility and have attempted to gain a good knowledge of the thermodynamics of the mixture (1,2). This approach provides a quantitative determination of the polymer–polymer interaction energies and makes it possible to predict whether a binary mixture will be miscible or immiscible. The interaction between polymer–polymer pairs affects the physical and mechanical properties of miscible systems and, for immiscible ones, it determines the nature and the width of the interface between the two polymers.

A large number of experimental methods have been used to investigate the miscibility in polymer blends. The morphology of the blend and the kinetics of the phase separation process can be investigated using static techniques such as light scattering and electron microscopy, whereas other methods such as small-angle x-ray scattering, small-angle neutron scattering, ellipsometry, and neutron reflectivity allow the determination of the interaction parameters in miscible and immiscible systems.

Only recently, detailed studies of the dynamics of polymer blends have been reported. Dynamic methods of analysis such as dielectric and mechanical relaxation spectroscopies have been extensively used to establish miscibility between two or more polymeric components. One of the most used criteria to define a polymer–polymer pair as miscible relies on the existence of a single glass transition that is intermediate between the glass transitions (T_gs) of the polymer components. The merging of the two T_gs into a single relaxation process signifies that there is intimate mixing of the polymer chains on a molecular level. On the other hand, an immiscible system will display two distinct T_gs that correspond to the transitions in the pure components, thus indicating the existence of two distinct phases.

Over the past decade there has been an increasing interest in the dynamics of polymer blends. The question that has been addressed is to what extent the dynamics of each of the components is altered by blending. The broad glass transitions observed for miscible blends by differential scanning calorimetry (DSC) compared to the T_gs of the components provide the main evidence of a dynamic effect due to blending (3–5). It is now widely accepted that the α-relaxation process associated with T_g and measured by mechanical or dielectric techniques is broader in a blend compared to the pure polymers (5–7). This broadening has been attributed to the presence of local concentration fluctuations in the blend (5). Because the local composition in a blend fluctuates about an average value, the chain relaxing segments will experience differing local environments and will therefore move at different rates. Thus the local heterogeneity results in a broadening of the relaxation process.

There are other aspects of the dynamic behavior of polymer blends that are not observed in the pure components. For example, a breakdown of time–temperature superpositioning has been observed for the segmental relaxation (6,8). A pronounced asymmetry of the relaxation spectrum with a low-frequency tail appears to be a common feature of many polymer blends. This behavior differs from the usual asymmetric shape toward higher frequencies that is observed in pure amorphous polymers (6).

The effect of blending on the dynamics of the components below the glass transition temperature has also been investigated in detail using various techniques including dielectric relaxation spectroscopy (DRS), dynamic mechanical thermal analysis (DMTA), NMR, and quasi-elastic neutron scattering. Because these secondary relaxations are believed to be correlated with the mechanical properties of the blends such as ductility, impact strength, and toughening, studies of sub-T_g blend dynamics are of considerable practical interest. Similarly to polymer–additive mixtures, the addition of a second polymer can be expected to suppress secondary relaxations as observed following the addition of low-molecular-weight plasticizers (9). In contrast to these studies, the effect of blending on the sub-T_g dynamics has been found to depend on the system investigated. A wide range of effects have been observed, from little or no evidence of antiplasticization (9) to a strong suppression of the β-relaxation (10–12).

It is well known that the physical and mechanical properties of polymeric materials change upon annealing below the glass transition temperature. This time-dependent behavior, which is referred to as physical aging, determines the material performance throughout its service life. The phenomenon of physical aging, which is particularly important when the material has a substantial amorphous component, is associated with volume contraction of the solid. This is believed to be due to small-scale relaxation processes in the glassy state. Aging is therefore strongly correlated to molecular motion, and it relies on a given amount of free volume that enables molecular motion to take place. Annealing causes free volume to decrease, and so molecular motion slows down.

Physical aging can be followed by enthalpy and volume relaxation or mechanical measurements such as creep, stress–relaxation, and dynamic mechanical methods. The aging process has been modeled in a number of ways, and some of these models are discussed in this chapter. In blends, the general observation is that aging rates differ from those of the individual components, often being intermediate between those of the pure components.

In this chapter we first review the dynamic properties of polymer blends in the region of the glass transition and discuss those models that have been developed to explain the dynamics of polymer blends. This is followed by discussion of the effect of blending on molecular motion below the glass transition. The last section deals with the long-term properties of polymer blends, i.e., the physical aging process in both homogeneous and heterogeneous blends (and composites).

II. RELAXATION PROCESSES IN BLENDS

The dynamics of polymer molecules covers a wide frequency range from the relaxation process, which is associated with the cooperative Brownian motion of the main-chain backbones at the glass transition T_g (α-relaxation), to vibrations and rotations of side groups occurring below the glass transition temperature (sub-T_g relaxations).

It may be expected that both the α-relaxation and any sub-T_g motions of a polymer molecule could be affected by the addition of a second polymer component. This can be justified considering that blending changes the environment surrounding the chain segments as well as any side groups. In some cases, the density of the blend has been found to differ from that expected by simple additivity rules (4), thus indicating that chain packing is different in a blend compared to the pure polymers. In other cases, the conformational properties of the components appeared to be altered by blending. These effects may all contribute to alter the dynamics of the individual components in a blend.

A large number of experimental techniques have been employed to investigate the dynamics of polymer blends. For example, DMTA and DRS have been used to establish the degree of homogeneity via measurements of the glass transition and the breadth of the mechanical and dielectric relaxation processes associated with T_g. One of the criteria that is widely used to establish miscibility in a binary blend is the presence of a single T_g intermediate between the T_gs of its constituents. This is a useful but limited criterion, and deeper insight into the degree of homogeneity of the system can be achieved by analyzing in detail the relaxation processes in the blend. It has been shown that the width of the DMTA or DRS loss curves provides a measure of concentration fluctuations in the blend as well as the degree of intermolecular coupling (7,8,13).

DRS presents a number of advantages over other techniques since it provides information on molecular motion occurring on a wide frequency range from 10^{-5} to 10^9 Hz compared to DMTA, which is limited to approximately 0.01 to 100 Hz. The capability of DRS to monitor motion on a wide frequency scale implies that a number of relaxation processes can be detected by DRS, i.e., molecular motion below the glass transition in addition to the α-relaxation.

Dielectric spectroscopy measures the response of dipoles to an electric perturbation, and it therefore relies on the chemical species possessing a permanent dipole. This feature of DRS has been used advantageously to trace the dynamics of a polar component in a blend with a second polymer that is nonpolar. In contrast to DRS, DMTA gives the global material response, and a distinction between the dynamics of each component in a blend cannot be generally made. The application of DRS to the study of miscible and immiscible blends has been recently reviewed (14).

In addition to DRS and DMTA, nuclear magnetic resonance has been extensively used to investigate the dynamics of amorphous polymers. The technique has also been applied to study molecular motion in miscible blends. For example, high-resolution solid-state ^{13}C-NMR (15,16) has been employed to identify the individual dynamics of the components in blends. By combining the techniques of magic angle spinning with high-power proton decoupling it is possible to observe well-resolved ^{13}C-NMR signals for specific C atoms. The dynamics of each component can then be followed through measurements of line broadening. This technique is highly selective, and it offers the advantage that the individual dynamics of the components can be observed simultaneously. Other NMR techniques, such as two-dimensional solid-state NMR, have been used to obtain information on one of the blend components using deuterium or ^{13}C-labeling (17,18).

In comparing data obtained using different techniques one needs to be aware that the degree of homogeneity that is detected is highly dependent upon the instrumental method used (19); DRS is believed to probe local segmental motions of a few units up to 10 bonds, whereas DMTA explores cooperative motions that may involve up to fifty bonds. The combination of these two techniques as well as additional information from other methods may prove useful in providing information on the level of heterogeneity on the different lengths and frequency scales.

A. The Glass Transition Region

One of the first detailed studies of compatibility in polymer blends was performed on miscible and immiscible blends of poly(2,6-dimethyl-1,4-phenylene oxide) (PPO) and copolymers of styrene and 4-chlorostyrene, poly(styrene-co-4-chlorostyrene) (4,5). Miscibility can be drastically changed in this system by varying the copolymer composition; blends of PPO and copolymers with compositions < 67.1 mol% of 4-chlorostyrene are compatible with PPO, while for higher 4-chlorostyrene content immiscibility is observed. Although the transition from one- to two-phase behavior is sharp as it occurs over a narrow copolymer composition range, subtle differences in blend homogeneity as a function of copolymer composition were detected on a wide composition range. The width of the calorimetric glass transition was found to increase with increasing 4-chlorostyrene composition in the copolymers, with a maximum occurring at 67.1 mol% prior to phase separation. The DSC data (4) were supported by dielectric relaxation measurements (5), which indicated that for miscible blends broad relaxations were observed compared to those measured in the pure copolymers; whereas for the copolymers, the width at half height of the dielectric spectra extended approximately over two decades, it increased to three decades in the miscible blends. Immiscible blends of PPO and copolymers with 4-chlorostyrene composition above 67.1 mol% showed a second α-relaxation peak. The width of the relaxation, even accounting for the presence of the second peak, was less than for the miscible blends (log $f/f_m = 2.5$). This result implies that the broadening is a characteristic of the residual heterogeneity in miscible blends that disappears upon phase separation.

Similar results were obtained by Alexandrovich et al. (20) on blends of polystyrene (PS) and poly(2-chlorostyrene) (PoC1S). These blends are miscible at room temperature but exhibit a lower critical solution temperature (LCST) behavior and phase separate 40 to 60° above their glass transition temperature. The dielectric spectra for a 40/60 (by weight) PS/PoC1S blend in the one- and two-phase regions are presented in Fig. 1. It is evident that the spectrum of the blend in the one-phase region is much broader compared to the phase separated sample. Because the polarity of PS is negligible compared to that of PoC1S, the dielectric measurements mainly reflect the dynamics of the PoC1S component. As shown in Fig. 1, the loss peaks of the phase-separated blend and the pure PoC1S sample coincide.

The broadening of the α-relaxation in miscible blends increases with the difference between the glass transitions of the components. This can be clearly seen

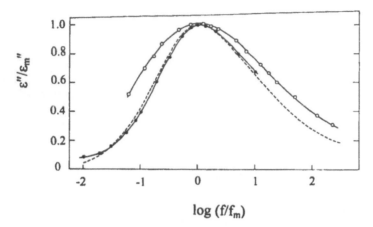

Figure 1 Normalized dielectric relaxation spectra for a 40/60 (by weight) PS/PoClS blend: (O) annealed at 150.3°C, below the LCST, and (●) annealed at 171.8°C, in the two-phase region, above the LCST. The dashed line is the relaxation spectrum of pure PoClS. (Adapted from Ref. 20.)

by comparing data for PS/PoC1S blends (Fig. 1) with the dielectric loss peaks for miscible blends of poly(vinyl methylether) (PVME) and PS (Fig. 2). Due to the considerable difference between the T_gs of PVME (255 K) and PS (380 K) compared to those of PS and PoC1S, which differ by only 30°, the DRS spectra of PS/PVME are much broader than those of pure PVME and the blends reported in Fig. 1. The α-relaxation was also measured for samples annealed in the two-phase region, above the LCST. The spectra indicated the presence of two relaxations, which were attributed to PVME- and PS-rich phases.

It has been observed that the dielectric spectra of PS/PVME as well as other miscible blends are strongly temperature dependent and present an asymmetric shape towards lower frequencies that increases as the temperature is lowered (6,7). This behavior is markedly different from that of pure amorphous polymers where a slight asymmetry towards higher frequencies is generally observed. Roland and Ngai (6) have explained the asymmetric DRS spectra of miscible blends in the framework of the coupling model, which will be discussed later.

The immediate consequence of this strong temperature dependence of relaxation time distribution is that the technique of master-curve construction, which is commonly used to describe the relaxation behavior of pure polymers, is no longer valid. For miscible blends, this leads to a failure of the time–temperature superposition principle (6,7).

Miscible blends of poly(tetramethyl-bisphenol-A-polycarbonate) (TMPC) and PS have also been studied by dielectric spectroscopy (21,22). Because the loss tangent maximum measured by DRS is ten times larger for TMPC than for pure

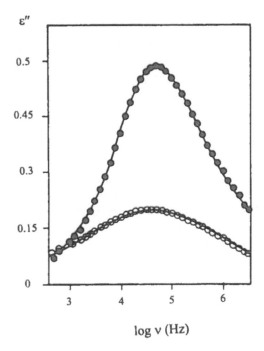

Figure 2 Dielectric α-relaxations ($\bullet$) in pure PVME at $T = 291.53$ K and ($\bigcirc$) for a 40/60 (by weight) PS/PVME blend at $T = 299.55$ K. The lines indicate fits to the experimental data using the Havriliak–Negami equation with $\alpha = 0.65$, $\beta = 0.68$, $\tau = 5.8 \cdot 10^{-6}$ s for PVME and $\alpha = 0.44$, $\beta = 1$, $\tau = 4 \cdot 10^{-6}$ s for the blend. (Adapted from Ref. 7.)

PS, the former dominates the dielectric response. The dynamics of the high T_g component, TMPC, is therefore monitored by DRS. These results are complementary to those of PS/PVME blends, where only the effect of blending on the low T_g component was detected, due to the negligible dipole moment of PS compared to PVME.

The effect of miscibility on chain dynamics was investigated by fitting the Williams–Landel–Ferry (WLF) equation to plots of the logarithmic frequency at the peak maximum, $\log f_m$, versus temperature. Both the fractional free volume and the β parameter of the Kohlrausch–Williams–Watts function (KWW) (23),

$$\phi(t) = \exp\left(-\frac{t}{\tau}\right)^{\beta} \tag{1}$$

were evaluated. The stretched exponential or KWW function is widely used to describe relaxations in complex condensed systems in the time domain t. According to Eq. (1), this is a function of two parameters, the characteristic time τ and the β

parameter, which is related to the width of the relaxation process (β varies from 0 to 1 as the width of the distribution decreases). For the TMPC/PC blends mentioned above, the β was found to be at a minimum at 60 to 80% TMPC content. This suggests a contraction of free volume on mixing, which is accompanied by the broadening of the distribution of relaxation times (lower β parameter). A similar behavior was observed by Mansour and Madbouly (22) in their study of PS/TMPC blends. The Cole–Cole distribution parameter for the α-relaxation showed a minimum (indicating a broadening of the distribution) at TMPC composition equal to 80 wt%.

A composition dependence of the broadening of the α-relaxation process has been reported in many miscible blends. However, it has been observed (24) that the position of the maximum is not simply a function of the T_g or the relaxation strength of the blend components. Whereas for binary liquid mixtures the maximum of the width of the relaxation frequency distribution was reported to be shifted towards the high T_g component, for both PS/PoClS and blends of bisphenol-A polycarbonate (BPA-PC) and tetramethyl bisphenol-A polycarbonate (TMPC) (24) this maximum was found to be closer to the lower T_g polymer.

It is interesting to note that, unlike Katana et al. (24), Mansour and Mabbouly (25) did not observe a similar change in the broadening of the dielectric α-relaxation of BPA-PC/TMPC blends, a result that is in agreement with DSC data on these blends (24,25). Although the two polymers have different T_gs and therefore different degrees of molecular interactions and constraints, the absence of any blending effect on the distribution of relaxation times was taken as evidence that blending did not cause any change in the local constraints.

Many of the experimental studies carried out on polymer blends have either been unable to resolve the dynamics of the individual components or have deliberately attempted to monitor only the motion of one of the constituents in the blend. For dielectric measurements, this has often relied on the high dipole moment of one species, PVME or poly(chloro-styrene) in the examples discussed above, compared to the other components in the blends (e.g., polystyrene). In these studies only the influence of local environment on the motion of one component was followed.

Some of the studies reported in the literature have made use of different techniques and labeled samples to extract information on the dynamics of the individual components in blends. For example, Chin et al. have employed two-dimensional solid-state NMR (17,18) to compare the chain motion of PS and PPO in miscible PS/PPO blends. The dynamics of each component in the blend was monitored by using either ^{13}C-NMR on ^{13}C-labeled poly(2,6-dimethyl-1,4-phenylene oxide) or ^{2}H-NMR on polystyrene samples deuterated in the chain backbone (PS-d_3). In both studies the distribution of correlation times was found to be broader in the blend compared to the pure polymers, in agreement with early DRS measurements carried out by Wetton et al. (4,5). The 2-D patterns were quantitatively analyzed, and neither a single-exponential correlation function nor a single stretched exponential function was capable of describing the experimental data. A multimodal distribution was invoked that consisted of two stretched exponentials for 25/75 PPO/PS blends (bimodal distribution) and three components for a 50/50 blend (trimodal distribution). The relative weighting of the components of the distribution was correlated with local concentration fluctuations that were determined from a lattice model based on purely random statistics.

As with this study, a combination of experimental methods was used by Alegria et al. (26) to resolve the dynamics of each component in poly(vinylethylene) (PVE)/polyisoprene (PIP) blends by exploiting the difference between the dielectric strength and the mechanical response of the two polymers. The dielectric strength of PVE is more than twice that of PIP, but PIP gives a sharper, more intense loss modulus peak making a stronger contribution to the mechanical loss modulus of the blend. The approach used by Alegria et al. (26) provides information on the dynamics of the individual components and therefore makes it possible to compare the experimental data with the results of theoretical models of blend dynamics.

The system used by Alegria et al. (26) is particularly interesting, since the two polymers PVE and PIP have different chemical structures and are expected to exhibit different mobility. PVE has inflexible vinyl units attached to the chain backbone that are responsible for strong intermolecular constraints to molecular motion. The dynamic mechanical spectrum of PVE is unusually broad compared to the spectra of most amorphous polymers (27), and this results in a large intermolecular coupling parameter ($n = 0.74$). Blends of PVE with PIP, a polymer with weaker intermolecular coupling, are therefore of interest in order to establish the relative contributions to blend dynamics due to the difference between the intrinsic mobility of the components.

PVE/PIP blends present unusually broad loss peaks in the dynamic mechanical spectra, particularly at high PVE concentrations (28). This made it difficult to characterize the blend by conventional mechanical spectrometers; ^{13}C-NMR (8) and dielectric spectroscopy (26) were used for this purpose. The results of these studies indicate that each component in the blend has its own intrinsic mobility. For the two blend components, both the distribution of the relaxation frequency and its temperature dependence were found to differ considerably, a result that reflects the different chemical structure of PVE and PIP. In addition, due to local composition fluctuations in the blends, the distribution of the relaxation frequency differs from those in the pure polymers.

Similar behavior has been observed in PS/PVME blends using high-resolution solid-state ^{13}C-NMR (15,16). Due to the high selectivity of this technique, the mobility of the PS and PVME components in the blend could be studied simultaneously by following the temperature dependence of the line widths of the carbons in the PS and PVME units (15). Both PS and PVME showed a large line broadening followed by a decrease of the line width with increasing temperature. Although for PS the aromatic and main-chain carbons had similar temperature dependence with maxima occurring at similar temperature, for PVME, the maximum line broadening occurred at much lower temperature. It was then concluded that, although the blends are compatible as indicated by the existence of a single T_g, the two components do not share the same dynamics at any given interval (T-T_g) (15,16). Theoretical attempts have been made to explain the dynamic properties of polymer blends. These will be discussed in Sect. II.c.

B.   Sub-T_g Relaxations

On cooling through the glass transition, segmental motion becomes increasingly hindered until it practically ceases at T_g. At lower temperature, only local dynamics such as side group reorientations takes place. For amorphous polymers, these sub-T_g motions are usually much broader compared to the α-process. This is believed to be due to the disordered structure of amorphous materials. In the glassy state, there exists a range of different environments surrounding the chemical units that undergo molecular motion. This heterogeneity generates a distribution of potential energy barriers hindering the reorientational motion, which is reflected experimentally in the broad distribution of relaxation frequencies.

In polymer blends, addition of a second component is expected to increase the level of heterogeneity, thus giving rise to observable blending effects on sub-T_g relaxations. In analogy to polymer–additive mixtures, where the presence of the low-molecular-weight component strongly suppresses the β-relaxation (9), an antiplastization effect could be expected. Whereas the glassy mixtures of polymers and additives are generally found to be harder and more brittle than the pure polymers, antiplasticization is not always observed in polymer blends. Fischer et al. (9) have suggested that whereas the local free volume fluctuations are responsible for the suppression of the β-relaxation in polymer–additive mixtures, a similar dynamic effect would not be expected in polymer blends, due to much lower volume fluctuations in these systems. Studies on BPA-PC/TMPC blends appear to support this hypothesis (9,24,25).

The β-relaxation in TMPC/BPA-PC blends has been investigated by various authors using dielectric spectroscopy (24,25,29). The local processes in TMPC and BPA-PC can be easily distinguished, as they are separated by more than 90° with the β-relaxation of TMPC occurring at the higher temperature. This difference indicates that changing the chemical structure of the polymers alters the β-process. Because this relaxation is attributed to the cooperative motion of the phenyl and carbonyl groups, it is evident that this motion should be hindered by the presence of the four bulky methyl groups in TMPC.

The dielectric loss spectra of TMPC/BPA-PC blends at various compositions are plotted in Fig. 3. The β-relaxation of both polymers can be clearly detected.

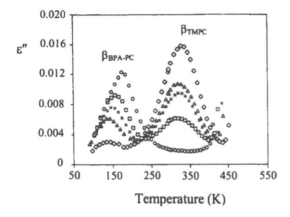

Figure 3 Dielectric loss spectra obtained at 10 Hz versus temperature for blends of BPA-PC and TMPC with TMPC volume fraction equal to (O) 0, (□) 0.22, (x) 0.38, (△) 0.53, and (◇) 0.82. (Adapted from Ref. 24.)

Although this sub-T_g motion is not suppressed in BPA-PC, it is noticed that the peak shifts towards lower temperatures as the TMPC content increases. At ϕ_{TMPC} = 0.82 the β-peak of BPA-PC has shifted by 40°. This result has been attributed to a suppression of the high-temperature tail of the relaxation process due to selective inhibition of the local motion of BPA-PC (24).

The β-relaxation in TMPC is unchanged by blending, as illustrated in Fig. 3, and the activation energy of this process is constant with blend composition. By contrast, the activation energy of the same process in BPA-PC decreases with increasing TMPC content from 54 kJ/mol for the pure polymer to 38 kJ/mol for an 80/20 TMPC/BPA-PC blend (24).

The strengths of the two β-relaxations are not affected by blending but vary according to dilution laws. This result is in contrast with data reported by Jho et al. (29) who observed that the β-process was fully suppressed above a critical TMPC concentration equal to 75%.

The absence of any blending effect on the local dynamics in BPA-PC/TMPC and TMPC/PS blends have brought Mansour and Madbouly (22,25) to conclude that the local environments in these blends are relatively unchanged by blending and consequently miscibility at the segmental level is not achieved.

Although Fischer et al. (9) have attributed the absence of any β-suppression in TMPC/BPA-PC to the small free volume fluctuations in blends compared to those in polymer–additive mixtures, other authors have highlighted the importance of specific interactions among the blend components and their effects on sub-T_g motions.

A negative volume of mixing has been observed for many polymer–polymer systems as a result of specific interactions. For blends of polystyrene (PS) and poly(2,6-dimethyl-1,4-phenylene oxide) (PPO) this effect may be related to the conformational changes of the PS units suggested by Mitchell and Windle (30). WAXS measurements carried out by these authors indicated that the introduction of PPO disrupts the stacked arrangement between the phenyl groups of pure PS. This microsegregation of the phenyl groups into stacks is believed to hinder molecular motion and bring about the poor mechanical properties of PS.

If the PPO chains effectively act as templates for the polystyrene ones as suggested by Mitchell et al., then it would be expected that molecular motion in PS be enhanced by blending. This effect was confirmed experimentally by Feng et al. (31) using ^{13}C-NMR. In this study, the spin relaxation times $T_1(C)$ of the aromatic rings in PPO and PS were found to be equal to 9.5 and 38.5 s, respectively. This indicates that the dynamics of PPO is five times faster than that of PS. For PS/PPO blends, $T_1(C)$ values of the two components were found to be almost the same, thus suggesting that the aromatic rings of the two polymers move cooperatively in the blend, probably due to a strong π–π interaction between them. The ability of the PS chains to undergo rapid conformational transitions in the PS/PPO blends could explain the improved impact strength of PS chains upon addition of PPO.

Although the NMR data discussed above appear to be consistent with the mechanical properties of PS/PPO blends, these results are not supported by other data reported in the literature. ^{2}H-NMR measurements reported by Chin et al. (12) have shown that the population of fast flipping PS phenyl groups is slightly reduced in the blend (32). It is suggested by the authors that the more pronounced β-process observed in the mechanical data of these blends may be a consequence of the higher glass transition of these mixtures compared to the pure polymers. Because the phenyl motion is almost unchanged, but the T_g occurs at higher temperature, the β-relaxation is better resolved from the α-process.

A suppression of the low-temperature local relaxation has been observed in BPA-PC blends with poly(methyl methacrylate) (PMMA) using dynamic mechanical spectroscopy and solid state ^{2}H-NMR (10). Although the spectra of 50/50 BPA-PC/PMMA blends in the two-phase region could be described as the superposition of those of the pure components, the spectra of blends in the one-phase region were found to be considerably different from those of pure BPA-PC and PMMA. For example, there was no evidence of the β-peak of BPA-PC in the dynamic mechanical spectra of a 50/50 blend. Deuterium NMR spectroscopy indicated that the distribution of correlation times for the ring motion in BPA-PC was broader in the blend compared to the pure polymer, thus suggesting that motion in BPA-PC is hindered by the presence of PMMA.

A similar effect would be expected for PMMA, but a reduction in the mobility of this polymer has not been observed experimentally (10). The β-process of

PMMA was found to move towards lower temperature with increasing BPA-PC content, thus suggesting that PMMA is less hindered by the presence of BPA-PC.

Antiplasticization was also observed in miscible blends of Chloral-PC and PMMA by de los Santos Jones et al. (11) using proton line shape and proton spin-lattice relaxation, $T_{1\rho}$, measurements. Because the PMMA component was fully deuterated, the effect of blending was investigated through the motion of the polycarbonate units. The proton $T_{1\rho}$ data showed a suppression of the π-flip process in Chloral-PC due to the addition of PMMA. At high PMMA compositions, a second process was detected that occurred at a temperature and time scale close to the ester group rotation in PMMA. This was interpreted as being due to coupling between the sub-T_g relaxations and the ester-CH_3 rotation of PMMA (11), a behavior similar to that observed in polymer–additive mixtures (33). The NMR data were interpreted in terms of a lattice model first developed for polymer–diluent mixtures as discussed in the following section.

C. Theoretical Models of Blend Dynamics

A few models have been developed to describe quantitatively the α-relaxation process observed in miscible blends. The models described below deal with the dynamics associated with the glass transition, although the lattice model originally developed for polymer–diluent mixtures (33) was also applied to describe sub-T_g motion in polymer blends (11).

The difference between the dynamics of a blend and that of its components, close to the glass transition, has been attributed to concentration fluctuations (4,5,7), and theoretical models that, by accounting for local heterogeneities, explain the origin of the distribution of relaxation times have been developed (34,35). The dynamics of polymer blends has also been described in the framework of the coupling model. Here it is assumed that the local environment of the relaxing segments affects the degree of intersegmental cooperativity and gives rise to a distribution of coupling parameters (6,8,13).

1. Lattice Model

It has been noted in Sect. II.B that the suppression of the β-relaxation in BPA-PC or Chloral-PC following the addition of a second polymeric component such as PMMA is similar to the effect observed after addition of a low-molecular-weight diluent (33). Considering this similarity, de los Santos Jones et al. (11) adapted the lattice model developed for polymer–additive mixtures (33) to the description of polymer blends.

Using a lattice with six nearest neighbors, these authors attempted to explain the antiplasticization effect observed in Chloral-PC/PMMA blends (11). A lattice site was defined considering the relative volume of the two polymer repeat units

as either two Chloral-PC or five PMMA units. The weight fraction of the blends
was then converted into sites fraction occupied by a given repeat unit. If p defines
the fraction of lattice sites occupied by Chloral-PC and d is the number of lattice
sites containing the second components, PMMA, then the fraction of Chloral-PC
units, F_i, where i represents the nearest neighbors lattice sites occupied by the second component, can be calculated by assuming a purely statistically random arrangement of the neighbouring lattice sites:

$$F_0 = p^4 \qquad F_1 = 4\,p^3\,d \qquad F_2 = 6\,p^2\,d^2$$
$$F_3 = 4\,p\,d^3 \qquad F_4 = d^4 \tag{2}$$

As for the polymer–diluent system, three types of polymer units could be distinguished (33,11): (a) those surrounded only by the same units, expressed as F_0, (b)
those in contact with only one diluent molecule or polymer component, F_1, and (c)
units surrounded by lattice sites occupied by more than one second component
unit (F_2, F_3, and F_4). It has been assumed (11) that units surrounded by the same
monomers relax as in the pure polymer system and therefore their dynamics is unchanged. Antiplasticization only occurs when one neighboring site is occupied by
the second component as a consequence of improved packing.

For the polymer–diluent system, units that are in contact with two or more
diluent units were considered to have increased mobility with respect to the pure
polymer, due to fast rotational diffusion of the diluent (33). This was supported by
the presence of an additional low temperature minimum in the proton T_{1p} data of
BPA-PC/diluent. The presence of the low temperature minimum was attributed to
those phenylene groups that moved more rapidly than the pure polymer.

The Chloral-PC/PMMA blend (11) showed a similar behavior with the de-
velopment of a new relaxation process at high PMMA concentrations attributed to
Chloral-PC units relaxing cooperatively with the ester side group rotation in
PMMA.

Although the lattice model gives a good description of polymer–diluent dynamics, the application to polymer blends was not found to be equally successful.
The discrepancies observed between the experimental data and the predictions
from the lattice model suggested that there are fewer contacts between Chloral-PC
and PMMA than expected from random mixing. This points out differences between polymer–diluent and polymer–polymer mixtures. It was suggested that the
reduced antiplasticization effect observed in blends is a result of the reduced ability of the polymer additive to fill local high–free volume regions compared to the
low-molecular-weight diluents.

The lattice model was also applied by Chin et al. (17,18) to the interpretation
of the NMR data of PS/PPO blends close to the glass transition. It has been already
discussed in Sect. II.B.1 that the dynamics of the PS chains in 25/75 PPO/PS
blends was described by a bimodal distribution consisting of two stretched exponential functions. The fast component accounted for 25% of the intensity, whereas

the slow component corresponded to the remaining 75%. This was found to be in close agreement with the lattice model where the fraction of the mobile PS units, F_0, is equal to 0.24, whereas the slow component is identified with those PS units that are in contact with the less mobile PPO units.

The good agreement between model predictions and blend dynamics close to T_g is in contrast to the disagreement found for the sub-T_g motion. A possible explanation is given by the different length scale of the dynamic processes involved. The π-flips that are correlated with low-temperature motion occur over a local scale, and effects due to nonrandom mixing may be important. On the contrary, the distance scale that is involved in the segmental motion close to T_g is much longer, and this may explain the success of the lattice model.

2. Fluctuation Model

The model developed by Fischer and Zetsche (34,35), which is based on the effect of concentration fluctuations on the segmental motion of one of the components in a blend, enables the determination of the temperature dependence of the concentration fluctuations, $<(\delta\phi)^2>$, from the temperature dependence of the dielectric spectra.

The "fluctuation model" relies on a number of assumptions. The sample is divided into subvolumes V_i each having a concentration ϕ_i of one of the components. The distribution $p(\phi_i)$ of the number of subvolumes with concentration ϕ_i is considered to be Gaussian. The concentration fluctuations are assumed to be stationary, i.e., the lifetimes of the concentration fluctuations are much longer than the relaxation times of the α-process.

The presence of concentration fluctuations leads to fluctuations of the time-dependent properties of the blend, and a composition-dependent glass transition $T_{gi}(\phi_i)$ is defined for each subvolume. This is transformed into a distribution of relaxation times via the empirical Williams–Landel–Ferry (WLF) equation, which describes the temperature dependence of the α-relaxation time, $\tau(T-T_g)$. A corresponding relaxation time $\tau_i(\phi_i, T)$ is defined. The Gaussian distributions of concentration $p(\phi_i)$ can then be written as a distribution of relaxation times $n(\tau_i, T)$.

Dielectric loss curves can be calculated for mixtures in which the relaxation behavior is dominated by the response of one of the components, for example PVME in PS/PVME blends (34,35). To do this, the shape of the relaxation spectrum is assumed to be the same in the pure polymer and in the blend, i.e., characterized by the same Havriliak–Negami parameters.

The model has been applied to the description of blend dynamics in PS/PVME blends. As shown in Fig. 4, fits with the model function are excellent at all concentrations and temperatures.

A number of features are reproduced by the fluctuation model, particularly the broadening towards lower frequencies that is observed with decreasing tem-

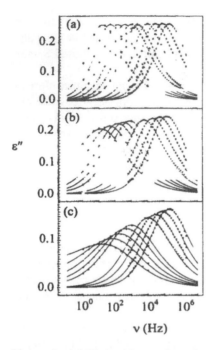

Figure 4 (a) Dielectric α-relaxation in pure PVME with fits performed using the Havriliak–Negami equation. The width of the loss curves slightly increases with decreasing temperature from 262.7 to 299.8 K. (b) Measured dielectric loss curves of a 20/80 PS/PVME blend and fits according to the fluctuation model. Measuring temperatures from 266.8 to 302.7. (c) Measured dielectric loss curves of a 40/60 PS/PVME blend and fits according to the fluctuation model. Measuring temperatures from 271.8 to 308.6. (Adapted from Ref. 35.)

perature. It has been pointed out that the HN function cannot describe the α-relaxation near T_g in miscible blends since, depending on the parameters used, it is either symmetric or asymmetric towards higher frequencies. The fluctuation model, however, is able to generate asymmetric shapes towards lower frequencies.

The mean square concentration fluctuations $<(\delta\phi)^2>$ can be extracted from the dielectric data as fitting parameters. The temperature dependence of $<(\delta\phi)^2>$ was correlated to the volume of the cooperatively rearranging domains, which has been discussed in theoretical treatments of the glass transition. For PS/PVME blends, this was estimated to have a radius of about 4 nm at 20 K above T_g.

A more detailed analysis of blend dynamics in terms of the concentration fluctuation model was recently performed by Katana et al. (36) on mixtures of PS and a statistical copolymer poly(cyclohexylacrylate-butylmethacrylate). Mean-square concentration fluctuations for this blend were obtained from the dielectric loss curves, as described for PS/PVME. Values of $<(\delta\phi)^2>$ are temperature dependent, but no composition dependence was observed for blends with ϕ between 0.2 and 0.5 when data were compared at equal $(T-T_g)$ intervals. This is consistent with small-angle neutron scattering (SANS) data on this system, which revealed negligible composition dependence of the interaction parameter $\chi(T)$ in the same composition range.

By combining the $<(\delta\phi)^2>$ with SANS measurements of the static structure factor $S(Q)$, the size of the domains characterized by cooperative dynamics was evaluated. It was found that the radius r of these domains varied from 6 to 10 nm in the temperature range from T_g to $T_g + 50$ K for blend compositions with $\phi = 0.2$–0.5. The radius obeyed the scaling law $r \propto (T - T_o)^{-\nu}$ with ν values ranging from 0.64 to 0.67, which is consistent with the theoretical predictions for the temperature dependence of the volume of cooperative relaxation given by Donth (37) and Adam and Gibbs (38).

3. Coupling Model

The segmental motion of polymer chains in the amorphous bulk phase can be described in terms of correlated conformational transitions that arise from both intra- and intermolecular constraints due to the connectivity of the monomer units. In pure polymers, the degree of the intermolecular coupling of segmental relaxations depends on the chemical structure of the monomers and, in the framework of the coupling model, it is expressed by the coupling parameter n (39).

The effective relaxation time τ^* is related to the coupling parameter by

$$\tau^* = \left[(1 - n)\omega_c{}^n\tau_0 \right]^{1/(1 - n)} \tag{3}$$

where τ_0 is the relaxation time in the absence of intermolecular coupling and $\omega_c{}^{-1}$ is a characteristic time defining the onset of coupling. This is typically of the order of 10^{-10} s (39).

The two components of any binary mixture will each be characterized by a relaxation time τ_0 and a coupling parameter n. Blending a polymer with a component of a different chemical structure will cause a change in the degree of intermolecular coupling. From a dynamic viewpoint, the local environment that surrounds the polymer segments changes upon blending, and chain mobility is therefore expected to vary. For example, if the added component has a stronger intermolecular coupling, it is expected that the coupling parameter of the polymer segments will increase compared to the pure polymer. In addition, because the coupling parameter depends on the chemical structure, the blend components are expected to experience different levels of intermolecular coupling even if surrounded by the same environment.

The presence of concentration fluctuations in miscible blends complicates this situation further. Because the degree of cooperativity of the segmental relaxation depends on the local environment as well as the chemical structure of the relaxing units, the distribution of local environments in blends leads to a distribution in the degree of intermolecular coupling and therefore to a distribution of coupling parameters.

The coupling model has been successfully applied to describe the dynamic properties of polymer blends such as PS/PVME (6), PIP/PVE (8,26), and TMPC/PS (13).

The dielectric relaxation spectra of miscible PS/PVME blends were fitted using a coupling model developed for blends (6). Data fitting indicated that the coupling parameter was a function of local composition and it therefore varied across the spectrum. For a PS/PVME blend (40/60) the coupling parameter n was found to range from 0.75 to 0.52 from low to high frequency, respectively (6). The mean coupling parameter for this blend was determined to be 0.67, substantially larger than $n = 0.56$ for pure PVME. This implies that there exists a considerable distribution in the degree of coupling of the PVME segmental relaxation due to changes of the local environment.

The coupling model successfully describes the asymmetry of the dielectric relaxation spectra observed in miscible blends (6). For PS/PVME (6) as well as other miscible blends (8,13,26), the response at higher frequencies arises primarily from those regions in the mixture rich in the low-T_g component having weaker intermolecular coupling (e.g., PVME), whereas environments rich in the high-T_g strongly intermolecularly coupled component will contribute predominantly to lower frequencies (longer times). As a result of this heterogeneous distribution in blends, the components with long τ^* and therefore large n will be shifted more rapidly towards longer times as the temperature is lowered. This has the effect of generating an asymmetric broadening of the distribution of relaxation times, with the well-known reversal of the asymmetric shape observed in pure amorphous materials.

As a consequence of concentration fluctuations in miscible blends, the dynamics of each component is expressed by a heterogeneous distribution of stretched exponential functions. Due to the presence of a distribution of coupling parameters, the blend dynamics no longer obey time–temperature superposition.

The dielectric data on PS/PVME blends only detect the relaxation of the highly polar PVME component. According to the coupling model, whereas the coupling parameter of PVME in the blends is always larger than n for pure PVME, the opposite effect would be expected for the PS component. By replacing PS with the more mobile PVME units a decrease in the PS coupling parameter is expected. This effect was indirectly observed using blends of TMPC/PS (13). It has been already observed that, for this system, the dielectric response is dominated by TMPC.

In TMPC/PS blends, TMPC plays the role of PS in PS/PVME blends, as it is the high-T_g component. The addition of PS to TMPC was found to lower the cou-

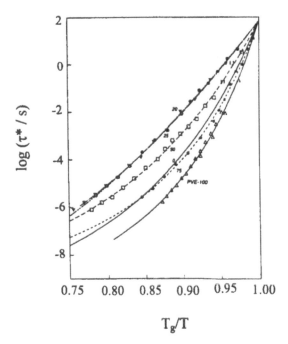

Figure 5 Cooperativity plot for the PVE component in PVE/PIP blends. The PVE content is indicated. T_g appearing in the label of the x-axis stands for the glass transition temperature T_{gPVE-x} of the PVE component in the PVE-x blend. (Adapted from Ref. 26.)

pling parameter, thus confirming that the coupling of the TMPC segmental relaxation is always weaker in the blend compared to the pure polymer (13).

More recently Alegria et al. (26) were able to follow the mobility of the individual components in PIP/PVE blends. The dielectric and dynamic mechanical data were interpreted using the coupling model. The relaxation frequencies of the segmental motion of the two blend components, converted to $\tau_i^* = (2\pi f_i)^{-1}$, are plotted versus T_g/T in Figs. 5 and 6 (cooperativity plots).

PVE has a steeper curve compared to PIP, in agreement with the larger coupling parameter of the former. On addition of PIP to PVE, a reduction of the steepness of the curve is observed and, when the PVE content reduces to 50%, this effect is so pronounced that the steepness of the PVE component is reduced compared to that of pure PIP.

No considerable differences in the coupling parameter of the PIP component in the blend and the pure polymer were observed up to 25% PVE content (Fig. 6). This is consistent with PIP being still the major component. A more pronounced blending effect would be expected at higher PIP content.

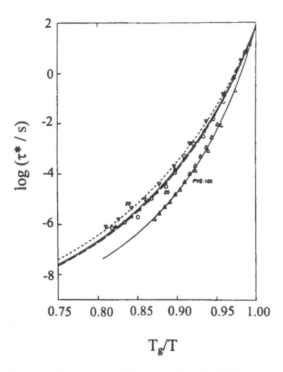

Figure 6 Cooperativity plot for the PIP component in PVE/PIP blends. The PVE content is indicated. The solid triangles represent data for pure PIP. T_g appearing in the label of the x-axis stands for T_{gPIP-x} of the PIP component in the PVE-x blend. (Adapted from Ref. 26.)

III. PHYSICAL AGING

The long-term stability of polymeric materials is a matter of considerable importance, both to materials scientists and to engineers. Two types of aging occur that lead to changes in the properties of polymers: chemical and physical. Chemical aging normally leads to modification of the polymer chain and may involve chain scission, oxidation, dehalogenation, loss of pendant groups, hydrolysis, and crosslinking, all of which are chemical reactions. Thus chemical aging usually leads to polymer degradation with a concurrent deterioration of the properties (e.g., coloration, embrittlement).

On the contrary, in the case of physical aging the polymer structure remains unchanged but the local packing of the chains alters. This leads to dimensional changes and alteration of physical properties such as brittleness, tensile strength, and glass transition temperature T_g. Thus physical aging is a reversible process, whereas chemical aging is not.

Physical aging is a manifestation of small-scale relaxation processes that take place in the amorphous region of a glassy polymer. When a polymer is cooled from the melt to a temperature below its glass transition, the glass that is formed is not in equilibrium with its surroundings. Physical aging is then the slow structural reorganization of the glass, on annealing at a given temperature T_a, as it attempts to achieve the equilibrium state. The continuous, slow relaxation of the glass from the initial nonequilibrium state towards a final thermodynamic equilibrium state produces time-dependent changes in the physical properties of the polymer. As the extent of physical aging increases, there are corresponding decreases in the enthalpy, the specific volume, and the fracture toughness, while increases in the glass transition temperature, the yield stress, and the tensile modulus of the material can also be observed. The volume and enthalpy relaxations that are features of this phenomenon, as are the time-dependent small-strain mechanical properties, can then be used to follow the progress of the physical aging process.

A. Fundamental Principles

1. Enthalpy Relaxation

Physical aging involves segmental relaxation processes that occur in the temperature range $T_\beta < T < T_g$ (40), where T_β is the temperature of any secondary relaxation process that can occur in the glassy state. This restricted temperature range is disputed (41–42), and it is believed that it is a phenomenon affecting all viscoelastic relaxation processes. As the annealing temperatures T_a drop further away from T_g [i.e., $(T_g - T_a)$ increases], the aging process slows down and the time scales involved become quite long. Consequently many studies are carried out under thermally accelerated conditions, and the relaxation of the enthalpy of the glass is a convenient parameter to follow when monitoring the physical aging process. Measurements are readily carried out at temperatures below T_g using DSC, if suitable precautions are taken.

It is important to define a reproducible thermal history for the sample before meaningful comparisons can be made. The first step is to ensure that the influence of any previous thermal history is erased by annealing the sample at a temperature in excess of $T_g + 50°C$, i.e., at the point A in Fig. 7a. The polymer is then cooled from the melt at temperature T_2 (A), at a rate q_1, into the glass at temperature T_a (B). The distance of the sample from its "equilibrium state" at T_a will depend on the rate of cooling of the sample, and so this thermal treatment should remain unchanged for all measurements that are to be compared. Annealing the sample at temperature T_a, for a specified time t_a, results in an enthalpy loss $H_B - H_C$ (along the line B–C), the extent of which depends on the magnitude of t_a. When measurements are carried out using DSC, the sample is then quenched from T_a after t_a, and on reheating the polymer at a rate q_2, the enthalpy overshoots the equilibrium curve, as shown in Fig. 7a

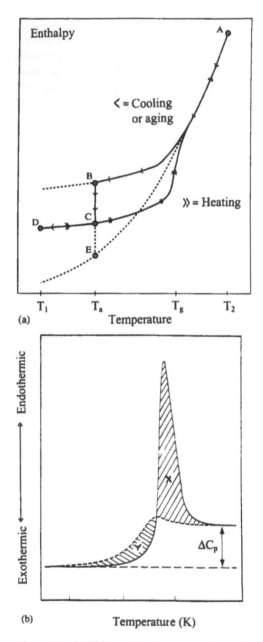

Figure 7 (a) Schematic diagram of cooling, aging, and heating cycles using a fixed thermal history. (b) Schematic diagram of DSC curves for an aged (full line) and unaged (broken line) polymer sample.

by an amount proportional to that lost during the aging process. This is represented by the line C–A. The enthalpy difference $H_B - H_C$ is then given by

$$(H_B - H_C) = \Delta H(t_a, T_a) = \int_{T_\alpha}^{T_\beta} \left\{ C_p \text{ (aged)} - C_p \text{ (unaged)} \right\} dT \qquad (4)$$

and represents the area X–Y in Fig. 7b. However, the methods of calculating enthalpy changes can depend on the model used to describe physical aging, as explained next.

2. The Multiparameter Phenomenological Models

A number of theoretical treatments have been developed that attempt to model the aging process in organic and inorganic glasses. Notable examples are the phenomenological models developed by Moynihan et al. (43) and Hodge et al. (44–46) and more recently by Gomez and Monleon (47). As the relaxation processes in the glassy state and glass-transition region are nonexponential and nonlinear, the theories must take account of the thermal history of glass formation and the asymmetry of the relaxations, which depend on how the system departs from equilibrium. These theories attempt to curve fit the heat capacity data from aging experiments and use integral equations to model the cumulative effects of a given thermal history imposed on the polymer starting from the equilibrium melt state. Use is then made of the convenient concept of a 'fictive' temperature T_f, first proposed by Tool (48) as the temperature at which the nonequilibrium value of a macroscopic property of the system would be an equilibrium one. This may be better understood by referring to Fig. 8, where the enthalpy difference $H(T_2)-H(T)$ can be expressed in terms of the heat capacity C_p and leads finally to

$$\int_T^{T_f} \left[C_p (T') - C_{p \text{ glass}} (T') \right] dT' - \int_{T_f}^{T_2} \left[C_{p \text{ liquid}} (T') - C_p (T') \right] dT' = 0 \qquad (5)$$

where T' is a dummy variable. From this a normalized heat capacity can be formulated in terms of experimental quantities as

$$\frac{dT_f}{dT} = \left[\frac{C_p (T) - C_{p \text{ glass}}(T)}{C_{p \text{ liquid}} (T_f) - C_{p \text{ glass}}(T_f)} \right] \qquad (6)$$

When the heat capacities of the glass and the liquid are defined by linear functions of temperature, the value of T_f can be calculated from numerical integration of the DSC curves to obtain $H(T_2)-H(T)$.

Gomez and Monleon (47) do not use the fictive temperature but consider instead the temporal evolution of the configurational entropy, S_c. This allows one to examine the raw C_p data without having to resort to normalization procedures and also removes the restriction of having to calculate the limiting C_p of the fully aged

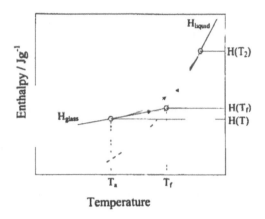

Figure 8 Representation of how the fictive temperature is defined.

glassy polymer from a linear extrapolation of the liquid C_p curve. In other words, they assume that the polymer may not be able to attain such a value because of the physical restraints imposed by chain entanglements and inefficient packing in the glass. This concept has also been used by Cowie and Ferguson (49).

The annealing of the glass at the aging temperature T_a results in a relaxation of the enthalpy towards the equilibrium value $H(\infty)$. This can be described by a relaxation function $\phi(t)$ defined by

$$\phi(t) = \frac{H(t) - H(\infty)}{H(0) - H(\infty)} \tag{7}$$

$\phi(t)$ can also be expressed in terms of a semiempirical function introduced originally by Kohlrausch and revived by Williams and Watts (23), abbreviated as the KWW equation:

$$\phi(t) = \exp\left[-\left(\frac{t}{\tau}\right)^{\beta}\right] \tag{8}$$

where β is related to the breadth of the distribution of relaxation times and has values $0 < \beta < 1$. A value of $\beta = 1$ would imply an infinitely sharp distribution with only one relaxation time. The latter parameter is represented by τ and the effects of physical aging are then analyzed in terms of β and τ.

The relaxation time τ can be expressed in terms of the fictive temperature T_f following Narayanaswamy (50) and Moynihan et al. (43), who used the form

$$\ln \tau = \ln A + \left[\frac{x\Delta h}{RT} + \frac{(1 - x)\Delta h}{RT_f}\right] \tag{9}$$

which assumes the system to be thermorheologically simple with only a single relaxation time. Here Δh is the activation energy, x is a structural parameter with values $0 < x = 1$, and A is a fitting constant. In this model Δh and x have no clear physical meaning, but Hodge (51) has used the Adam–Gibbs theory (38) to show that $x = T_f/T_2$, where T_f is the value of the fictive temperature in the glass and T_2 is the "equilibrium" T_g value postulated by Gibbs and Di Marzio (52). A similar approach is used by Gomez and Monleon, only T_f is replaced by S_c in their model.

The parameters in Eq. (9) can be estimated by assuming that Δh has a fixed value and allowing $\ln A$, x, and β to vary until the best fit to the experimental, normalized, C_p data is obtained. A more informative assessment can be made by comparing $\Delta H(t_a, T_a)$ values from the model with experimental data.

Alternative analytical methods are the KAHR isothermal single relaxation time model (53) and the peak shifting technique (54). None of these attempt to use the data to predict long-term aging effects, for which a different approach is necessary.

3. Predictive Models for Long-Term Aging

The descriptions of physical aging given in the previous section cannot predict long-term aging effects, so for this one must use either the Petrie–Marshall (P–M) (55) or the Cowie–Ferguson (C–F) (49,56) models. In both, the enthalpy lost on aging is given by

$$\Delta H(t_a, T_a) = \Delta H(\infty, T_a)\{1 - \Phi(t_a)\} \tag{10}$$

where the P–M approach uses Eq. (8) to define $\Phi(t_a)$ with $\beta = 1$, but C–F express $\Phi(t_a)$ as

$$\Phi(t_a) = \exp\left[-\left(\frac{t}{t_c}\right)^{\beta}\right] \tag{11}$$

where t_c is a characteristic time such that $t_a = t_c$ when the polymer glass has aged to 63.2% of the fully aged glass. The method of determining $\Delta H(\infty, T_a)$ also differs. In the P–M model this is estimated by a linear extrapolation of the heat capacity (liquid) into the glassy state. The relaxation time is then related to the departure from equilibrium of the enthalpy (δH) by

$$\frac{1}{\tau} = -\frac{1}{t_a} \ln\left[1 - \frac{\Delta H(t_a, T_a)}{\Delta H(\infty, T_a)}\right] \tag{12}$$

where

$$\delta H = \Delta H(\infty, T_a) - \Delta H(t_a, T_a). \tag{13}$$

The C–F approach treats $\Delta H(\infty, T_a)$ as an adjustable parameter, as it is considered that the linear C_p extrapolation is inaccurate. Thus a direct measure of the areas of the aged and unaged samples is used to calculate $\Delta H(t_a, T_a)$. The data are then analyzed by curve fitting plots of $\Delta H(t_a, T_a)$ against $\log_{10} t_a$, to assess the thermodynamic aspects from the $\Delta H(\infty, T_a)$ parameter, and the kinetic aspects, embodied in $\Phi(t_a)$, both of which are obtainable from this approach. Also considered in the C–F approach is the prediction of t_e, which is the time to reach 99.9% of the thermodynamic equilibrium state of the infinitely aged glass. This can be predicted from short-term experiments and is illustrated in the following section.

B. Physical Aging in Polymer Blends

Remarkably few studies of physical aging in polymer blends have appeared in the open literature, considering the importance of the phenomenon. There have been many investigations of single-polymer systems, and this is useful as it provides a means of comparing blend behavior with that of the original components. While the data described in detail here will refer to blend systems only, the relevant homopolymer data will also be mentioned. Some of the work reported is purely qualitative and is summarized in Table 1 for blends and composite systems. When an attempt at quantitative analysis of the data has been made, these references are gathered in Table 2 for blends and composites.

1. Homogeneous Blends—Thermal Methods

One of the most commonly used criteria for establishing the phase behaviour in amorphous binary polymer blends is the presence of one or more T_gs. If the blend

Table 1 Physical Aging in Blends, Copolymers, and Composites: Qualitative Treatment

System	Polymer	Technique	Reference
Blend	ABS/PC	G(t)	(74)
Blend	Aromatic Polyamides	DSC	(61)
Blend	Nylon 6,6/PPO	E(t)	(81)
Blend	PMMA/SAN blends	DSC	(67)
Blend	Poly(Lactide) blends	DSC	(58)
Blend	PS/P(2-VP)	DSC	(59)
Blend	PSF/CPSF	DSC	(82)
Blend	PVC/PiPMA	DSC	(60)
Blend	PVC/PMMA	DSC	(60)
Copoly	PS-b-PMMA	DSC	(83)
Comp	PE	Creep	(84)
Comp	PEEK	Impact toughness	(79)
Comp	PEEK	Creep	(77)
Comp	PPS	Impact toughness	(79)

Table 2 Physical Aging in Blends: Quantitative Treatment

Polymer blend	Technique	Model	Reference
PEEK/PEI	DSC	W. Watt	(70)
PES/Epoxy	DSC	CF	(71)
PMMA/PVDF	E(t)	W. Watt	(72)
PMMA/SAN	DSC	Hodge	(68)
PMMA/SAN	DSC	Hodge	(85)
PMMA/SAN	E(t)	W. Watt	(72)
PMMA/SAN	E(t)	Stress Rel.	(68)
PS/PPO	E(t)	W. Watt	(72)
PS/PVME	DSC	CF	(56)

is one-phase, a single T_g lying between the values for each component is detected and characterizes the mixture. If the blend is two-phase, then two T_gs are observed close to or matching those of the two components. On occasions the component glass transition temperatures are so close together that it is difficult to discriminate between the presence of one or two T_gs in a DSC measurement at normal scan rates. Both Bosma et al. (57) and Jorda and Wilkes (58) demonstrated that the use of isothermal aging experiments could overcome this problem. It was argued that in a homogeneous one-phase blend the kinetics of the aging process would be an average representing the blend and as such would exhibit only one enthalpy recovery peak. If, on the other hand, the blend was two-phase, the components would age at their own individual rates, and this would be manifest in the appearance of two enthalpy recovery peaks on the DSC trace. The technique was demonstrated for several systems. Blends of poly(vinyl chloride), PVC (T_g = 80°C), and poly(isopropyl methacrylate), PiPMA (T_g = 82.5°C), are believed to be immiscible, but because of the closeness of the T_g values this is difficult to confirm. A 50/50 blend was annealed first at 195°C, to erase previous thermal history, then quenched to 60°C, i.e., ($T_g - T_a$) ~20°C, and aged for various times t_a (57). The thermograms shown in Fig. 9a that were obtained demonstrate clearly the development of two distinct enthalpy recovery peaks that increase with increasing aging time t_a. Good separation of the peak maxima are obtained as T_{max} tends to increase with aging time for both components but at different rates. This indicates that phase separation has occurred and the enthalpy recovery peaks are characteristic of each distinct phase in the mixture. A miscible blend of PVC and atactic polymethylmethacrylate, PMMA, was also treated in a similar fashion, but only one enthalpy recovery peak could be detected, Fig. 9b, indicating a single-phase system.

This work was extended by Grooten and ten Brinke (59) to include the immiscible blend of polystyrene, PS, with poly(2-vinyl pyridine), P2VP. The authors

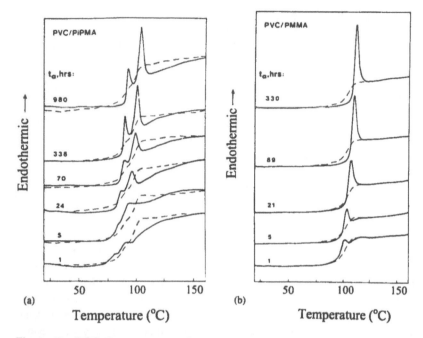

Figure 9 DSC thermograms for aged polymer blends: (a) poly(vinyl chloride)/poly(isopropyl methacrylate), immiscible blend, aged at a temperature of 60°C, and (b) poly(vinyl chloride)/poly(methyl methacrylate), miscible blend, aged at 80°C. Time of aging, t_a in hours, is shown alongside each curve. Broken lines represent the unaged samples for comparison. (Adapted from Ref. 57.)

concluded that the most appropriate range of aging temperature for this type of experiment was T_g to ~$(T_g - 20°C)$. The method was also applied to distinguish the two-phase nature of PS-P2VP diblock copolymers (60).

The sensitivity of this method was demonstrated by Jorda and Wilkes (58) who showed, after aging, that blends of racemic polylactide and its optically active L-form were two-phase, a convincing demonstration that two stereo-regular forms of a polymer may be immiscible. Similar levels of sensitivity were reported by Ellis (61) during studies on the phase behavior of blends of aromatic polyamides. Having predicted the phase behavior of some of these blends using the mean field binary interaction model (62,63), proof was required for blends with components having similar values of their T_g. The aging experiments confirmed the predicted phase behavior.

The first comprehensive study of physical aging in a miscible blend system using enthalpy relaxation was reported by Cowie and Ferguson (56), who followed the enthalpic relaxation in a series of blends of PS and poly(vinyl methyl ether), PVME. Comparison of the blend behavior with that of the two components by analyzing the data on the basis of both P–M and the C–F models led to the conclusion that the blends aged more slowly than PVME when aging was carried out at a comparable temperature below T_g; hence the component with the lower T_g, i.e., the more mobile component in the blend, PVME, was responsible for most of the aging effects seen in this blend (see Table 3). The PS component did not appear to contribute significantly to the total aging because the relaxation processes were much slower than PVME at the aging temperature. This suggests that even though the blend can be regarded as a miscible, one-phase system, the components can largely relax independently, although the relative rates of each will be influenced by the second component. In the PS/PVME blends the T_gs of the two components are quite different, and the blends can be regarded as comprising a flexible polymer (PVME) mixed with a relatively stiff polymer (PS), a situation that is not favored thermodynamically. It is interesting to note that if an even more rigid analog, poly(α-methyl styrene), replaces PS in the blend, a two-phase system is obtained. Thus PS/PVME may not be a typical miscible blend system, as its glass transition region is broad, indicating a mixture close to phase separation.

Oudhuis and ten Brinke (64) have addressed these points by examining the aging of blends of PS with poly(2,6-dimethyl-1,4-phenylene ether), PPE, where now both components are relatively rigid, although there is still about 100°C difference between the T_g values. These authors found that while the amount of enthalpy relaxation observed in the blends was lower than that for either component, there was no evidence for faster relaxation by the component with the lower T_g, viz., PS (Table 4). This observation has been confirmed by Elliot (65) for this sys-

Table 3 PS/PVME Blends

Petrie Model

wt% PVME	$\ln(A/\text{min}^{-1})$	E_h kJ mol^{-1}	C/g J^{-1}
0	366.6	1192	2.92
50	388.2	929	4.08
100	192.3	407	1.14

C–F Model

Blend	T_a/K	T_g-T_a/K	$\Delta H_\infty(T_a)$/Jg^{-1}	$\text{Log}_{10}(t_c/\text{min})$	β	$\text{Log}_{10}(t_e/\text{min})$
PS/PVME	270.0	11.8	1.08	2.15	0.449	3.83
	265.0	16.8	1.39	2.46	0.316	5.12
	260.0	21.8	1.65	2.47	0.373	4.72
	255.0	26.8	1.88	2.74	0.412	4.77
	250.0	31.8	2.09	3.25	0.364	5.56

Source: Ref. 56.

Table 4 Enthalpy Relaxation in Blends of PS/PPE and SAN/PMMA Analyzed Using the Hodge Model (N Parameters)

Blend	$w2^a$	$\ln(A/\text{sec})$	$\Delta h^b/R$, kK	x	β	T_f/K	Reference
SAN/PMMA[b]	100	−328.9	125.0	0.221	0.48	380.05	(67)
	80	−358.5	134.7	0.137	0.46	375.73	(67)
	60	−370.8	138.6	0.147	0.42	373.79	(67)
	40	−392.6	146.2	0.253	0.35	372.39	(67)
	20	−377.6	140.3	0.227	0.35	371.56	(67)
	0	−359.8	132.2	0.338	0.26	367.43	(67)
PS/PPE[c]	0	−288.1	140	0.37	0.58	485.94	(64)
	50	−316.6	135	0.33	0.56	426.41	(64)
	100	−328.8	126	0.28	0.57	383.21	(64)

[a] $w2$ = weight % first homopolymer component in blend.
[b] Values for unaged.
[c] Values for t_a = 120 minutes.

tem using three different blend compositions. Oudhuis and ten Brinke suggested that since the enthalpic definition of T_g was used as proposed by Cowie and Ferguson (56) and the blend showed a broad glass transition region covering about 23°C, the use of the onset T_g instead of the mid-range temperature might be a more accurate reference for selection of the aging temperatures. When a broad transition region is present, some aging temperatures close to the enthalpic T_g might impinge on the onset region of the glass transition process and show an accelerated aging for the more flexible component. This illustrates some of the problems associated with accelerated aging studies where the precise location of T_g for a weakly miscible blend may be difficult to define.

However, an alternative explanation may be found in the sequential aging theory proposed by Chai and McCrum (66). The authors postulated that at a given T_a and t_a, the viscoelastic elements with relaxation times equivalent to t_a will be aging, but that elements with $t < t_a$ will already have reached equilibrium and those with $t > t_a$ will not yet have begun to move towards equilibrium. Thus in the PVME/PS case the more flexible PVME, which will also be closer to its own T_g, will possess more elements with shorter t than the PS. Consequently the PVME relaxation spectrum will tend to move more rapidly towards equilibrium than the PS and so would age more rapidly. This idea will require further testing.

Blends of PMMA with a commercial sample of poly(styrene-*stat*-acrylonitrile), SAN, containing 25 wt% AN, have been studied using enthalpy (67) and stress relaxation (68) measurements. Data are shown in Table 4 and 5. The authors observed, in common with all other workers, that aging is faster at higher temperatures. No comparison was made with the component polymers, but it was ob-

Table 5   Physical Aging in Polymer Blends, Studied by Stress
Relaxation, Mijovic Equation[a] Parameters

System	w_2[b]	C_W	C_T	C_t	Reference
SAN/PMMA	0	−7.72	3.325	1.013	(68)
	20	−7.65	3.325	1.013	(68)
	40	−7.49	3.325	1.013	(68)
	60	−7.25	3.325	1.013	(68)
	80	−7.02	3.325	1.013	(68)
	100	−6.68	3.325	1.013	(68)
PS/PPE	0	−8.08	3.61	0.357	(72)
	20	−10.34	4.24	0.683	(72)
	40	−14.55	5.15	0.885	(72)
	60	−14.35	5.15	0.936	(72)
	80	−10.42	4.31	0.838	(72)
	100	−8.07	3.68	0.849	(72)

[a] Mijovic equation: $E(t)/E_0 = \exp[-(t/\tau)^\beta]$ and $\ln \tau = C_W + C_T \ln(T_g - T_a) + C_t \ln t_a$.
[b] w_2 = weight % first homopolymer component in blend.
Note: $C_t = \partial(\ln \tau)/\partial(\ln t_a) = \mu$.

served that blends rich in SAN relaxed faster than PMMA-rich blends. The aging times used in that work were no more than 150 min.

A more comprehensive study of this system has been carried out by Cowie and Ferguson (69). These authors followed the enthalpy and stress relaxation of a series of PMMA/SAN blends with SAN compositions spanning the miscibility window, i.e., from 13.3 to 30 wt% AN. It was found that the blends relaxed faster than either of the components, when aging temperatures were $(T_g - T_a) = 10°C$, but that this was no longer the case at $(T_g - T_a) = 20°C$, where blend aging was intermediate to both components. The data were analyzed using both P–M and C–F approaches, and examples of $\Delta H(t_a, T_a)$ vs. log t_a plots are shown in Fig. 10. The C–F model provided a better estimate of long-term aging effects where over estimates of the time taken to reach equilibrium were made by the P–M model. The parameters derived from each analysis are shown in Table 6 for PMMA/SAN (26.6 wt% AN) at $T_a = (T_g - 10°C)$.

Enthalpy relaxation studies have also been used to assess the aging of polyether ether ketone blends with polyetherimide, PEEK/PEI = 50/50 (70). The preparation of the blend produced an amorphous system with T_g ~215°C, but crystallization of the PEEK occurred after raising the temperature above T_g. Enthalpic relaxation could only be observed in the temperature range T_g to $(T_g - 50)$, and no aging could be detected at temperatures below 150°C (Table 7). The system was analyzed using the KWW, Eq. (8), that yielded values of $\beta = 0.4$, intermediate between those of the component polymers. When the blends were examined using dielectric relaxation measurements that probe the dipole relaxation

$\Delta H\ (t_a,\ T_a)\ /\ Jg^{-1}$

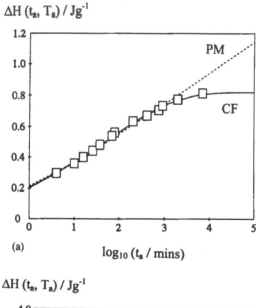

(a) $\log_{10}\ (t_a\ /\ mins)$

$\Delta H\ (t_a,\ T_a)\ /\ Jg^{-1}$

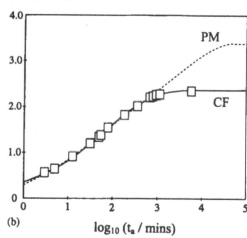

(b) $\log_{10}\ (t_a\ /\ mins)$

Figure 10 Comparison of the experimental enthalpy change on aging for t_a minutes for (a) PMMA, (b) SAN containing 26.6 wt% AN, and (c) a 50/50 blend of these two polymers, with the theoretical curves derived from the Petrie–Marshall (P–M) and Cowie–Ferguson (C–F) models. (Adapted from Ref. 69.)

$\Delta H\,(t_a, T_a)\,/\,Jg^{-1}$

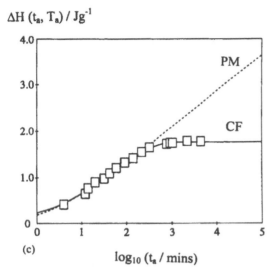

(c)

$\log_{10}(t_a\,/\,mins)$

Figure 10 Continued

spectrum, values of β were found to be much lower (0.1–0.22). This was interpreted as indicating the development of heterogeneity at the molecular level caused probably by the crystallization of the PEEK component.

Polyether sulphone, PES, can be blended with epoxy resins in certain combinations that do not lead to phase separated systems, and Breach et al. (71) have investigated the aging characteristics of Epikote 828 (Shell) and Victrex 5003P. Comparison of aged and unaged samples allowed the enthalpy relaxation to be calculated from the peak areas. Data are collected in Table 8. It was found that the blends aged at a faster rate than the components, and that this increased with increasing PES content. It was observed that β also increased, and these effects were

Table 6 Aging Parameters Calculated from the C–F and P–M Analysis of Data Obtained from PMMA, SAN (26.6 wt% AN) and Their (50/50) Blend PMMA/SAN at $T_a = (T_g - 10K)$

	C–F Model				P–M Model	
Polymer	$\Delta H(\infty)$	$\log t_c$	β	$\log t_e$	$\Delta H(\infty)$	$\log t_e$
PMMA	0.820	1.780	0.290	4.678	2.381	11.479
SAN (26.6 wt % AN)	2.365	1.864	0.422	3.853	3.386	4.697
PMMA/SAN (50/50)	1.766	1.678	0.498	3.362	4.142	6.008

Table 7 PEEK/PEI Blends Analyzed Using the C–F Model

wt% PEEK	T_a/K	$\Delta H\infty(T_a)$/Jg^{-1}	$\log_{10}(t_c$/min)	β
0	457–482	not given	not given	0.35
50	429–445	not given	not given	0.40
100	389–410	not given	not given	0.55–0.60

Source: Ref. 70.

explained by considering that the incorporation of PES loosened the epoxy resin network, thereby increasing the free volume. As this would change the size distribution of the free volume "holes," the presence of larger holes would accelerate the relaxation process and narrow the relaxation time distribution.

2. Homogeneous Blends—Mechanical Methods

Since there is a shift in the viscoelastic relaxation spectrum to longer times with aging time, aging can be followed using stress relaxation, creep, or volume relaxation measurements. It has been shown that for aging experiments (40), momentary creep curves have a universal shape, and a master curve can be constructed using either time–aging (t-t_a) or time–temperature (t-T) superposition. The stress relaxation modulus can then be expressed in terms of the KWW and the equation proposed originally by Struik for analyzing creep compliance:

$$G(t) = G_0 \exp\left[-\left(\frac{t}{\tau}\right)^{\beta}\right] \tag{14}$$

where G_o and $G(t)$ are the stress relaxation moduli at zero time and time t respectively. Hence $\phi(t) = G(t)/G_o$.

As long as the measuring time is short compared with the relaxation or retar-

Table 8 Epoxy Resin Blends with Polyethersulphone Data Derived Using the C–F Model from Enthalpy Relaxation Measurements

Blend	T_a/K	T_g-T_a/K	$\Delta H\infty(T_a)$/Jg^{-1}	$\log_{10}$ (t_c/min)	β	$\log_{10}$ (t_a/min)
E828DDS/PES[a] 0	453.2	—	3.35	3.672	0.26	6.900
E828DDS/PES 20	453.2	—	3.13	3.515	0.36	5.846
E828DDS/PES 30	453.2	—	2.52	2.881	0.52	4.495

[a]Epikote 828 resin (Shell) cured with 4,4'-diaminodiphenylsulphone (DDS) and blended with polyethersulphone (Victrex 5003P).
Source: Ref. 71.

dation times the aging process can be studied effectively, and short-term stress–relaxation measurements have been carried out on blends of SAN/PMMA and PS/PPE (72). Equation (14) was used to fit the data, and the authors found that t could be expressed by

$$\ln \tau = A + C_T \ln(T_g - T_a) + C_t \ln t_a \tag{15}$$

where A is a function of the weight fraction of one of the components in the blend, while C_T and C_t are adjustable temperature and time coefficients. Data are listed in Table 5.

Ho et al. (72) also observed that the values for β fell within a narrow range centered on what was called a "universal value" $\beta = 0.41$. While many results suggest β does not vary significantly, it is more likely that β is an increasing function of (T_g-T_a), but that it may eventually reach a plateau for large values of (T_g-T_a).

3. Positron Annihilation Lifetime Spectroscopy (PALS)

Three blends have been studied by Chang et al. (73) using stress relaxation measurements and PAL spectroscopy. It was observed that for blends of polystyrene (PS)/poly(2,6-dimethyl-1,4-phenylene oxide) (PPO) and PS/poly(vinyl methyl ether) (PVME) the stress relaxation rates were faster for the blends in comparison with PS alone, whereas the opposite was true for a poly(methyl methacrylate) (PMMA)/poly(ethylene oxide) (PEO) blend when compared with pure PMMA.

As aging tends to lead to changes in the packing density of the system, PALS was used to study the blends, as it will provide a qualitative estimate of the free volume in the system. The stress relaxation results were confirmed qualitatively using PALS, and it was found that PS/PPO and PS/PVME blends were less dense than PS, while PMMA/PEO was denser than PMMA, thereby making chain relaxation easier in the former and more difficult in the latter.

4. Heterogeneous Systems

Maurer et al. (74) have examined the two-phase blend of acrylonitrile-butadiene-styrene copolymer, ABS ($T_g = 110°C$) and polycarbonate of bisphenol-A, PC ($T_g = 151°C$) using stress–relaxation measurements. Four regimes of behavior were found. Below 70°C both (t-T) and (t-t_a) superpositions were possible, because the aging rates μ defined as

$$\mu = \frac{d(\log \tau)}{d(\log t_a)} \tag{16}$$

of both components were equal. Between 70°C and 100°C only (t-t_a) superposition worked. However, close to the T_g(ABS), and between the T_gs of both components, neither were valid. While Eq. (14) has been used by other workers Booij and Palmen (75) found that a more precise form was

$$G(t, \tau) = G_0 \left(\frac{t}{\tau}\right)^\alpha \exp\left[-\left(\frac{t}{\tau}\right)^\beta\right] \tag{17}$$

where $\beta = 0.4$ was regarded as a constant (although this might be structure dependent) and α was normally a very small number. This form was used to analyse the ABS/PC blend data.

The behavior of the two-phase systems is complex, and responses on aging can be affected by the thermal history and aging temperature. This was illustrated by aging ABS/PC blends in regime three and quenching from regime four. Quenching to 95°C followed by aging gave curve a in Fig. 11, whereas slow cooling to 95°C then heating to 115°C and quenching back to 95°C produced curve b. For the first (rapid quench) both components age from the glass at a similar rate, but in the second case the thermal history ensures that ABS is largely relaxed, whereas PC still has to age fully from the glassy state. This change in the relative aging rates of the two components allows a modulus plateau to develop in the modulus–time curve and illustrates the need for care with thermal histories (75).

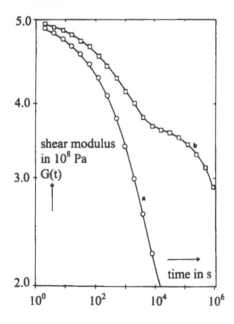

Figure 11 The shear modulus plotted as a function of aging time (in seconds) at 95°C, for a (34/66) ABS/PC blend, prepared using two different thermal histories: (a) quenched at 160°C and annealed at 95°C, and (b) heated to 160°C, cooled slowly to 95°C, heated to 115°C, and then quenched to 95°C. (Adapted from Ref. 74.)

Polycarbonate has been found to lose its toughness when aged at temperatures below its T_g, and impact modifiers are often added to counteract this effect of physical aging. Blends of PC with several core-shell methacrylate-butadiene-styrene impact modifiers have been studied after aging at 125, 130, and 135°C (76). Tests for changes in impact strength, tensile strength, dynamic mechanical response, and fracture morphology were complicated by the simultaneous chemical degradation of the samples when aged in air, but the modifiers did appear to slow down the sample embrittlement caused by physical aging. The increase in the toughness lifetime of the modified PC depended on the amount and nature of the impact modifier used, and the aging environment conditions. The effectiveness of the modifiers was reduced by their chemical degradation at elevated temperatures, particularly in the region of the sample surface.

A number of composites of carbon fiber with engineering polymers have become increasingly important in the aerospace industry. Physical aging is a crucial feature in their long-term stability, and this has stimulated studies of this effect in PEEK, polyphenylene sulphide, PPS, and polyamideimide, PAI, based composites. D'Amore et al. (77) reported that while the crystalline regions in PEEK and the carbon fibers themselves do not participate in the aging process, they are likely to constrain the mobility of chains in the amorphous regions near the inclusion boundaries. This led to an increase in the composite T_g and lowered the material sensitivity to physical aging. Creep compliance (J) measurements were made at various aging times, but the curves were modeled using the Findley relation (78):

$$J = J_0 + A \left(\frac{t}{t_0} \right)^n \tag{18}$$

rather than Eq. (14). The time–temperature shift factor a_T was calculated from the curves at various aging times t_a and was found to increase with increasing t_a. The rate of aging was obtained from

$$\mu = \frac{d(\log a_T)}{d(\log t_a)} \tag{19}$$

It was observed that crystallinity in PEEK slowed the aging rate. Close to T_g the rates of aging for the amorphous, semicrystalline PEEK and PEEK composites were similar but, as the aging temperature dropped further into the glassy state, retardation of the aging process in the composite and the semicrystalline PEEK became more obvious. Thus

$$\mu_{amorph} > \mu_{cryst} > \mu_{comp} \tag{20}$$

indicating that the aging process becomes less self-retarding as the amount of filler increases.

The influence of physical aging on the comparative and intrinsic toughness of PEEK and PPS carbon-fiber composites was determined by Ma et al. (79). Both types exhibited a loss of toughness with increase in aging time and temperature. The PEEK composites tended to retain more impact toughness but after aging were less ductile than the PPS composites.

Nichols et al. (80) examined carbon-fiber composites where a PAI was used as the thermoplastic matrix (Amoco A1696/T650-2). At high loadings the nonrecoverable creep strain in the composite depended on the length of time the material remained under load, and this effect was attributed primarily to physical aging of the specimen. These viscoplastic creep measurements were made after aging for 14 h at 225°C ($T_g = 250°C$), and it was observed that the plateau strain is lower in the aged sample. Differences were also exemplified in the steady state creep rates that could be up to 50 times greater in the aged, compared with unaged, samples. This was explained by assuming that the increase in stiffness due to aging leads to development of a smaller strain under load for the aged material compared with the unaged sample where the chains slip past one another more easily.

These authors also highlighted the fact that the reduction in free volume on aging must lead to shrinkage of the sample during a creep measurement. Consequently, this effect opposes the deformation, and negative creep may be seen in low-stress tests. When these were carried out, the master curve constructed from the data at various T_a deviated from that expected. Because aging effects led to an overprediction of the strain when short-term data were extrapolated to long-term responses, Nichols et al. (80) concluded that while physical aging reduced the undesirable creep effects in thermoplastics at elevated temperatures, the reduction in fracture toughness was too great and tended to negate any advantages a thermoplastic might have over a thermoset in these composite materials.

IV. SUMMARY AND CONCLUSIONS

Relaxation processes in polymer blends can be investigated using a number of techniques, for example DRS, DMTA, and NMR. These methods of analysis do not simply provide experimental evidence that allows the characterization of the blend phase behavior but also offer additional information on the mobility of the individual components. While in two-phase systems the molecular motion of each component is unaltered, a pronounced "blending effect" has been observed in many miscible systems. The degree of this effect has been discussed theoretically in terms of concentration fluctutations and intermolecular coupling. Both the fluctuation model and the coupling model have successfully reproduced a number of features of the relaxations in miscible blends.

Sub-T_g relaxations have been often associated with the mechanical properties of polymeric materials such as toughness and impact strength. The effect of blending on these motions, although of important practical interest, is less well understood compared to the α-relaxation in miscible blends. Experimentally, the addition of a second component causes diverse effects on the molecular motion below T_g of a polymer. In some cases, no "blending effect" has been observed, whereas in others a suppression of the β-relaxation process has been reported. It is also possible that the addition of a second polymer may enhance molecular motion of the other component. Changes in the free volume or the free volume distribution upon mixing may be responsible for this behavior.

Physical aging has been used to determine the phase behavior of blends where the T_gs of the individual components are very close and therefore it is not possible to discriminate between miscibility and immiscibility using conventional DSC analysis. Quantitative analysis of the physical aging process in a miscible blend suggests that even though the blend can be regarded as a miscible one-phase system, the components can sometime relax independently, but blending usually alters the rate at which the components relax when compared with the pure polymers. Miscible blends relaxing faster than either of the components have been reported (71). This behavior has been related to free volume changes in blends. The presence of larger holes in some systems would accelerate the relaxation process and narrow the relaxation time distribution.

The study of blends by physical aging is still largely unexplored and requires close attention from both the experimental and the theoretical standpoint. One major advance would be the ability to cross-correlate mechanical and thermodynamic physical aging data, but while there is a theory that links the two this is imperfect and requires refinement. One problem is that the aging processes, when followed by these different methods, proceed at different rates. If conditions could be established that led to these proceeding at equal rates then this cross-correlation might be possible.

REFERENCES

1. D. R. Paul and S. Newman. Polymer Blends. New York: Academic Press, Vols. I and II, 1978.
2. O. Olabisi, L. M. Robeson, and M. T. Shaw. Polymer–Polymer Miscibility. New York: Academic Press, 1979.
3. S. F. Lau, J. Pathak, B. Wunderlich. Study of phase separation in blends of polystyrene and poly(alpha-methylstyrene) in the glass transition region using quantitative thermal analysis. Macromolecules 15:1278–1283, 1982.
4. J. R. Fried, F. E. Karasz, and W. J. MacKnight. Compatibility of poly(2,6-dimethyl-1,4-phenylene oxide) (PPO)/poly(styrene-co-4-chlorostyrene) blends. I. Differential scanning calorimetry and density studies. Macromolecules 11:150–158, 1978.
5. R. E. Wetton, W. J. MacKnight, J. R. Fried, and F. E. Karasz. Compatibility of poly(2,6-dimethyl-1,4-phenylene oxide) (PPO)/poly(styrene-co-4-chlorostyrene) blends. II. Dielectric study of the critical composition region. Macromolecules 11:158–165, 1978.
6. C. M. Roland and K. L. Ngai. Segmental relaxation and the correlation of time and temperature dependencies in poly(vinyl methyl ether)/polystyrene mixtures. Macromolecules 25:363–367, 1992.

7. A. Zetsche, F. Kremer, W. Jung, and H. Schulze. Dielectric study on the miscibility of binary polymer blends. Polymer 31:1883–1887, 1990.

8. K. J. McGrath and C. M. Roland. Concentration fluctuations and dynamic heterogenity in PIP/PVE blends. J. Non-Cryst. Solids 172–174:891–896, 1994.

9. E. W. Fischer, G. P. Hellmann, H. W. Spiess, F. J. Horth, U. Ecarius, and M. Wehrle. Mechanical properties, molecular motions and density fluctuations in polymer–additive mixtures. Makromol. Chem.-Macromol. Chem. Phys. S12:189–214, 1985.

10. C. J. T. Landry and P. M. Henrichs. The influence of blending on the local motions of polymers: studies involving polycarbonate, poly(methyl methacrylate) and a polyester. Macromolecules 22:2157–2165, 1989.

11. H. de los Santos Jones, Y. Liu, P. T. Inglefield, A. A. Jones, C. K. Kim, and D. R. Paul. Coupled relaxations in a blend of PMMA and a polycarbonate. Polymer 35:57–65, 1994.

12. Y. H. Chin, Y. Liu, A. A. Jones, P. T. Inglefield, and R. P. Kambour. Secondary relaxations in polymer blends by solid state NMR. Polym. Prepr. (Am. Chem. Soc., Div. Polym. Chem.) 35:66–67, 1994.

13. K. L. Ngai, C. M. Roland, J. M. O'Reilly, and J. S. Sedita. Trends in the temperature dependency of segmental relaxation in TMPC/PS blends. Macromolecules 25:3906–3909, 1992.

14. G. P. Simon. Dielectric relaxation spectroscopy of thermoplastic polymers and blends. Materials Forum 18:235–264, 1994.

15. C. Le Menestrel, A. M. Kenwright, P. Sergot, F. Laupretre, and L. Monnerie. ^{13}C NMR investigation of local dynamics in compatible polymer blends. Macromolecules 25:3020–3026, 1992.

16. K. Takegoshi and K. Hikichi. Effects of blending on local chain dynamics and glass transition: polystyrene/poly(vinyl methyl ether) blends as studied by high-resolution solid-state ^{13}C nuclear magnetic resonance spectroscopy. J. Chem. Phys. 94:3200–3206, 1991.

17. Y. H. Chin, C. Zhang, P. Wang, P. T. Inglefield, A. A. Jones, R. P. Kambour, J. T. Bendler, and D. M. White. Glass transition dynamics in a compatible blend by two-dimensional solid state NMR. Macromolecules 25:3031–3038, 1992.

18. Y. H. Chin, P. T. Inglefield, and A. A. Jones. Dynamic Study of poly(styrene-vinyl-d₃) in a glassy blend by two-dimensional solid state NMR. Macromolecules 26:5372–5378, 1993.

19. S. Spall, A. A. Goodwin, M. D. Zipper, and G. P. Simon. Molecular dynamics in a miscible polyester blend. J. Polym. Sci.: Part B: Polym. Phys. 34:2419–2431, 1996.

20. P. S. Alexandrovich, F. E. Karasz, and W. J. MacKnight. Dielectric study of polymer compatibility: blends of polystyrene/poly-2-chlorostyrene. J. Macromo. Sci., Phys. B17:501–516, 1980.

21. J. M. O'Reilly and J. S. Sedita. Dielectric and enthalpic relaxation behavior of polyesters, polycarbonates and polystyrene and their blends near T_g. J. Non-Cryst. Sol. 131–133:1140–1144, 1991.

22. A. A. Mansour and S. A. Madbouly. Dielectric investigation of the molecular dynamics of blends (II): polymers with dissimilar molecular architecture (PS/TMPC blend). Polymer International 37:267–276, 1995.

23. G. Williams and D. C. Watts. Non-symmetrical dielectric relaxation behavior arising from simple empirical decay function. Trans. Faraday Soc. 66:80–85, 1970.

24. G. Katana, F. Kremer, E. W. Fischer, R. Plaetschke. Broadband dielectric study on binary blends of bisphenol-A and tetramethylbisphenol-A polycarbonate. Macromolecules 26:3075–3080, 1993.

25. A. A. Mansour and S. A. Madbouly. Dielectric investigation of the molecular dynamics in blends: polymers with similar molecular architecture (TMPC/PC blend). Polym. Int. 36:269–277, 1995.

26. A. Alegria, J. Colmenero, K. L. Ngai, and C. M. Roland. Observation of the component dynamics in a miscible polymer blend by dielectric and mechanical spectroscopies. Macromolecules 27:4486–4492, 1994.

27. K. J. McGrath, K. L. Ngai, and C. M. Roland. Temperature dependence of segmental motion in polyisobutylene and poly(vinylethylene). Macromolecules 25:4911–4914, 1992.

28. C. M. Roland and K. L. Ngai. Dynamic heterogenity in a miscible polymer blend. Macromolecules 24:2261–2265, 1991.

29. Y. J. Jho and A. F. Yee. Secondary relaxation motion in bisphenol-A polycarbonate. Macromolecules 24:1905–1913, 1991.

30. G. R. Mitchell and A. H. Windle. The local structure of blends of polystyrene and poly(2,6-dimethylphenylene oxide). J. Polym. Sci.: Polym. Phys. Ed. 23:1967–1974, 1985.

31. H. Feng, Z. Feng, H. Ruan, and L. Shen. A high-resolution solid-state NMR study of the miscibility, morphology, and toughnening mechanism of polystyrene with poly(2,6-dimethyl-1,4-phenylene oxide) blends. Macromolecules 25:5981–5985, 1992.

32. J. Zhao, Y. H. Chin, Y. Liu, A. A. Jones, P. T. Inglefield, R. P. Kambour, and D. M. White. Deuterium NMR study of phenyl group motion in glassy polystyrene and a blend of polystyrene with polyphenylene oxide. Macromolecules 28:3881–3889, 1995.

33. A. A. Jones, P. T. Inglefield, Y. Liu, A. K. Roy, and B. J. Cauley. A lattice model for dynamics in a mixed polymer-diluent glass. J. Non-Cryst. Solids 131–133:556–562, 1991.

34. E. W. Fischer, and A. Zetsche. Molecular dynamics in polymer mixtures near the glass transition as measured by dielectric relaxation. Polym. Prepr. (Am. Chem. Soc., Div. Polym. Chem.) 33:78–79, 1992.

35. A. Zetsche, and E. W. Fischer. Dielectric studies of the α-relaxation in miscible polymer blends and its relation to concentration fluctuations. Acta Polymer. 45:168–175, 1994.

36. G. Katana, E. W. Fischer, Th. Hack, V. Abetz, and F. Kremer. Influence of concentration fluctuations on the dielectric α-relaxation in homogeneous polymer mixtures. Macromolecules 28:2714–2722, 1995.

37. E.-J. Donth. Characteristic length of glass-transition. J. Non-Cryst. Solids 131–133:204–206, 1991.

38. G. Adam and J. H. Gibbs. On the temperature dependence of cooperative relaxation properties in glass-forming liquids. J. Chem. Phys. 43:139–146, 1965.

39. K. L. Ngai, A. K. Rajagopal, and S. J. Teitler. Slowing down of relaxation in a complex system by constraint dynamics. J. Chem. Phys. 88:5086–5094, 1988.

40. L. C. E. Struik. Physical aging in amorphous polymers and other materials. Amsterdam: Elsevier, 1978.

41. G. P. Johari. Effect of annealing on the secondary relaxations in glasses. J. Chem. Phys. 77:4619–4626, 1982.

42. N. G. McCrum. The interpretation of physical ageing in creep and DMTA from sequential ageing theory. Plast. Rubb. Comp. Proc. Appl. 18:181–191, 1992.

43. C. T. Moynihan, P. B. Macedo, C. J. Montrose, P. K. Gupta, M. A. DeBolt, J. F. Dill, B. E. Dom, P.W. Drake, A.J. Easteal, P.B. Elterman, R.P. Moeller, H. Sasabe, and J.A. Wilder, Ann. N.Y. Acad. Sci. 279:15–35, 1976; C.T. Moynihan. Analysis of structural relaxation in glass using rate heating data. J. Am. Ceram. Soc. 59:16–21, 1976.

44. I.M. Hodge and A.R. Berens. Calculation of the effect of annealing on sub-Tg endotherms. Macromolecules 14:1598–1599, 1981.

45. I.M. Hodge, and A.R. Berens. Effects of annealing and prior history on enthalpy relaxation in glassy polymers. 1. Mathematical modeling. Macromolecules 15:762–770, 1982.

46. I.M. Hodge, and G.S. Huvard. Effects of annealing and prior history on enthalpy relaxation in glassy polymers. 3. Experimental and modeling studies of polystyrene. Macromolecules 16:371–375, 1983.

47. J.L. Gomez-Ribelles and M. Monleon-Pradas. Structural relaxation of glass-forming polymers based on an equation for configurational entropy. 1. DSC experiments on polycarbonate. Macromolecules 28:5867–5877, 1995.

48. A.Q. Tool. Relation between inelastic deformability and thermal expansion of glass in its annealing range. J. Amer. Ceram. Soc. 29:240–253, 1946.

49. J.M.G. Cowie and R. Ferguson. The ageing of poly(vinyl methyl ether) as determined from enthalpy relaxation measurements. Polym. Commun. 27:258–260, 1986.

50. O.S. Narayanaswamy. A model of structural relaxation in glass. J. Amer. Ceram. Soc. 54:491–498, 1971.

51. I.M. Hodge. Effects of annealing and prior history on enthalpy relaxation in glassy polymers. 6. Adam-Gibbs formulation of nonlinearity. Macromolecules 20:2897–2908, 1987.

52. J.H. Gibbs and E.A.D. Marzio. Nature of the glass transition and the glassy state. J. Chem. Phys. 28:373–383, 1958.

53. A.J. Kovacs, J.J. Aklonis, J.M. Hutchinson, and A.R. Ramos. Isobaric volume and enthalpy recovery of glasses. II. A transparent multiparameter theory. J. Polym. Sci. Polym. Phys. Ed. 17:1097–1162, 1979.

54. J.M. Hutchinson. Interpretation of structural recovery of amorphous polymers from DSC data. Progr. Colloid Polym. Sci. 87:69–73, 1992.

55. A.S. Marshall and S.E.B. Petrie. Rate-determining factors for enthalpy relaxation of glassy polymers. Molecular weight. J. Appl. Phys. 46:4223–4230, 1975.

56. J.M.G. Cowie and R. Ferguson. Physical aging studies in poly(vinyl methyl ether). 1. Enthalpy relaxation as a function of aging temperature. Macromolecules 22:2307–2312, 1989.

57. M. Bosma, G. ten Brinkle and T.S. Ellis. Polymer–polymer miscibility and enthalpy relaxations. Macromolecules 21:1465–1470, 1988.

58. R. Jorda and G.L. Wilkes. A novel use of physical aging to distinguish immiscibility in polymer blends. Polym. Bull. 20:479–485, 1988.

59. R. Grooten and G. ten Brinke. Enthalpy relaxations in blends of polystyrene and poly(2-vinylpyridine). Macromolecules 22:1761–1766, 1989.

60. G. ten Brinke and R. Grooten. Enthalpy relaxations in polymer blends and block copolymers: influence of domain size. Colloid Polym. Sci. 267:992–1001, 1989.

61. T.S. Ellis. Aromatic polyamide blends: enthalpy relaxation and its correlation with phase phenomena. Macromolecules 23:1494–1503, 1990.

62. G. ten Brinke, F.E. Karasz, and W.J. MacKnight. Phase-behavior in co-polymer blends—poly(2,6-dimethyl-1,4-phenylene oxide) and halogen-substituted styrene co-polymers. Macromolecules 16:1827–1832, 1983.

63. D.R. Paul, and J.W. Barlow. A binary interaction model for miscibility of copolymers in blends. Polymer 25:487–494, 1984.

64. A.A.C.M. Oudhuis, G. ten Brinke. Enthalpy relaxations and concentration fluctuations in blends of polystyrene and poly(oxy-2,6-dimethyl-1,4-phenylene). Macromolecules 25:698–702, 1992.

65. S. Elliot. The physical ageing of homopolymers and blends. Ph.D. thesis, Heriot-Watt University, Edinburgh, 1990.

66. C.K. Chai and N.G. McCrum. Mechanism of physical aging in crystalline polymers. Polymer 21:706–712, 1980.

67. J. Mijovic and T.K. Kwei. Physical aging in poly(methyl methacrylate)/poly(styrene-co-acrylonitrile) blends. Part II: Enthalpy relaxation. Polym. Eng. Sci. 29:1604–1610, 1989.

68. J. Mijovic, S.T. Devine, and T. Ho. Physical aging in poly(methyl methacrylate)/poly(styrene-co-acrylonitrile) blends. Part I: Stress relaxation. J. Appl. Polym. Sci. 39:1133–1151, 1990.

69. J.M.G. Cowie and R. Ferguson. Physical ageing studies in a series of SAN copolymers and a blend of PMMA/SAN. Paper presented at IUPAC, Montreal (1991).

70. J.N. Hay. Enthalpy relaxation in amorphous and crystalline polymers. Polyaryl ether ether ketone blends. Prog. Colloid Polym. Sci. 87:74–77, 1992.

71. C.D. Breach, M.J. Folkes, and J.M. Barton. Physical ageing of an epoxy resin/polyethersulphone blend. Polymer 33:3080–3082, 1992.

72. T. Ho, J. Mijovic, and C. Lee. Effect of structure on stress relaxation of polymer blends in glassy state. Polymer 32:619–627, 1991.

73. G.-W. Chang, A. Janneson, Z. Yu, and J.D. McGervey. Physical aging in the mechanical properties of miscible polymer blends. J. Appl. Polym. Sci. 63:483–496, 1997.

74. F.H.J. Maurer, J.H. M. Palmen, and H.C. Booij. Generalized stress relaxation behaviour and physical aging in heterogeneous amorphous polyblends. Rheol. Acta 24:243–249, 1985.

75. H.C. Booij, and J.H.M. Palmen. Viscoelasticity of ABS samples differing in thermal history. Polym. Eng. Sci. 18:781–787, 1978.

76. T.W. Cheng, H. Keskkula, and D.R. Paul. Thermal aging of impact-modified polycarbonate. J. Appl. Polym. Sci. 45:531–551, 1992.

77. A. D'Amore, F. Cocchini, A. Pompo, A. Apicella, and L. Nicolais. The effect of physical aging on long-term properties of poly-ether-ketone (PEEK) and PEEK-based composites. J. Appl. Polym. Sci. 39:1163–1174, 1990.

78. W.N. Findley, S.S. Lai, and K. Onaran. Creep relaxation of non linear viscoelastic materials. Amsterdam: North Holland, 1976.

79. C.M. Ma, C.L. Lee, M.J. Chang, and N.H. Tai. Effect of physical aging on the tough-

ness of carbon fiber reinforced poly(ether ether ketone) and poly(phenylene sulfide) composites. I. Polymer Composites 13:441–447, 1992.

80. M.E. Nichols, S.S. Wang, and P.H. Geil. Creep and physical aging in a polyamideimide carbon fiber composite. J. Macromol. Sci. -Phys. B29:303–336, 1990.

81. J.J. Laverty. Effect of heat aging on the properties of a nylon 6,6/poly(phenylene oxide) blend. Polym. Eng. Sci. 28:360–366, 1988.

82. W.W.Y. Lau, Y.G. Jiang, and P.P.K. Tan. Miscibility and enthalpy relaxations in polysulfone-carboxylated polysulfone blends. Polymer International 31:163–167, 1993.

83. C. Tsitsilianis and G. Staikos. Phase behavior in PS-b-PMMA block copolymer by enthalpy relaxation. Macromolecules 25:910–916, 1992.

84. J. Kubat, M. Rigdahl, and M. Welander. Ageing effects and internal stresses in quenched unfilled and clay-filled high density polyethylene. Colloid Polym Sci. 266:509–517, 1988.

85. J. Mijovic, and T. Ho. Proposed correlation between enthalpic and viscoelastic measurements of structural relaxation in glassy polymers. Polymer 34:3865–3869, 1993.

5

Thermoplastic Rubbers via Dynamic Vulcanization

József Karger-Kocsis
University of Kaiserslautern, Kaiserslautern, Germany

I. DEFINITION

Thermoplastic dynamic vulcanizates (TDV), also called elastomer alloy thermoplastic vulcanizates (EA TPV), belong to a novel family of thermoplastic rubbers. TDVs are produced by dynamic cross-linking (curing/vulcanization) of blends composed of thermoplastic resins and thermoset elastomers. The term "dynamic" or "in-situ" cross-linking means the selective curing of the thermosetting rubber component and its fine dispersion in a molten thermoplastic resin via an intensive mixing and kneading process. This process yields a fine dispersion of partially or fully cross-linked, micron-size rubber particles in a thermoplastic matrix. Since dynamic vulcanization is a versatile and innovative process, the above definition and morphological description do not always hold, as will be shown later. The main difference between TDVs and other thermoplastic elastomers produced from rubber/plastic blends is that the rubber component in the latter systems is not cross-linked. Therefore a clear difference is made between thermoplastic olefin elastomers (TPO) produced without (TPO-O) and with dynamic curing (TPO-V). There are several comprehensive reviews on TDVs in the literature (1–5), but the overlap between some of them is quite significant.

II. HISTORY

The author believes that the appearance of TDV was triggered by R&D activities with impact-modified thermoplastics and especially by the development of rubber-modified or high-impact isotactic polypropylenes (iPP). It was reported in the mid 1970s that incorporation of saturated (EPM or EPR) or nonsaturated (EPDM) rubbers (R) consisting of ethylene (E), propylene (P), and diene (D) monomers (M) highly improves the impact resistance of iPP homopolymers, especially below the glass transition temperature (Tg) of iPP ($Tg{\sim}0°C$). It was also recognized that the impact performance of iPP/elastomer blends (the content of which was usually below 20 wt%) strongly depends on the dispersion state of the rubber phase [e.g., (6–7)]. This resulted in a generalized criterion (viz., critical matrix ligament distance) for rubber-toughened thermoplastics (8). To overcome difficulties related to agglomeration, recombination, and coalescence phenomena in the dispersion, the partial curing of the rubber was also recommended and practiced (9). Recently, impact-modified thermoplastics with dynamically cured inclusions were selected to study the effects of interfacial adhesion on the impact strength (10).

The invention of dynamic vulcanization is credited to Gessler (U.S. Patent 3 037 954), who suggested that a high-tensile-strength semirigid thermoplastic compound can be produced by dynamic curing of a blend of 5–50 parts chlorinated butyl rubber (CIIR) and 95–50 parts iPP with conventional curatives that do not contain peroxides. It should be noted here that iPP decomposes in the melt in the presence of peroxides by β-scission. This process (known as viscosity breaking or "visbreaking") is widely used to change the molecular weight distribution (MWD) and melt flow characteristics of iPP (11).

Further pioneering work on this topic was done by Fischer (U.S. Patents 3,758,643, 3,806,558 and 3,835,201) who showed that elastoplastic compositions can be obtained by dynamic curing of 1-olefin copolymer elastomers (like EPM and EPDM) in molten polyolefins. However, the real breakthrough was achieved by coworkers of Monsanto, who filed several patents (U.S. Patents 4,104,210, 4,130,535, and 4,271,049 by Coran and Patel, and U.S. Patent 4,311,628 by Abdou-Sabet and Fath) at the end of the 1970s and in the early 1980s. This development culminated in commercialization of the first TDV (Santoprene® by Monsanto) in 1981 and initiated vigorous research for analogous systems both in industry and in academia. One of the next targets was to reduce the oil absorption swelling of Santoprene® (produced and marketed now by Advanced Elastomer Systems, AES) composed of iPP and EPDM. This was reached by the dynamic curing of nitrile rubber (NBR)/iPP blends, which were released into the market in 1985 under the trade name Geolast® (Monsanto, now AES) (12). The subsequent R&D activity worldwide resulted in several proprietary processes and additional TDV grades.

What is the beauty of dynamic vulcanization and of the resulting thermoplastic elastomers? First, the compounds behave like rubbers but can be processed and reprocessed as thermoplastics. This is a great advantage over conventional thermoset rubbers, which can be converted into rubber goods only with considerable waste (which is neither soluble nor melt fusible) as by-products. Second, the method of dynamic curing can be adopted for a large variety of rubber/plastic blends, and thus the property profile of the resulting TDV can be tailored accordingly. Third, this development, along with the appearance of other thermoplastic elastomers, allowed plastic processors to enter in some market segments that were occupied and controlled by the rubber industry. Last but not least, the processing of thermoplastic elastomers is energy efficient compared to thermoset stocks.

The approach of dynamic curing can also be adopted for thermoset rubbers. During the mixing processes of a rubber/rubber blend one of the rubber components will be dynamically vulcanized and thus dispersed in the other rubber, which becomes the matrix phase. This is achieved by the action of selective curatives, which induce solely the vulcanization of the target rubber. The resulting stock is vulcanizable via curing of the matrix-forming rubber phase. By this technique some properties of the rubber can be improved (ozone-, heat-, and chemical resistance) in comparison to the traditionally blended and static cured rubber blends (13). This method allows us also to use cured rubber waste, as was shown recently by the example of fluororubbers (14).

In spite of the fast development in this field it is worth emphasizing that the practice (how?) of TDVs was far more advanced than the theory (why?) behind it, and this is still more or less the case.

III. MORPHOLOGY AND ITS CONTROL

A. Morphology

Before going into details on the morphology of TDV, it is worthwhile to give a short overview of the microstructure of other thermoplastic elastomers (TPE) (15). Figure 1 shows various thermoreversible or physical network structures. The term "physical" is introduced in order to distinguish the thermoplastic rubbers from the traditional thermoset ones in which a chemical or covalently bonded network (cross-linked structure) is formed. Note that the matrix in these TPEs is always rubbery. The rubbery or "soft" matrix is given by a flexible chain polymer that may also be present as flexible segments or blocks. The "knots" shown in Fig. 1 are amorphous domains formed by phase separation in styrenic TPE. These styrenic TPEs are di-, tri- or multiblock polymers composed of polystyrene (S) and polybutadiene (B) or polyisoprene (I) segments, which are not compatible with each other. Beyond the softening point of the polystyrene (PS) domains this TPE can be processed in the melt. The physical network structure is restored by cooling.

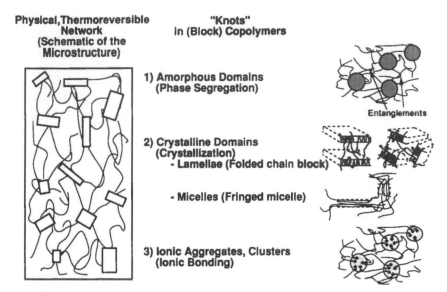

Figure 1 Alternative physical network structures in thermoplastic elastomers of (block) copolymer type.

In case of semicrystalline TPE the "knots" are given by crystalline domains and aggregates (15). In their formation several polymer chains or segments of more than one polymer molecule may participate. Depending on the length of the crystallizable segments either lamellae (folded chain blocks) or micelles (in analogy to the early proposed "fringed micelle" model) are formed. The most well-known variants of this type of TPE are polyurethane-, polyester-, and polyamide-based (15). All of them are block copolymers composed of "soft" (rubbery, noncrystallizable) and "hard" (crystallizable) segments. Above the melting point of the crystallites this TPE behaves like the usual thermoplastic resins. The physical network structure is formed during cooling via crystallization.

The third possibility of network generation occurs by ionic interaction (15). Clusters are held together by ionic bonds in the so-called ionomers (the invention with ionomers is credited to Rees: U.S. Patents 3,264,272 and 3,404,134). The ionomers are usually produced by neutralization of the acidic groups (saponification by adding salts) in copolymers or functionalized polymers.

It should be emphasized here that in the soft rubbery or amorphous phase a further network structure may be present: the entanglement network (cf. Fig. 1). Prerequiste for the appearance of an entanglement network is that the MW of the noncrystallizable polymer molecules or molecular segments be sufficiently high.

It is obvious that all the microstructures displayed in Fig. 1 differ substantially from that of the TDV. In case of TDV a semicrystalline thermoplastic is usually the matrix-giving polymer, so none of the alternatives in Fig. 1 holds for TDV. TDV exhibits an "inverse" character in respect to the TPEs of multiblock structure, since its continuous phase is the hard one. The thermoplasticity of TDV is obvious, but how can the rubbery performance, resilience (a very basic dynamic property measured in rebound tests) be explained? Why is the tension set (recovery after removal of an applied tensile strain) of a TDV with iPP matrix considerably less than that of the plain iPP (see Fig. 2), although the continuous phase is given by the same polymer, viz., iPP?

Surprisingly, works were devoted to this fundamental question only recently. The outcome of a two-dimensional (2D) finite element (FE) calculation of the Inoue group (16) is straightforward, therefore it will be described briefly. For this FE modeling (FEM) the following boundary conditions were set:

1. The true stress vs. true strain behavior of the plain iPP and that of the iPP matrix in the TDV are identical.
2. There is an infinite adhesion between the rubber particles (modeled as octahedral disc inclusions) and the matrix.
3. The shear yielding performance obeys the von Mises criterion.

In the FEM study it was found that in the TDV the yielding (i.e., plastic deformation reducing the recovery potential of the TDV) starts in the equatorial direction (i.e., perpendicular to the load) of the rubber inclusions (cf. Fig. 3). This prediction is in concert with findings of theoretical and experimental studies focused on rubber-toughened thermoplastics (17,18). It was found that crazing starts in the

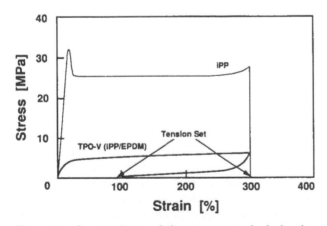

Figure 2 Comparison of the stress–strain behavior and tension set (after 300% deformation) of plain iPP and TDV composed of iPP/EPDM, schematically.

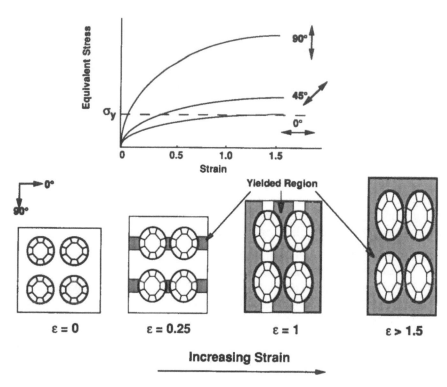

Figure 3 Illustration of the tensile strain recovery in TDVs based on an FEM model. (Redrawn from Ref. 16.)

equatorial region of the particles, the E-modulus of which is below that of the matrix (soft inclusion). At further increase of the true strain (cf. Fig. 3) the ligaments between the particles in ±45° direction yield, and only the polar ligaments remain intact. The nonyielded polar ligaments act as "elastic domains" and guarantee the shape recovery even after considerable strain. When all polar ligaments (termed by the authors "in-situ formed adhesives" in Ref. 16) undergo yielding, no strain recovery will take place. In respect to Fig. 3 it should be emphasized that the cavitation process caused by the difference in the Poisson number between the rubber (close to 0.5) and the iPP (at about 0.35) is not considered. Recall that debonding between particle and matrix was excluded by the infinite adhesion condition. The above elastoplastic FEM analysis also worked well for a TDV composed of poly(butylene terephthalate) (PBT) and poly(ethylene-co-glycidylmethacrylate) copolymer rubber (19). In the cited work it was also shown experimentally that the strain recovery is hampered by cavitation of the rubbery particles. The reliability of the aforementioned 2D FEM treatise (16,19) seems to be supported by other FEM works (20). Furthermore, Anderlik (21) showed that the strength response of a TDV can well be estimated by a 2D model based on a surface-centered cubic cell model adopted for the "packing" of the dispersed rubber in the iPP matrix.

These results, along with some direct evidence achieved mostly by transmission electron microscopy (TEM), substantiate the earlier concluded morphology for TDV: fine dispersion of the cured rubber in the thermoplastic matrix. Nevertheless, the question arises: is this morphological picture always correct? If this morphological description is true, a TDV at the same composition should exhibit a more pronounced elasticity as the rubber particles are more finely dispersed in the matrix. This is associated with the formation of a small uniform ligament between the particles. In the first paper of their legendary paper series, Coran and Patel (22) have shown how the stress–strain curve is extended in both directions when the mean size of the cured EPDM rubber is reduced (Fig. 4). Although this curve appears smooth, only the finest dispersion was produced by dynamic curing. All other blends, indicated by the dotted line in the curve in Fig. 4) were produced by dilution of rubber powders of various particle size with iPP. Therefore a direct comparison of the stress–strain response between the "diluted" and in-situ vulcanized blends may be questionable. The development of a fine rubber dispersion in a thermoplastic matrix is not obvious if one considers that the thermoplastic:rubber ratio is usually between 25 : 75 and 75 : 25 parts in TDV formulations. In this compositional range both rubber and thermoplastic resin may form a continuous phase (dual phase or cocontinuous structure or interpenetrating network, IPN). Fig. 5 makes clear that the presence of the IPN morphol-

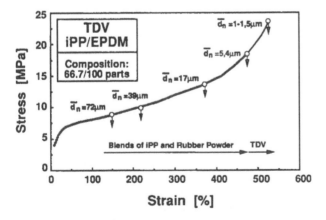

Figure 4 Stress–strain behavior of vulcanized rubber particle filled iPP and an iPP-based TDV at the same composition. d_n = number-average particle size; arrow indicates ultimate failure. (Redrawn from Ref. 22.)

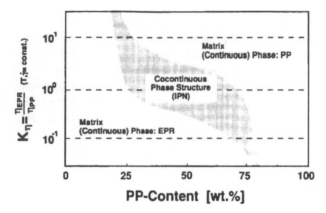

Figure 5 Phase diagram (viscosity ratio K_η vs. blend composition) for EPR/iPP blends. Change in the viscosity ratio is due to various MW characteristics of the components.

ogy strongly depends on the viscosity ratio of the components. The viscosity ratio K_η (see Sect. III.B), defined as the viscosity relation of the rubber to the thermoplastic measured at the same thermal and rheological history, can be influenced by both material selection (products of higher or lower MW, cf. Fig. 5) and processing conditions (effects of temperature and shear rate). It should be pointed out that K_η varies for the same composition and at the same temperature as a function of the shear rate (γ). This is because the shear thinning (viscosity reduction by increasing shear rate) behavior of the rubber differs from that of the thermoplastic polymer. It is well known that the dispersion morphology of polymeric blends depends on the viscosity ratio. This is learnt from the early works of Rayleigh (23), Taylor (24,25) and Tomotika (26). Although their model systems, used in the cited works, differed from polymers, the basic rules seem to be valid for polymer blends (27,28).

Referring to Fig. 6 the previous question can be more precisely formulated: is the initial IPN structure fully broken up? The TEM pictures in Fig. 6 compare the phase structure of the same rubber/thermoplastic blend before and after dynamic curing. This TDV was produced according to our invention (Hungarian Patent 197 338) containing 50 parts polyolefins (20 parts LDPE and 30 parts EVA) and 50 parts styrene/butadiene rubber (SBR). Dynamic curing occurred by a sulphur-donor system in a Banbury type mixer without external heating. The cocontinuous phase structure in the uncured blend (Fig. 6a), in which the rubber phase was made visible by staining, is obvious. Although in the TEM picture taken from the TDV (Fig. 6b), rubber particles in the size range ≤0.5 μm are clearly visible, it cannot be claimed that the initially continuous rubber phase was fully transformed into particles. On the contrary, the TEM picture in Fig. 6b substantiates the presence of an IPN, however, restricted likely for smaller vol-

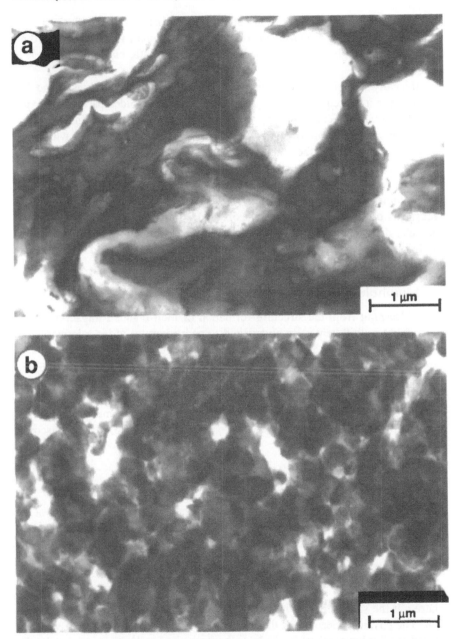

Figure 6 TEM photomicrographs taken from a polyolefin/SBR blend (a) before and (b) after dynamic curing (recipe according to Hungarian Patent 197 338). The rubber phase was contrasted by OsO_4 and RuO_4.

ume elements ("micro-IPN"). Interestingly and in accordance with the afore-mentioned, Sperling (29) grouped some dynamic vulcanizates (those according to the patents of Fischer; see Sect. II) into thermoplastic elastomers with IPN structure.

If locally some IPN structure is present, it will be partially broken up and dis-torted during a subsequent molding operation. Thus the molded part should expe-rience some mechanical anisotropy. Remember that no or only marginal anistropy [due to some skin-core morphology (30)] is expected in case of a TDV with a fine dispersion structure. On the other hand, quite pronounced mechanical anisotropy was reported for TDVs (31), and related information can be found also in product brochures (e.g., Sarlink® grades of DSM).

Summing up, it should be emphasized that the morphology of the TDV is not necessarily of the dispersion type. The open question in this respect is whether the non-fully-dispersed morphology is a result of the formulation, of the processing conditions, or of both.

B. Morphology Development and Control

Again, learning from the structure formation of melt-blended impact-modified thermoplastics will help us to understand some basic rules of the microstructure development in TDVs. Since mixing occurs mostly under shear and a finely dis-persed structure is targeted, it is reasonable to focus on the deformation and breakup of an isolated droplet in a fluid under shear stresses. When both the dis-perse (subscript D) and matrix phase (subscript M) are purely viscous (i.e., New-tonian fluids) the size and deformation of the droplet depend mainly on two pa-rameters [Refs. 27, 32 and references therein including (23–26)]: (a) the viscosity ratio

$$K_\eta = \left(\frac{\eta_D}{\eta_M}\right)_{T,\gamma = \text{const}} \tag{1}$$

where η_D and η_M are the viscosities of the dispersed and continuous phase at the same processing conditions (shear rate γ and temperature T are constants), and (b) capillary (capillarity) or Weber number

$$K_{We} = \frac{\tau \cdot d}{\sigma} = \frac{\eta_M \cdot \gamma \cdot d}{\sigma} \tag{2}$$

where τ is the shear stress, d is the droplet diameter, and σ is the interfacial ten-sion. The physical meaning of the Weber number (the inverse term of which is termed the Taylor number) is a balance between the local acting and droplet de-forming shear stress ($\tau = \eta_M \cdot \gamma$) and the deformation resistance of the droplet (σ/d). The deformation of the droplet becomes irreversible, and thus the elongated thread will break up when K_{We} exceeds a critical value. This critical K_{We} depends, on the other hand, mostly on K_η. In simple shear flow the necessary condition for breakup is $K_\eta < 4$ (32). For polymers of viscoelastic rather than viscous nature, K_{We} has to be corrected by terms considering the elasticity of the phases, or other descriptions should be used [e.g., (27,28,32)].

According to the predictions it was found for several binary polymer blends that the finest dispersion forms when K_η is approaching 1 (6,27,28,32). Above and below this value a coarser dispersion was resolved. The often observed shift for the finest dispersion from $K_\eta = 1$ was attributed to the viscoelastic, elastic nature of the components. It should be mentioned here that the criteria for elongational flow conditions are different from those for shear. Therefore a superposition of both shear and elongational flow may also cause a shift in K_η assigned to the finest dispersion.

The above simple treatise cannot be used for describing the morphology development in TDV, since during dynamic curing the viscosity of the rubber phase is not constant but strongly increasing. On the other hand, η_M can be considered as constant. The increase in K_η should result in a more coarse structure per se (27,28,32). Furthermore, it is not known whether the interfacial tension is changing due to curing [see Eq. (2)]. These are the possible reasons why locally an IPN structure may be preserved, as supposed above. As the cross-linking reaction occurs, both the viscosity of the rubber and the viscosity of the whole system increase. This is due to the initial dual-phase continuity of the blend. Thus according to Eq. (2) the increase in η_M is accompanied by increasing shear stresses, which break up the continuous rubber phase more efficiently. It should be kept in mind that in the above simplified scenario other important effects, like heat development due to cross-linking and its transfer, rate of the cross-linking reaction, thermal degradation etc. have not been considered. Though model descriptions taking into consideration the complexity of this problem for both discontinuous (33) and continuous (34) processing were proposed recently, further work is needed to get a deeper understanding of the dynamic curing process.

Based on the works of Coran and coworkers (35,36), the best mechanical performance of a TDV is guaranteed when the following criteria are met (37):

1. The mismatch in the surface free energy between the rubber and the thermoplastic resin is small (or more properly the interfacial tension of the components in the molten stage is small).
2. The rubber component is densely cross-linked.
3. A crystalline thermoplastic is the matrix-forming polymer.

The first two requirements seem to be coherent. Small interfacial tension should result in a more homogeneous, more "technologically compatible" blend, the effect of which can only be compensated for by cross-linking in order to insure the dispersion type morphology required. Furthermore, in case of thermoplastics and rubbers with very different surface energetics, the compatibilization problem has to be solved (38,39). This can be done in two ways: (1) adding a third polymer of block or graft type, which shows compatibility toward both thermoplastic resin and rubber, and (2) generation of the compatibilizer in-situ or on-line.

The target of the in-situ compatibilization is the "functionalization" of the thermoplastic polymer or the rubber, which is often termed "reactive processing." The interested reader is referred to excellent reviews on this topic focusing on both rubbers and polymers (40–42).

The third requirement related to the crystallinity also requires some explanation. The author suspects that this is due to the difference in the deformation and failure behavior between amorphous (tendency for shear yielding) and semicrystalline thermoplastics (failure onset by crazing). The failure mode may affect the deformation of the matrix ligaments between the rubbery particles (see the model explanation in Fig. 3) and thus impedes the elastic recovery.

IV. PRODUCTION OF THERMOPLASTIC DYNAMIC VULCANIZATES

A. Mixing Technology

Conventional masticating equipment (Banbury, Farrel, and inner mixers of various types) widely used in the rubber industry can also be used for the production of TDVs, provided that the heat generation in these mixers is sufficient to melt the thermoplastic polymers. Since the majority of rubber blending facilities have no external heating (during compounding a rubber recipe, cooling is generally required), the selection of the thermoplastic resins is limited for polyethylenes (PE) and ethylene copolymers (e.g., ethylene-vinylacetate copolymers, EVA). The onset and progress of the vulcanization can well be monitored by the change in the mixing torque or energy uptake during the dynamic curing process. Fig. 7 schematically demonstrates the change of the mixing torque and blend temperature for the in-situ cross-linking of a thermoplastic rubber composed of low-density PE (LDPE), EVA, and a diene rubber (SBR) (43). In the first step, thermoplastics, rubbers, and other components (e.g., processing aids, ZnO, stearic acid) are introduced and mixed at a temperature higher than that of the melting (when semicrystalline) or softening temperature (when amorphous) of the most thermal resistant thermoplastic resin in the blend. After obtaining a homogeneous blend, characterized by a minimum torque value (cf. Fig. 7), the curatives are added under steady kneading and mixing. The mixing torque first rises and then goes through a maximum before reaching a constant level. The constant torque of the ready compound prior to dump usually lies beyond that of the homogeneous blend. This torque value may be, however, also below that of the "homogenized" blend when the temperature difference between the homogenization and discharge is high (affecting the melt viscosity of the TDV and thus the torque).

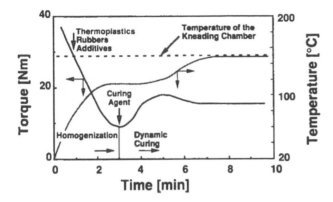

Figure 7 Characteristic change in the mixing torque and temperature during dynamic curing of a thermoplastic/rubber blend. Mixing in a Brabender PlastiCorder; recipe according to Hungarian Patent 197 338.

The maximum in the torque–mixing time curve is due to the superposition of the rubber cross-linking and breakup of the continuous rubber phase (torque increases in both cases) and temperature rise (torque decrease). Recall that the temperature is rising due to the exothermic cross-linking reaction and mixing-related internal friction.

It is obvious that both position and shape of the maximum strongly depend on the quality and amount of the curing agents used if all other parameters are kept constant. The faster is the rate of vulcanization, the more intensive the mixing must be to insure the breakup of larger rubber domains and transform them into small particles. The nearly constant torque after completing dynamic curing already implies some information about the melt flow behavior of the TDV. Based on the above scenario one can anticipate that the quality (controlled by the morphology of the TDV) of the TDV does depend on the mixing conditions. The TDV produced in such a discontinuous or batchwise operation can be sheeted (as practiced with thermoset rubbers), pelletized or granulated (e.g., by a granulating extruder), or even processed to a final product on-line (e.g., warm strip feeding of an extruder that produces profiles).

A higher output and a more constant quality can be, on the other hand, achieved by a continuous process using single- or twin-screw extruders. According to a study using different methods (one- or two-step; the latter includes the production of a master batch) and techniques (twin-screw extruder, internal mixer, and open mill) it was claimed that the performance of the TDV produced by twin-screw extrusion is superior to all other variants (44).

There are further benefits of the continuous extrusion process:

1. The screw configuration (metering, mixing, and kneading zones and elements) and residence time can be adjusted for requirements of the blend.

2. The dosage (by gravimetric or volumetric control) of the ingredients can occur at any section of the extruder (side-feeding).

3. The TDV produced can be granulated on-line.

All large TDV producers favor the extrusion techniques. Furthermore, the reactive extrusion opens a new horizon for TDV production. Strong benefits are expected for formulations from thermoplastics and rubbers with very dissimilar thermal resistance and compatibility. According to a research concept, reactive extrusion will be used to convert poly(ethylene terephthalate) (PET) waste derived from discarded or one-way soft drink bottles into a thermoplastic elastomer (45). According to the scheme in Fig. 8 the first extruder section serves for grafting (by glycidylmethacrylate, GMA) of the rubber (NR or EPDM), and the melt blending of this functionalized rubber with PET will occur in the second section of the extruder (or alternatively in a second extruder by the cascade method). The functionalization and dynamic curing of the rubber can hardly be done when PET is fed together with the rubber: this is due to the high melting temperature of PET (T_m = 255°C) affecting the related chemical reactions (grafting, cross-linking, degradation) of the rubber component.

B. TDV Grades and Formulations

1. Hardness Range

The dynamic curing of rubber/thermoplastics blends is a very versatile process. By changing the compositional range of rubber to plastics the product hardness (considered by colleagues in the rubber industry as the first relevant information from a rubber) can cover a very broad range. Although the works of Coran et al. (35,36,46) demonstrate that semicrystalline polymers should be preferred as the thermoplastic component, the dynamic vulcanization works also with amorphous polymers like atactic polypropylene (aPP) (47). The Tauropren TDV produced by coworkers of the author in the mid-1980s according to Hungarian Patent 189 266 exhibited very low hardness (Shore A 35) in an unfilled version and was recommended for use as an impact modifier for iPP (48,49). The low hardness was reached by dynamic curing of aPP of high MW with EPM or EPDM rubber by action of peroxides. The upper hardness threshold of TDVs is likely when linear polyesters, e.g., poly(butylene terephthalate) (PBT) or polyamides (PA), are compounded with suitable rubbers. Fig. 9 displays the service temperature and hardness range of TDVs compared to other thermoplastic elastomers of block copolymer type.

2. TDV Formulations

In the open literature there are many papers dealing with various TDV formulations. A further invaluable source of information is the related patent literature.

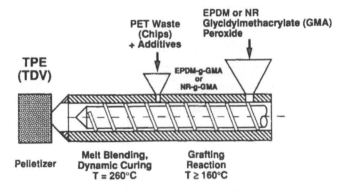

Figure 8 Production of a TDV containing PET and functionalized rubber via reactive extrusion shown schematically. (From Ref. 45.)

Though the reader can get a fine overview of thermoplastic/rubber blends from review papers, book chapters, and books, e.g., Refs. 1–5, 36, it is useful to list the most important combinations also here. In the updated list below, the blends are grouped according to their thermoplastic component, regardless of whether this was the major component in the recipe (generally it was not). Apart from the blend composition the cited papers also inform of the curing agents and conditions used.

Polyethylenes and ethylene copolymers

PE/NR (3,31,50,51)
PE/SBR (3)
PE/EPDM (3,51)
PE/NR, CPE (chlorinated polyethylene) (51)
PE/NR, EPDM (50)
PE/NR, S-EPDM (sulfonated EPDM) (50)
PE/ENR (epoxidized NR) (51)
LLDPE (linear low density PE)/NR (52)
LDPE/EPDM (53)
LDPE, EVA/SBR (43)
HDPE/NR with carbon fiber (CF) (54)
HDPE, LDPE/IIR (butyl rubber) (13)
PE/SI (silicon rubber) (3)

Polypropylenes (homopolymers: isotactic, atactic; copolymers: random, block copolymer, or heterophasic system)

PP/NR (3,31)
PP/NBR (3,35,39)

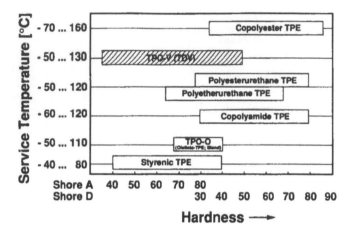

Figure 9 Hardness and service temperature range of TDV compared to thermoplastic elastomers of (multi)block structure.

PP/EPDM (22,33,35,55)
PP homo- and copolymers/EPDM (56)
PP/EPDM with organosilane cross-linking (34,57,58)
PP/EVA (35)
aPP/EPR, EPDM (48,49)
PP/SBS (styrene-butadiene-styrene block copolymer) (59,60)

Polystyrenes (PS) and styrenic polymers

PS/IIR, EPDM, NR, SBR, CPE, NBR, BR (butadiene rubber) (3)
PS, SAN (styrene-acrylonitrile copolymer)/EPDM (35)
PS, SAN/EVA (35)
PS, SAN/NBR (35)

Polyamides (PA) and related copolymers

PA-6/CPE (38)
PA-6, 6.6 copolymer/NBR, EVA (3)
PA-6, 6.9 copolymer/NBR (3,35)
PA-6, 6.6, 6.10 terpolymer/CPE (3)
PA-6, 6.6, 6.10 terpolymer/PUR (polyurethane) (46)
PA-11/EVA (3,35)
PA-11/NBR (61)
Aromatic PA/HNBR (hydrogenated NBR) (62,63)

Linear polyesters and related copolymers

PBT/EGMA, poly(ethylene-co-glycidylmethacrylate) (19)
PBT/EPDM-g-GMA (GMA-grafted EPDM) (64)
Copolyester elastomer/NBR (44,65)

High-temperature resistant thermoplastics

PEEK poly(ether ether ketone), PPS poly(phenylene sulfide)/FKM (fluororubber)
(66)

The above nonexhaustive list hints that the R&D activity is nowadays focused on
TDV systems containing engineering thermoplastics rather than high-volume
commodity, thermoplastic resins.

V. PROPERTIES OF TDVS

To date TPO grades are most widely used. Since many are simple blends (TPO-O),
DSM's Keltan®TP, AES's VistaFlex®, Montell's Hifax® (and similar products
are offered by others), they should be distinguished from the TDVs. Comparing the
properties of TPO-O and TPO-V at the same constituents and hardness, the benefits given by the fully cured rubber in TPO-V become clear: improved set properties, fluid and chemical resistance, service temperature, and tensile strength. These
are achieved, however, at the cost of some other properties (e.g., reduced melt
flow, tear strength, and elongation at break). The limits of the service temperature
of TPO-V are linked to the glass transition (T_9) of the rubber (ca. $-50°C$) and to
the melting point of the plastic phase (ca. 130°C in the case of PP), respectively.
    The properties of TPO-V vary in a broad range depending on the hardness and
thus on the composition range of rubber and PP (see Table 1). Keeping in mind
that hardness increases with increasing amount of PP, the related changes in the
properties can be easily recognized. It is worthwhile to note that although the permanent set (both tensile and compression) values of TDVs are good, they are always inferior to those of traditionally vulcanized rubbers. The most widespread
TPO-V grades are offered under the tradename Santoprene® (Advanced Elastomers Systems, USA), Sarlink® (DSM, Holland) and Forprene® (DTR, Italy).
    Academic interests concentrate now on the assessment and description of the
mechanical properties of TDVs. The earlier phenomenological description of the
ultimate tensile properties as a function of surface tension, crystallinity, and cross-
link density of the rubber (see Refs. 35 and 36) will be replaced by others. It was
recently shown (59) that the yield stress (defined when not obvious in the
stress–strain curve as the stress value at which the related slope becomes 0) obeys

Table 1 Comparison of the Properties of TPO-V Grades of the Same Components for the Lowest and Highest Hardness Range, Respectively

Properties	Unit	Standard	Types	
Shore hardness (5 s)	[deg.]	ISO 868	45–50A	50–55D
Density	[gcm^{-3}]	ISO 1183	0.95	0.96
In flow direction		ISO 37 (II)		
Tensile strength	[MPa]		2.8	20.7
Modulus at 100%				
elongation	[MPa]		2.7	17.6
Elongation at break	[%]		≈150	≈500
Transverse to flow direction				
Tensile strength	[MPa]		3.9	22.0
Modulus at 100%				
elongation	[MPa]		1.2	13.5
Elongation at break	[%]		≈600	≈650
Tear strength		ISO 34A		
(trouser)	[kNm^{-1}]		5	64
Compression set		ISO 815		
22 h, 70°C	[%]		44	75
22 h, 100°C	[%]		57	81
Volume swell		ISO 1817		
72 h, 100°C water	[%]		+5	+2
72 h, 100°C ASTM oil 1	[%]		+60	+9
Hot air aging		ISO 188		
(28 d, 125°C)				
Hardness change	[deg.]		+2	−5
Retention tensile				
strength	[%]		123	85
Retention				
elongation at break (%)	[%]		118	90

Source: Data based on Sarlink® types of DSM.

a Nicolais–Narkis type power law function recommended for filled polymers (17):

$$\sigma_y = \sigma_{y,0} (1 - 1.21 V_f^{2/3}) \tag{3}$$

where σ_y, and σ_y, 0 are the yield strength of the TDV and the plain thermoplastic polymer, respectively, and V_f is the volume fraction of the rubber filler. The weighting factor proved to be below the value indicated in Eq. (3) (viz., 1.02). This means a lower stress concentration due to the rubber particles compared to hard spherical particles. Though further analysis is needed to figure out whether

Eq. (3) holds for higher V_f values, the beauty of this description is given by the analogy with the TDV microstructure.

Another treatise is based on an elastoplastic model in which both the entropy elasticity of the rubber (according to classical rubber theory) and the plastic flow behavior of the PP matrix are taken into account (67).

In one of the pioneering works on the rheological behavior of TDV the authors concluded (68) that TDVs behave like highly filled fluids. That is the reason why their die-swell is small and no Newtonian plateau at low deformation rate can be observed. Figure 10 compares the course of the complex (Fig. 10a) and apparent shear viscosity (Fig. 10b) as a function of the angular frequency and shear rate, respectively. The missing Newtonian plateau in both cases is obvious, especially at higher rubber contents. This can be attributed to matrix immobilization by the rubber dispersion (34). Chung et al. (69) assumed that at low shear rates a 3D network structure is formed in the TDV and modeled it as a filled polymer. The alternative explanation would be the assumption of a microscale IPN structure, as proposed by the author of this chapter, which may form via coalescence and recombination processes. In the case of partial curing of the rubber component in the blend, such arguments may be straightforward. In Figs. 10a and b one can see how the viscosity is changing with increasing rubber content (decreasing hardness) in the TDVs. This effect is smaller under shear flow (cf. Fig. 10b). Based on the power law behavior of the melt (shear thinning), TDVs should be processed preferentially by extrusion and injection molding (cf. Fig. 10b) (68,70). The beneficial low die swell behavior is a further argument for extrusion processing.

It was also early recognized that TDVs can compete with thermoset rubbers only when their resistance to environmental attack is comparable. Therefore work was and is in progress on their heat aging (53,71), outdoor and ozone exposure (72), as well as on their sorption behavior (73).

In several application fields the TDV product is finished [e.g., decorated, painted (74)] or assembled by adhesive bonding. During the development of a shoe sole material the author's group was faced with the problem of how to guarantee the gluing ability of the shoe sole (TDV) to the shoe upper (leather)—see Fig. 11 (43). The philosophy of this R&D work was to develop a shoe sole compound that met the standardized physicomechanical properties and had good melt processability and gluing capability via dynamic curing of inexpensive commodity thermoplastic resins and rubbers. It was early recognized that the major problem was related to the gluing of the TDV compound to the shoe upper by using polyurethane-based adhesives. On the other hand, to fulfil the requirements in respect to physicomechanical and rheological properties by the TDV compounds was a much easier task. The problem with the adhesion was solved at the end by adding a polymer of high polarity (EVA) as a further thermoplastic component to the recipe consisting of LDPE and SBR (43).

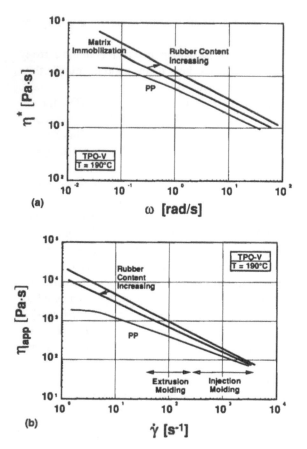

(a)

(b)

Figure 10 (a) Complex viscosity η^* vs. angular frequency ω and (b) apparent shear viscosity η_{app} vs. shear rate γ at the same temperature $T = 190°C$ for TPO-V grades, schematically.

VI. PROCESSING

Most commercial TDVs are fully formulated and tailored for given molding operations. The melt viscosities of TDVs at low shear rates are considerably above those of the uncured blends. Their viscosity decays with increasing shear rate (shear thinning or pseudoplastic feature); see Fig. 10. On the other hand, the viscosity change with temperature is notably less than for plain polymers. Due to this beneficial shear thinning effect, TDV is shaped mostly by injection molding, which represents a molding operation at high shear rates (cf. Fig. 10b). In order to get a uniform cavity filling, the mold layout should meet the following criteria: short sprues and runners, runners of identical length, cavities arranged in a bal-

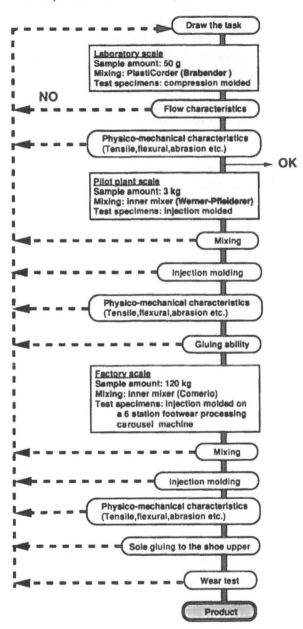

Figure 11 R&D strategy to produce a low-cost shoe sole material of TDV type according to Hungarian Patent 197 338. (From Ref. 43.)

anced pattern. It is also essential to position the weld lines in nonfunctional regions of the parts. Molds produced according to these guidelines yield an optimum melt flow, flash-free molding, short production cycles, and last but not least products of uniform quality. TPO-V is successfully molded on reciprocating screw injection molding machines, provided they have a proper clamping force (in respect of the projected surface of the parts a clamping pressure of 50–80 MPa is required). The mass of the product should not exceed 70% of the maximum shot size of the injection molding machine. Coinjection, two-component molding, or overmolding (both by thermoplastics and a similar TDV of different hardness) can also be practiced. Now Santoprene® grades have become available with good adhesion to PAs and acrylonitrile-butadiene-styrol (ABS) resins. The hot runner system, which avoids the use of sprues and runners, is also gaining acceptance.

For extrusion and extrusion blow molding, the machinery designed for PP processing (i.e., length-to-diameter ratio of the screw >20, compression ratio ≈3) is suitable. Tubes and hoses can be produced with and without size calibration. It is noteworthy that both injection and extrusion molding may result in some flow-induced mechanical anistropy (cf. Table 1). The extrusion foaming by means of both physical and chemical blowing agents is also viable and carried out for TPO-V compounds. By coextrusion a soft-touch (low-hardness) TDV can be combined with a higher hardness one to produce a part with improved gripping and flex resistance (e.g., the new Honda cars are equipped with such glazing strips).

Irrespective of the higher melt viscosity at low shear rates (cf. Fig. 10), TDVs are available also for processing at low shear rates, like calendering or compression molding.

TPO-V processing scraps can be easily reprocessed by adding to the virgin material; generally with up to 10 wt% addition of recycled material no deterioration in the mechanical and rheological properties can be observed. Usually no change in the mechanical properties of TPO-V occurs until ca. five reprocessing cycles; this is in harmony with the thumbnail rules known for PP. The recycling of TPO-V parts after their service lifetime is more problematic when they suffer oxidative and/or fluid attack during service. In a recent report it was shown that TPO-V waste taken from car wrecks can be recycled via remelting and blending (75). Interestingly, the dynamic curing process was adopted to convert cured latex waste (e.g., gloves) to a thermoplastic elastomer (76).

VII.   MARKET AND APPLICATIONS

The total world demand for thermoplastic elastomers, including all types, will reach ca. 1.5 million tons in the year 2000. The annual growth rate of polyolefin-based thermoplastic elastomers was 9.1% between 1985 and 1995, and for the next five years an increase of about 7.6%/year was forecast (perhaps optimistically) by the Freedonia Group (77). The overall market share of the thermoplastic elastomers is ca. 10% in the nontire market.

Most applications of TPO-V (ca. 40%) are in the automotive sector (airbag covers, axle sleeves, bumper fascia, under hood cables and hoses, sealings), where it replaces mostly thermoset rubbers and PVC. For bumpers, TPO-O grades are preferentially used. The second big market (ca. 30%) for TPO-V is building and construction (window glazing, weather seal, expansion joint, roofing membranes, etc.), followed by electric and electronic applications (ca. 15%; covering wire and cable jacketing, electric plugs). TPO-V is also gaining acceptance for medical goods, like tubes (dialysis, blood collection) and for sealing (safety needle sealing, medical stoppers). In case of technical rubber goods (seals, gaskets, bushings, etc.), TDVs are replacing conventional rubbers. Since their set properties are, however, inferior to those of the traditional rubbers, the replacement should be accompanied with some "redesign" (i.e., taking into consideration the property profile of the TDV) of the product. Forecasts for the consumption and applications, including replacement trends for TPEs, can be taken from recent reviews (78–80).

VIII. OUTLOOK AND FUTURE TRENDS

Further development with TDVs is likely interlinked with that of metallocene synthesis. The rationale behind this prophecy is based on the potential and recent results achieved by this revolutionary technology. The fine structure and thus the mechanical and rheological behavior of polyolefins can be tailored by this method (81). The TDV technology should and will make profit of this development in order to keep market share. It should be emphasized here that metallocene synthesis has an impact also on the synthesis of olefin rubbers (81). Several Nordel® grades of DuPont are produced by metallocene technology and show excellent flow behavior. In a very recent paper the synthesis of a new TPO-O (via sequential copolymerization) was reported (82). The ultimate properties of this product after dynamic curing were superior to the reference TPO-V, the components of which were prepared by traditional Ziegler–Natta catalysts.

On the other hand, metallocene synthesis is the biggest challenge TDVs have ever faced. By this technology elastomeric polyolefins can also be produced, which will compete with TPO-O and TPO-V types in the future. At present polyolefin elastomers (Engage®) are marketed by DuPont Dow Elastomers (81,83). The hardness, ultimate tensile strength, and ultimate tensile strain of the Engage® products are in the range of Shore A60–Shore D50, 4–33 MPa, and >1000–700%, respectively. The low temperature resistance of Engage®, being an ethylene/1-octene copolymer, is also excellent. A further metallocene product is the elastomeric PP (ELPP) (81,84) produced by PCD (Linz, Austria) under licence from DuPont. This ELPP is a multiblock polypropylene homopolymer in which stereoregular (and thus crystallizable, isotactic) and stereoirregular (and thus amorphous, atactic) sequences are present. The ultimate tensile strength and tensile strain values are in the range of 1–30 MPa and >1000–700%, respectively (85). Unfortunately, the T_g of the ELPP is rather high (as that of the PP: $T_g \approx 0°C$), which is a major limitation for its application.

In the aforementioned olefinic TPEs the physical network structure is given by crystalline domains. These domains are formed by cocrystallization in which several macromolecular chains or segments participate (the crystalline domains can thus be viewed using the "fringed micelle" model; cf. Fig. 1). Beside the development with metallocene thermoplastics and rubbers, it is worthwhile to pay attention also to the development potential of the dynamic curing technology and processing of TDV products. From the viewpoint of the dynamic curing the most promising trends are multicomponent TDV formulations and reactive processing (functionalization). This will definitely yield new strategies for the curing reactions [e.g., transesterification (86)]. The processing developments will likely focus on 3D blow molding (combined eventually with a coextrusion technology) and foaming. A further open question is whether the TDV performance can be improved by fiber reinforcement (added separately or formed in-situ from a polymer with adequate properties). Studies are still ongoing to understand the morphology-related elastic recovery of TDVs. In a recent paper the authors found that the elastomer is built in in the PP matrix and that PP crystals were also occluded in the cured rubber particles in a commercial TPO-V compound (87). This partial phase mixing, which improved the elastic tensile recovery, was attributed to shear effects that emerged during dynamic vulcanization.

ACKNOWLEDGMENT

This work was done in the framework of an Inco-Copernicus project (45) and partially also supported by the German Science Foundation (DFG—Ka 1202/2).

REFERENCES

1. A. Y. Coran. Thermoplastic elastomers based on elastomer–thermoplastic blends dynamically vulcanized. In: N.R. Legge, G. Holden, and H. E. Schroeder, eds. Thermoplastic Elastomers. München: Hanser, 1987, pp. 133–161.
2. C. P. Rader. Elastomeric alloy thermoplastic vulcanizates. In: B. M. Walker and C. P. Rader, eds. Handbook of Thermoplastic Elastomers. 2nd ed. New York: Van Nostrand Reinhold, 1988, pp. 85–140.
3. S. K. De and A. K. Bhowmick, eds. Thermoplastic Elastomers from Rubber-Plastic Blends. Chichester: Ellis Horwood, 1990.
4. A. Y. Coran and R. P. Patel. Thermoplastic elastomers by blending and dynamic vul-

canization. In: J. Karger-Kocsis, ed. Polypropylene: Structure Blends and Composites. London: Chapman and Hall, 1995, vol. 2, pp. 162–201.

5. A. Y. Coran and R. P. Patel. Thermoplastic elastomers based on elastomer/thermoplastic blends dynamically vulcanized. In S. Al-Malaika, ed. Reactive Modifiers for Polymers. London: Blackie Academic Prof., 1997, pp. 349–394.

6. J. Karger-Kocsis, A. Kalló, and V. N. Kuleznev. Phase structure of impact-modified polypropylene blends. Polymer 25:279–286, 1984.

7. E. Martuscelli. Structure and properties of polypropylene–elastomer blends. In: J. Karger-Kocsis, ed. Polypropylene: Structure, Blends and Composites. London: Chapman and Hall, 1995, vol. 2, pp. 95–140.

8. S. Wu. A generalized criterion for rubber toughening: the critical matrix ligament thickness. J. Appl. Polym. Sci. 35:549–561, 1988.

9. K. C. Dao: Mechanical properties of polypropylene/crosslinked rubber blends. J. Appl. Polym. Sci. 27:4799–4806, 1982.

10. T. Inoue and T. Suzuki. Selective crosslinking reaction in polymer blends. III. The effects of the crosslinking of dispersed EPDM particles on the impact behavior of PP/EPDM blends. J. Appl. Polym. Sci. 56:1113–1125, 1995.

11. K. Hammerschmid and M. Gahleitner. Controlled rheology polypropylene. In: J. Karger-Kocsis, ed. Polypropylene: An A–Z Reference. London: Chapman and Hall/Kluwer Academic, 1998, pp. 95–103.

12. N. R. Legge: Thermoplastic elastomers—three decades of progress. Rubb. Chem. Technol. 62:529–547, 1989.

13. A. Y. Coran: New elastomers by reactive processing, Part I. Vulcanizable precured alloys from NBR and ACM. Rubb. Chem. Technol. 63:599–612, 1990.

14. N. N. Petrova, I. D. Khodzhaeva, O. A. Adrianova, and I. N. Cherskii. Dynamic vulcanization of elastomer mixtures based on synthetic rubbers SKI-3 and SKF-32. Kauchuk i Rezina (2), 40–42, 1997.

15. N. R. Legge, G. Holden, and H. E. Schroeder, eds. Thermoplastic Elastomers. München: Hanser, 1987.

16. Y. Kikuchi, T. Fukui, T. Okada, and T. Inoue. Elastic-plastic analysis of the deformation mechanism of PP-EPDM thermoplastic elastomer: origin of rubber elasticity. Polym. Eng. Sci. 31:1029–1032, 1991.

17. L. E. Nielsen and R. F. Landel. Mechanical Properties of Polymers and Composites. 2d ed. New York: Marcel Dekker, 1994, pp. 435–436.

18. G. H. Michler. Kunststoff-Mikromechanik. Munich: Hanser, 1992.

19. M. Okamoto, K. Shiomi, and T. Inoue. Structure and mechanical properties of poly(butylene terephthalate)/rubber blends prepared by dynamic vulcanization. Polymer 35:4618–4622, 1994.

20. S. Kawabata, S. Kitawaki, H. Arisawa, Y. Yamashita, and X. Guo. Deformation mechanism and microstructure of thermoplastic elastomer estimated on the basis of its mechanical behavior under finite deformation. J. Appl. Polym. Sci., Appl. Polym. Symp. 50:245–259, 1992.

21. R. Anderlik: Thermoplastische Elastomere: Berechnung der Festigkeitseigenschaften. Kautschuk, Gummi, Kunststoffe 45:814–817, 1992.

22. A. Y. Coran and R. P. Patel. Rubber-thermoplastic compositions. Part I. EPDM-polypropylene thermoplastic vulcanizates. Rubb. Chem. Technol. 53:141–150, 1980.

23. J. W. S. Rayleigh. On the capillary phenomena in jets. Proc. Roy. Soc. London 29:71–97, 1879.

24. G. I. Taylor. The viscosity of a fluid containing small drops of another fluid. Proc. Roy. Soc. London A 138:41–48, 1932.

25. G. I. Taylor. The formation of emulsions in definable fields of flow. Proc. Roy. Soc. London A 146:501–523, 1934.

26. S. Tomotika. On the instability of a cylindrical thread of a viscous liquid surrounded by another viscous fluid. Proc. Roy. Soc. London A 150:322–337, 1935.

27. L. A. Utracki. Polymer Alloys and Blends. Müchen: Hanser, 1990, pp. 162–167.

28. V. N. Kuleznev. Smesy polimerov (Mixture of Polymer). Moscow: Znanie, 1984.

29. L. H. Sperling. Interpenetrating polymer networks: now thermoplastics. Mod. Plast. Oct., pp. 74–78, 1981.

30. M. Fujiyama. Processing-induced morphology. In: J. Karger-Kocsis, ed. Polypropylene: An A–Z Reference. London: Chapman and Hall/Kluwer Academic, 1998, pp. 668–677.

31. D. S. Campbell, D. J. Elliott, and M. A. Wheelans. Thermoplastic natural rubber blends. NR Technology 9(Part 2):2–12, 1978.

32. K. Søndergaard and J. Lyngaae-Jørgensen. Shear flow-induced structural changes in polymer blends and allyos. In: K. Søndergaard and J. Lyngaae-Jørgensen, eds. Rheo-Physics in Multiphase Polymer Systems. Lancaster: Technomic, 1996, pp. 445–530.

33. H.-J. Radusch and T. Pham. Morphologiebildung in dynamisch vulkanisierten PP/EPDM-Blends. Kautschuk, Gummi, Kunststoffe 49:249–257, 1996.

34. H.-G. Fritz. Neue thermoplastische Elastomere: Rezeptierung, Aufbereitung und Werkstoffeigenschaften. Chem.-Ing. Tech. 67:560–569, 1995.

35. A. Y. Coran and R. Patel. Rubber-thermoplastic compositions. Part IV. Thermoplastic vulcanizates from various rubber-plastic combinations. Rubb. Chem. Technol. 54:892–903, 1981.

36. A. Y. Coran and R. Patel. Rubber-thermoplastic compositions. Part V. Selecting polymers for thermoplastic vulcanizates. Rubb. Chem. Technol. 55:116–136, 1982.

37. M. Akiba and A. S. Hashim. Vulcanization and crosslinking in elastomers. Progr. Polym. Sci. 22:475–521, 1997.

38. A. Y. Coran and R. Patel. Rubber-thermoplastic compositions. Part VII. Chlorinated polyethylene rubber-nylon compositions. Rubb. Chem. Technol. 56:210–225, 1983.

39. A. Y. Coran and R. Patel. Rubber-thermoplastic compositions. Part VIII. Nitrile rubber polyolefin blends with technological compatibilization. Rubb. Chem. Technol. 56:1045–1060, 1983.

40. D. J. Burlett and J. T. Lindt. Reactive processing of rubbers. Rubb. Chem. Technol. 66:411–434, 1993.

41. M. Xanthos, ed. Reactive Extrusion, München: Hanser, 1992.

42. S. Al-Malaika, ed. Reactive Modifiers for Polymers. London: Blackie Academic Prof., 1997.

43. M. Binet-Szulman. Manufacturing, characterization and processing of a dynamically vulcanized thermoplastic shoe sole material. J. Polym. Eng. 12:121–153, 1993.

44. F. Cai and A. I. Isayev. Dynamic vulcanization of thermoplastic copolyester elastomer/nitrile rubber alloys: I. Various mixing methods. J. Elast. Plast. 25:74–89 (1993).

45. J. Karger-Kocsis, S. Al-Malaika, B. Pukánszky, and M. Binet-Szulman. Recycling of

postconsumer PET bottles by conversion into thermoplastic elastomers (Inco-Copernicus project, PL 964056), 1997–1999.

46. A. Y. Coran. Vulcanization:conventional and dynamic. Rubb. Chem. Technol. 68:351–375, 1995.

47. J. Karger-Kocsis: Amorphous or atactic polypropylene. In: J. Karger-Kocsis, ed. Polypropylene: An A–Z Reference. London: Chapman and Hall/Kluwer Academic, 1998, pp. 7–12.

48. J. Karger-Kocsis, B. Kozma, and M. Schober. Tauroprene—a new versatile polyolefinic thermoplastic rubber. Kautschuk, Gummi Kunststoffe 38:614–616, 1985.

49. J. Karger-Kocsis. Tauroprene—a new thermoplastic rubber. Plasticheskiye Massy (8), 58–59, 1986 (in Russian).

50. N. R. Choudhury and A.K. Bhowmick. Compatibilization of natural rubber-polyolefin thermoplastic elastomeric blends by phase modification. J. Appl. Polym. Sci. 38:1091–1109, 1989.

51. N. R. Choudhury and A. K. Bhowmick. Strength of thermoplastic elastomers from rubber-polyolefin blends. J. Mater. Sci. 25:161–167, 1990.

52. K.-H. Kim, W.-J. Cho, and C.-S. Ha. Properties of dynamically vulcanized EPDM and LLDPE blends. J. Appl. Polym. Sci. 59:407–414, 1996.

53. P. Ghosh, B. Chattopadhyay, and A. K. Sen. Thermal and oxidative degradation of PE-EPDM blends vulcanized differently using sulfur accelerator systems. Eur. Polym. J. 32:1015–1021, 1996.

54. D. Roy, S. K. De, and B. R. Gupta. Anisotropy in hysteresis loss and tension set in short-carbon-fibre-filled thermoplastic elastomers. J. Mater. Sci. 29:4113–4118, 1994.

55. C. S. Ha, D. J. Ihm, and S. C. Kim. Structure and properties of dynamically cured EPDM/PP blends. J. Appl. Polym. Sci. 32:6281–6297, 1986.

56. M. D. Ellul. Novel low temperature resistant thermoplastic elastomers for specialty applications. Plast. Rubb. Compos., Process. Appl. 26:137–142, 1997.

57. R. Anderlik and H.-G. Fritz. Herstellung thermoplastischer Elastomere durch dynamisches Vernetzen silangeprofter EP-Elastomere. Kautschuk, Gummi, Kunststoffe 45:527–530, 1992.

58. H.-G. Fritz and R. Anderlik. Elastische Eigenschaften von thermoplastischen Elastomeren auf der Basis dynamisch vernetzter PP/EPDM-Blends. Kautschuk, Gummi, Kunststoffe 46:374–379, 1993.

59. M. Saroop and G. N. Mathur. Studies on the dynamically vulcanized polypropylene (PP)/butadiene styrene block copolymer (SBS) blends: mechanical properties. J. Appl. Polym. Sci. 65:2691–2701, 1997.

60. M. Saroop and G. N. Mathur. Studies on the dynamically vulcanized polypropylene (PP)/butadiene styrene block copolymer (SBS) blends: melt rheological properties. J. Appl. Polym. Sci. 65:2703–2713, 1997.

61. M. Mehrabzadeh and R. P. Burford. Effect of crosslinking on polyamide 11/butadiene-acrylonitrile copolymer blends. J. Appl. Polym. Sci. 64:1605–1611, 1997.

62. A. K. Bhowmick and T. Inoue. Structure development during dynamic vulcanization of hydrogenated nitrile rubber/nylon blends. J. Appl. Polym. Sci. 49:1893–1900, 1993.

63. A. K. Bhowmick, T. Chiba, and T. Inoue. Reactive processing of rubber-plastic blends: role of chemical compatibilizer. J. Appl. Polym. Sci. 50:2055–2064, 1993.

64. A. J. Moffett and M. E. J. Dekkers. Compatibilized and dynamically vulcanized thermoplastic elastomer blends of poly (butylene terephthalate) and ethylene propylene diene rubber. Polym. Eng. 32:1–5, 1992.

65. F. Cai and A. I. Isayev. Dynamic vulcanization of thermoplastic copolyester elastomer/nitrile rubber alloys: II. Rheology, morphology and properties. J. Elast. Plast. 25:249–265, 1993.

66. J. Karger-Kocsis and K. Friedrich. Solvent- and heat resistant thermoplastic elastomeric vulcanized/crosslinked blends of polymers and rubber. Hungarian Patent Application CA: 114:104, 080 (1991).

67. Th. Lüpke, H.-J. Radusch, M. Sandring, and N. Nicolai. Spannungs-Dehnungsverhalten dynamischer Vulkanisate. Kautschuk, Gummi, Kunststoffe 45:91–94, 1992.

68. L. A. Goettler, J. R. Richwine, and F. J. Wille. The rheology and processing of olefinbased thermoplastic vulcanizates. Rubb. Chem. Technol. 55:1448–1463, 1982.

69. O. Chung, A. Y. Coran, and J. L. White. Melt rheology of dynamically vulcanized rubber/plastic blends. SPE ANTEC 43:3455–3460, 1997.

70. S. Abdou-Sabet, R. C. Puydak, and C. P. Rader. Dynamically vulcanized thermoplastic elastomers. Rubb. Chem. Technol. 69:476–494, 1996.

71. M. T. Payne. Olefinic thermoplastic vulcanizates: impact on rubber recycling. In: C. P. Rader, S. D. Baldwin, D. D. Cornell, G. D. Sadler, and R. F. Stockel, eds. Plastics, Rubber, and Paper Recycling. ACS Symp. Ser. 609. Washington: Am. Chem. Soc., 1995, pp. 221–236.

72. S.-G. Hong and C.-M. Liao. The surface oxidation of a thermoplastic olefin elastomer under ozone exposure: ATR analysis. Polym. Degr. Stab. 49:437–447, 1995.

73. T. M. Aminabhavi and H. T. S. Phayde. Sorption, desorption, diffusion, and permeation of aliphatic alkanes into Santoprene thermoplastic rubber. J. Appl. Polym. Sci. 55:17–37, 1995.

74. R. J. Clark. TPO paintability enhancement. SPE ANTEC 42:2736–2738, 1996.

75. C. P. Rader and R. C. Wegelin. Thermoplastic elastomers—a major recycling opportunity. SPE ANTEC 40:3026–3031, 1994.

76. R. S. George and R. Joseph. Studies on thermoplastic elastomers from polypropylene and latex waste products. Kautschuk, Gummi, Kunststoffe 47:816–819, 1994.

77. N. N.: TPE demand to reach 1.5 million tonnes, Eur. Plast. News., Nov., 1996, p. 14.

78. R. School. Markets for thermoplastic elastomers into the 21st century. Kautschuk, Gummi, Kunststoffe 48:811–816, 1995.

79. H. M. J. C. Creemers. Thermoplastische Elastomere (TPE). Kunststoffe—German Plastics 86:1845–1851, 1996.

80. W. P. Lauhus, E. Haberstroh, and F. Ehrig. Technische Elastomere. Kunststoffe—German Plastics 87:706–716, 1997.

81. R. Mülhaupt. Metallocene catalysis and tailor-made polyolefins. In: J. Karger-Kocsis, ed. Polypropylene: An A–Z Reference. London: Chapman and Hall/Kluwer Academic, 1998, pp. 454–475.

82. G. Collina, V. Braga, and F. Sartori. New thermoplastic polyolefin elastomers from the novel multicatalyst reactor granule technology: their relevant physical-mechanical properties after crosslinking. Polym. Bull. 38:701–705, 1997.

83. A. Batistini. New polyolefin plastomers and elastomers made with the Insite® technology: structure-property relationship and benefits in flexible thermoplastic appli-

cations. Macromol. Symp. 100:137–142, 1995.

84. W. J. Gauthier. Elastomeric polypropylene homopolymers using metallocene cata-
    lysts. In: J. Karger-Kocsis, ed. Polypropylene: An A–Z Reference. London: Chapman
    and Hall/Kluwer Academic, 1998, pp. 178–185.
85. D. E. Mouzakis, M. Gahleitner, and J. Karger-Kocsis. Toughness assessment of elas-
    tomeric polypropylene (ELPP) by the essential work of fracture method. J. Appl.
    Polym. Sci. 70:873–881, 1998.
86. A. Y. Coran and S. Lee. New elastomers by reactive processing. Part II. Dynamic vul-
    canization of blends by transesterification. Rubb. Chem. Technol. 65:231–244, 1992.
87. Y. Yang, T. Chiba, H. Saito, and T. Inoue. Physical characterization of a polyolefinic
    thermoplastic elastomer. Polymer 39:3365–3372, 1998.

6

Thermosetting Polymer Blends: Miscibility, Crystallization, and Related Properties

Qipeng Guo
University of Science and Technology of China, Hefei, People's Republic of China

I. INTRODUCTION

Thermosetting polymers are among the most important materials in many diverse industries and are being used increasingly in structural engineering applications. The term "thermosetting resins" covers a wide range of cross-linking polymers, which are principally epoxy resins, unsaturated polyester resins, phenol-formaldehyde resins, and amino resins.

Thermosetting polymers are generally amorphous, highly cross-linked polymers, and this structure results in these materials possessing various desirable properties such as high tensile strength and modulus, easy processing, good thermal and chemical resistance, and dimensional stability. However, it also leads to low toughness and poor crack resistance, and the materials are normally brittle at room temperature. This is the basic reason for toughening thermosetting polymers for many end-use applications. Thermosetting polymer blends are usually multiphase polymers; the dispersed phase consists of rubbery or thermoplastic domains, and the continuous phase is a cross-linked thermosetting polymer matrix. One of the most successful methods of improving the toughness of a thermosetting polymer is to incorporate a second phase of dispersed rubbery particles or thermoplastic domains into the cross-linked polymer. The developments in the area of toughening thermosetting resins with elastomers or thermoplastics have been included in several excellent books (1–6) and hence will not be covered in detail here.

155

However, relatively few systematic studies have been paid to the miscibility, phase behavior and crystallization in blends of thermosetting resins with linear polymers. Since the resulting morphology and extent of phase separation is known to affect the optical and mechanical properties of the cured blends, the need for an understanding of the miscibility and phase behavior in thermosetting blends is of great practical importance. Furthermore, it is also of much academic interest to examine the phase behavior of such thermosetting polymer blends. Particularly, in the thermosetting polymer blends where the linear component is crystallizable, crystallization will be greatly affected by the miscibility and phase behavior of the blends. The interrelationship between the miscibility, phase behavior, crystallization, and composition is very complicated in such thermosetting polymer blends and has been the subject of a research program in this laboratory during the past decade. The intention of this chapter is to highlight some of the progress that has been made in the areas, but for reasons of space not all are referenced.

II. EPOXY RESIN

A. Epoxy Resin/Elastomer Blends

Epoxy resins in particular are commonly used. They are generally the strongest thermosetting materials and are widely employed as molding compounds, laminates, structural adhesives in aerospace technology, and matrices for fiber-composite materials (7–11).

Elastomers are widely employed to modify epoxy resins (12–21). The miscibility requirement of blending an elastomer or a thermoplastic with thermoset resins is rather different from that of blending two linear polymers. The low-molecular-weight liquids of elastomers are usually miscible enough to dissolve and disperse in the epoxy resin monomers due to entropy contribution, but they phase-separate out during cure of the resin, owing to unfavorable enthalpy effect. It is the morphology formed during the phase separation that results in the enhancement in toughness.

Visconti and Marchessault (22) first reported a qualitative description of the phase-separation process, in a homogeneous mixture composed of a cycloaliphatic epoxy resin, an anhydride hardener, and a carboxyl-terminated butadiene-acrylonitrile (CTBN) random copolymer. They found that at a certain point of the cross-linking reaction small spherical domains of an elastomeric phase were segregated. The variation in size and shape of the rubbery domains was analyzed by using small-angle light scattering. Phase separation was found to take place well before gelation, and a phase inversion occurred beyond 20 wt% CTBN content.

The mechanisms of phase separation during the processing of epoxy resins have been extensively studied by Williams and coworkers (23–25). They reported the miscibility and cloud point curves of mixtures comprising monomers of bisphenol-A type epoxy resin and CTBN random copolymers, and, eventually, amines as hardeners (24). It was found that the miscibility of the mixtures is very sensitive to the molecular weight of the epoxy molecule, and the increase in the molecular weight decreases the miscibility. Based on the thermodynamic description through a Flory–Huggins equation and constitutive equations for the polymerization and phase-separation rates, they proposed a theoretical model to predict the fraction, composition, and particle size distribution of dispersed domains (25).

It was also reported that a variety of different morphologies could be obtained from a single rubber-modified epoxy formulation by varying the cure temperature (26,27). The phase separation process is controlled by both thermodynamic and kinetic factors as pointed out Bucknall and Partridge (28–30). For example, in epoxy resin modified with CTBN rubber, the increase in molecular weight during curing, for both the rubber and the epoxy, reduced the entropy of mixing. This, together with the positive (endothermic) enthalpy, increases the likelihood of a two-phase morphology. The overall resultant morphology is also influenced by reaction rate at a given temperature and the period over which phase separation can take place.

In the mixing of two relatively low molecular weight components with nearly matched solubility parameters (δ) of CTBN rubber and epoxy, the negative entropy contribution dominates the free energy of mixing (28–30).

B. Epoxy Resin/Thermoplastic Blends

Because addition of rubbery materials to epoxy resins has been shown to lower their glass transition temperatures (T_g) and their thermal and oxidative stability, high-performance thermoplastics have been employed to toughen epoxy resins in recent years. Owing to the high modulus and high T_g of these thermoplastics, the modulus and T_g of the modified epoxy resin can reach or even surpass those of the pure epoxy resin (31–41). A recent review of the developments in the area of thermoplastic-modified epoxy resins is available (42).

Many uncured mixtures of thermoplastic polymers with epoxy resins form homogeneous blends, but most of these homogeneous liquid blends would turn into multiphase systems after they undergo cross-linking reactions with curing agents or hardeners (35,43,44). It has been shown that the modifications necessitate a fine phase-separated structure and a good interfacial adhesion between the two separated phases. A thermoplastics-dispersed phase structure or a cocontinuous phase structure in thermoplastics-modified epoxy resins usually yields greater fracture toughness. An understanding of the phase diagram and its evolution during the curing process is very important. Reaction-induced phase separation was usually observed, and spinodal decomposition was proven to be the dominant mechanism of the phase separation by Inoue and coworkers (45–47).

We have reported that uncured bisphenol-A type epoxy resin, i.e., diglycidyl ether of bisphenol A (DGEBA), is miscible with several thermoplastic polymers, including phenolphthalein poly(ether ether sulfone) (PES-C) (48,49), poly(hydroxyether of bisphenol A) (phenoxy) (50), polysulfone (PSF) (51), phenolphthalein poly(ether ether ketone) (PEK-C) (52), poly(vinyl acetate) (40), and poly(styrene-co-acrylonitrile) (41).

We also successfully obtained the homogeneous 4,4'-diaminodiphenylmethane (DDM)-cured blends of bisphenol-A type epoxy resin (ER) with PES-C (48,49), phenoxy (50), PSF (51), and PEK-C (52). Homogeneous blends have also been reported by Chen et al. in the ER/polyetherimide system (53) and by Jayle et al. in the ER/polycarbonate system (54). However, it was noted that incorporation of these thermoplastics does not effectively increase either the fracture toughness or the flexural properties. The lacking of toughening in the homogeneous blends has been proposed to be due to the reduced cross-link density of the epoxy network. In thermoset/thermoplastic blends, the significance of entropy factors is greatly diminished, and thus the miscibility has been considered to be due primarily to the enthalpic contributions.

The morphology and phase behavior of the cured blends were found to be greatly dependent on the choice of curing agent. For example, it was shown that PES-C exhibits more overall miscibility with the aromatic amine-cured epoxy resins than with the anhydride-cured epoxy resins. For the blends cured with DDM and 4,4'-diaminodiphenylsulphone (DDS), no phase separation occurred as indicated by either dynamic mechanical analysis (DMA) or scanning electron microscopy (SEM). However, for the blends cured with maleic anhydride (MA) and phthalic anhydride (PA), both DMA and SEM clearly showed evidence of phase separation. This has been considered to be due mainly to the presence of hydroxyl groups in the amine-cured system, which offer an excellent potential for hydrogen-bonding interaction with side ester groups and/or ether oxygens of PES-C in blends. For the ER/phenoxy blends cured with DDM and aliphatic anhydrides, i.e., MA, and hexahydrophthalic anhydride (HHPA), no phase separation occurred. However, phase separation occurred for the ER/phenoxy blends cured with DDS and PA.

The ER/PEK-C blends cured with anhydrides (MA, PA, and HHPA) showed a two-phase structure (55), whereas the phase behavior of the DDM-cured ER/PEK-C blends was dependent on the curing condition (52,56). The homogeneous DDM-cured ER/PEK-C blends could be obtained at a low initial curing temperature of 80°C (52), whereas the DDM-cured ER/PEK-C blends prepared at a higher initial curing temperature of 150°C were found to be heterogeneous (56).

The phase behavior and morphology of a thermosetting polymer blend is determined by both thermodynamic and kinetic factors. If the systems containing one cross-linked and one linear component are still miscible in the highly cross-linked cases, a semi-interpenetrating network (semi-IPN) with homogeneous phase structure may form. In this semi-IPN or IPN, the components that are initially immiscible may exhibit some characteristics of miscibility, i.e., the semi-IPN or IPN can be considered as a compatibilized immiscible polymer blend. On the other hand, the resulting increase in molecular weight causes a decrease in the configurational entropy of mixing as the cure proceeds. Therefore curing may cause phase separation in an initial miscible blend. The phase separation process is governed by not only thermodynamic but also kinetic factors. Raising curing temperature increases the rate of the curing reaction and reduces the time for gelation. If gelation occurs prior to phase separation, no domains appear. On the other hand, raising curing temperature enhances the mobility of molecules, so that phase separation takes place more easily. Curing reaction and phase separation are competitive. In the DDM-cured ER/PEK-C system, heterogeneous blends were obtained where the initial curing temperature was 150°C (56), whereas only homogeneous blends were obtained at an initial curing temperature of 80°C (52). Elevation of curing temperature is more beneficial for phase separation.

Teng and Chang (57,58) reported that both homogeneous and inhomogeneous DDS-cured phenoxy/epoxy blends can be prepared through kinetic control of the curing rate by varying the amount of accelerator in the system.

Chemical reaction between the epoxy resin and the modifier is often desirable for effective toughening. Indeed, reactions or strong specific interactions between epoxy resin and the thermoplastic polymer, if taking place at an appropriate temperature, might prevent or lessen the extent of phase separation at the fully cured state. Recently, Nichols and Robertson investigated blends of DGEBA with poly(butylene terephthalate) (PBT) and found that thermoreversible gelation occurred (59,60). However, reaction-induced phase separation was observed in the DDS-cured blends of DGEBA and PBT (61,62). Homogeneous blends of DGEBA and poly(ethylene terephthalate) (PET) were obtained by solution casting in this laboratory (63). Interchange reactions between DGEBA and PET at elevated temperatures occurred, resulting in the formation of copolymers based on the blend components, and there also existed a self-cross-linking reaction among the DGEBA molecules. The miscibility and chemical interactions in hardener-free epoxy/polycarbonate blends were investigated by several authors (64–69), and it was proven that chemical exchange reactions occurred and cross-linked structure formed in the blends after heat treatment. Polycarbonate–epoxy semi-IPNs were also reported recently by Rong and Zeng (70,71).

Coleman and coworkers (72) investigated the effect of cross-linking on the degree of miscibility in a miscible polymer blend and the IPN derived from the blend. Kim et al. (73) found that the extent of hydrogen bonding in the blends and IPNs was controlled by changing the temperature of the cross-linking reaction and by reducing the density of the hydrogen bond acceptor groups in the chain.

Several theoretical attempts have been made to explain the complex interaction between phase behavior and cross-linking. Binder and Frisch (74) considered the phase behavior of a blend in which the thermoset had formed a complete "weakly" cross-linked network before phase separation could take place. Donatelli et al. (75) semiempirically derived phase domain sizes by using a similar method. Teng and Chang (57) have proposed a model for phase separation prior to network formation in which only the average molecular weight of the thermoset during curing is considered. Clarke et al. (76,77) calculated the growth rate of concentration fluctuations in star/star and star/linear polymer blends and concluded that the dynamics of entangled branched polymers are very important in the early stages of spinodal decomposition. They considered that the cluster size distribution is dependent only on cure time. Tanaka and Stockmayer (78) have considered thermoreversible gels in blends, and the interaction between the gel point and phase separation. Jo and Ko (79) investigated the structure development during the cure of epoxy resin modified with rubber or thermoplastic polymer by a Monte Carlo simulation technique.

C. Epoxy Resin/Poly(ethylene oxide) Blends

The study of miscibility for polymer blends where one component is crystallizable and another is highly cross-linked has received relatively little attention (80–86). From the point of view of thermodynamics, an increase in molecular weight for either component of a miscible blend could reduce the entropy of mixing. This, together with the positive (endothermic) enthalpy, would result in phase separation. Therefore the occurrence of miscibility or even partial miscibility in such polymer blends containing one component with an infinite molecular weight (i.e., highly cross-linked) requires a negative (exothermic) enthalpy.

We reported our studies on bisphenol-A type epoxy resin (ER)/poly(ethylene oxide) (PEO) blends (82,83) and, in particular, the role of cross-linking in influencing the miscibility of ER with PEO and the phase structure of the resulted blends. Tetraethylenepentamine (TEPA) was used as cross-linking agent. The blend samples could be fully cured at only 110°C, and interchange reactions could be avoided at such a low cure temperature. The miscibility and morphology of the ER/PEO blends were investigated by differential scanning calorimetry (DSC).

We found that PEO is miscible with uncured epoxy resin, DGEBA, as shown by the existence of a single glass transition temperature in each blend. However, PEO was judged to be immiscible with the highly TEPA-cross-linked ER. The miscibility and morphology of the ER/PEO blends was remarkably affected by cross-linking. It was observed that the phase separation in the ER/PEO blends occurred as cross-linking proceeded.

All the DGEBA/PEO blends were transparent just above the melting point of PEO. The thermal transitions obtained from the DSC thermograms of the

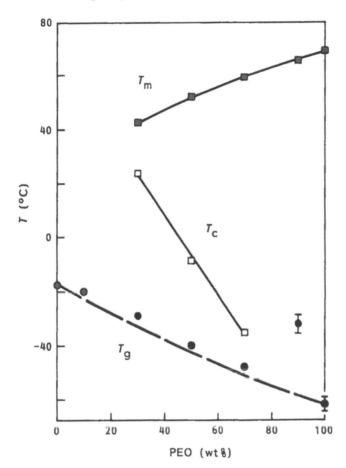

Figure 1 Dependence of T_g (●), T_c (□), and T_m (■) of the quenched DGEBA/PEO blends on the weight percent of PEO. – – –, Fox equation prediction. (From Ref. 83.)

quenched samples are plotted in Fig. 1 as functions of blend composition. The full circles in Fig. 1 display a single glass transition temperature (T_g) varying with overall blend composition. The appearance of a single T_g strongly suggests that the blend presents a homogeneous single amorphous phase, i.e., the two components are miscible in the amorphous phase.

The broken curve in Fig. 1 is as predicted by the Fox equation (84), fitting the experimental T_g data quite well. The deviation of experimentally obtained T_g data from the Fox equation at high PEO content (90 wt% PEO) is because of the crystallization of PEO in the blend during the quenching.

Figure 1 shows also the T_c and T_m of the quenched samples as functions of blend composition. For the pure PEO and 10/90 DGEBA/PEO blend, no crystallization exotherm was observed, since crystallization was sufficiently rapid to occur completely during the quenching. However, for the blends with a higher DGEBA content up to 70 wt%, T_c increases with increase of DGEBA content. This phenomenon indicates that crystallization of PEO in the blend becomes progressively more difficult with increase of DGEBA content. Furthermore, the blend containing 90 wt% DGEBA did not show any crystallization exotherm. These results support the idea that DGEBA is completely miscible with PEO over the entire composition range in the melt. The decrease in the crystallinity of PEO with increase of DGEBA content is attributed both to the higher T_g of DGEBA than that of PEO and to the interactions between these two components.

The T_m depression with increase of DGEBA content shown in Fig. 1 is more substantial than those usually observed for miscible blends having one crystallizable component. This is because both enthalpic and entropic effects contribute to the melting point depression in the DGEBA/PEO blends owing to the small molecular weight of the DGEBA, whereas in the case that the molecular weights of both components are adequately large, only enthalpic contribution to the melting point depression is not negligible (85).

Figure 2 shows the DSC thermograms of the quenched samples of the ER/PEO blends containing 50 wt% PEO and cured with different TEPA contents. A crystallization exotherm occurred for blends with TEPA/DGEBA $\leqslant$ 0.08, while no exotherm occurred for those with TEPA/DGEBA $\geqslant$ 0.12, which implies that crystallization was so rapid that it was completed during the quenching. The more TEPA content the blend had, the more rapidly the crystallization occurred and the higher the crystallinity of PEO. For the blend without TEPA, i.e., 50/50 DGEBA blend, the crystallization exotherm equaled the melting endotherm; this means that no crystallization occurred during the quenching. In the blends of low TEPA-content (TEPA/DGEBA $\leqslant$ 0.08), the T_g is observed to shift to lower temperature with increasing TEPA content. This is because the greater the TEPA content of the blend, the more the cured ER separated out and the more PEO in the uncured phase the blend had. For the blends with TEPA/DGEBA $\geqslant$ 0.12, two T_gs are observed, as shown in Fig. 2. One (T_{g1}) around 108°C is attributed to the cross-linked ER remaining almost invariant. That is to say, a highly cross-linked ER was formed and separated out during the cross-linking. The other one (T_{g2}) is higher than that of the 50/50 DGEBA/PEO blend and stays almost constant, indicating that the related phase might be composed of PEO and partially cured ER. Additionally, the melting point of PEO is shifted to higher temperature and approaches the T_m of the pure PEO as a result of increasing cross-link density. It can be seen that PEO is miscible or partially miscible with insufficiently cross-linked ER but immiscible with highly cross-

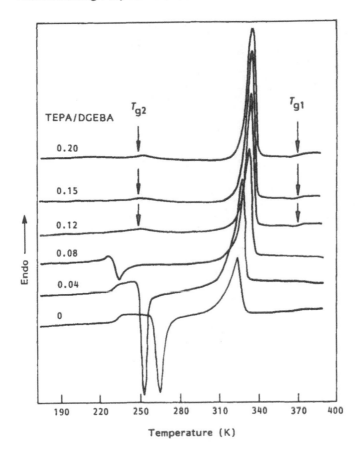

Figure 2 The DSC thermograms of the ER/PEO blends containing 50 wt% PEO and cured with various amount of TEPA. (From Ref. 83.)

linked ER. The higher the cross-linking degree, the more completely the phase separation occurred. It was also noticed that the degree of crystallinity of the PEO phase of the cross-linked 50/50 ER/PEO blends increases with TEPA content.

All the ER/PEO blends cured with TEPA/DGEBA = 0.12 were quite opaque at room temperature and did not become clear above the T_m of PEO. The TEPA-cured ER/PEO blends exhibited two separate T_gs, corresponding to the cured ER phase and the PEO-rich phase, respectively. The T_m for PEO was observed to lie between the two T_g values, and the T_m depression was slight. PEO is immiscible with the highly TEPA-cross-linked ER even though it may be miscible with the partially TEPA-cured ER.

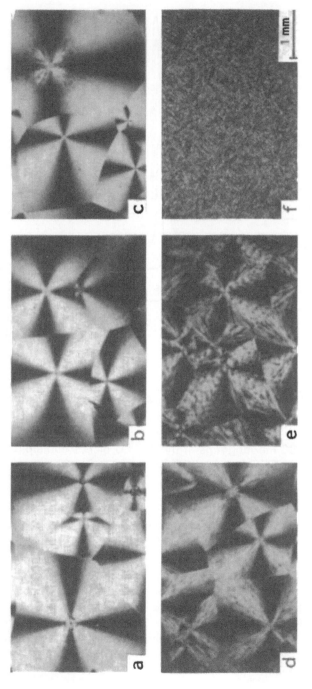

Figure 3 Optical micrographs for the thin films of (a) 0/100, (b) 10/90, (c) 30/70, (d) 50/50, (e) 70/30, and (f) 90/10 DGEBA/PEO blends. (From Ref. 83.)

Figure 3 shows optical micrographs for the DGEBA/PEO blends. In the optical microscope with crossed polars, the spherulites display a 'Maltese cross' birefringence pattern and have a regular shape with defined borders. The following observations can be made: first, there is no evidence that the noncrystallizable component segregates in large domains in intraspherulitic regions or in interspherulitic contact zones exceeding the dimensions corresponding to the resolving power of the technique used. Secondly, the DGEBA/PEO blend samples are always completely volume filled with PEO spherulites up to a composition of 90/10. Finally, spherulites of 90/10 DGEBA/PEO blend show a less regular texture; this may be caused by the coarseness of the crystalline lamellae due to the presence of uncrystallized material in interlamellar regions. All these observations suggest that PEO is miscible with DGEBA in the molten state; and during the crystallizing process of the blends, the DGEBA is incorporated in the interlamellar regions of PEO spherulites.

Optical micrographs for the ER/PEO blends containing 50 wt% PEO and cured with various amount of TEPA are shown in Fig. 4. The spherulites first become smaller, which may be attributed to the enhanced nucleation rate owing to the cross-linked ER phase, and then gradually become larger again and clearer as a result of increasing cross-link density. The spherulites of the highly cross-linked blends tend to become like those of plain PEO as the cross-link density increases.

It was shown that PEO may be miscible with partially cured ER or partially miscible with more completely but not yet sufficiently cured ER. However, it was concluded that PEO is immiscible with the highly cross-linked ER although it is miscible with both uncured ER, i.e., DGEBA, and poly(hydroxyether of bisphenol A) (phenoxy) (86–88), which may be considered to be a model linear polymer of epoxy resin. A brief FTIR study (89) has revealed that there is hydrogen-bonding interaction between PEO and phenoxy that is even stronger than that in the pure phenoxy.

Following the above work, ER/PEO blends cured with DDM and with phthalic anhydride (PA) were investigated by Ma and coworkers (90–93). The miscibility was established on the basis of thermal analysis and DMA results. Single composition-dependent glass transition temperatures were observed for both the DDM-cured and the PA-cured blends. However, the dilution effect of the PEO component and the participation of PEO in the cure reaction resulted in incomplete cross-linking, i.e., the formation of imperfect cross-linking network structures. It was speculated that the PA-cured blend is characterized by a negative Flory–Huggins interaction parameter (χ_{12}) (90). A short FTIR study suggested the presence of intermolecular interaction occurring in the PA-cured blends, and its strength was found to be higher than that in the pure PA-cured epoxy resin (90).

Bates and coworkers (94) recently investigated the PA-cured systems of bisphenol-A type epoxy resin with a set of amphiphilic poly(ethylene oxide)-poly(ethyl ethylene) (PEO-PEE) and poly(ethylene oxide)-poly(ethylene-alt-

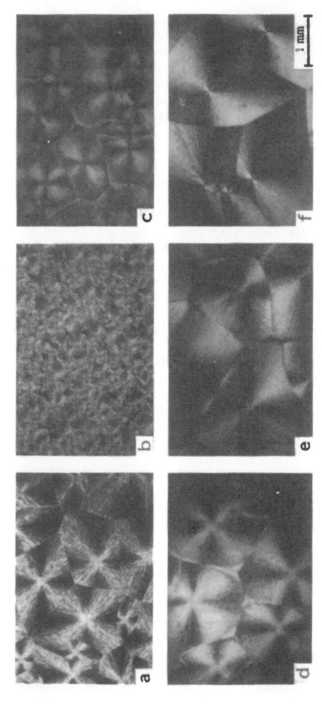

Figure 4 Optical micrographs of the ER/PEO blends containing 50 wt% PEO and cured with various amount of TEPA. TEPA/DGEBA: (a) 0, (b) 0.04, (c) 0.08, (d) 0.12, (e) 0.15, and (f) 0.20. (From Ref. 83.)

propylene) (PEO-PEP) block copolymers. The epoxy resin could swell the PEO domains in the block copolymers without dissolving the polyalkane blocks and thus lead to an ordered composite material. They found that the matrix epoxy underwent a dramatic change during the gelation reaction in which its molecular weight diverged. This resulted in a significant decrease in the miscibility of the PEO in the epoxy matrix. Cross-linking of epoxy matrix without macroscopic phase separation of the block copolymers led to stable structures with nanoscopic features.

D.   Epoxy Resin/Poly(ε-caprolactone) Blends

Poly(ε-caprolactone) (PCL) has been shown to be miscible with many other polymers due to its high potentiality to form hydrogen bonds with these polymers. In particular, it has been proven that there is hydrogen-bonding interaction between PCL and phenoxy (89), a linear model epoxy resin of high molecular weight, and PCL was found to exhibit miscibility with phenoxy (86).

Noshay and Robeson (80) examined the miscibility of a range of anhydride-cured epoxy resins with PCL of various molecular weights and with different end groups in 1974. They concluded that above a critical molecular weight (3,000–5,000) of PCL, the blends had a two-phase structure and that the PCL end groups reacted with the anhydride curing agent to produce a type of block copolymer.

In another work by Clark et al. (81), PCL blends with amine-cured epoxy resins were examined with attention to the opportunity that existed for hydrogen bonding in amine-cured epoxy resin/PCL blends. They found that PCL with average molecular weight about 20,000 was partially miscible with amine-cured epoxy resins although it was largely immiscible with anhydride-cured epoxy resins as reported by Noshay and Robeson (80). The different miscibility with PCL between amine-cured and anhydride-cured systems was considered to be due to the presence of hydroxyl groups in the amine-cured system, which offer an excellent potential for hydrogen-bonding interaction with ester groups of PCL in blends.

However, in these earlier studies, the miscibility of the ER/PCL system was established only based on the DMA and infrared results; and the molecular weight of PCL they used was relatively low. In order to obtain a better understanding on this thermosetting polymer blend system, a further examination was conducted in this laboratory. The PCL we used was of rather high molecular weight (70,000–100,000), and 2,2-bis(4-(4-aminophenoxy)phenyl propane (BAPP) was used as curing agent. The miscibility, crystallization, and thermal properties were investigated by DSC, while intermolecular hydrogen-bonding interaction was revealed by FTIR spectroscopy. Solid-state NMR was employed to characterize homogeneity at the molecular level in the BAPP-cured ER/PCL blends. Here, we present some preliminary results of our investigation (95–97).

Our DSC studies revealed that both the uncured and the BAPP-cured blends exhibited a single T_g varying with blend composition. PCL was found to be completely miscible with both DGEBA and the BAPP-cured ER. A detailed FTIR study revealed the existence of the hydrogen-bonding interaction in the BAPP-cured ER/PCL blends, i.e., the hydrogen-bonding interaction occurred between the carbonyl groups of PCL and the hydroxyls of the BAPP-cured ER.

The homogeneity of the BAPP-cured ER/PCL blends at the molecular level was examined by solid-state ^{13}C CP/MAS spectra of the BAPP-cured ER/PCL blends could not be taken as the superposition of the spectra of both the pure BAPP-cured ER and the pure PCL; and there were some significant changes in chemical shifts and line width of resonances, especially for the resonance of the carbons involving the intermolecular (or intramolecular) hydrogen-bonding interactions. One narrow resonance peak was observed around 174 ppm in the ^{13}C CP/MAS spectra for the blends containing 50 wt% PCL and more, resulting from the combination of both the mobile, amorphous phase and the well-ordered crystalline phases of PCL. It was noted that the 1.3 ppm high-field shift was seen for the ^{13}C resonance of the carbonyl carbons. As the FTIR studies had revealed the presence of the intermolecular hydrogen-bonding interaction, these results were considered as indicative of the intimate mixing of the blend components.

In order to obtain information about the scale of miscibility and phase structure, the dynamic relaxation experiment of protons was conducted from the ^{13}C CP/MAS spectroscopy. The single values of proton spin-lattice relaxation time in the laboratory frame ($T_1(H)$) were found for all the BAPP-cured ER/PCL blends, intermediate between the relaxation times of the two pure components. The results indicate that a fast spin diffusion, carried out among all the protons in the blends, averaged out the whole relaxation process. Therefore the blends are homogeneous on the scale where the spin diffusion occurs within $T_1(H)$. The mixing scale was estimated to be 20–30 nm.

To examine the homogeneity of the BAPP-cured ER/PCL blends at the molecular level, we further measured the proton spin-relaxation time in the rotating frame ($T_{1\rho}(H)$). For the pure BAPP-cured ER, the fitting of the experimental ^{13}C intensity data with signal exponential decay function was quite good, indicating that a fast spin-diffusion has been carried out among all the protons in the pure BAPP-cured ER, which averaged out the whole relaxation process. However, the relaxation of protons for the pure PCL displays a typical two-stage process. The shorter represents the relaxation of protons in the crystalline region, whereas the longer indicates that in the amorphous region. The relaxation of protons in the BAPP-cured ER/PCL blends is greatly dependent on the blend composition. For the 70/30 BAPP-cured ER/PCL blends, all protons can relax in the form of single-exponential decay, i.e., all $T_{1\rho}(H)$ value are identical. The results show that this blend was homogeneous on the scale of 20–30 Å. However, for the 50/50 and

30/70 BAPP-cured ER/PCL blends, the $T_{1\rho}(H)$ values are dependent on the selected carbons, indicating an inefficient dipolar coupling and hence the presence of structural heterogeneity at the molecular level.

The isothermal crystallization kinetics of PCL in both the uncured DGEBA/PCL blends and the BAPP-cured ER/PCL blends were investigated by DSC (96). It was shown that the cross-linking caused a considerable increase in the overall crystallization rate and hence dramatically influenced the mechanism of nucleation and growth of the PCL crystals in the blends. The phenomenon of T_m depression was observed for both the DGEBA/PCL blends and the BAPP-cured ER/PCL blends. The analysis of kinetic data according to nucleation theories showed that the surface energy of extremity surfaces dramatically increased with content of the amorphous component for the DGEBA/PCL blends but decreased for the BAPP-cured ER/PCL blends.

III. NOVOLAC RESIN

Phenol-formaldehyde resins are a major class of thermosetting polymers that are widely employed as molding compounds, laminates, adhesives, and shell molds for metals and electrical insulation due to their low manufacturing cost, dimensional stability, resistance to aging, and high tensile strength (98,99). Blending novolacs with other polymers is of great interest for industry. In fact, it has been found that novolac resins exhibit miscibility with a number of polar polymers containing carbonyl or carbonate groups (100–106). In such blends, intermolecular hydrogen bonding acts as the dominant driving force for miscibility. However, as most novolac resins are used in cured forms, studies of cross-linked novolac blends are of more practical importance. Some literature has reported the effect of cross-linking on the novolac blends and analogous blends (72,73,107).

The miscibility and phase structures of initially miscible novolac/poly (methyl methacrylate) (PMMA) blends after full curing of the novolac were investigated by Zhang and Solomon (107) using solid-state NMR, FTIR, and DSC techniques. It was proven that there are strong intermolecular hydrogen-bonding interactions between novolac and PMMA, resulting in the blends being miscible at the molecular level. Curing novolac/PMMA blends with hexamine (HMTA) reduced the intermolecular hydrogen-bonding significantly, but a considerable amount of residual intermolecular hydrogen bond still remained in the cured blends. The phase structures of the fully HMTA-cured novolac/PMMA blends were found to show composition dependence. Fully HMTA-cured novolac-rich blends resulted in semi-IPNs; novolac formed a highly cross-linked network through the whole blend while PMMA chains distributed uniformly in the network with domain sizes on 2–3 nm scales. The segmental motion of the PMMA chains was frozen by the network, and the glass transition was not observed. For the PMMA-rich blends, cross-linked novolac chains distributed in the PMMA matrix. Most fully HMTA-cured blends were partially miscible or microphase-separated with domain sizes around a scale of 20–30 nm.

We have reported our studies on the miscibility and crystalline morphology of uncured novolac/PCL blends as well as highly HMTA cross-linked novolac/PCL blends with attention focused on the role of cross-linking in influencing the miscibility and phase behavior of novolac resin with PCL and the morphology and crystallization of the resulted blends (108). The miscibility and crystalline morphology of novolac/PCL blends before and after curing were investigated by optical microscopy, DSC, and FTIR techniques. DSC study revealed that there existed a single T_g in each blend, and PCL is miscible with uncured novolac.

Remarkable changes took place after the novolac/PCL blends were cured with HMTA, which can be considered to be due to the dramatic changes in chemical and physical nature of novolac resin during the cross-linking. Phase separation in the initially miscible novolac/PCL blends occurred after curing with 15 wt% HMTA (relative to novolac content). The phase structures of the cured blends show composition dependence. The cured novolac/PCL blends with novolac content up to 70 wt% were observed to be partially miscible, whereas the 90/10 cured novolac/PCL blend was found to be miscible.

An FTIR study by Coleman and coworkers revealed the existence of hydrogen-bonding interaction between PCL and poly(p-vinyl phenol) (PVPh), a high-molecular-weight analogue of novolac, and PCL was judged to exhibit miscibility with PVPh (109). In the novolac/PCL blends (108), the miscibility was proposed to be due to the formation of intermolecular hydrogen bonds between PCL and novolac. Indeed, our FTIR studies proved that hydrogen-bonding interaction in novolac/PCL blends occurred between the hydroxyl groups of novolac and the carbonyl groups of PCL.

The curing reduces the intermolecular hydrogen bonding significantly, but there still exists a considerable amount of residual intermolecular hydrogen bond in the cured blends with a strength much lower than that in the uncured blends. Figure 5 shows the FTIR spectra of novolac/PCL blends cured with 15 wt% HMTA in the region of 2790 to 3710 cm^{-1}. The band at 3498 cm^{-1} is attributed to the free hydroxyl groups stretching vibration and the band at 3383 cm^{-1} corresponds to the self-associated hydroxyl groups of cured novolac. The position of the free hydroxyl band at 3498 cm^{-1} remains constant, while the hydrogen-bonded hydroxyl band at 3383 cm^{-1} slightly shifts to high frequency with the increase of PCL content. That is to say, there is still a considerable amount of intermolecular hydrogen bond in the HMTA-cured novolac/PCL blends, although it is very small relative to the uncured novolac/PCL blends. This phenomenon is more clear as we examine the FTIR spectra of the HMTA-

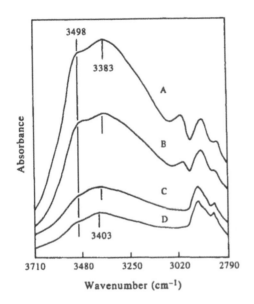

3498

3383

A

Absorbance

B

C

D

3403

| 3710 | 3480 | 3250 | 3020 | 2790 |

Wavenumber (cm⁻¹)

Figure 5 FTIR spectra in the 2790–3710 cm^{-1} region of the novolac/PCL blends cured with 15 wt% HMTA. Novolac/PCL: (A) 100/0, (B) 90/10, (C) 70/30, and (D) 50/50. (From Ref. 110.)

cured novolac/PCL blends in the region of 1605 to 1821 cm^{-1} in Fig. 6. Compared with the spectra of the uncured novolac/PCL blends, the following results were obtained. First, since the frequency difference between the free hydroxyl absorption and those of the hydrogen bonding species (Δv) is a measure of the average strength of the intermolecular interactions (110,111), the average strength of the hydrogen bond in pure HMTA-cured novolac ($\Delta v = 115$ cm^{-1}) is much lower than that in pure uncured novolac ($\Delta v = 217$ cm^{-1}). Second, the average strength of the hydrogen bond between the hydroxyl groups of novolac and the carbonyl groups of PCL in the cured blends ($\Delta v = 95$ cm^{-1} for the 50/50 HMTA-cured novolac/PCL blend) is also much lower than that in the uncured blends ($\Delta v = 154$ cm^{-1} for the 50/50 uncured novolac/PCL blend). Third, as can be seen from the relative intensities of the absorption bands, the ratio of free hydroxyl groups to the self-associated hydroxyl groups in cured novolac/PCL blends is almost constant and is much higher than that in uncured blends. This phenomenon can be explained as follows: novolac was cured with a high amount of cross-linking agent, and the high cross-link density makes the segmental motion of novolac chains very difficult. Therefore it is difficult to form hydrogen bonds even in the pure cured novolac resin. In other words, the average strength of hydrogen bond in cured samples is very low, and the ratio of

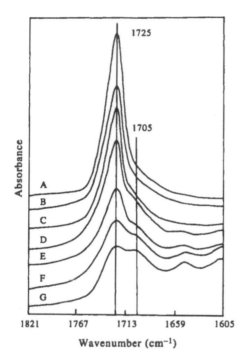

Figure 6 FTIR spectra in the 1605 to 1821 cm^{-1} region of the novolac/PCL blends cured with 15 wt% HMTA. Novolac/PCL: (A) 0/100, (B) 10/90, (C) 20/80, (D) 30/70, (E) 40/60, (F) 50/50, and (G) 70/30. (From Ref. 110.)

nonassociated hydroxyl groups to the self-associated hydroxyl groups in a cured system is larger than that in an uncured system.

Figure 6 shows the FTIR spectra of novolac/PCL blends cured with 15 wt% HMTA in the region of 1605 to 1821 cm^{-1}. The absorption band of non-hydrogen-bonded carbonyl groups stretching vibration of PCL is observed to center at 1725 cm^{-1}. However, with increasing novolac content, a second band at 1705 cm^{-1} appears; this can be attributed to the hydrogen-bonded carbonyl groups stretching vibration of PCL. Compared with the uncured blends, one can find that the average strength of the hydrogen bond between the hydroxyl groups of novolac and the carbonyl groups of PCL in the cured blends ($\Delta v = 20$ cm^{-1} for the HMTA-cured novolac/PCL blend, Fig. 6) is lower than that in the uncured blends ($\Delta v = 23$ cm^{-1} for the uncured novolac/PCL blend). It is also noticed that the relative intensity of the band at 1705 cm^{-1} in the cured blends (Fig. 6) is relatively lower than that at 1702 cm^{-1} for the uncured blends, suggesting that there is much less amount of intermolecular hydrogen bond in the cured novolac/PCL blends than in the uncured novolac/PCL blends.

Blends of novolac and PEO were prepared by solution casting from N,N-dimethylformamide (DMF) (112). We found that PEO is miscible with uncured novolac over the entire composition range, as was shown by the existence of a single composition-dependent T_g. The existence of hydrogen-bonding interaction between PEO and PVPh has been revealed by Coleman and coworkers (109–111). Belfiore and coworkers (113) confirmed this result by a further FTIR study and established the miscibility of PEO/PVPh blends on the basis of DSC results.

Our FTIR studies revealed that hydrogen-bonding interactions exist between the hydroxyl groups of novolac and the ether oxygens of PEO. The relative amount and the average strength of the hydrogen bonds in the blends were higher than those in the pure novolac resin. The curing with 15 wt% HMTA (relative to novolac content) resulted in the disappearance of a detectable T_g in both the neat novolac and the novolac-rich blends, due to the reduced mobility of the novolac chain segments. An analysis of the reduction in T_m and crystallization rate with increasing novolac content revealed that the HMTA-cured blends remained completely miscible. After curing with HMTA, considerable hydrogen-bonding interaction between the components still existed, which is the driving force for the miscibility of the HMTA-cured blends. The relative amount and the average strength of hydrogen bonds in the cured blends were lower than those in the uncured blends.

The complexation interaction of an aqueous phenolic resin with PEO and poly(acrylamide-co-ethylene glycol) was recently studied by Xiao et al. (114). However, we have reported cross-linkable interpolymer complexes of novolac and PEO prepared by mutually mixing ethanol solutions of novolac and PEO (115). The morphology and thermal properties of the complexes before and after curing were investigated by optical microscopy and DSC. We found that the uncured novolac/PEO complexes had a single composition-dependent T_g. The curing with 15 wt% HMTA (relative to novolac content) resulted in the disappearance of T_g for both the neat novolac and the novolac-rich complexes, owing to less mobility of the novolac chain segments. The T_m and crystallization rate of the HMTA-cured novolac/PEO complexes decreased with increasing novolac content, and no T_m was observed for the cured complexes with PEO content less than 50 wt%.

The uncured novolac/PEO complexes exhibited typical 'Maltese cross' birefringent pattern at high-PEO containing compositions, whereas obvious characteristic of dendritic texture was observed for the 30/70 and 40/60 novolac/PEO compositions. The crystalline morphology of the HMTA-cured novolac/PEO complexes at high-PEO containing compositions is similar to that of the uncured complexes.

We have conducted a comparative study on crystallization kinetics of PEO between both the novolac/PEO blends and the novolac/PEO complexes (116). Results of an investigation of isothermal crystallization and thermal behavior were obtained for both the uncured and the HMTA-cured samples. The crystallization process from the melt and the melting behavior of PEO were found to be strongly influenced by factors such as composition, crystallization temperature, complexation, and cross-linking. The dependence of the relative degree of crystallinity on time deviates the Avrami equation (117) at low conversion. The cured complexes display an obvious two-stage crystallization (primary crystallization and crystal perfection), and this is more remarkable at higher crystallization temperature. The inclusion of noncrystallizable components into PEO causes a depression in both the overall crystallization rate and the T_m. The influence of complexation and curing on the overall crystallization rate is rather complicated. In general, complexation and curing results in an increase of the overall crystallization rate. Experimental data on the overall kinetic rate constant K_N were analyzed by means of the kinetic theory (118–120). The surface free energy of folding (σ_e) shows an increase with increasing novolac content for both the uncured blends and the uncured complexes, whereas σ_e displays a maximum with the variation of composition for both the cured blends and the cured complexes. The uncured complexes have larger σ_e and preexponential factor A_o than those of the uncured blends.

The complexation interaction between novolac resin and poly(N-vinyl-2-pyrrolidone) (PVP) in various solvents was studied (121). Interpolymer complexes were formed from methanol, ethanol, 2-butanone, cyclohexanone, and DMF, whereas only miscible blends were obtained from N-methyl-2-pyrrolidone. It has been found that the nature of the solvent assumes a profound influence on the degree of the interpolymer association. The formation of 1/1 and 2/1 ([novolac]/[PVP]) interpolymer complex in DMF solution was shown by measurements of reduced viscosity, conductivity, and clear point. DSC studies revealed a single-phase nature of both the blends and the complexes of uncured novolac/PVP. The complexes had higher T_gs than those of the blends with the same composition, and the strength of the interactions between the components in the complex is much higher than that in the blend. The driving force in the formation of the interpolymer complexes between novolac and PVP is the hydrogen bonding interaction between the hydroxyls of novolac and the proton-accepting groups of PVP. Both the blends and the complexes were cured with 15 wt% HMTA relative to novolac content. The cured novolac/PVP blends were only partially miscible, whereas there still existed remarkable complexation in the cured complexes. FTIR studies revealed that the interaction between the components was rather strong for both the uncured and the cured blends.

IV. UNSATURATED POLYESTER RESIN

In comparison with the large amount of literature relating to epoxy resins, only a few researchers concentrate on unsaturated polyester resins (UPs). The curing kinetics and the mechanism of low-profile behavior are the main topics in this field (122–144). As UPs are a major type of practical thermosetting plastic in industry, systematic investigation of them and their blends is of practical significance. One of the major drawbacks for UPs is their brittleness and poor resistance to crack propagation (145), which has confined the application of UPs to situations where the stress is relatively low and preferably static. Although the technique of blending has solved many practical problems in polymer science and engineering, it cannot develop its great potential in UPs due to the poor solubility between UPs and the blending components.

In order to improve the mechanical properties of cured UP materials, Martuscelli and coworkers (146,147) made an attempt chemically to modify commercial liquid rubbers to enhance their reactivity toward the UP matrix. Mucha (148) investigated blends of PEO with a cross-linked polyester resin (PER) and concluded that the PEO is partially miscible with oligoester and cross-linked polyester resins. As PEO has the potential to form hydrogen-bonding interaction with other proton donor polymers, we inferred that PEO/PER polymer blends can be miscible if the oligoester, i.e., the precursor of PER, is terminated with hydroxyl, and that the toughness of PER can be improved in low PEO content with little loss of rigidity due to the plasticization of PEO. Under the guidance of this idea, a systematic investigation of the thermodynamics, crystallization behavior, and mechanical properties of PEO/PER blends has been conducted. Here, we summarize some results on the miscibility and crystallization of an unsaturated polyester resin with PEO and PCL (149–153).

The miscibility and thermal properties of PEO/oligoester resin (OER) blends and PEO/cross-linked polyester resin (PER) blends were studied by DSC. The DSC thermograms of all the quenched blend samples gave single composition-dependent T_gs. For both PEO/OER blends and PEO/PER blends, the variation of T_g with composition fits the Fox equation (84) quite well, indicating that the two systems are miscible in the amorphous state at all compositions.

The effect of a quenching process on the crystallization behavior of PEO for these two systems was investigated. It was found that the quenching process is a greater hindrance to the crystallization of PEO/OER blends than to that of PEO/PER blends.

With decreasing PEO content, the variation of crystallinity degree, Xc(PEO), for PEO/PER quenched samples, PEO/OER as-cast samples, and PEO/OER quenched samples shows the following tendency: PEO/OER quenched samples > PEO/PER quenched samples > PEO/OER as-cast samples.

The interaction parameter χ_{12} for PEO/OER blends and that for PEO/PER blends are found to be -1.29 and -2.01, respectively, implying that both PEO/OER blends and PEO/PER blends are miscible in the amorphous state.

For miscible blends having one crystallizable component, the interaction parameter χ_{12} for the system can be estimated from the T_m depression. For the

PEO/PER blends, the molecular weights of both components are adequately large, but enthalpic contribution to the T_m depression is not negligible (85). The interaction parameter χ_{12} was found to be -2.01 by the Nishi–Wang plot (154). The negative value of χ_{12} confirmed that PEO/PER blends are miscible in the molten state.

In the PEO/OER blends, both enthalpic and entropic effects contribute to the T_m depression owing to the small molecular weight of the OER; the interaction parameter χ_{12} cannot be simply obtained by the conventional Nishi–Wang plot (154). Thus the Flory–Huggins theory (155) is employed, which can be expressed as

$$\frac{\Delta H_{2U} V_{1U}}{(RV_{2U})(1/T_m - 1/T_m^0)} + \ln \frac{V_2}{m_2} + \left(\frac{1}{m_2} - \frac{1}{m_1}\right) V_1 = -\chi_{12} V_1^2 \tag{1}$$

where T_m^0 and T_m are the melting points of pure crystallizable component and that in the blend, respectively; V_{1U}, V_{2U} are the molar volumes of the repeat units of OER and PEO, respectively; m_1, V_1 and m_2, V_2 are the degree of polymerization and the volume fractions of the noncrystallizable (OER) and crystallizable (PEO) polymers, respectively. ΔH_{2U} is the heat of fusion per mole 100% crystalline PEO. The details of the evaluation will be published elsewhere (146); here we show a plot of Eq. (1) in Fig. 7 for the PEO/OER blends. A least-squares fit of the experimental data yields a straight line with an intercept of 0.17 and a slope of 1.29, which gives $\chi_{12} = -1.29$. The negative value of χ_{12} (-1.29) strongly indicates that in the melt at T_m, PEO and OER are miscible. The relative small value of the intercept (0.17) implies that this approach is reasonable.

FTIR studies revealed that there is hydrogen-bonding interaction between the two components in PEO/OER blends and in PEO/PER blends. Its strength is approximately as strong as the self-association of hydroxyl groups in either the pure OER or the pure PER. The hydrogen-bonding interaction plays an important role in the miscibility of PEO/OER blends and PEO/PER blends.

The crystallization kinetics and crystalline morphology of PEO in PEO/PER blends was remarkably affected by cross-linking. The curing of amorphous component OER increases the crystallization rate of PEO. It was found that the overall crystallization rate of PEO in PEO/PER blends is larger than that in PEO/OER blends at the crystallization temperature investigated (41°C), which was considered to be the result of nucleation controlling mechanism.

With decreasing PEO content, the regular shape of PEO spherulites in PEO/OER blends gradually turns irregular, which is the result of the inclusion of amorphous OER and uncrystallized PEO between lamellae of spherulites. In PEO/PER blends, birefringent spherulites of PEO disappear gradually and dendritic structures are substituted. They are concerned with the mechanism of spherulitic nucleation and growth in cross-linked polymer blends. The relatively more difficult diffusion process of PEO due to cross-linked PER compared with that in PEO/OER blends makes even the growing of PEO spherulites impossible

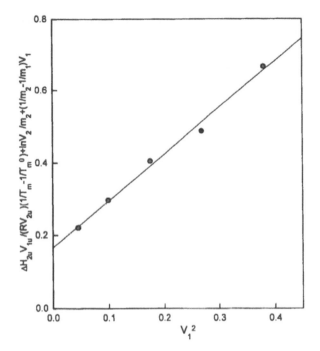

Figure 7 T_m depression analysis by using Flory–Huggins theory to obtain interaction parameter χ_{12} for the PEO/OER blends.

in PEO/PER blends, especially at high PER content. Dendrites with many edges and corners give a favorable morphology in this cross-linked system. Raising the crystallization temperature favors the formation of dendrites in PEO/PER blends.

It was also found that the addition of PEO had a great effect on the mechanical properties of cross-linked PER (152). The elongation at break first increased greatly and then decreased slightly, while the tensile modulus and the tensile strength first decreased and then increased slightly with increasing PEO content in the blends. The variation of tensile properties was considered to be due to the plasticization effect of PEO in the blends. The impact strength remains almost unchanged with PEO content up to 40 wt%. Thermal stability for PER/PEO blends did not dramatically decrease until the blend containing 30 wt% PEO.

PCL was found to be miscible with OER, as indicated by the appearance of a single T_g in each blend (153). The interaction parameter χ_{12} between PCL and OER was evaluated to be -0.48 by using the Flory–Huggins theory (155). The entropy contribution to the miscibility was found to be not negligible in the PCL/OER blends. However, phase separation occurred in the blends after crosslinking, and PCL was judged to be only partially miscible with PER.

FTIR studies revealed that hydrogen-bonding interaction is an important driving force for both the miscibility of the PCL/OER blends and the partial miscibility of the PCL/PER blends. The average strength of the hydrogen bond between PCL and either OER or PER is stronger than that between hydroxyls in either of the pure OER and PER. The average strength of hydrogen bond between PCL and PER is higher than that between PCL and OER.

The morphological studies revealed that only birefringent spherulites can be seen in the uncured PCL/OER blends. However, a distinct pattern of extinction rings is apparent in the partially miscible PCL/PER blends, which was considered as a result of the rejection of PER in the PCL-rich phase from crystalline lamellae to the interlamellar amorphous region.

The overall crystallization rate of PCL in the cross-linked PCL/PER blends was greatly enhanced due to the phase separation. With decreasing PCL concentration, the kinetic rate constant K_N decreased sharply for the PCL/OER blends, whereas it decreased evenly for the cross-linked PCL/PER blends. In the miscible PCL/OER blends, amorphous OER was incorporated in the interlamellar or interfibrillar regions of PCL spherulites, which greatly hindered the crystallization of PCL. However, in the partially miscible PCL/PER blends, the small concentration of PER in the PCL rich phase showed relatively little hindrance to the crystallization of PCL. The mechanism of nucleation and geometry of the growing PCL crystals was not affected by the incorporation of OER but changed gradually with the addition of PER; the cross-linking of OER had great influence on the crystallization of PCL in the blends.

V. CONCLUDING REMARKS

As this chapter has shown, there has been significant progress in understanding the miscibility, crystallization, and related properties in blends of thermosetting resins with linear polymers during the past decade.

In the blending of the low-molecular-weight liquids of elastomer modifiers with the monomers of thermoset resins, the negative entropy contribution is important to the free energy of mixing. The increase in molecular weight during curing reduces the entropy of mixing. This, together with the positive (endothermic) enthalpy, increases the likelihood of a two-phase morphology.

Initially miscible blends containing one cross-linkable and one linear component could remain miscible in the highly cross-linked cases; a semi-IPN with homogeneous phase structure may form. On the other hand, the resulting increase in molecular weight causes a decrease in the configurational entropy of mixing as the cure proceeds; as a result, curing may cause phase separation in an initially miscible blend. The phase separation process is governed by not only thermodynamic factors but also kinetic factors. Curing reaction and phase separation are competitive. If gelation occurs prior to phase separation, no domains appear.

In thermosetting polymer blends where the linear component is crystallizable, the miscibility and crystallization is greatly influenced by cross-linking. Furthermore, crystallization is greatly affected by the miscibility and phase structures of the resulted blends.

In thermoset/thermoplastic blends, the significance of entropy factors is greatly diminished, and thus the miscibility has been considered to be due primarily to the enthalpic contributions. The occurrence of miscibility or partial miscibility in thermosetting polymer blends where one component is highly cross-linked requires a negative (exothermic) enthalpy. Hydrogen-bonding interaction is an important driving force for the miscibility and the partial miscibility of thermosetting polymer blends.

The influence of curing on the overall crystallization rate is rather complicated. In general, curing results in a considerable increase in the overall crystallization rate and hence dramatically influences the mechanism of nucleation and growth of the crystalline component in the blends.

The thermal and mechanical properties of thermosetting polymer blends are also dependent on the miscibility and phase morphology. In particular, the toughening effect is greatly dependent on the phase structures of the blends.

ACKNOWLEDGMENT

The author wishes to express his appreciation to the National Natural Science Foundation of China for awarding a 'Premier Grant' for Outstanding Young Investigators (No. 59525307).

REFERENCES

1. C. K. Riew and J. K. Gillham, eds. Rubber-Modified Thermoset Resins. Adv. Chem. Ser. No. 208. Washington DC: ACS, 1984.
2. C. K. Riew, ed. Rubber-Toughened Plastics. Adv. Chem. Ser. No. 222. Washington DC: ACS, 1989.
3. C. K. Riew and A. J. Kinloch, eds. Toughened Plastics I: Science and Engineering. Adv. Chem. Ser. No. 233. Washington DC: ACS, 1993.
4. R. J. Young. Fracture of thermosetting resins. In: E. H. Andrews, ed. Developments in Polymer Fracture-1. London: Applied Science Publishers, 1979, Chap. 6.
5. C. B. Bucknall. Toughened Plastics. London: Applied Science Publishers, 1977.
6. A. J. Kinloch. Multiphase thermosetting polymers. In: D. J. Walsh, J. S. Higgins, and A. Maconnochie, eds. Polymer Blends and Mixtures. Dordrecht: Martinus Nijhoff, 1985, pp. 393–412.

7. H. Lee and N. Neville. Handbook of Epoxy Resins. New York: McGraw-Hill, 1967.
8. W. G. Potter. Epoxide Resins. New York: Springer-Verlag, 1970.
9. C. A. May and G. Y. Tanaka, eds. Epoxy Resin Chemistry and Technology. New York: Marcel Dekker, 1973.
10. R. S. Bauer, ed. Epoxy Resin Chemistry. Adv. Chem. Ser. No. 114. Washington DC: ACS, 1979.
11. R. S. Bauer, ed. Epoxy Resin Chemistry II. Adv. Chem. Ser. No 201. Washington DC: ACS, 1983.
12. E. H. Rowe, A. R. Siebert, and R. S. Drake. Toughening thermosets with liquid butadiene/acrylonitrile polymers. Modern Plastics 47:110, 1970.
13. J. M. Sultan and F. McGarry. Effect of rubber particle size on deformation mechanism in glassy epoxy. Polym. Eng. Sci. 13:29, 1973.
14. R. S. Raghava. Role of matrix-particle interface adhesion on fracture toughness of dual phase epoxy-polyethersulfone blends. J. Polym. Sci., Polym. Phys. Ed. 25:1017, 1987.
15. J. B. Enns and J. K. Gillham. Time-temperature-transformation (TTT) cure diagram: modeling the cure behavior of thermosets. J. Appl. Polym. Sci. 28:2567, 1985.
16. R. A. Pearson and A. F. Yee. Toughening mechanisms in elastomer-modified epoxies. Part 3. The effect of cross-link density. J. Mater. Sci. 24:2571, 1989.
17. D. Verchère, H. Sautereau, J. P. Pascault, S. M. Moschiar, C. C. Riccardi, and R. J. J. Williams. Rubber-modified epoxies. I. Influence of carboxyl-terminated butadiene-acrylonitrile random copolymers (CTBN) on the polymerization and phase separation processes. J. Appl. Polym. Sci. 41:467, 1990.
18. D. Verchère, J. P. Pascault, H. Sautereau, S. M. Moschiar, C. C. Riccardi, and R. J. J. Williams. Rubber-modified epoxies. II. Influence of the cure schedule and rubber concentration on the generated morphology. J. Appl. Polym. Sci. 42:701, 1991.
19. S. M. Moschiar, C. C. Riccardi, R. J. J. Williams, D. Verchère, H. Sautereau, and J. P. Pascault. Rubber-modified epoxies. III. Analysis of experimental trends through a phase separation model. J. Appl. Polym. Sci. 42:717, 1991.
20. A. C. Grillet, J. Galy, and J. P. Pascault. Influence of a two-step process and of different cure schedules on the generated morphology of a rubber-modified epoxy system based on aromatic diamines. Polymer 33:34, 1992.
21. S. Zheng, H. Wang, Q. Dai, X. Luo, D. Ma, and K. Wang, Morphology and structure of organosilicon polymer-modified epoxy resins. Macromol. Chem. Phys. 196:269, 1995.
22. S. Visconti and R. H. Marchessault. Macromolecules 7:913, 1974.
23. D. Verchère, H. Sautereau, J. P. Pascault, S. M. Moschiar, C. C. Riccardi, and R. J. J. Williams. Rubber-modified epoxies: analysis of the phase-separation process. In: C. K. Riew and A. J. Kinloch, eds. Toughened Plastics I: Science and Engineering. Adv. Chem. Ser. No. 233. Washington DC: ACS, 1993, pp. 335–363.
24. D. Verchère, H. Sautereau, J. P. Pascault, S. M. Moschiar, C. C. Riccardi, and R. J. J. Williams. Miscibility of epoxy monomers with carboxyl-terminated butadiene-acrylonitrile random copolymers. Polymer 30:107, 1989.
25. A. Vazquez, A. J. Rojas, H. E. Adabbo, J. Borrajo, and R. J. J. Williams. Rubber-modified thermosets: prediction of particle size distribution of dispersed domains. Polymer 28:1156, 1987.

26. L. T. Manzione, J. K. Gillham, and C. A. McPherson. Rubber-modified epoxies. I. Transitions and morphology. J. Appl. Polym. Sci. 26:889, 1981.

27. L. T. Manzione, J. K. Gillham, and C. A. McPherson. Rubber-modified epoxies. II. Morphology and mechanical properties. J. Appl. Polym. Sci. 26:907, 1981.

28. C. B. Bucknall and I. K. Partridge. Phase separation in epoxy resins containing polyethersulfone. Polym. Eng. Sci. 24:639, 1983.

29. C. B. Bucknall and I. K. Partridge. Phase separation in crosslinked resins containing polymeric modifiers. Polym. Eng. Sci. 26:54, 1986.

30. W. H. Lee. Toughened polymers. In: M. J. Folkes, ed. Polymer Blends and Alloys, London: Chapman and Hall, 1993, pp. 163–194.

31. J. L. Hedrick, I. Yilgor, M. Jurek, J. C. Hedrick, G. L. Wilkes, and J. E. McGrath. Chemical modification of matrix resin networks with engineering thermoplastics. 1. Synthesis, morphology, physical behavior and toughening mechanisms of poly(arylene ether sulfone) modified epoxy networks. Polymer 32:2020, 1991.

32. J. A. Cerere and J. E. McGrath. Morphology and properties of amine terminated poly(arylether ketone) and poly(aryl ether sulphone) modified epoxy resin systems. Am. Chem. Soc. Div. Polym. Chem. Polym. Prepr. 27:299, 1986.

33. T. Iijima, T. Tochimoto, and M. Tomoi, Modification of epoxy resins with poly(aryl ether ketone)s. J. Appl. Polym. Sci. 43:1685, 1991.

34. Z. Fu and Y. Sun. Epoxy resin toughened by thermoplastics. Chinese J. Polym. Sci. 7:367, 1989.

35. C. B. Bucknall and A. H. Gilbert. Toughening of tetrafunctional epoxy resins using polyetherimide. Polymer 30:213, 1989.

36. S. C. Kim and H. R. Brown. Impact-modified epoxy resin with glassy second component. J. Mater. Sci. 22:2589, 1987.

37. M. C. Chen, D. J. Hourston, and W. B. Sun. Morphology and fracture behavior of an epoxy resin-bisphenol-A polycarbonate blend. Eur. Polym. J. 28:1471, 1992.

38. C. B. Bucknall, I. K. Partridge, L. Jayle, I. Nozue, A. Fernyhough, and J. N. Hay. Toughenability studies in ternary blends of epoxy, rubber and polycarbonate. Am. Chem. Soc. Div. Polym. Chem. Polym. Prepr. 33:378, 1992.

39. M. C. Chen, D. J. Hourston, F. U. Schafer, and T. N. Huckerby. Miscibility and fracture behavior of epoxy resin-nitrated polyetherimide blends. Polymer 36:3287, 1995.

40. S. Zheng, Y. Hu, Q. Guo, and J. Wei. Miscibility, morphology and fracture toughness of epoxy resin/poly(vinyl acetate) blends. Colloid Polym. Sci. 274:410, 1996.

41. S. Zheng, J. Wang, Q. Guo, J. Wei, and J. Li. Miscibility, morphology and fracture toughness of epoxy resin/poly(styrene-co-acrylonitrile) blends. Polymer 37:4667, 1996.

42. R. A. Pearson. Toughening epoxies using rigid thermoplastic particles: a review. In: C. K. Riew and A. J. Kinloch, eds. Toughened Plastics I: Science and Engineering. Adv. Chem. Ser. No. 233. Washington DC: ACS, 1993, pp. 405–425.

43. R. A. Pearson and A. F. Yee. Toughening mechanisms in thermoplastics-modified epoxies: 1. Modification using poly(phenylene oxide). Polymer 34:3658, 1993.

44. C. B. Bucknall, C. M. Gomez, and I. Quintard. Phase separation from solutions of poly(ether sulfone) in epoxy resins. Polymer 35:353, 1994.

45. T. Inoue. Reaction-induced phase decomposition in polymer blends. Prog. Polym. Sci. 20:119, 1995.
46. K. Yamanaka and T. Inoue. Structure development in epoxy resin modified with poly(ether sulphone). Polymer 30:662, 1989.
47. B. S. Kim, T. Chiba, and T. Inoue. Phase separation and apparent phase dissolution during cure process of thermoset/thermoplastic blend. Polymer 36:67, 1995.
48. Q. Guo, Phase behaviour in epoxy resin containing phenolphthalein poly(ether ether sulphone). Polymer 34:70, 1993.
49. Z. Zhong, S. Zheng, J. Huang, X. Cheng, Q. Guo, and W. Zhu. Phase behavior and fracture mechanics of epoxy resin containing phenolphthalein poly(ether ether sulfone). J. Mater. Sci., submitted.
50. Q. Guo. Effect of curing agent on the phase behaviour of epoxy resin/phenoxy blends. Polymer 36:4753, 1995.
51. P. Huang, S. Zheng, J. Huang, Q. Guo, and W. Zhu. Miscibility and mechanical properties of epoxy resin/polysulfone blends. Polymer 38:5565, 1997.
52. Z. Zhong, S. Zheng, J. Huang, X. Cheng, Q. Guo, and J. Wei. Phase behavior and mechanical properties of epoxy resin containing phenolphthalein poly(ether ether ketone). Polymer 39:1075, 1998.
53. M. C. Chen, D. J. Hourston, and W. B. Sun. The morphology and fracture behavior of a miscible epoxy resin-polyetherimide blend. Eur. Polym. J. 31:199, 1995.
54. L. Jayle, C. B. Bucknall, I. K. Partridge, J. N. Hay, A. Fernyhough, and I. Nozue. Ternary blends of epoxy, rubber and polycarbonate blends: phase behavior, mechanical properties and chemical interactions. Polymer 37:1897, 1996.
55. Q. Guo, J. Huang, L. Ge, and Z. Feng. Phase separation in anhydride-cured epoxy resin containing phenolphthalein poly(ether ether ketone). Eur. Polym. J. 28:405, 1992.
56. Q. Guo, J. Huang, B. Li, T. Chen, H. Zhang, and Z. Feng. Blends of phenolphthalein poly(ether ether ketone) with phenoxy and epoxy resin. Polymer 32:58, 1991.
57. K. C. Teng and F. C. Chang. Single-phase and multiple-phase thermoplastic/thermoset polyblends: 1. Kinetics and mechanisms of phenoxy/epoxy blends. Polymer 34:4291, 1993.
58. K. C. Teng and F. C. Chang. Single-phase and multiple-phase thermoplastic/thermoset polyblends: 2. Morphologies and mechanical properties of phenoxy/epoxy blends. Polymer 37:2385, 1996.
59. M. E. Nichols and R. E. Robertson. Preparation of small poly(butylene terephthalate) spheres by crystallization from solution. J. Polym. Sci., Polym. Phys. Ed. 32:573, 1994.
60. M. E. Nichols and R. E. Robertson. Crystallization and thermoreversible gelation of poly(butylene terephthalate)-epoxy mixtures. J. Polym. Sci., Polym. Phys. Ed. 32:1607, 1994.
61. P. A. Oyanguren, P. M. Frontinie, R. J. J. Williams, E. Girard-Reydet, and J. P. Pascault. Reaction-induced phase separation in poly(butylene terephthalate)-epoxy systems: 1. Conversion-temperature transformation diagrams. Polymer 37:3079, 1996.
62. P. A. Oyanguren, P. M. Frontinie, R. J. J. Williams, G. Vigier, and J. P. Pascault. Reaction-induced phase separation in poly(butylene terephthalate)-epoxy systems: 2. Morphologies generated and resulting properties. Polymer 37:3087, 1996.

63. P. Huang, Z. Zhong, S. Zheng, and Q. Guo. Miscibility and interchange reactions in blends of bisphenol-A-type epoxy resin and poly(ethylene terephthalate). J. Appl. Polym. Sci., in proof, 1998.

64. Y. Yu and J. P. Bell. Chemistry of epoxide-polycarbonate copolymer network networks. J. Polym. Sci., Polym. Chem. Ed. 26:247, 1988.

65. M. Abbate, E. Martuscelli, P. Mosuto, G. Ragosta, and G. Scarini. Toughening of a highly cross-linked epoxy resin by reactive blending with bisphenol A polycarbonate. I. FTIR spectroscopy. J. Polym. Sci., Polym. Phys. Ed. 32:395, 1994.

66. V. Di Liello, E. Martuscelli, P. Mosuto, G. Ragosta, and G. Scarini. Toughening of a highly cross-linked epoxy resin by reactive blending with bisphenol A polycarbonate. II. Yield and fracture behavior. J. Polym. Sci., Polym. Phys. Ed. 32:409, 1994.

67. E. M. Woo and M. N. Wu. Chemical interactions in blends of bisphenol-A polycarbonate with tetraglycidyl-4,4'-diaminodiphenylmethane epoxy. Macromolecules 28:6779, 1995.

68. E. M. Woo and M. N. Wu. Blends of a diglycidylether epoxy with bisphenol-A polycarbonate or poly(methyl methacrylate): case of miscibility with or without specific interactions. Polymer 37:2485, 1996.

69. M. S. Li, C. C. M. Ma, J. L. Chen, M. L. Lin, and F. C. Chang. Epoxy-polycarbonate blends catalyzed by a tertiary amine. 1. Mechanism of transesterification and cyclization. Macromolecules 29:499, 1996.

70. M. Rong and H. Zeng. Polycarbonate-epoxy semi-interpenetrating polymer network: 1. Preparation, interaction and curing behavior. Polymer 37:2525, 1996.

71. M. Rong and H. Zeng. Polycarbonate-epoxy semi-interpenetrating polymer network: 2. Phase separation and morphology. Polymer 38:269, 1997.

72. M. M. Coleman, C. J. Serman, and P. C. Painter. Effect of cross-linking on the degree of molecular level mixing in a polymer blend. Macromolecules 20:226, 1987.

73. H. I. Kim, E. M. Pearce, and T. K. Kwei. Miscibility control by hydrogen bonding in polymer blends and interpenetrating networks. Macromolecules 22:3374, 1989.

74. K. Binder and H. L. Frisch. Phase stability of weakly crosslinked interpenetrating polymer networks. J. Chem. Phys. 81:2126, 1984.

75. A. A. Donatelli, L. H. Sperling, and D. A. Thomas. A semiempirical derivation of phase domain size in interpenetrating polymer networks. J. Appl. Polym. Sci. 21:1189, 1977.

76. N. Clarke and T. C. B. McLeish. Kinetics of concentration fluctuations and spinodal decomposition in star/star and star/linear polymer blends. J. Chem. Phys. 99:10034, 1993.

77. N. Clarke, T. C. B. McLeish, and S. D. Jenkins. Phase behavior of linear/branched polymer blends. Macromolecules 28:4650, 1995.

78. F. Tanaka and W. H. Stockmayer. Thermoreversible gelation with junctions of variable multiplicity. Macromolecules 27:3943, 1994.

79. W. H. Jo and M. B. Ko. Effect of reactivity on cure and phase separation behavior in epoxy resin modified with thermoplastic polymers: a Monte Carlo simulation approach. Macromolecules 27:7815, 1994.

80. A. Noshay and L. M. Robeson. Epoxy/modifier block copolymers. J. Polym. Sci., Polym. Chem. Ed. 12:689, 1974.

81. J. N. Clark, H. J. Daly, and A. Garton. Hydrogen bonding in epoxy resin/poly(ε-caprolactone) blends. J. Appl. Polym. Sci. 9:3381, 1984.

82. Q. Guo, X. Peng, and Z. Wang. Phase separation during crosslinking of epoxy resin/poly(ethylene oxide) blends. Polym. Bull. 21:593, 1989.

83. Q. Guo, X. Peng, and Z. Wang. The miscibility and morphology of epoxy resin/poly(ethylene oxide) blends. Polymer 32:53, 1991.

84. T. G. Fox. Influence of diluent and of copolymer composition on the glass temperature of a polymer system. Bull. Am. Phys. Soc. 1:123, 1956.

85. R. L. Imaken, D. R. Paul, and J. W. Barlow. Transition behavior of poly(vinylidene fluoride)/poly(ethyl methacrylate) blends. Polym. Eng. Sci. 16:593, 1976.

86. L. M. Robeson, W. F. Hale, and C. N. Merriam. Miscibility of the poly(hydroxy ether) of bispheno A with water-soluble polyethers. Macromolecules 14:1644, 1981.

87. M. Iriarte, J. I. Iribarren, A. Etxeberria, and J. J. Iruin. Crystallization and melting behavior of poly(hydroxyether of bisphenol A)/poly(ethylene oxide) blends. Polymer 30:1160, 1990.

88. M. Iriarte, E. Espi, A. Etxeberria, M. Valero, M. J. Fernandez-Berridi, and J. J. Iruin. Miscible blends of poly(ethylene oxide) and the poly(hydroxyether) of bisphenol A (phenoxy). Macromolecules 24:5546, 1991.

89. M. M. Coleman and E. J. Moskala. FTIR studies of polymer blends containing the poly(hydroxy ether of bisphenol A) and poly(ε-caprolactone). Polymer 24:251, 1983.

90. X. Luo, S. Zheng, N. Zhang, and D. Ma. Miscibility of epoxy resins/poly(ethylene oxide) blends cured with phthalic anhydride. Polymer 35:2619, 1994.

91. S. Zheng, N. Zhang, X. Luo, and D. Ma. Epoxy resin/poly(ethylene oxide) blends cured with aromatic amine. Polymer 36:3609, 1995.

92. S. Zheng, H. Wang, X. Luo, N. Zhang, D. Ma, C. Zhu, and J. Hu. Ultrasonic behavior of epoxy resins/poly(ethylene oxide) blends cured with phthalic anhydride. Chinese J. Polym. Sci. 13:20, 1995.

93. X. Luo, S. Zheng, D. Ma, and K. Hu. Mechanical relaxation and intermolecular interaction in epoxy resins/poly(ethylene oxide) blend cured with phthalic anhydride. Chinese J. Polym. Sci. 13:144, 1995.

94. M. A. Hillmyer, P. M. Lipic, D. A. Hajduk, K. Almdal, and F. S. Bates. Self-assembly and polymerization of epoxy-resin amphiphilic block-copolymer nanocomposites. J. Am. Chem. Soc. 119:2749, 1997.

95. Q. Guo et al. Miscibility in thermosetting polymer blends of epoxy resin and poly(ε-caprolactone) characterized by DSC, FTIR and solid-state NMR. Unpublished work.

96. Q. Guo et al. Morphology and crystallization kinetics in thermosetting polymer blends of epoxy resin and poly(ε-caprolactone). Unpublished work.

97. S. Zheng. In situ polymerization preparation, phase behavior, properties and solid-state NMR characterization of polymer blends. Ph.D. Diss., University of Science and Technology of China, Hefei, 1997.

98. A. Knop and L. A. Pilato. Phenolic Resins. Springer-Verlag, Heidelberg, 1979.

99. A. Knop and W. Scheib. Chemistry and Application of Phenolic Resins. Springer-Verlag, Heidelberg, 1985.

100. S. R. Fahrenholtz and T. K. Kwei. Compatibility of polymer mixtures containing novolac resins. Macromolecules 14:1076, 1981.

101. T. K. Kwei. The effect of hydrogen bonding on the glass transition temperatures of polymer mixtures. J. Polym. Sci., Polym. Lett. Ed. 22:307, 1984.

102. J. R. Pennacchia, E. M. Pearce, T. K. Kwei, B. J. Bulkin, and J. P. Chen. Compatibility of substituted phenol condensation resins with poly(methyl methacrylate). Macromolecules 19:973, 1986.

103. A. K. Kalkar and N. K. Roy. Thermal and dynamic mechanical properties of polycarbonate/poly(p-t-butylphenol formaldehyde) blend films. Eur. Polym. J. 29:1391, 1993.

104. N. Mekhilef and P. Hadjiandreou. Miscibility behavior of ethylene vinyl acetate/novolac blends. Polymer 36:2165, 1995.

105. T. P. Yang, E. M. Pearce, and T. K. Kwei. Complexation of poly(N,N-dimethylacrylamide) and phenol-formaldehyde resins. Macromolecules 22:1813, 1989.

106. P. Lin, C. Clash, E. M. Pearce, and T. K. Kwei. Solubility and miscibility of poly(ethyl oxazoline). J. Polym. Sci., Polym. Phys. Ed. 26:603, 1988.

107. X. Zhang and D. H. Solomon. Phase structures of hexamine cross-linked novolac blends. 1. Blends with poly(methyl methacrylate). Macromolecules 27:919, 1994.

108. Z. Zhong and Q. Guo. The miscibility and morphology of hexamine cross-linked novolac/poly(ε-caprolactone) blends. Polymer 38:279, 1997.

109. E. J. Moskala, D. F. Varnell, and M. M. Coleman. Concerning the miscibility of poly(vinyl phenol) blends—FTIR study. Polymer 26:228, 1985.

110. M. M. Coleman, J. F. Graf, and P. C. Painter. Specific Interactions and The Miscibility of Polymer Blends. Lancaster, PA: Technomic, 1991.

111. M. M. Coleman and P. C. Painter. Hydrogen bonded polymer blends. Prog. Polym. Sci. 20:1, 1995.

112. Z. Zhong and Q. Guo. Miscibility and morphology of thermosetting polymer blends of novolac resin with poly(ethylene oxide). Polymer 39:517, 1998.

113. C. Qin, A. T. N. Pires, and L. A. Belfiore. Morphological and physicochemical interactions in semicrystalline polymer-polymer blends. Polym. Commun. 31:177, 1990.

114. H. Xiao, R. Pelton, and A. Hamielec. The association of aqueous phenolic resin with poly(ethylene oxide) and poly(acrylamide-co-ethylene glycol). J. Polym. Sci., Polym. Chem. Ed. 33:2605, 1995.

115. Z. Zhong and Q. Guo. Crosslinkable interpolymer complexes of novolac resin and poly(ethylene oxide). J. Polym. Sci., Polym. Chem. Ed. 36:401, 1998.

116. Z. Zhong and Q. Guo. Crystallization kinetics of thermosetting polymer blends and hexamine-crosslinked complexes of novolac resin and poly(ethylene oxide). Polymer, submitted.

117. M. Avrami. Kinetics of phase change. I. General theory. J. Chem. Phys. 7:1103, 1939.

118. J. D. Hoffman. Soc. Plast. Eng. Trans. 4:315, 1960.

119. J. D. Hoffman and J. J. Weeks. Melting process and the equilibrium melting temperature of polychlorotrifluoroethylene. J. Res. Natl. Bur. Stand. Sect. A 66:13, 1962.

120. J. D. Hoffman. Region III crystallization in melt-crystallized polymers: the variable cluster model of chain folding. Polymer 24:3, 1983.
121. Z. Zhong and Q. Guo. Interpolymer complexes and miscible blends of poly(N-vinyl-2-pyrrolidone) with novolac resin and the effect of crosslinking on the related behaviors. Polym. Int. 41:315, 1996.
122. K. E. Atkins. Low-profile behavior. In: D. R. Paul and S. Newman, eds. Polymer Blends. New York: Academic Press, 1978, Vol. 2, pp. 391–414.
123. E. J. Bartkus and C. H. Kroekel. Low shrinkage reinforced polyester systems. Appl. Polym. Symp. 15:113, 1970.
124. V. A. Pattison, R. R. Hindersinn, and W. T. Schwartz. Mechanism of low profile behavior in unsaturated polyester systems. J. Appl. Polym. Sci. 18:2763, 1974.
125. V. A. Pattison, R. R. Hindersinn, and W. T. Schwartz. Mechanism of low profile behavior in single-phase unsaturated polyester systems. J. Appl. Polym. Sci. 19:3045, 1975.
126. A. Siegmann, M. Narkis, J. Kost, and A. T. DiBenedetto. Mechanism of low-profile behavior in unsaturated polyester systems. Inter. J. Polym. Mater. 6:217, 1978.
127. K. Wailem and C. D. Han. Chemorheology of thermosetting resins. III. Effect of low profile additive on the chemorheology and curing kinetics of unsaturated polyester resin. J. Appl. Polym. Sci. 28:3207, 1983.
128. S. C. Ma, H. L. Lin, and T. L. Yu. Glass transition temperature, free volume, and curing kinetics of unsaturated polyester. Polym. J. 25:897, 1983.
129. K. W. Lem and C. D. Han. Thermokinetics of unsaturated polyester and vinyl ester resins. Polym. Eng. Sci. 24:175, 1984.
130. C. B. Bucknall, P. Davies, and I. K. Partridge. Phase separation in styrenated polyester resin containing a poly(vinyl acetate) low-profile additive. Polymer 26:109, 1985.
131. L. Suspene, D. Fourquier, and Y. Yang. Application of phase diagrams in the curing of unsaturated polyester resins with low-profile additives. Polymer 32:1593, 1991.
132. Y. J. Huang and C. C. Sun. Effects of poly(vinyl acetate) and poly(methyl methacrylate) low-profile additives on the curing of unsaturated polyester resins: rheokinetics and morphological changes up to gelation. Polymer 35:2397, 1994.
133. B. Sun and T. L. Yu. Effects of low-profile additives on the curing reaction of unsaturated polyester resins. J. Appl. Polym. Sci. 57:7, 1995.
134. B. Sun and T. L. Yu. Study of microgelation of unsaturated polyester resins in presence of low-profile additives by dynamic light scattering. Macromol. Chem. Phys. 197:275, 1996.
135. M. Kinkelaar, B. Wang, and L. J. Lee. Shrinkage behavior of low-profile unsaturated polyester resins. Polymer 35:3011, 1994.
136. E. L. Rodriguez. The effect of reactive monomers and functional polymers on the mechanical properties of an unsaturated polyester resin. Polym. Eng. Sci. 33:115, 1993.
137. L. Suspene and J. P. Pascault. Incompatible ternary blends on unsaturated polyester resins. I. Phase diagrams. J. Appl. Polym. Sci. 41:2665, 1990.
138. M. Mucha. Morphology and optical properties of liquid crystals embedded in polyester resin matrix. Colloid Polym. Sci. 269:1111, 1991.

139. M. S. Lin, R. J. Chang, T. Yang, and Y. F. Shin. Kinetic study on simultaneous interpenetrating polymer network formation of epoxy resin and unsaturated polyester. J. Appl. Polym. Sci. 55:1607, 1995.

140. Y. C. Chou and L. J. Lee. Mechanical properties of polyurethane-unsaturated polyester interpenetrating polymer networks. Polym. Eng. Sci. 35:976, 1995.

141. Y. C. Chou and L. J. Lee. Reaction-induced phase separation during the formation of a polyurethane-unsaturated polyester interpenetrating polymer network. Polym. Eng. Sci. 34:1239, 1994.

142. C. B. Bucknall, I. K. Partridge, and M. J. Phillips. Morphology and properties of thermoset blends made from unsaturated polyester resin, poly(vinyl acetate) and styrene. Polymer 32:786, 1991.

143. G. A. Crosbie and M. G. Phillips. Toughening of polyester resins by rubber modification. Part 1. Mechanical properties. J. Mater. Sci. 20:182, 1985.

144. G. A. Crosbie and M. G. Phillips. Toughening of polyester resins by rubber modification. Part 2. Microstructures. J. Mater. Sci. 20:563, 1985.

145. C. K. Riew, E. H. Rowe, and A. R. Siebert. Rubber toughened thermosets. In: R. D. Deanin and A. M. Crugnola, eds. Toughness and Brittleness of Plastics. Adv. Chem. Ser. No. 154. Washington DC: ACS, 1976, pp. 326–343.

146. E. Martuscelli, P. Musto, G. Ragosta, G. Scarinzi, and E. Bertotti. Reactive rubbers as toughening agents for thermoset polyester resins. Molecular analysis by FTIR and fracture behavior of the resulting blends. J. Polym. Sci., Polym. Phys. Ed. 31:619, 1993.

147. M. Abbate, E. Martuscelli, P. Musto, G. Ragosta, and G. Scarinzi. Maleated polyisobutylene: a novel toughener for unsaturated polyester resins. J. Appl. Polym. Sci. 58:1825, 1995.

148. M. Mucha. Poly(ethylene oxide) blends with crosslinking polyester resin. Colloid Polym. Sci. 272:1090, 1994.

149. H. Zheng. Miscibility, crystallization and properties of thermosetting polymer blends of unsaturated polyester resin. M.Sc. thesis, University of Science and Technology of China, Hefei, 1997.

150. H. Zheng, S. Zheng, and Q. Guo. Thermosetting polymer blends of unsaturated polyester resin and poly(ethylene oxide). I. Miscibility and thermal properties. J. Polym. Sci., Polym. Chem. Ed. 35:3161, 1997.

151. H. Zheng, S. Zheng, and Q. Guo. Thermosetting polymer blends of unsaturated polyester resin and poly(ethylene oxide). II. Hydrogen-bonding interaction, crystallization kinetics, and morphology. J. Polym. Sci., Polym. Chem. Ed. 35:3169, 1997.

152. H. Zheng, S. Zheng, and Q. Guo. Phase behavior, mechanical properties, and thermal stability of thermosetting polymer blends of unsaturated polyester resin and poly(ethylene oxide). J. Mater. Sci., submitted.

153. Q. Guo and H. Zheng. Miscibility and crystallization of thermosetting polymer blends of unsaturated polyester resin and poly(ε-caprolactone). Polymer 40:637, 1999.

154. T. Nishi and T. T. Wang. Melting point depression and kinetic effects of cooling on crystallization in poly(vinylidene fluoride)-poly(methyl methacrylate) mixtures. Macromolecules 8:909, 1975.

155. P. J. Flory. Principles of Polymer Chemistry. Ithaca: Cornell Univ. Press, 1953.

7

Computer Simulation of Spinodal Decomposition in Polymer Mixtures

Takaaki Matsuoka
Toyota Central Research and Development Laboratories, Inc., Nagakute, Aichi, Japan

I. INTRODUCTION

Polymer mixtures, called polymer alloys and blends, are very attractive as structural and functional materials because of their variable properties. A number of polymer alloys and blends have been developed in polymer industries and are now commercially available for use in practical products. Because it is well known that their variable properties are caused by morphology or phase structure, morphological study is important for developing polymer alloys and blends. Experimental studies of phase separation and morphology have been made by digital image analysis (1) and light scattering (2,3).

Recently, computer simulations have been used to investigate the phase separation of polymer mixtures. Monte Carlo simulations have been performed for studying the spinodal decomposition of polymer systems (4–7). An approach called cell dynamical systems has been proposed for studying phase separation dynamics (8).

The dynamics of concentration fluctuations are phenomenologically described by the time-dependent Langevin equation. Cahn and Hilliard first developed a diffusion equation for the spinodal decomposition of metal alloys (9), and Cook added thermal fluctuations to the equation (10). Petschek and Metiu (11) applied the Ginzburg–Landau model to the numerical simulation of the spinodal de-

189

composition in two dimensions by using a finite difference method and discussed the results of the probability of the concentration for binary mixtures. Chakrabarti et al. (12) presented a numerical study of a dimensionless Cahn–Hilliard model without the thermal fluctuation term for the late stage of spinodal decomposition of binary system in three dimensions. Ariyapadi et al. (13) predicted the domain size in polymer blends during phase separation by solving a modified form of the Cahn–Hilliard equation, in which the mobility was described as a function of the concentration. Chen et al. (14) performed a computer simulation of the early-stage spinodal decomposition of polymer solutions in two-dimensional Fourier space considering a composition-dependent mobility and diffusivity. The effect of polymer characteristics on the phase separation was investigated by a computer simulation of the late-stage phase separation (15). These studies indicate that the Cahn–Hilliard–Cook equation can be used for the prediction of phase separation in binary polymer systems.

In this chapter, computer simulations of the phase separation due to the spinodal decomposition described by the Cahn–Hilliard model are represented by considering polymer characteristics.

II. THEORY AND NUMERICAL METHODS

A. Phase Separation Model

The time evolution of concentration fluctuation in the spinodal decomposition is described by the nonlinear Langevin equation:

$$\frac{\partial \phi}{\partial t} = m\nabla^2 \frac{\delta F}{\delta \phi} + \eta(r, t) \tag{1}$$

where $\phi (r, t)$ is the order parameter related to the concentration of one of components, $F(t)$ is the free energy functional, $\eta (r, t)$ is the thermal noise, m is the mobility, t is the time, and r is the spatial position vector.

The free energy functional is assumed to be given by the Ginzburg–Landau expansion:

$$F(t) = k_B T \int dr \left[\frac{1}{2} K(\nabla\phi)^2 + \frac{1}{2!} \left(\frac{\partial^2 f}{\partial c^2} \right) \phi^2 + \frac{1}{4!} \left(\frac{\partial^4 f}{\partial c^4} \right) \phi^4 \right] \tag{2}$$

where k_B is the Boltzmann constant, T is the temperature, $f(c)$ is the free energy density, c is the concentration, and K is the gradient energy parameter. From Eqs. (1) and (2), the time-dependent Ginzburg–Landau equation is obtained as follows:

$$\frac{\partial \phi}{\partial t} = k_B T m \nabla^2 (-b\phi + u\phi^3 - K\nabla^2\phi) + \eta(r, t) \tag{3}$$

where

$$b = \frac{\partial^2 f}{\partial c^2} \tag{4}$$

$$u = \frac{\partial^4 f}{\partial c^4} \tag{5}$$

Eq. (3) is also known as the Cahn–Hilliard–Cook equation.

The thermal noise plays a role in the early stage of phase separation.

$$\langle \eta(r, t)\, \eta(r', t') \rangle = -2k_B Tm \nabla^2 \delta(r - r') \delta(t - t') \tag{6}$$

where $\langle \rangle$ denotes an ensemble average. In the late stage, since the thermal noise does not affect the phase structure, it is negligible. Then Eq. (3) becomes

$$\frac{\partial \phi}{\partial t} = k_B Tm \nabla^2 (-b\phi + u\phi^3 - K\nabla^2 \phi) \tag{7}$$

Assuming that T, m, b, u, and K are constant, the equation becomes the following dimensionless form in a simple fashion for the late stage of the phase separation (12):

$$\frac{\partial \phi^*}{\partial t^*} = \frac{1}{2}\nabla^2(-\phi^* + \phi^{*3} - \nabla^2\phi^*) \tag{8}$$

where

$$r^* = \left(\frac{b}{K}\right)^{1/2} r \tag{9}$$

$$\phi^* = \left(\frac{u}{b}\right)^{1/2} \phi \tag{10}$$

$$t^* = \left(\frac{2k_B Tmb^2}{K}\right) t \tag{11}$$

B. Polymer Characteristics

For binary polymer mixtures, the free energy density is given by Flory–Huggins lattice theory:

$$f(c) = \frac{c_1}{N_1}\ln(c_1) + \frac{c_2}{N_2}\ln(c_2) + \chi c_1 c_2 \tag{12}$$

where $c_1 + c_2 = 1$ and $c = c_2$. N is the number of segments and χ is the interaction parameter. Subscripts 1 and 2 denote polymer 1 and polymer 2, respectively. Substituting Eq. (12) into Eqs. (4) and (5), b and u become

$$b = -\frac{1}{N_1c_1} - \frac{1}{N_2c_2} + 2\chi \tag{13}$$

$$u = \frac{1}{3}\left(\frac{1}{N_1c_1^3} + \frac{1}{N_2c_2^3}\right) \tag{14}$$

K is given by (16)

$$K = \frac{a^2}{18c_1c_2} \tag{15}$$

where a is the characteristic length related to the segment size s (17):

$$a = (s_1^2c_1 + s_2^2c_2)^{1/2} \tag{16}$$

The concentration is given from the order parameter

$$c = \frac{\phi/\phi_e + 1}{2} \tag{17}$$

where

$$\phi_e = \left(\frac{b}{u}\right)^{1/2} \tag{18}$$

In the late stage, order parameters approach the ϕ_e or $-\phi_e$, region except in the boundary of domains.

Polymer characteristics required for obtaining coefficients b, u, and K are the number of segments N, the segment size s, and the interaction parameter χ. If the monomer is defined as a segment, these parameters are estimated by the following relationships.

$$N = \frac{M_w}{M_u} \tag{19}$$

$$s = \frac{r_0}{M^{1/2}} M_u^{1/2} \tag{20}$$

$$\chi = \frac{v}{RT}(\delta_1 - \delta_2)^2 \tag{21}$$

$$v = \frac{M_u}{\rho} \tag{22}$$

where M_w is the weight-averaged molecular weight, M_u is the molecular weight of monomer, $r_0/M^{1/2}$ is the mean-square end-to-end distance, v is the molar volume, R is the gas constant, δ is the solubility parameter, and ρ is the density. Table 1 shows some values of parameters typical for polymers.

Table 1   Characteristic Values for Polymers Related to the Phase Separation

	PP	PA$_{66}$	PE	PMMA	PS
p	0.91	1.14	0.95	1.19	1.05
M_u	42	226	28	100	104
M_w	3.5×10^5	3.3×10^4	1.8×10^5	3.5×10^6	2.7×10^5
$r_0/M^{1/2}$[a]	0.0685	0.0890	0.1320	0.0670	0.0690
δ[a]	9.30	13.60	8.00	9.30	8.85

[a] From Ref. 25.
PP: polypropylene; PA$_{66}$: nylon66; PE: polyethylene; PMMA: polymethylmetacrylate; PS: polystyrene; p: density [g/cm^3]; M_u: molecular weight of monomer [−]; M_w: weight-averaged molecular weight [−]; $r_0/M^{1/2}$: mean-square end-to-end distance [nm]; δ: solubility parameter [(cal/cm^3)$^{1/2}$].

C.   Numerical Method

The equation for the phase separation can be numerically solved by using a finite difference method, which is the easiest to use. The derivatives are described with finite difference formulations. A lattice grid is constructed over an analysis domain in a rectangular coordinate system, as shown in Fig. 1. The grid provides dis-

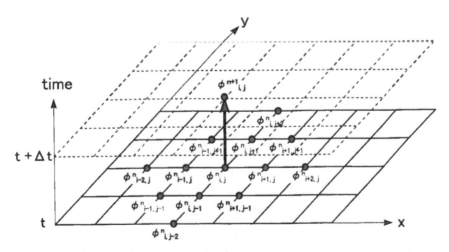

Figure 1   Lattice grid of an analysis domain for a finite difference method in two-dimensional space.

crete points or nodes, where the order parameter will be calculated. The locations of the nodes are identified by using the indices i, j, and k in the x, y, and z directions, respectively. The spatial derivatives are replaced by central difference expressions as follows:

For two dimensions (x, y):

$$\nabla^2 \phi = h^{-2}(\phi_{i+1,j} + \phi_{i-1,j} + \phi_{i,j+1} + \phi_{i,j-1} - 4\phi_{i,j}) \tag{23}$$

$$\nabla^2(\nabla^2\phi) = h^{-4}[20\,\phi_{i,j} - 8(\phi_{i+1,j} + \phi_{i-1,j} + \phi_{i,j+1} + \phi_{i,j-1})$$
$$+2(\phi_{i+1,j+1} + \phi_{i-1,j+1} + \phi_{i+1,j-1} + \phi_{i-1,j-1}) \tag{24}$$
$$+ \phi_{i+2,j} + \phi_{i-2,j} + \phi_{i,j+2} + \phi_{i,j-2}]$$

For three dimensions (x, y, z):

$$\nabla^2\phi = h^{-2}(\phi_{i+1,j,k} + \phi_{i-1,j,k} + \phi_{i,j+1,k} + \phi_{i,j-1,k} \tag{25}$$
$$+ \phi_{i,j,k+1} + \phi_{i,j,k-1} - 6\phi_{i,j,k})$$

$$\nabla^2(\nabla^2\phi) = h^{-4}[42\,\phi_{i,j,k}$$
$$-12(\phi_{i+1,j,k} + \phi_{i-1,j,k} + \phi_{i,j+1,k} + \phi_{i,j-1,k}$$
$$+ \phi_{i,j,k+1} + \phi_{i,j,k-1})$$
$$+2(\phi_{i+1,j+1,k} + \phi_{i-1,j+1,k} + \phi_{i+1,j-1,k} + \phi_{i-1,j-1,k})$$
$$+2(\phi_{i+1,j,k+1} + \phi_{i-1,j,k+1} + \phi_{i+1,j,k-1} + \phi_{i-1,j,k-1}) \tag{26}$$
$$+2(\phi_{i,j+1,k+1} + \phi_{i,j-1,k+1} + \phi_{i,j+1,k-1} + \phi_{i,j-1,k-1})$$
$$+ \phi_{i+2,j,k} + \phi_{i-2,j,k} + \phi_{i,j+2,k} + \phi_{i,j-2,k}$$
$$+ \phi_{i,j,k+2} + \phi_{i,j,k-2}]$$

where h is the equal grid size ($h = \Delta x = \Delta y = \Delta z$).

The time derivative is approximated with a forward differential scheme.

$$\frac{\partial \phi}{\partial t} = \frac{\phi^{n+1} + \phi^n}{\Delta t} \tag{27}$$

where Δt is the time increment. Superscripts n and $n + 1$ indicate the old and new times, respectively. The new time is a time level greater by Δt than the old time.

The partial differential equation is transformed into a set of finite difference equations using the above formulations. Since the time scheme is fully explicit, the spatial derivatives are evaluated at the old time level. The nodal order parameters at the new time, or unknowns, are explicitly calculated from their known values at the old time. However, this requires small time steps for stable computations. The maximum allowable time increment is related to the polymer properties and the grid size.

A periodic boundary condition is assumed around the analysis domain to model an infinite space. When the thermal noise is neglected, the initial nodal order parameters are chosen to be randomly distributed with small fluctuations at the average concentration. The average of the fluctuations must equal 0.

III. CALCULATIONS

A. Phase Structures

Figure 2 shows phase structures during phase separation calculated in two dimensions for a volume fraction of 0.5. The volume fraction f equals the average concentration of polymer 2. A square lattice grid with 128×128 cells is used and the thermal noise is neglected for the computation. Two components are dissolved at the initial time $t = 0$, and then are quenched in an immiscible region. The calculation conditions are as follows:

Boltzmann constant	3.3×10^{-24}	cal/K
Temperature	573	K
Mobility	4.19×10^{-11}	$cm^2mol/(cal \ s)$
Number of segments	1000	—
Segment size	1×10^{-6}	cm
Volume fraction	0.5	—
Molar volume	100	cm^3/mol
Difference of solubility parameters	1	$(cal/cm^3)^{1/2}$
Grid size	1×10^{-6}	cm
Time increment	1×10^{-4}	s

They are the standard conditions for the simulations below.

The darkness indicates the concentration of polymer 2; that is, white areas are fully occupied by polymer 1 and black areas by polymer 2. In gray areas, polymer 2 is dissolved in polymer 1 at the composition shown by the level of darkness. At the initial time, the whole domain seems to be uniform, although there are small fluctuations around the average concentration, but with increasing time, the fluctuation increases. However, the morphology does not change visibly until 0.001 s. The fluctuation starts to be recognized at 0.002. At 0.005, a fine percolated structure is clearly observed. After that, the pattern becomes larger with time, owing to coalescence coarsening in the late stage of the spinodal decomposition.

The order parameters of the cells located on the horizontal center line in Fig. 2 are plotted against position, as shown in Fig. 3. The order parameters are scaled by ϕ_e. In (a), as the initial fluctuation grows, a wave appears clearly, and the amplitude of the wave increases with time without a change in frequency. This period is called the early stage of spinodal decomposition. After that, there is an in-

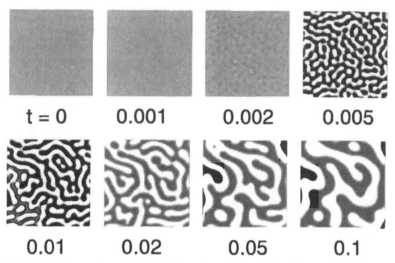

t = 0   0.001   0.002   0.005

0.01   0.02   0.05   0.1

Figure 2 Phase structures during phase separation for $f = 0.5$, $\Delta\delta = 1.0$, and $N = 1000$. The darkness denotes the concentration of polymer 2 and the unit of the time is seconds.

termediate stage when both the amplitude and the frequency change, as shown in (b). As the order parameter draws closer to 1 or -1, the domain is completely separated into two components. At the late stage (c), the frequency increases with time because of the coarsening.

By changing the volume fraction, phase structures for volume fractions 0.3 and 0.7 can be generated and are shown in Fig. 4. Other conditions are the same as those for the previous calculation. The times when the separation begins to be clearly observed are later than for volume fraction 0.5, and the morphology becomes droplet/matrix. When the volume fraction is 0.3, polymer 2 is dispersed as droplets in the matrix of polymer 1 because the volume fraction of polymer 2 is less than that of polymer 1. For $f = 0.7$, the structure type is the same as that for $f = 0.3$, but the components are altered, the droplets being polymer 1 and the matrix polymer 2. Both structures coarsen as the droplets become larger with time.

The statistical dispersion, $\Phi = \langle \phi^2 \rangle - \langle \phi \rangle^2$, is shown in Fig. 5 as a function of the time for various volume fractions. The results for volume fractions 0.6 and 0.7 are almost the same as those of 0.4 and 0.3, respectively. There is no phase separation when $f \leq 0.2$ or $f \geq 0.8$. The phase separation time, which is defined as the time when Φ rises up rapidly from zero, is delayed by decreasing or increasing volume fraction below or above 0.5. The curve is shifted in time by changing the volume fraction.

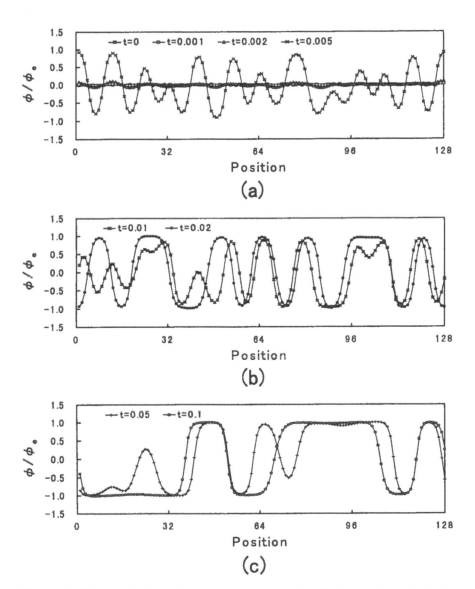

Figure 3 Time evolution of order parameters of cells located on the horizontal center line. The calculation conditions are standard and the unit of the time is seconds.

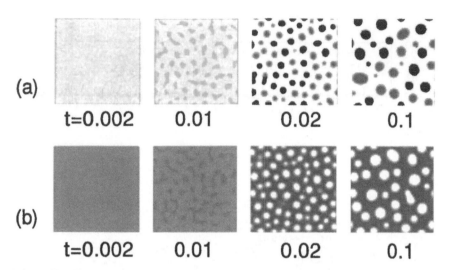

Figure 4 Phase structures during phase separation for $\Delta\delta = 1.0$, $N = 1000$, and (a) $f = 0.3$ and (b) $f = 0.7$. The unit of the time is seconds.

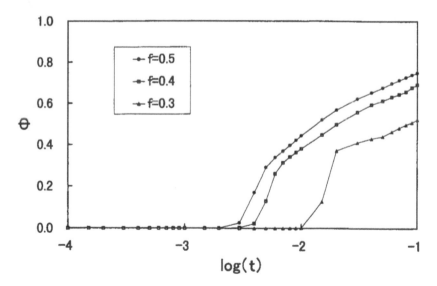

Figure 5 Effect of the volume fraction on Φ.

B. Effects of Polymer Characteristics

The molar volume and the segment size within the polymer characteristics are constant for each polymer. Since no change of the polymers is assumed in the calculations, the effects of only the number of segments and the solubility parameter are discussed here.

Figure 6 demonstrates phase structures for $N = 20$ and $N = 100$. Figure 7 shows the effect of the number of segments on Φ. The beginning of the phase separation is delayed by decreasing the number of segments. However, the phase structure of $N = 100$ is not much different from that of $N = 1000$, although the number of segments is greater by an order of magnitude. If the number of segments is decreased to 20, the phase separation still occurs, but the separated structures become unclear. To avoid the phase separation, the number of segments must be less than 12, which means that one of components must be an oligomer, not polymeric. The mobility is assumed to be constant in the calculation, but it actually depends on the number of segments. By decreasing the number of segments, the mobility will increase and the time to phase separation will be decreased. Therefore it should be considered that there are two opposing effects of the number of segments on the phase separation in the actual behavior.

Phase structures for $\Delta\delta = 0.5$ and 1.5 are shown in Fig. 8. The volume fraction is 0.5. For $\Delta\delta = 0.5$, the phase separation is not observed until $t = 0.02$. The structure at the longest time is also a coarse, percolated type but its pattern is less

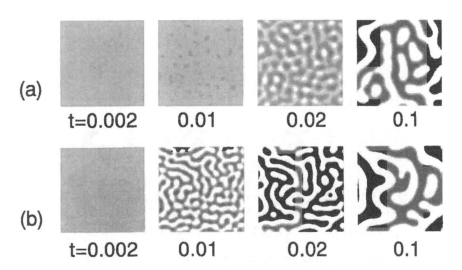

Figure 6 Phase structures during phase separation for $f = 0.5$, $\Delta\delta = 1.0$, and (a) $N = 20$ and (b) $N = 100$. The unit of the time is seconds.

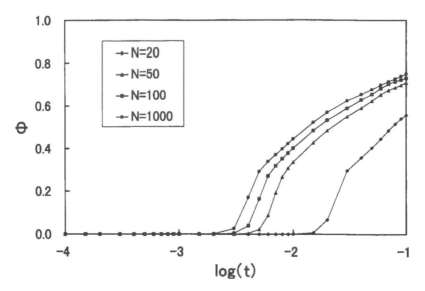

Figure 7   Effect of the number of segments on Φ.

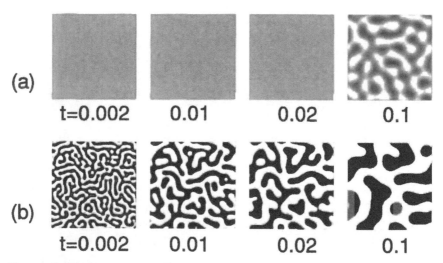

Figure 8   Phase structures during phase separation for $f = 0.5$, $N = 1000$ and (a) $\Delta\delta = 0.5$ and (b) $\Delta\delta = 1.5$. The unit of the time is seconds.

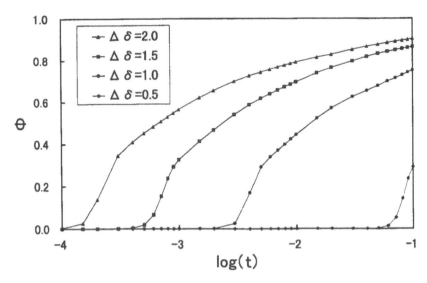

Figure 9 Effect of the difference of solubility parameters on Φ.

clear than for $\Delta\delta = 1.0$. For $\Delta\delta = 1.5$, the phase separation occurs immediately as a fine percolated structure at $t = 0.002$. The phase structure is very clear from the beginning and coarsens with time. By increasing the difference of the solubility parameter, the phase separation is faster and more distinct. The solubility parameter dramatically changes the evolution of Φ, as shown in Fig. 9 with a decrease in $\Delta\delta$ causing the phase separation time to increase. When $\Delta\delta \leqq 0.3$, the phase separation does not occur within the calculation conditions. For $\Delta\delta = 1.5$, the phase separation time is the shortest of all the different calculations. Φ increases with time because of the coarsening of the phase structure.

C. Interfacial Tension

The interfacial tension is important when discussing the phase structure. It is related to coefficients K, b, and u of Eq. (3) as follows (18):

$$\sigma \propto \left(\frac{1}{3}\right)\frac{K^{1/2}(2b)^{3/2}}{u} \tag{28}$$

The phase separation is theoretically caused when the following relation is satisfied:

$$\varepsilon = -b + 3u\,\phi_a^2 < 0 \tag{29}$$

where, ε is the criterion of phase separation and ϕ_a is the average order parameter.

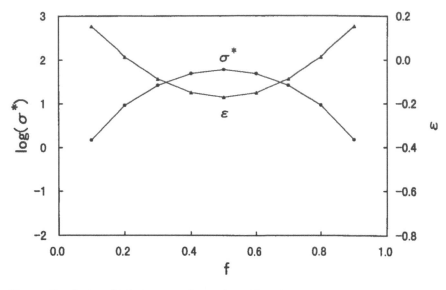

Figure 10   Plots of σ* and ε vs. the volume fraction.

σ*, which equals the value of the right hand side of Eq. (28), and ε are estimated from the calculation conditions. They are plotted versus the volume fraction as shown in Fig. 10. The interfacial tension is a maximum at $f = 0.5$ and decreases with deviation from the value of 0.5. The curve of the criterion shows an opposite tendency to that of the interfacial tension. The plot of ε shows that the phase separation is caused in the volume fraction range of 0.3 to 0.7 because ε is negative. On the other hand, when $f \leq 0.2$ or $0.8 \leq f$, the phases will not separate because $\varepsilon > 0$. The volume fraction to prevent phase separation agrees with that predicted by the previous computer simulation. Fig. 11 shows σ* and ε versus the number of segments. As the number of segments decreases, the interfacial tension is decreased towards 50 and then dramatically drops after that. ε is, however, almost constant until $N = 100$ and then increases while remaining negative. The change of ε corresponds to the shift of the phase separation time due to the number of segments in the simulation. Fig. 12 shows the effect of the solubility parameter on σ* and ε. By decreasing the difference between solubility parameters, the interfacial tension decreases and the criterion increases and approaches zero. If Δδ is very small, the phase separation may not appear for a long time.

Summarizing the above three figures, the relation between σ* and ε is shown in Fig. 13. As is evident from the figure, a large interfacial tension causes the phase separation. We can reduce the interfacial tension by changing the number of segments and the solubility parameter. However, the occurrence of the phase separation cannot be predicted only from the interfacial tension. It could be that

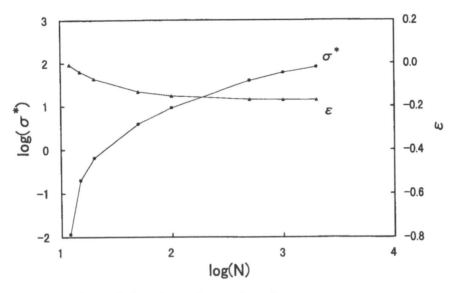

Figure 11   Plots of σ* and ε vs. the number of segments.

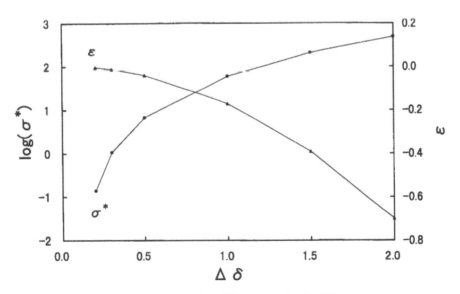

Figure 12   Plots of σ* and ε vs. the difference of solubility parameters.

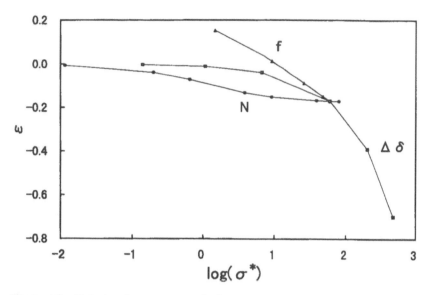

Figure 13 Relation between ε and σ*.

the only factor that makes the criterion positive is the volume fraction, and this corresponds with the phase diagram. For the number of segments and the solubility parameters, the criterion only approaches zero with a negative sign, although interfacial tension decreases. In this case, the criterion indicates that the phase separation is not solely prevented by changing the number of segments and solubility parameters. The phase separation disappears as these parameters are decreased, as described in the previous simulation. This discrepancy is due to the difference of time scale in that the criterion is referred to infinite time but the simulation is finite. If a computer could be run infinitely, the phase separation predicted by simulation might also occur any time, as the criterion indicates. The computer simulation is useful to predict the occurrence of the phase separation within a given time and the phase structure after that.

D. Flow-Induced Phase Structures

The morphology of polymer mixtures strongly depends on processing due to transport phenomena. The effect of the flow is added to the dimensionless equation of phase separation:

$$\frac{\partial \phi^*}{\partial t^*} = \frac{1}{2}\, \nabla^2(-\phi^* + \phi^{*3} - \nabla^2 \phi^*) - \nabla u \phi^* \tag{30}$$

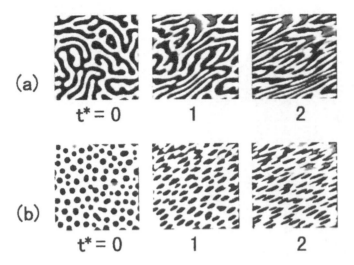

Figure 14 Deformation of phase structures due to a simple shear flow of $\dot{\gamma}$ = 1. The volume fractions are (a) 0.5 and (b) 0.3.

where u is the velocity. Here, we consider a simple shear flow:

$$u = (\dot{\gamma}\,y, 0, 0) \tag{31}$$

where $\dot{\gamma}$ is the shear rate in the rectangular coordinates, x-y-z.

When a simple shear flow, $\gamma = 1$, is applied to separated structures, they are deformed by the flow as shown in Fig. 14, where (a) $f = 0.5$ and (b) $f = 0.3$. The flow is caused by moving the upper side left to right in the calculation. For $f = 0.5$, the structure pattern is deformed by the flow, but the percolation still remains and appears as a lamination. For $f = 0.3$, the droplets are stretched and aligned in the flow direction, and the orientation of extended droplets results in anisotropy induced into the macroscopic properties. The behavior of the domain growth under steady shear flow has been investigated in detail by Ohta et al. (19). The effect of the flow-induced phase structure on mechanical properties of mixtures can be predicted by using a finite element method (20).

E. Three-Dimensional Simulation

Three-dimensional phase structures at a late stage of t* = 1000 are demonstrated in Fig. 15. The volume fractions are (a) 0.5 and (b) 0.3, respectively. The calculations are made by using a cubic grid of 64 × 64 × 64 in dimensionless scale. Only cells in which the concentration of polymer 2 is greater than 0.5 are displayed in

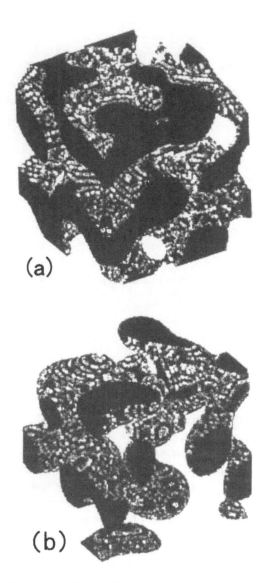

Figure 15 Three-dimensional phase structures at $t^* = 1000$ for (a) $f = 0.5$ and (b) $f = 0.3$. Cells with a positive order parameter are displayed. The solid corresponds to polymer 2 and the cavity to polymer 1. The darkness indicates the concentration of polymer 2.

the snapshots. The cavity is almost totally occupied by polymer 1. The dark region represents the concentration of polymer 2. The percolated structure is also predicted inside and outside the same as the two-dimensional simulation for $f = 0.5$. For $f = 0.3$, droplets appear to be independent of each other from examination of the surface of the analysis domain. On the cross section, the morphology appears to be droplet/matrix. However, the droplets are connected with other ones inside and thus are part of the percolated structure. The results of the three-dimensional simulation are different from those of the two-dimensional one. By calculating the phase separation in the three-dimensional space, the percolation can be predicted for the nonsymmetric 30 : 70 blend. Because of limited computer resources, the two-dimensional simulation is more useful in practice at this time. However, a three-dimensional simulation would be more desirable.

IV. SUMMARY

Phase separation due to spinodal decomposition is numerically simulated by solving the time-dependent Langevin equation with Flory–Huggins free energy using a finite difference method. The amplification of concentration fluctuation, the structure coarsening during the phase separation process, and the flow-induced phase structure are simulated for binary polymer mixtures. Effects of the volume fraction and polymer characteristics such as the number of segments and the difference of solubility parameter on the phase separation are numerically predicted. The phase separation time, at which the statistic dispersion of order parameter begins to rise rapidly from nearly zero, decreases when the volume fraction differs from 0.5 and with decreasing the number of segments and the difference of solubility parameter. The difference of the solubility parameter has the largest influence on the phase separation time of all the variables. Effects of the polymer characteristics are qualitatively explained by the change of the interfacial tension.

The computer simulation has been applied to studies of phase separation of block copolymers (21), diblock copolymers (22), and ternary polymer systems (23,24), which are nearer to materials practically used than simple binary systems. It is expected that computer simulation will become a useful tool for designing and developing new polymer alloys and blends in the near future.

REFERENCES

1. H. Tanaka, T. Hayashi, T. Nishi. Application of digital image analysis to pattern formation in polymer systems. J Appl Phys 59:3627–3643, 1986.
2. H. L. Snyder, P. Meakin. Details of phase separation processes in polymer blends. J Polym Sci Polym Symp 73:217–239, 1985.
3. T. Hashimoto, M. Itakura, N. Shimidzu. Late stage spinodal decomposition of a binary polymer mixture. II. Scaling analyses on $Q_m(\tau)$ and $I_m(\tau)$. J Chem Phys 85:6773–6786, 1986.

4. A. Levy, S. Reich, P. Meakin. The shape of clusters on rectangular 2D lattices in a simple "phase separation" computer experiment. Phys Letters 87A:248–252, 1982.
5. A. Baumgärtner, D. W. Heermann. Spinodal decomposition of polymer films. Polymer 27:1777–1780, 1986.
6. A. Sariban, K. Binder. Critical properties of the Flory–Huggins lattice model of polymer mixtures. J Chem Phys 86:5859–5873, 1987.
7. P. Cifra, F. E. Karasz, W. J. Macknight. Computer simulation of a binary polymer mixture in three dimensions. J Polym Sci B 26:2379–2383, 1988.
8. Y. Oono, S. Puri. Study of phase-separation dynamics by use of cell dynamical systems. I. Modeling. Phys Rev A 38:434–453, 1988.
9. J. W. Cahn, J. E. Hilliard. Free energy of a nonuniform system. I. Interfacial free energy. J Chem Phys 28:258–267, 1958.
10. H. E. Cook. Brownian motion in spinodal decomposition. Acta Met 18:297–306, 1970.
11. R. Petschek, H. Metiu. Computer simulation of the time-dependent Ginzburg–Landau model for spinodal decomposition. J Chem Phys 79:3443–3456, 1983.
12. A. Chakrabarti, R. Toral, J. D. Gunton. Late stages of spinodal decomposition in a three-dimensional model system. Phys Rev B 39:4386–4394, 1989.
13. M. V. Ariyapadi, E. B. Nauman, J. W. Haus. Spinodal decomposition in polymer–polymer systems: a two-dimensional computer simulation. In: R. J. Roe, ed. Computer Simulation of Polymers. Englewood Cliffs, NJ: Prentice-Hall, 1991, pp. 374–384.
14. Y. Chen, K. Solc, G. T. Caneba. Analysis of polymer membrane formation through spinodal decomposition. III: Two-dimensional simulations of early-stage behavior. Polym Eng Sci 33:1033–1041, 1993.
15. T. Matsuoka, S. Yamamoto. Computer simulation of phase separation in binary polymer mixtures. J Appl Polym Sci 57:353–362, 1995.
16. P. G. de Gennes. Dynamics of fluctuations and spinodal decomposition in polymer blends. J Chem Phys 72:4756–4763, 1980.
17. K. Binder. Collective diffusion, nucleation and spinodal decomposition in polymer mixtures. J Chem Phys 79:6387–6409, 1983.
18. M. Doi, A. Onuki. Polymer Physics and Dynamics of Phase Transition. Tokyo: Iwanami, 1992, pp. 127–148.
19. T. Ohta, H. Nozaki, M. Doi. Computer simulations of domain growth under steady shear flow. J Chem Phys 93:2664–2675, 1990.
20. Matsuoka, S. Yamamoto. Computer simulation of phase structure and mechanical properties of polymer mixtures. J Appl Polym Sci 68:807–813, 1998.
21. M. Bahiana, Y. Oono. Cell dynamical system approach to block copolymers. Phys Rev A 41:6763–6771, 1993.
22. L. A. Molina, A. L. Rodriguez, J. Freire. Monte Carlo study of symmetric diblock copolymers in nonselective solvents. Macromolecules 27:1160–1165, 1994.
23. A. Sariban. Monomer distribution functions of a ternary polymer system: a Monte Carlo investigation. Macromolecules 24:1134–1144, 1991.
24. N. Vasishtha, D. Q. He, E. B. Nauman. Self-similarity and scaling transitions during spinodal decomposition. Comput Polym Sci 4:1–6, 1994.
25. J. Brandrup, E. H. Immergut, eds. Polymer Handbook, 3rd Ed., New York: John Wiley, 1989.

8

Interactions and Phase Behavior of Polyester Blends

Thomas S. Ellis*

General Motors Research and Development Center, Warren, Michigan

I. INTRODUCTION

Polyesters are arguably the most versatile and important class of polymers used today. Their uses include engineering thermoplastics, fibers, films, textiles, packaging, and biomedical applications. Recent commercial innovations, leading to economic production of naphthalene-based polyesters such as poly(ethylene naphthalate), PEN, and poly(trimethylene terephthalate), PTT, will allow several new additions to the family of polyesters that promise to extend the range of uses. Although the properties of polyesters have many desirable attributes, many applications involve their use as blends or alloys (1) with polymers such as polycarbonate, or as copolymers, modified for example with polyether and polyurethane blocks, such as those used in thermoplastic elastomers (2). Manipulation of polymer materials by blending or incorporation of other polymer backbones remains a simple and cost-effective method for extending the property profile of polymers. To achieve these objectives in a rational manner and to understand the factors necessary to achieve useful multicomponent materials require knowledge of their fundamental mixing behavior.

---

* Current affiliation: Delphi Automotive Systems Research and Development Center, Warren, Michigan.

Notwithstanding rheological and processing issues, the latter is heavily dependent on the thermodynamic enthalpic interaction between the constituent polymers. Accordingly, the focus of this chapter is to examine one particular strategy that has been directed to evaluating or estimating interactions in blends of polyesters with other polymers and to review the experimentally reported behavior in the light of predicted behavior.

The mutual solubility of two polymers produces a molecularly homogeneous, or miscible, blend. This state of "compatibility" (which is used here as a qualitative term that describes a beneficial outcome resulting from the blending of polymers) can also be applied to heterogeneous blends. Miscibility is often confused with compatibility, and in some instances miscibility may not be a desirable feature in a blend. For example, the rate of crystallization in a semicrystalline polymer is usually inhibited or slowed dramatically when a miscible higher-T_g amorphous polymer is present.

A measure of the thermodynamic compatibility can be obtained by assessing the magnitude of the unfavorable interaction. For example, when this is minimized, as in the case of blends of bisphenol A polycarbonate, PC, and acrylonitrile-butadiene-styrene, ABS, polymers, by manipulation of the acrylonitrile content in the copolymer with styrene, optimum mechanical performance is attained (3). Mechanical compatibility may be obtained for some polymer mixtures, without the aid of compatibilizers, even though they exhibit gross thermodynamic incompatibility, e.g., PC and polyethylene (4,5).

Polyesters have been shown to form miscible or partially miscible blends with a wide variety of polymers; however, the major emphasis of this chapter will be restricted to reviewing some of the recent advances in quantifying interactions and explaining observed phase behavior in blends of polyesters with other polyesters and those containing polycarbonate, chlorinated polymers, and polyamides, respectively. Determination of the interaction in any polymer blend is a formidable and difficult task. Solubility parameter theories, although often applied, lack the necessary complexity to provide reliable information. In this review the approach adopted will be that described by a binary interaction model (6–8). Much of the information and methodology introduced here has been extracted from a similar application of the model to account for mixing behavior in polyamide blends (9–16).

The premise on which mixing behavior, and the analysis, is founded is summarized in Fig. 1 (9), which illustrates how a polymer A can be miscible, as denoted by the interaction, in a series of polymers or copolymers that differ in a systematic manner, such as homologous (CH_2) structure of an aliphatic polyester, or comonomer content in a random addition copolymer. The miscibility window exists only for a finite range of composition, and if a small change of chemical constitution is introduced into polymer A, then the window of miscibility can shift along the composition ordinate (curve 2) or often will disappear completely as depicted by curve 3. All the information presented below describes some of the re-

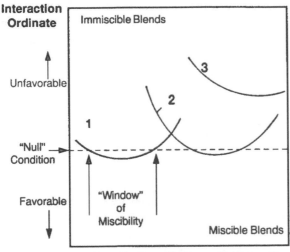

Composition Ordinate Polymer B

Figure 1 Schematic representation of the interaction of a polymer A in a series of polymers B that possess a systematic compositional variation, e.g., —(CH$_2$)$_n$—. Curves 1–3 represent a structural change of polymer A, such as the introduction of a chemically different segment. (See Ref. 9.)

cent progress towards establishing a quantitative representation for blends involving polyesters that can be placed in the context of this figure.

II. THEORETICAL BACKGROUND

Although lacking in completeness, the Flory–Huggins interpretation of the thermodynamics of polymer mixtures, given as Eq. (1) remains a versatile tool for describing the free energy of mixing, ΔG. When expressed using a lattice volume element as a reference, typically 100 cm^3 mole^{-1}, N is the weight average degree of polymerization, ϕ is the volume fraction of polymer in the mixture, R is the gas constant, T is absolute temperature, and χ_{12} is the Flory interaction parameter between polymers 1 and 2.

The theory fails to account for all behavior, but it has mathematical simplicity, and the data required to obtain quantitative information by it are accessible.

$$\frac{\Delta G}{RT} = \frac{\phi_1 \ln \phi_1}{N_1} + \frac{\phi_2 \ln \phi_2}{N_2} + \chi_{12}\, \phi_1\, \phi_2 \tag{1}$$

As a first approximation, polymer–polymer solubility occurs for $\Delta G < 0$; however, a closer inspection reveals that homogeneous mixing or miscibility can also occur when the interaction is slightly positive but below a critical value χ_{crit}. This is because the combinatorial entropy, represented by the terms in parentheses in Eq. (1), is always negative. A simple expression for approximating χ_{crit}, which depends on molecular mass of the blend constituents, is shown as Eq. (2).

$$\chi_{crit} = \frac{(N_1{}^{0.5} + N_2{}^{0.5})^2}{2} \qquad (2)$$

In the limit of high molecular mass polymers ($N_1 = N_2 \approx 2 \times 10^3$; $\chi_{crit} \approx 0.001$), such as that typically encountered in polyolefins and other addition type polymers, the sign of the interaction parameter χ_{12} often has the decisive influence on phase behavior.

The interaction in polymer mixtures is usually positive or unfavorable for miscibility. This turns out to be true for many blends involving polyesters, however, polyesters of very high molecular mass are difficult to obtain. Therefore for blends involving these polymers of only moderate chain length, the entropy of mixing assumes a more important role. In other words the value of χ_{crit} may be relatively large ($\chi_{crit} \approx 0.008$).

A conceptually simple, and accessible, route to obtaining an estimation of the interaction has been developed through the use of a binary interaction model. Shown as Eq. (3), where $\chi_{12} \equiv \chi_{blend}$, the polymers are defined in terms of segments or monomer units, and each pair of segments possess an interaction χ_{ij} that contributes to the overall interaction.

$$\chi_{blend} = \sum_{i,j} \psi_i^1 \psi_j^2 \chi_{ij} - \left[\sum_{i,j} \psi_i^1 \psi_j^1 \chi_{ij} + \sum_{i,j} \psi_i^2 \psi_j^2 \chi_{ij} \right] \qquad (3)$$

The first term summarizes all the unlike contacts between the two polymers and the second term in brackets summarizes all the unlike contacts within each polymer. The model assigns zero value to interactions between like segments. This is sometimes referred to as the "repulsion" model because an overall negative interaction can be obtained even when all the segmental interactions are positive or unfavorable for mixing. It is important to note that χ_{blend} is not a pure enthalpic term but really a free energy parameter that contains both enthalpic and entropic components.

The respective volume fraction ψ_i of each of the segments can be calculated using tabulated values of their group contribution v_i to the molar volume of the polymer (18). For addition polymers, monomer units are a sensible choice for segments; however, for the polymers discussed here, a fundamental scheme, based on elementary units such as ester, amide, and methylene groups, has been adopted. This allows a wider application of the predictive potential of the segmental interaction parameters obtained. Unfortunately, there are associated disadvantages in that more parameters are usually required for calculating the overall behavior, especially if the calculation has to account for the configuration or the arrangement of segments and different isomeric species (14,17). Two examples of how poly-

$$-CH_2 - CH_2 - CH_2 - CH_2 - CH_2 - CO\ O -$$ Aliphatic Polyester

A   A   A   A   A   D   Segments

Polycarbonate

C   A   C   F   Segments

mers are represented for the model are shown here.

Thus with reference to bisphenol A polycarbonate, PC, we obtain $\psi_A = 0.2441$, $\psi_C = 0.5954$, and $\psi_F = 0.1605$ using $\nu_A = -(CH_2)-$ 16.45 cm^3 mol^{-1}, $-C$ $(CH_3)_2-$50.35 cm^3 mol^{-1}; $\nu_C = 61.4$ cm^3 mol^{-1}; $\nu_D = 24.6$ cm^3 mol^{-1}; $\nu_F = 33.1$ cm^3 mol^{-1} (18).

Note that although ester (D) and carbonate (F) groups are similar, the scheme adopted has distinguished between the two segments; however, the alkyl fragment in PC, shown above, and methylene groups (A) are assumed to be equivalent in their interactions. The latter simplification has proven to be more than adequate in describing phase behavior of blends of semiaromatic polyamides and aliphatic polyamides (14).

An attractive feature of the model is that copolymers can be used to define experimental limits of miscibility so that boundary conditions, where $\chi_{blend} = 0$, can be established. The mathematical solution of the simultaneous equations, from which the unknown segmental interactions are obtained, is a simple task; however, at least one of the equations must contain a nonzero value of χ_{blend} in order to scale the model correctly. A summary of the relevant segmental interaction parameters currently available is shown in Table 1.

III. CHEMICAL REACTIVITY IN POLYESTER BLENDS

A recognized feature of investigations of polyester blends, and sometimes the pur-

Table 1   Values of Segmental Interaction Parameters

Segments	Segmental interaction parameter, χ_{ij}	Value
—CH$_2$—/—NHCO—	χ_{AB}	8.53
—CH$_2$—/—C$_6$H$_4$—	χ_{AC}	0.10
—CH$_2$—/—CO O—	χ_{AD}	2.23
—CH$_2$—/—O CO O—	χ_{AF}	2.78
—CH$_2$—/—CH Cl—	χ_{AH}	0.50
—NHCO—/—C$_6$H$_4$—	χ_{BC}	7.97
—NHCO—/—CH Cl—	χ_{BH}	6.75
—NHCO—/—CO O—	χ_{BD}	3.88
—C$_6$H$_4$—/—CO O—	χ_{CD}	1.69
—C$_6$H$_4$—/—O CO O—	χ_{CF}	2.53
—C$_6$H$_4$—/—CH Cl—	χ_{CH}	0.68
—CO O—/—O CO O—	χ_{DF}	0.30
—CO O—/—CH Cl—	χ_{DH}	0.04

Source: Refs. 15, 17, 36, 65.

veyor of confusion concerning fundamental mixing behavior, is the relative ease with which they can react chemically at temperatures typically encountered during melt processing >260°C). The purpose of this chapter is to describe phase behavior in the absence of chemical reactions; however it is important to note that this can have a profound effect on mixing behavior.

In blends of only polyesters, the mechanism is one of catalyzed transesterification (19–21). Often strenuously avoided, for the determination of phase behavior, such reactions are tolerated, and sometimes promoted (22), in the expectation of enhanced mechanical performance in the resulting blends. The resulting block copolymers, produced by these reactions, have the ability to transform immiscible or partially miscible blends of polyesters to seemingly miscible mixtures; however, in some instances it has been concluded that a high degree of conversion is required to effect this transformation. The relatively low molecular mass, inherently typical of this class of polymers, is also a critical factor in allowing trans-reaction-induced miscibility to occur.

Putting aside the basic chemical reactivity of functional groups and the associated dependence on structure and temperature, the feature that affects the ability of transreactivity to influence phase behavior in immiscible polymer blends is the relative frequency that reactive species come into contact with each other. This will depend upon processing conditions, e.g., temperature and the interaction parameter. The latter is perhaps the most important, since it can affect directly the width of the interfacial region and hence the contact between reactive species.

When evaluating fundamental mixing behavior of these materials, it is preferable, for the reasons noted above, that the least intrusive method of blend preparation should be sought. This usually involves removal of catalysts and melt mixing under carefully controlled conditions of time and temperature, or solution methods such as coprecipitation from a common solvent or solvent evaporation from a mixed solution. Casting films using solvents can lead to erroneous conclusions of immiscibility because of the different χ values between the respective polymers and the solvent. Moreover, coprecipitation of blends containing crystallizable components requires a preliminary melting step in order to establish a liquid–liquid equilibrium prior to analysis. In this regard, even a low-level exposure to elevated temperature can sometimes be sufficient to promote reactivity and frustrate determination of fundamental behavior.

The latter thermal treatment also masks any underlying phase transition, e.g., lower critical solution temperature, LCST, that may be present below the melting point of the respective component. When all of these concerns are taken into account, it is not surprising that the literature is replete with contradictory observations and conclusions concerning phase behavior of polyester blends (20,21).

IV. BLENDS OF POLYESTERS AND POLYAMIDES

An experimental and theoretical investigation of blends of aliphatic polyesters, polyamides, and their copolymers has indicated that the interaction in these mixtures will be invariably positive and that miscibility occurs only when the favorable entropic contribution overcomes the unfavorable interaction, i.e., $0 < \chi_{blend} < \chi_{crit}$ (15).

It was found that in a blend of PA-6 and a caprolactam–caprolactone copolymer of typical molecular mass ($M_w \approx 40,000$), the miscible blend could tolerate approximately 15 mole percent of lactone in the copolymer before phase separation occurred. A similar situation was also found to exist at the opposite end of the composition scale involving blends of polycaprolactone and caprolactone–caprolactam copolymers (16).

Although entropically driven miscibility can occur, provided that the compositions of the blend constituents are not too dissimilar and that the respective molecular masses are not too large, calculations predict that for a mixture of aliphatic polyester and polyamide homopolymers this practically never occurs. These conclusions are summarized schematically in Fig. 2 and are exemplified by the interaction in blends whose constituents contain 5 ($\chi_{blend} = 0.207$) and 12 ($\chi_{blend} = 0.055$) methylene groups, respectively. The values of χ_{blend} are far greater than a reasonable critical value ($\chi_{crit} \approx 0.008$) below which miscible

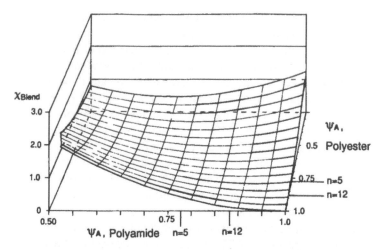

Figure 2 Calculated values of χ_{blend} for mixtures of aliphatic polyesters —(CH$_2$)$_n$—CO.O— and aliphatic polyamides —(CH$_2$)$_n$—NHCO. (See Ref. 15.)

blends may be anticipated. As the methylene contents of both polymers increase, the figure indicates that χ_{blend} approaches zero, representing the mixing of pure polymethylene.

Although binary blends of aliphatic polyesters, and polyesters with aliphatic polyamides, may be of interest for biomedical applications, blends of semiaromatic polyesters, e.g., poly(ethylene terephthalate), PET, and aliphatic polyamides, e.g., PA-66, are considered the most attractive from an engineering thermoplastic viewpoint.

The blending of these polymers has been investigated on numerous occasions (23–28), indicating the formation of phase-separated mixtures with poor interfacial adhesion; a situation characterized by a highly unfavorable interaction between the components. Miscibility in blends of semiaromatic polyamides, such as poly(hexamethylene isophthalamide), PA-6I, and random aliphatic copolyesteramides, derived from ε-caprolactam and ε-caprolactone, has been used to determine the segmental interactions that contribute to this situation (29).

Dilution of ε-caprolactam by ε-caprolactone in PA-6 transforms miscible blends based on PA-6I and poly(trimethylhexamethylene terephthalamide), PA-TMDT, respectively to immiscible mixtures at approximately 35 mole% and 46 mole%, respectively, of ε-caprolactone in the copolymer. These findings together with model calculations are reproduced as Fig. 3. The shaded area represents the regions where $\chi_{blend} < 0$ and miscible blends will occur.

The results have been used (29) to develop a picture, shown as Fig. 4, of interactions in blends of semiaromatic polyesters, such as PET, and aliphatic

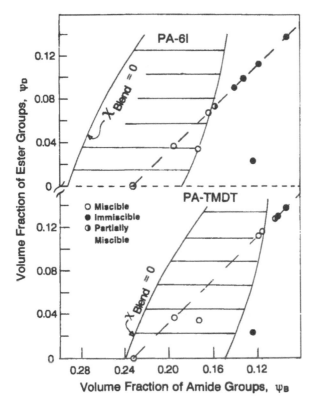

Figure 3 Correlation of phase behavior with calculated values of χ_{blend} for blends of aliphatic copolyesteramides with semiaromatic polyamides PA-6I and PA-TMDT, respectively (See Ref. 29.)

polyamides. The behavior of poly(butylene terephthalate), PBT, will be intermediate between that shown for PET and poly(hexamethylene terephthalate), PHT. The figure also shows that similar results are obtained when the analysis extended to a polyarylate, PAr, formed from the condensate of bisphenol A and a mixture of iso- and terephthalic acids.

Blends of semiaromatic polyesters and semiaromatic polyamides can also be expected to form immiscible blends with a strongly unfavorable interaction (29) and lead to the general conclusion that heterogeneous mixing will always be encountered when pure polyesters are blended with pure polyamides. Alloying opportunities may still exist, provided that some form of reactive processing or compatibilizer is involved to promote adhesion at the interface. Most attempts currently focus on transreaction of the components themselves in order to effect compatibility (28).

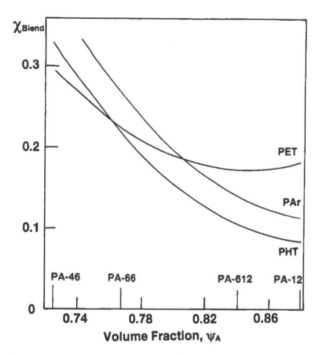

Figure 4 Calculated values of χ_{blend} for blends of semiaromatic polyesters, e.g., PET, PHT and PAr, respectively, with aliphatic polyamides. (See Ref. 29.)

V. BLENDS OF POLYESTERS AND BISPHENOL A POLYCARBONATE

The most widely used commercial blends containing PC and a semiaromatic polyester are those based on PBT, and to a lesser extent PET. A critical feature of the utility of the PBT/PC blend is that the components are partially miscible, i.e., the interaction is positive, but only slightly greater than χ_{crit}. For a blend containing constituents of typical commercial molecular mass (e.g., PC $M_w \approx 30,000$; PBT $M_w \approx 50,000$), this situation promotes compatibility between the components while maintaining the beneficial ability of the polyester to crystallize rapidly in the blend. Indeed, a blend of low-molecular-mass PBT and PC is completely miscible (30). Not surprisingly, the practical success of these blends has promoted a great deal of interest in blends of additional semiaromatic polyesters and copolyesters, such as those based on cyclohexane dimethanol, which have also indicated miscibility with PC.

Attempts to understand why this occurs can be traced back to earlier publications describing phase behavior of aliphatic polyesters in blends with PC, that identified a correlation between miscibility and the ratio of methylene to ester groups in the polyester (31–33). Additional calorimetric studies indicated that there was a negative interaction in this "window" of miscibility (34). More recent attention (35,36) to this subject has confirmed that favorable interactions can also occur in blends with semiaromatic polyesters.

The foregoing has been captured by model calculations as shown in Fig. 5. The figure indicates a region of complete miscibility (shaded area) for PC in polyesters and a region of expected miscibility or partial miscibility, defined here as $0 < \chi_{blend} < 0.008$, that would depend upon the molecular mass of the constituents. Also shown in the figure is experimentally reported behavior for polyesters and copolyesters (20,21); including PET/Caprolactone copolymers (37–39).

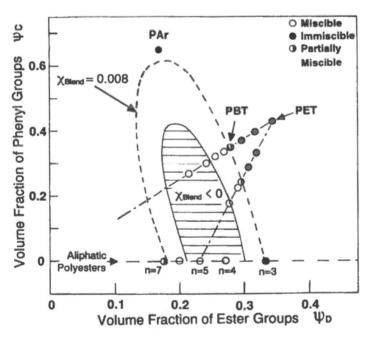

Figure 5 Comparison of experimental data of phase behavior of blends of bisphenol A polycarbonate with aliphatic (—(CH₂)ₙ COO—) and semiaromatic polyesters with binary interaction model calculations. Shaded area is where $\chi_{blend} < 0$. The broken lines (1–2) represent the locus of composition of copolyesters of PCL-PET and semiaromatic polyesters and copolyesters —OCO phenyl COO—(CH₂)ₘ—, respectively (See Ref. 35.)

In some of the data points given, the blends involved copolyesters containing cyclohexyl structures and mixed isomers of the phenyl group. In all cases, the cyclohexyl group has been simplified as an ensemble of methylene (A) groups, and no distinction is made between the different isomers. These assumptions appear to be reasonable in the case of semiaromatic polyesters; however, in aliphatic copolyesters there appear to be some exceptions to this rationalization of the alicyclic structure (36). The figure reveals that PC will be miscible with semiaromatic polyesters containing between approximately 6 and 10 methylene groups, or their equivalent.

There is considerably less interest in the economic exploitation of aliphatic polycarbonates; however, the parameters in Table 1 may be extended (36) to include predictions of phase behavior in blends of aliphatic polycarbonates with aliphatic polyesters, semiaromatic polyesters, and polycarbonate, respectively. The overall trends, exemplified by Fig. 6, which summarizes results for blends of aliphatic polyesters and two aliphatic polycarbonates, show that for most practi-

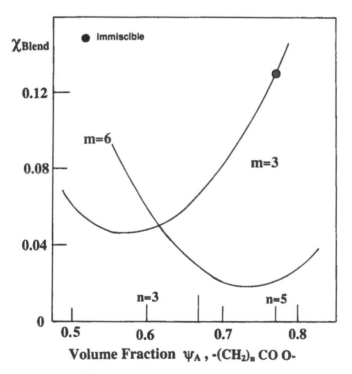

Figure 6 Calculated values of χ_{blend} for blends of aliphatic polycarbonates —(CH$_2$)$_m$ O CO O— for $m = 3$ and $m = 6$, respectively, in aliphatic polyesters. (See Ref. 36.)

cal polymers heterogeneous mixing will prevail. There is little published information with which to compare these results with experimental behavior; however, the reported (40) phase separation in a block copolymer of PCL and poly(trimethylene carbonate) lends support to the calculated behavior.

The commercial availability of the aromatic polyester polyarylate, PAr, has created considerable interest involving blends with PC. The extension of the nomenclature, and the associated segmental interaction parameters currently available, to include the PAr/PC blend is somewhat optimistic. Nevertheless, a calculated value of $\chi_{blend} = 0.011$, which is probably slightly underestimated, supports the observed heterogeneity (36) that may be easily transformed to a miscible blend in the presence of a small amount of transreaction (41).

The parameters in Table 1 also allow an examination of the interaction, presented as Fig. 7 for PET and PBT, respectively of blends of semiaromatic polyesters and polyestercarbonates, which in this instance are composed of random copolymers of polycarbonate and polyarylate. The response of χ_{blend} to com-

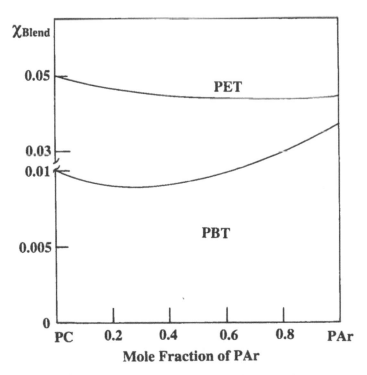

Figure 7 Calculated values of χ_{blend} for blends of semiaromatic polyesters PET and PBT in semiaromatic copolyestercarbonates of polycarbonate, PC, and polyarylate, PAr.

position is only slightly concave to the composition ordinate. This is because of the relatively small repulsion effect between the polycarbonate and polyarylate segments ($\chi_{blend} = 0.011$); however, an underestimation of the latter interaction would signify that the concavity is too shallow and illustrates how miscible blends of polyestercarbonates and PBT may been obtained (42). At 40 mole% PAr, the calculated value of $\chi_{blend} = 0.008$ would be reduced even further and support entropically driven miscibility. Although the interaction in blends containing PET would also be reduced, it is doubtful that this would be sufficient to account for reported (43,44) miscibility in these blends.

The foregoing demonstrates both the utility and the shortcomings of the currently available data. Only six segmental interaction parameters have been implemented to characterize interactions in blends of structurally and compositionally complex polymers.

VI. BLENDS OF COPOLYESTERAMIDES AND POLYCARBONATE

Aliphatic copolyesteramides, derived from caprolactone and caprolactam, can be used in a similar way to that noted above to determine a critical lactone content for miscibility with PC, above which the miscible blend will be transformed to an immiscible mixture. This reasoning is based on the known miscibility of PC with PCL and conversely the immiscibility of PC and PA-6. Studies have yet to be concluded to determine this; however, there are experimental data to complement exploratory calculations concerning blends of PC and semiaromatic copolyesteramides derived from PBT and the polyamide analogue, PA-4T. Both the model, illustrated by Fig. 8, and the experimental data (45) indicate a more unfavorable interaction as the polyamide content increases. A tentative estimation of the amide–carbonate segmental interaction parameter ($\chi_{BF} = 3.00$), obtained from scaling relationships in Table 1, has been applied to compute the behavior shown.

VII. BINARY BLENDS OF POLYESTERS

Both the experimental and theoretical aspects of binary aliphatic polyester blends have been addressed in the literature (16,46,47). For blends ($A_y\ D_{1-y}/A_x\ D_{1-x}$), mixing is exclusively endothermic, as indicated by Eq. (4); however, because of the relatively small ester–methylene interaction, these blends can tolerate a relatively large dissimilarity of alkyl content before the entropically driven miscibility is overcome by the unfavorable enthalpic interaction.

$$\chi_{blend} = (x - y)^2 \chi_{AD} \tag{4}$$

When the methylene, or alkyl content, difference of the polyesters becomes sufficiently large, a heterogeneous blend is obtained (35,48). The essential features de-

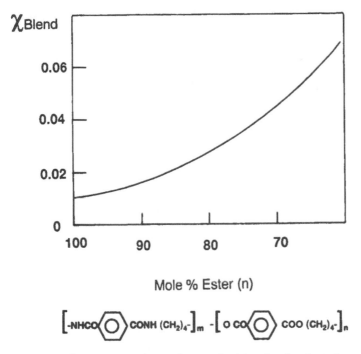

Figure 8   Calculated values of χ_{blend} for blends of polycarbonate in blends of semiaromatic copolyesteramides of PBT and PA-4T.

scribed above are summarized in Fig. 9, which contains a comparison of experimentally reported behavior and a theoretical description of expected phase behavior. The loci shown for χ_{crit} are relatively large because of the low molecular mass of the polyesters.

A selection of data (36,49,50) and some calculated behavior of binary blends of aliphatic polyesters with semiaromatic polyesters is shown in Fig. 10. The predicted trends show that χ_{blend} is always positive, with no athermal condition (χ_{blend} = 0), and unfavorable for homogenous mixing. Although the experimental data support this result, entropically driven miscibility may be expected as the alkyl (methylene) content of both components increase.

The heterogeneous blend of the copolyester Kodar A150, which is a copolyester containing cyclohexyl groups, and poly(butylene adipate) represents an example where χ_{blend} (≈ 0.005) is predicted to be close to zero and where entropically driven miscibility may be expected to occur. The confounding effects of cyclohexyl groups in some polyester blends has been observed on several occasions (36) and is almost certainly caused by sensitivity of the overall interaction to the simplification of alicyclic structures as a polymethylene sequence.

A distinctly more favorable comparison of experimental data with calcula-

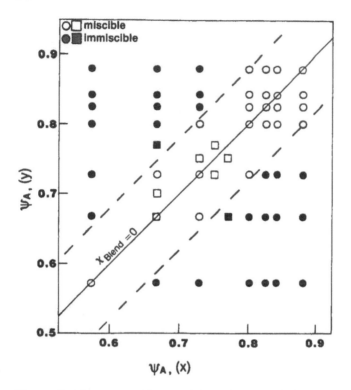

Figure 9 Summary of experimental phase behavior of binary aliphatic polyester blends A_xD_{1-x}/A_yD_{1-y}: lines (---) represent the loci of $\chi_{blend} = 0.0145$ ($N_1 = N_2 \approx 140$). (See Ref. 16.)

tions is provided by blends of PBT-PCL random copolymers with PCL (51). The pure homopolymers have a significant unfavorable interaction ($\chi_{blend} = 0.028$) for homogeneous mixing. As the amount of PBT copolymerized with PCL is decreased, the interaction of the random copolymer with pure PCL will diminish to zero. Calculations indicate that entropically driven miscibility would be expected when the copolymer contains below 60 wt% PBT, a result in accordance with the conclusions of the authors. Phase behavior of binary blends of homologous semi-aromatic polyesters and copolyesters, such as PET and PBT, can be summarized as shown in Fig. 11. The calculations, centered on PBT, indicate the athermal condition when in a blend with an equivalent alkyl or methylene content. Miscibility would be found to occur under these circumstances as evidenced (52) by a blend of PBT with a copolyester of similar alkyl content ($\psi_A = 0.36$) to that of PBT ($\psi_A = 0.373$); however, mixing is usually endothermic. Investigations of PET/PBT

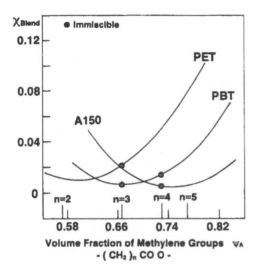

Figure 10 Comparison of experimentally observed phase behavior and calculated values of χ_{blend} for blends of aliphatic polyesters and semiaromatic polyesters, PET, PBT, and Kodar A150 polyester (condensation polymer of cyclohexanedimethanol and iso-/terephthalic acid). (See Ref. 36.)

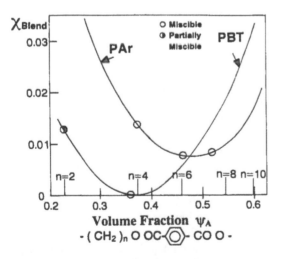

Figure 11 Calculated values of χ_{blend} for binary blends of semiaromatic polyesters, —OCO phenyl COO— $(CH_2)_n$—, centered on PBT ($n = 4$; $\chi_{blend} = 0$) and polyarylate, PAr, as a function of methylene (alkyl) content ψ_A. (See Ref. 36.)

blends have reported differing opinions concerning phase behavior (53–55). A calculated value of $\chi_{blend} = 0.013$ suggests that partial misciblity or complete immiscibility may occur with a sensitivity to molecular mass. Enthalpy recovery studies (56) applied to these blends indicate partial miscibility (57).

Figure 11 also includes projected behavior of polyarylate in semiaromatic polyesters. Polyarylates have attracted considerable attention in blend studies, most notably with PET and PBT, respectively. Although transesterification can occur quite easily, it is concluded that PAr is immiscible with PET and miscible with PBT (21). Copolyesters based on PET with cyclohexyl dimethanol have also been reported to be miscible with PAr (41,58). The calculated large unfavorable interaction of PAr with PET ($\chi_{blend} = 0.045$) agrees with experimental observations. Although the miscible blends involving PBT ($\chi_{blend} = 0.013$) and the copolyester Kodar A150 ($\chi_{blend} = 0.008$), are close to the boundary, the figure illustrates how miscibility can occur with polyesters in this range of composition. Calculations (36) indicate that PAr will be immiscible with all aliphatic polyesters, e.g., with PCL ($\chi_{blend} = 0.026$).

The imminent widespread commercial availability of poly(propylene terephthalate), PPT, suggests that blends of PPT/PET and PPT/PBT will attract some attention in the near future. Calculations indicate only slightly unfavorable interactions (0.004 and 0.003, respectively) and a prognosis for miscibility.

VIII. BLENDS OF POLYESTERS AND CHLORINATED POLYMERS

It has been known for a long time that many aliphatic polyesters are compatible (miscible) with chlorinated polymers, in particular polyvinyl chloride, PVC, with the result that plasticizers based on polyesters are commonly used in PVC applications. Early work on blends of PVC and aliphatic polyesters established a structure–miscibility correlation that was again dependent upon the ratio of methylene to ester groups. The reported window of miscibility for PVC was between approximately 4 to 10 methylenes per ester group (59). These findings were also examined in the framework of a binary interaction model and the segmental interactions determined. Not surprisingly, studies were expanded (60–64) to include additional chlorinated polymers such as chlorinated polyethylenes, CPE, chlorinated PVC, CPVC, and PVC/vinylidene chloride copolymers, as well blends of the latter with semiaromatic polyesters, e.g., PBT.

More recent studies have been extended using copolyesteramides (65), again derived from caprolactam and caprolactone. In addition to re-examining the dependence of chemical structure on miscibility, the objective was also to establish structure–interaction relationships between polyamides, aliphatic and semiaromatic, and chlorinated polymers. Figures 12 and 13 illustrate a fit of the model to the experimentally observed behavior. It has been found that miscible blends of

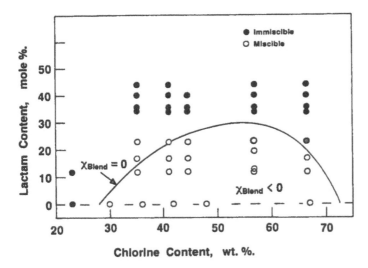

Figure 12 Comparison of experimentally observed phase behavior with model calculations for blends of PCL and caprolactone–caprolactam copolymers respectively with chlorinated polymers. The region enclosed by the locus of $\chi_{blend} = 0$ denotes where χ_{blend} polyethylenes < 0. (See Ref. 65.)

CPE and PVC, respectively with aliphatic copolyesteramides can tolerate a relatively large proportion of lactam, in a copolymer of a lactone, without inducing phase separation.

For the calculations shown in the figures, the chlorinated polymers are defined as copolymers of —CH_2— (A) and —CH Cl— (H) segments. Using this nomenclature, PVC is a perfectly alternating copolymer of A and H segments. There was some difficulty in trying to satisfy all the experimental data with just a single value of the parameters involving the chloromethylene segment, a finding that was found to be consistent with the experience of other researchers who have attempted to apply similar methodologies involving blends containing PVC and CPE. The configuration of segments in CPE and CPVC has been found to play a significant role on the magnitude of interactions in these blends, and the incorporation of an adjustable value for the χ_{AH} segmental interaction can overcome some of these difficulties.

Blends of chlorinated polymers with semiaromatic polyesters, such as PBT, have received much less attention than their aliphatic counterparts. Nevertheless, evidence suggests that blends of these materials are also capable of displaying a window of miscibility that depends upon the chlorine content (62–64). It has been possible to accommodate such behavior in the same way as that described above

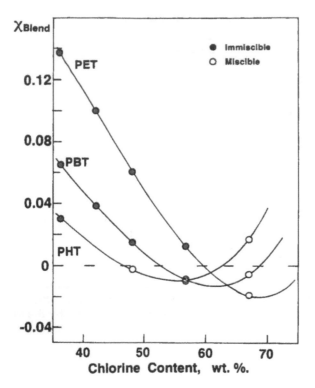

Figure 13 Comparison of calculated values of χ_{blend} and experimentally reported behavior for blends of chlorinated polyethylenes and semiaromatic polyesters. (See Ref. 65.)

by use of the appropriate segmental interaction parameters given in Table 1. Shown as Fig. 13, the estimated values of χ_{blend} compare favorably with reported behavior (62–64). As the methylene content of the polyester increases, the model describes quite accurately a shift of the miscibility window to polymers with a lower chlorine content.

IX. CONCLUDING COMMENTS

Care should be exercised when attempting to place physical significance to the sign and magnitude of segmental interactions, because they are easily influenced by reference quantities used to formulate the model; however, there are several conclusions that can be drawn. The segmental interactions derived from the application of the model to all of the blends described above are positive and therefore may be termed "repulsive." This includes those blends in which favorable "specific interactions" have been proposed as a possible contributor towards the formation of a miscible blend.

A specific interaction between the carbonyl of the ester group and the chlorine atom of chlorinated polymers is often proposed as the leading cause for miscibility in these systems; however, the ester–chloromethylene segmental interaction is slightly positive and unfavorable (Table 1). Such a result does not exclude the possibility of an "attractive" or associative carbonyl–chlorine interaction, but it does imply that it may have little or no influence on the resulting phase behavior. This situation may be the result of the combined effects of a repulsive component, arising from the adjoining —CO— and —CH— fragments, and an attractive component from the —Cl and —C=O fragments.

Interactions between extremely polar and nonpolar species, e.g., amide and hydrocarbon segments, are invariably large and unfavorable. In contrast, interactions between similar segments, e.g., hydrocarbon–hydrocarbon and ester–carbonate, are usually found to be small. Additionally, interactions of similar segments with a particular segment, e.g., methylene with ester and carbonate groups, respectively are also quantitatively similar. This is a sensible and consistent outcome of a now extensive tabulation of data on different polymer blends.

The remarkable success of this simple treatment to blends of these kinds of polymers can be attributed to the fact that small changes of chemical composition of the blend constituents has a far greater influence on their resultant phase behavior than does any of the additional thermodynamic factors that are not accounted for in the simple theoretical framework. A major drawback to applying the model using a fundamental segment classification scheme is the lack of information concerning the influence of microstructure and configuration, i.e., the effects of adjacent segments on segmental interactions. There are subtle effects, related to microstructure, that are often not accounted for; however, they usually tend not to dominate the phase behavior of these blends. This is quite unlike that encountered for example in blends of high-molar-mass polyolefins, where subtle changes of composition and structure can have a decisive influence.

Some of the projections contained in this chapter have been constructed using only three segmental interaction parameters whose principal deficiency is an inability to account for configuration of the segments. Accordingly, some of the quantitative aspects should be viewed only tentatively. Calculations directed to estimating the distribution of electric charge in polymer structures (67) illustrate why the arrangement of segments can influence their respective interactions with other segments. The quantitative inclusion of this feature into model calculations would represent a major advance.

REFERENCES

1. L. A. Utracki, ed. Encyclopedic Dictionary of Commercial Polymer Blends. Toronto: Chemtec, 1994.
2. N. R. Legge, G. Holden, and H. E. Schroeder, eds. Thermoplastic Elastomers: A Comprehensive Review. Munich: Hanser. 1987.
3. D. R. Paul. Effects of polymer–polymer interactions in multi-phase blends or alloys. Macromol. Symp. 78:83, 1994.
4. A. F. Yee. The yield and deformation behavior of some polycarbonate blends. J. Mater. Sci. 12:757, 1977.
5. E. Baer. The provision of useful mechanical properties in polymeric materials. Brit. Polym. J. 10:84, 1978.
6. P. Kambour, J. T. Bendler, and R. C. Bopp. Phase behavior of polystyrene, poly (2,6-dimethyl-1,4-phenylene oxide) and their brominated derivatives. Macromolecules 16:753, 1983.
7. G. ten Brinke, F. E. Karasz, and W. J. MacKnight. Phase behavior in copolymer blends: poly(2,6-dimethyl-1,4-phenylene oxide) and halogen substituted styrene copolymers. Macromolecules 16:1827, 1983.
8. D. R. Paul and J. W. Barlow. A binary interaction model for miscibility of copolymers in blends. Polymer 25:487, 1984.
9. T. S. Ellis. Blending of nylons. In: M. I. Kohan, ed. Nylon Plastics Handbook. Munich: Hanser, 1995, pp. 268–283.
10. T. S. Ellis. Miscibility and immiscibility of polyamide blends. Macromolecules 22:742, 1989.
11. T. S. Ellis. The role of repulsive interactions in polyamide blends. Polym. Eng. Sci 30:998, 1990.
12. T. S. Ellis. Critical miscibility limits in blends of aliphatic polyamides containing an aromatic polyamide. Polymer 31:1058, 1990.
13. T. S. Ellis. Aromatic polyamide blends: enthalpy relaxation and its correlation with phase phenomena. Macromolecules 23:1494, 1990.
14. T. S. Ellis. The influence of structure on phase behavior of polyamide blends. Macromolecules 24:3845, 1991.
15. T. S. Ellis. Polyamide–polyester blends: an estimation of the amide-ester interaction. J. Polym. Sci., Poly. Phys. 31:1109, 1993.
16. T. S. Ellis. Miscibility of blends of aliphatic main-chain polyesters. Macromolecules 28:1882, 1995.
17. T. S. Ellis. Miscibility of polyamide blends: effects of configuration. Polymer 36:3919, 1995.
18. D. W. Van Krevelen. Properties of Polymers. 2d ed. New York: Elsevier, 1976.
19. A. M. Kotliar. Interchange reactions involving condensation polymers. J. Polym. Sci., Macromol. Rev. 16:367, 1981.
20. R. S. Porter, J. M. Jonza, M. Kimura, C. R. Desper, and E. R. George. Polyesters 11: a review of phase behavior in binary blends: amorphous, crystalline, liquid crystalline and on transreaction. Polym. Eng. Sci. 29:55, 1989.
21. R. S. Porter and L. Wang. Compatibility and transesterification in binary polyester blends. Polymer 33:2019, 1992.

22. J. Devaux, P. Godard, and J. P. Mercier. The transesterification of bisphenol-A poly-carbonate and polybutylene terephthalate: a new route to block polycondensates. Polym. Eng. Sci. 22:229, 1982.

23. S. Varma and V. K. Dhar. Studies of nylon 6/PET polymer blends: structure and some physical properties. J. Appl. Polym. Sci. 33:1103, 1987.

24. R. Kamal, M. A. Sahto, and L. A. Utracki. Some solid state properties of blends of polyethylene terephthalate and polyamide-6,6. Polym. Eng. Sci. 22:1127, 1982.

25. Z. Pillon and L. A. Utracki. Compatibilization of polyester/polyamide blends via ester–amide interchange reaction. Polym. Eng. Sci. 24:1300, 1984.

26. I. Eguiazabal and J. J. Iruin. Polyarylate/polyamide 6 blends: a calorimetric study. Polym. Bull. 24:641, 1990.

27. M. Nadkarni, V. L. Shingankuli, and J. P. Jog. Blends of thermoplastic polyesters with amorphous polyamide. 1. Thermal and crystallization behavior. Polym. Eng. Sci. 28:1326, 1988.

28. Fakirov, M. Evstatiev, and J. M. Schultz. Microfibrillar reinforced composite from drawn poly(ethylene terephthalate)/nylon-6 blend. Polymer 34:4669, 1993.

29. T. S. Ellis. Phase behavior of polyamide–polyester blends: the influence of aromatic-ity. Polymer 38:3837, 1997.

30. D. G. Hamilton and R. R. Gallucci. The effects of molecular weight on polycarbon-ate–poly(butylene terephthalate) blends. J. Appl. Polym. Sci. 48:2249, 1993.

31. C. A. Cruz, D. R. Paul, and J. W. Barlow. Polyester–polycarbonate blends. IV. Poly(ε-caprolactone). J. Appl. Polym. Sci. 23:589, 1979.

32. C. A. Cruz, D. R. Paul, and J. W. Barlow. Polyester–polycarbonate blends. V. Linear aliphatic polyesters. J. Appl. Polym. Sci. 24:2101, 1979.

33. C. A. Cruz, J. W. Barlow, and D. R. Paul. The basis for miscibility in polyester–poly-carbonate blends. Macromolecules 12:726, 1979.

34. C. H. Lai, D. R. Paul, and J. W. Barlow. Group contribution methods for predicting polymer–polymer miscibility from heats of mixing of liquids. 2. Polyester containing binary blends. Macromolecules 22:374, 1989.

35. T. S. Ellis. Estimating interactions in blends of polyamides, polyesters and polycar-bonate using copolymers. Macromol Symposia 112:47, 1997.

36. T. S. Ellis. Phase behavior of blends of polyesters and polycarbonates. Polymer, 39:4741, 1998.

37. M. Dezhu, Z. Ruiyun, and L. Xiaolie. Miscibility of ethylene terephthalate–caprolac-tone blends with polycarbonate. Polymer 36:3963, 1995.

38. R. Zhang, X. Luo, and D. Ma. Multiple melting endotherms from ethylene tereph-thalate–caprolactone copolyesters. Polymer 36:4361, 1995.

39. M. Dezhu, L. Xiaolie, Z. Ruiyun, and T. Nishi. Miscibility and spherulites in blends of poly (ε-caprolactone) with ethylene terephthalate–caprolactone copolyester. Poly-mer 37:1575, 1996.

40. M. C. Luyten, E. J. F. Bogels, G. O. R. Alberda van Ekenstein, G. ten Brinke, W. Bras, B. E. Komanschek, and A. J. Ryan. Morphology in binary blends of poly(vinyl methyl ether) and ε-caprolactone-trimethylene carbonate diblock copolymer. Poly-mer 38:509, 1997.

41. L. M. Robeson. Phase behavior of polyarylate blends. J. Appl. Polym. Sci. 30:4081, 1985.

42. J. L. Rodriguez, J. I. Eguizabal, and J. Nazabal. Phase behavior and interchange reactions in poly(butylene terephthalate)/poly(ester-carbonate) blends. Polym. J. 28:501, 1996.

43. S. M. Aharoni. Aromatic poly(ester carbonate)/poly(ethylene terephthalate) alloys. J. Macromol Sci.-Phys. B22:813, 1983.

44. N. S. Murthy and S. M. Aharoni. X-ray and neutron scattering studies of poly(ester carbonate)/poly(ethylene terephthalate) alloys. Polymer 28:2171, 1987.

45. A. C. M. van Bennekom, D. T. Pluimers, J. Bussink, and R. J. Gaymans. Blends of amide modified poly(butylene terephthalate) and polycarbonate: transesterification and degradation. Polymer 38:3017, 1997.

46. D. Braun, D. Leiss, M. J. Bergmann, and G. P. Hellman. Miscibility behavior of various polymers made of alkyl and carbonate groups. Eur. Polym. J. 29:225, 1993.

47. J. S. Yoon, M. C. Chang, M. N. Kim, E. J. Kang, C. Kim, and I. J. Chin. Compatibility and fungal degradation of poly(R-3-hydroxybutyrate)/aliphatic copolyester blend. J. Poly. Sci.: Part B Poly. Phys. 34:2543, 1996.

48. M. Grimaldi, B. Immirzi, M. Malinconico, E. Martuscelli, G. Orsello, A. Rizzo, and M. Grazia Volpe. Reactive processing-property relationships in biodegradable blends useful for prosthesis application. J. Mater. Sci. 31:6155, 1996.

49. H. B. Tsai, H. C. Li, S. J. Chang, and H. H. Yu. Block copolyesters of poly(butylene terephthalate) and poly(butylene adipate). Polym. Bull. 27:141, 1991.

50. G. Montaudo, M. S. Montaudo, E. Scamporrino, and D. Vitalini. Mechanism of exchange in polyesters. Composition and microstructure of copolymers formed in the melt mixing process of Poly(ethylene terephthalate) and Poly(ethylene adipate). Macromolecules 25:5099, 1992.

51. M. Dezhu, X. Xiang, L. Xiaolie, and T. Nishi. Transesterification and ringed spherulites in blends of butylene terephthalate-ε-caprolactone copolyester with poly(-ε-caprolactone). Polymer 38:1131, 1997.

52. C. P. Papadopoulou and N. K. Kalfoglou. Blends of an amorphous copolyester with poly(butylene terephthalate). Eur. Polym. J. 33:191, 1997.

53. I. Avramov and N. Avramova. Calorimetric study of poly(ethylene terephthalate)/poly(butylene terephthalate) blends. J. Macromol. Sci. Phys B30(4):335, 1991.

54. G. O. Shonaike. Studies on miscibility of glass fibre reinforced blends of poly(ethylene terephthalate) with poly(butylene terephthalate). Eur. Polym. J. 28:777, 1992.

55. Y. Yu and K. J. Choi. Crystallization in blends of poly(ethylene terephthalate) and poly(butylene terephthalate). Polym. Eng. Sci. 37:91, 1997.

56. M. Bosma, G. ten Brinke, and T. S. Ellis. Polymer–polymer miscibility and enthalpy relaxations. Macromolecules 21:1465, 1988.

57. T. S. Ellis, unpublished results.

58. A. Golovoy and M. F. Cheung. Phase behavior studies of polyarylate/copolyester blends by thermal and dynamic mechanical analysis. Polym. Eng. Sci. 29:85, 1989.

59. E. M. Woo, J. W. Barlow, and D. R. Paul. Thermodynamics of the phase behavior of poly(vinyl chloride)/aliphatic polyester blends. Polymer 26:763, 1985.

60. G. Belorgey and R. E. Prud'Homme. Miscibility of polycaprolactone/chlorinated polyethylene blends. J. Polym. Sci.: Poly. Phys. Edn. 20:191, 1982.

61. G. Belorgey, M. Aubin, and R. E. Prud'Homme. Studies of polyester-chlorinated poly(vinyl chloride) blends. Polymer 23:1051, 1982.

62. J. J. Ziska, J. W. Barlow, and D. R. Paul. Miscibility in PVC-polyester blends. Polymer 22:918, 1982.
63. L. M. Robeson. Miscible blends of poly(vinyl chloride) and poly(butylene terephthalate). J. Poly. Sci. Polym. Lett. Ed. 16:261, 1978.
64. M. Aubin and R. E. Prud'Homme. A study of aromatic polyester/chlorinated polymer blends. Polym. Eng. Sci. 24:350, 1984.
65. G. O. R. Alberda van Ekenstein, H. Deuring, G. ten Brinke, and T. S. Ellis. Blends of caprolactam/caprolactone copolymers and chlorinated polymers. Polymer 38:3025, 1997.
66. A. C. Balazs, F. E. Karasz, W. J. MacKnight, H. Ueda, and I. C. Sanchez. Copolymer/copolymer blends: effect of sequence distribution on miscibility. Macromolecules 18:2784, 1985.
67. S. Ziaee and D. R. Paul. Polymer–polymer interactions via analog calorimetry. 1. Blends of polystyrene with poly(2,6,-dimethyl-1,4-phenylene oxide). J. Poly Sci.: Part B Poly Phys. 34:2641, 1996.

9

Miscibility of Nylon 66/Santoprene Blends

Gabriel O. Shonaike
Himeji Institute of Technology, Himeji, Hyogo, Japan

I. INTRODUCTION

In the early stage of polymer blends, research activities were focused on semicrystalline and amorphous thermoplastics. However, recent developments incorporate blending of rigid thermoplastics with elastomers or thermoplastic elastomers (TPEs). TPEs that can combine the processing characteristics of rigid thermoplastics with physical properties of vulcanized rubber are now receiving greater attention within the academic and industrial communities. The choice of TPEs over conventional elastomers is due to the following advantages (1–8):

Simple processing with fewer steps
Short molding cycle
No curing is needed during blending process
Damage tolerance
Unlimited shelf life
Thermal stability
Material quality consistence and higher economic return in recycling
Lower density in most cases, thereby giving additional cost savings

In most cases, when TPEs are a minor component in the blend, they act as impact modifiers or to improve some of its physio/mechanical or rheological characteristics (9). Nylon 66 on the other hand is chosen for its balance of properties in various applications where medium-to-high performance is required. However, properties of nylon such as impact can be improved by blending with thermoplastic elastomer (1). In this article, some of the properties of both components will be highlighted before discussing some effects of blend compositions on mechanical properties.

A. Nylon 66

Nylon 66 is a trade name for polyhexamethylene adipamide, which is one of the major polyamides manufactured by a condensation reaction between hexamethylene diamine and adipic acid (10–12). Its popularity as an engineering polymer is due to its superior balance of properties, including toughness over a wide range of temperatures, impact and abrasion resistance, and good resistance to organic and petroleum products (11). However, all polyamides are water sensitive owing to the hydrogen bonding character of the amide group. Water absorption of nylon 66 in ambient temperatures can be as high as 9%. The presence of water in the structure is one of the major disadvantages of nylon because it acts as a plasticizer, thereby reducing the tensile strength and modulus while increasing both the elongation at break and the toughness (14). It is also attacked by strong acids, oxidizing agents, and concentrated solutions of certain salts (15,16). One of the major advantages of nylon 66 is that it can retain its properties to a useful degree at high temperatures (15), e.g., it can withstand short-term exposure to temperatures exceeding 200°C. Application of heat stabilizers and other chemical modifications during processing increases long-term oxidative and hydrolytic stability (15). Nylon 66 has a melting point T_m of 265°C and a glass transition temperature T_g of around 50°C when dry. β- and γ-transitions occur at about $-80°C$ and $-140°C$, respectively (16).

Since its discovery, it has been subjected to various research activities in both industrial and academic communities. Its unique properties have made it one of the most successful engineering polymers suitable for various applications in all sectors. Since nylon is not a new polymer, a review on the work done would occupy the whole volume; more information can be obtained in standard textbook (16).

B. Santoprene

Santoprene is the first brand of vulcanizate TPE developed by Monsanto in 1981. It is a family of advanced thermoplastic elastomers that can combine the processing characteristics of thermoplastics with the physical properties of vulcanized rubber, such as heat resistance and low compression set (17). All grades of commercially available Santoprene are polymerized polyolefin compounds (polypropylene) with ethylene-propylene-diene monomer (EPDM) and require no post curing or annealing to attain their full range of performance and properties.

The morphology of Santoprene indicates that it contains a fine dispersion of highly vulcanized EPDM rubber in a matrix of polypropylene (PP) as the continuous phase. Miscibility of the two components may be due to nearly equivalent solubility parameters (18), this gives a fine rubber dispersion that provides the desired material properties. The process leading to a fine dispersion of EPDM in PP matrix is by dynamic vulcanization. Dynamic vulcanization means vulcanizing the elastomer with nonvulcanizable molten thermoplastics. Excellent works on dynamic vulcanization of EPDM-PP (19) and other combinations has been carried out by Coran and collaborators (20–28) at Monsanto. Akiba and Hashim (29) have shown that during dynamic vulcanization of EPDM-PP, small rubber droplets are vulcanized to give vulcanized rubber particles of stable domain morphology that are then dispersed in the molten PP. Recent studies (30) have shown that systems for dynamic vulcanization should have the following features:

1. The surface energy difference between the rubber phase and the resin should be small.
2. Crystalline resin should be used.
3. The rubber component should be densely entangled.

The morphology of Santoprene, which is typical of thermoplastic vulcanizates, is depicted in Fig. 1, which shows a fine dispersion of EPDM rubber uniformly dis-

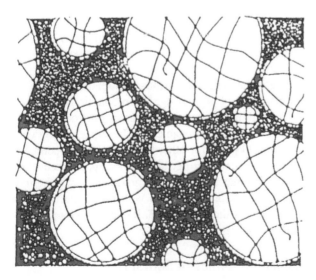

Figure 1 Morphology of dynamically cross-linked compound. (From Ref. 29.)

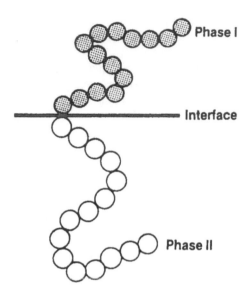

Figure 2 Distribution of compatibilizing block copolymer between rubber and plastic phases. (From Ref. 31.)

tributed in a matrix of polypropylene as the continuous phase (31). Coran and collaborators (20) have shown that in Santoprene, both EPDM and polypropylene are adequately miscible with fine rubber dispersion to provide the desired material properties. However, attainment of good dispersion is aided by incorporation of a compatibilizer that provides interaction between the two phases, as shown in Fig. 2.

Santoprene is increasingly in demand for various applications due to its unique combination of properties, ease of processing, low production costs, consistent quality, and greater production performance. In terms of production, its thermoplasticity and melt flow characteristics enable it to be processed on conventional thermoplastic equipment like that for other thermoplastics. Operations such as injection molding, extrusion, calendering, blow molding, and recycling can all be carried out with the efficiency and economy associated with thermoplastic materials. Cost vs. performance of Santoprene with other thermoplastic elastomers and thermoset elastomers is shown in Fig. 3. It can be seen in the figure that Santoprene fared well in both cases.

C. Physical and Mechanical Properties of Santoprene

Santoprene is attractive for its combination of properties that make it suitable for various applications. All grades of Santoprene have excellent physical properties. The specific gravity of around 0.98 compares favorably with that of other vulcan-

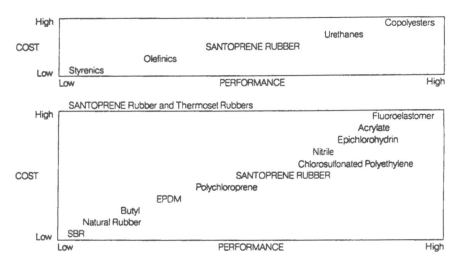

Figure 3 Cost vs. performance of Santoprene with other elastomers.

ized rubbers such as polychloropene chlorosulfonated polyethylene and EPDM (31). Santoprene has an excellent chemical resistance such as to acids and bases. It is regarded as one of the main engineering thermoplastic elastomers used for advanced applications in various sectors. The mechanical properties of various grades of Santoprene as obtained from Monsanto are shown in Table 1. The properties increase with increasing grades, e.g., the tensile strength of Grade 203-50 is almost four times higher than that of 201-73. Higher grades indicate a stiffer structure, i.e., an increase in PP content increases the mechanical properties. The hardness of Santoprene varies from 55A to 50D (90A) and this range permits the design of articles ranging from flexible to semirigid.

II. MISCIBILITY OF NYLON/SANTOPRENE BLENDS

It is well documented in the literature that blending of multicomponent polymeric systems generates a new material with advantages over either of the individual components. However, for this to occur, the system must be miscible or partially miscible. Miscibility is the tendency or capacity of the components to form a uniform blend (32), i.e., the blend is homogeneous down to the molecular level. In most cases, polymers exhibit immiscible behavior and require a compatibilizer. A compatibilizer is added to a blend system in order to introduce specific interactions between the components (33). Within the last couple of years, several investigations on compatibilization of polymeric materials have been carried out (34–43).

Table 1   Mechanical Properties of Santoprene

Properties	ASTM test method	Test units U.S. (SI)	Test temp. °F (°C)	Santoprene rubber grades						
				201-55 101-55	201-64 101-64	201-73 101-73	201-80 101-80	201-87 101-87	203-40 103-40	203-50 103-50
Hardness	D2240	5 sec Shore	77 (25)	55A	64A	73A	80A	87A	40D	50D
Specific gravity	D297	—	77 (25)	0.97	0.97	0.98	0.97	0.96	0.95	0.94
Tensile strength	D412	psi (MPa)	77 (25)	640 (4.4)	1000 (6.9)	1200 (8.3)	1600 (11.0)	2300 (15.9)	2750 (19.0)	4000 (27.6)
Ultimate elongation	D412	%	77 (25)	330	400	375	450	530	600	600
100% modulus	D412	psi (MPa)	77 (25)	290 (2.1)	340 (2.3)	470 (3.2)	700 (4.8)	1000 (6.9)	1250 (8.6)	1450 (10.0)
Tear strength	D624	pli (kN/m)	77 (25)	108 (19)	140 (24.5)	159 (27.8)	194 (34.0)	278 (48.7)	369 (64.6)	514 (90.0)
		pli (kN/m)	212 (100)	42 (7.3)	58 (10.2)	76 (13.3)	75 (13.1)	133 (23.3)	203 (35.5)	364 (63.7)
Tension set	D412	%	77 (25)	6	10	14	20	33	48	61
Compression set,168 h	D395	%	77 (25)	23	23	24	29	29	32	41
		%	212 (100)	25	36	36	41	45	49	81
Flex fatigue		Mega cycles to fail	77 (25)	<3.4 (<3.4)	<3.4 (<3.4)	<3.4 (<3.4)	— —	— —	— —	— —
Brittle point	D746	°F (°C)	— —	<-76 (-60)	<-76 (-60)	-81 (-63)	-81 (-63)	-78 (-61)	-71 (-57)	-29 (-34)

Nylon 66 is one of the most successful medium-range engineering polymers and has been blended with various polymers (44–51). However, its miscibility with Santoprene has not been clearly addressed. Several years ago, Coran and collaborators (20–27) reported the effects of blending vulcanizate rubbers with various thermoplastics. In one of their findings, a significant interaction occurred between modified polyolefin and nylon due to the formation of compatibilizing segments or blocks of both nylon and polyolefin sequences.

In this article, recent studies on nylon–Santoprene blend in our laboratory will be discussed. The blends were mixed to give various compositions and blended by using a single-screw extruder.

A. Melting Temperature

Table 2 shows the results obtained from differential scanning calorimetry (DSC) thermograms shown in Figs. 4 and 5. The melting temperature of nylon irrespec-

Table 2 Thermal Properties of Nylon 66/Santoprene Blends

FIRST SCANS						
Nylon 66/ Santoprene (%) (°C)	Nylon 66			Santoprene		
	T_m (°C)	ΔH (J/g)	T_c (°C)	T_m (°C)	ΔH (J/g)	T_c
100/0	264	70.06	208	—	—	—
90/10	264	55.09	226	153	1.54	102
75/25	265	53.47	225	153	3.57	102
50/50	260	36.22	224	153	11.09	102
25/75	259	18.41	224	152	21.17	101
0/100	—	—	—	154	28.61	101

SECOND SCANS						
Nylon 66/ Santoprene (%) (°C)	Nylon			Santoprene		
	T_m (°C)	ΔH (J/g)	T_c (°C)	T_m (°C)	ΔH (J/g)	T_c
100/0	261	51.43	208	—	—	—
90/10	260	61.85	226	152	1.30	102
75/25	261	53.11	225	153	4.33	102
50/50	258	34.39	225	153	12.48	102
25/75	257	17.80	225	152	21.17	102
0/100	—	—	—	153	28.61	101

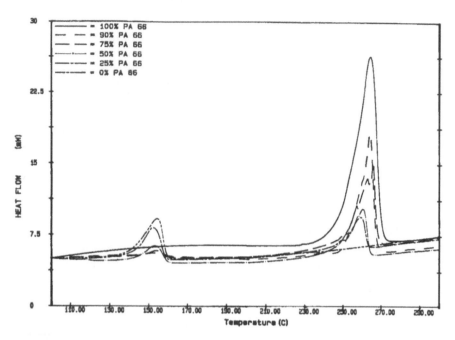

Figure 4   DSC thermograms of nylon 66/Santoprene blends (first scans).

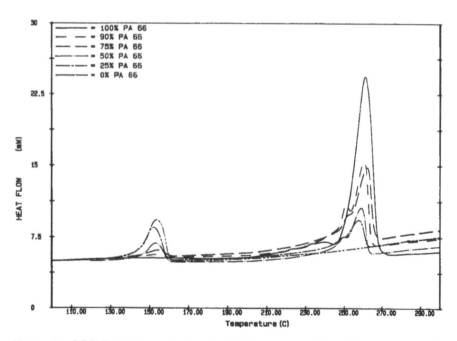

Figure 5   DSC thermograms of nylon 66/Santoprene blends (second scans).

tive of blend composition was slightly lower in second scans than in first scans because of the melting away of imperfections or the rearrangement of molecules.
Similar behavior has been observed in other blend combinations such as poly(hydroxyether of phenolphthalein) (PHP) with poly(ethylene terephthalate) (PET)
(52) and PET with poly(butylene terephthalate) (PBT) (53). A closer look at the
second scan DSC endotherm reveals small peaks that occurred around 248°C in
all the blends. This second peak however, is a consequence of reorganization and
the presence of different crystal structures owing to the heating process (54). In
Fig. 4, the first peak around 153°C corresponds to the melting temperature of PP
in Santoprene, whilst the second peak around 264°C is that of nylon. It is widely
acknowledged that a miscible blend will show a single T_g whilst an immiscible
blend will contain dual T_gs (55) due to phase segregation of individual components. In a simple way, this may be accepted, but it is not completely certain. In
most cases, some partially miscible blends will display double T_g, but the evidence
of partial miscibility is also reflected in other physical or mechanical properties.

The melting temperature of the blends indicates that Santoprene is partially
miscible with nylon in the Santoprene-rich region, because of a slight reduction of
the melting point of nylon in 50/50 and 75/25% nylon–Santoprene blend. Above
50% Santoprene content, the melting point is not affected, and this may indicate a
lack of significant interaction between the two components. The concept of melting temperature depression is well documented in the literature. Various investigators (56–60) have derived several equations to explain the depression of T_m as
a result of blending crystalline with amorphous polymers. A simple expression for
the analysis of melting point depression (based on the evaluation of interaction parameter) has been given by Ellis as follows (61):

$$\frac{1}{T^\circ_m \text{ blend}} - \frac{1}{T^\circ_m} = -\frac{RV_2}{\Delta h_2 V_1} \chi_{12}\,(1-\phi)^2 \tag{1}$$

(by ignoring entropic contributions to χ_{12}),

$$\chi_{12} = \frac{BV_1}{RT} \tag{2}$$

where B is the interaction energy density. Combining Eqs. 1 and 2 putting $T = T^\circ_{m\text{blend}}$ it is possible to obtain

$$1 - \frac{T^\circ_{m \text{ blend}}}{T^\circ_m} = \frac{-BV_2}{\Delta h_2}\,(1-\phi_2)^2 \tag{3}$$

where T°_m and $T^\circ_{m \text{ blend}}$ are the equilibrium melting points of crystallizable component in the pure state and the blend, respectively. R is the gas constant, V_1 is the
molar volume of the respective component, Δh_2 is the heat of fusion per mole of
repeat unit, and ϕ_2 is the volume fraction of component 2 present in the blend.
Thus, if values of T°_m and $T^\circ_{m \text{ blend}}$ are obtained, a suitable graphical procedure fa-
cilitates an estimation of B and therefore χ_{12}, i.e., χ_{blend} hr.

However, the melting point depression has been attributed to the diluting effect of the noncrystallizable polymers when the two components are compatible in the melt (62,63). Recently, Lee and collaborators (64) reported the effect of temperature depression on blends containing nylon 6 and poly(maleic anhydride-co-vinyl acetate) (poly(MAH-co-VAc) and polyhydroxylated poly(maleic anhydride-co-vinyl acetate) (MAH-co-VAc)H. A comparison of the melting temperatures of the two blends prompted the authors to conclude that miscibility of nylon 6 with poly(MAH-co-VAc)H was better due to a rapid depression of melting temperature. In some blend systems, depression of the melting temperature does not occur with increasing composition. Martuscelli and collaborators (62) have shown that the lack of a melting point depression is due to noninteraction between the crystallizable and noncrystallizable components at the melting point. The melting point depression in the Santoprene-rich compositions may be explained as follows:

$$\begin{array}{ccc} \text{NYLON} & & \text{SANTOPRENE} \\ \text{(semicrystalline)} & + & \text{(semicrystalline PP + amorphous EPDM)} \end{array}$$

During melt mixing, interaction of the two components is assumed to have taken place in amorphous domains. Thus the depression of the melting point of nylon will depend on the content of the amorphous phase, i.e., the higher the amorphous content, the greater the effect on the melting point. In the nylon-rich compositions, the amorphous content is lower than in the Santoprene-rich region, and as a result the melting temperature of nylon is not affected. On the other hand, in the Santoprene-rich region, a significant reduction of melting temperature of nylon occurred due to high amorphous content. In some cases, T_m may increase when the defective molecules of one component are selectively dissolved into the other (65–68). Lower values of T_m in blends containing nylon and other thermoplastics may result in lower crystallinity (69). This is analogous to Santoprene/nylon blends, because the degree of crystallinity of the blend was reduced above 50/50 blend composition. This is in agreement with temperature depression above in the Santoprene-rich region. The degree of crystallinity as a function of Santoprene content (Fig. 6) was obtained by analyzing the heat of fusion of nylon and taking note of the weight fraction of the blend. Figure 6 suggests that little interaction did occur between the two phases, especially in the Santoprene-rich region.

B. Crystallization Temperature

Crystallization (T_c) is the process of forming a crystalline material after cooling from the melt. It occurs below the melting point of semicrystalline polymers and generates an ordered structure in both crystalline and amorphous phases. The DSC thermogram of T_c is shown in Fig. 7. In this case, neat nylon crystallizes at 208°C

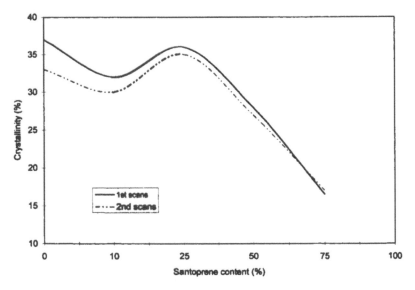

Figure 6 Degree of crystallinity of nylon in nylon/Santoprene blends.

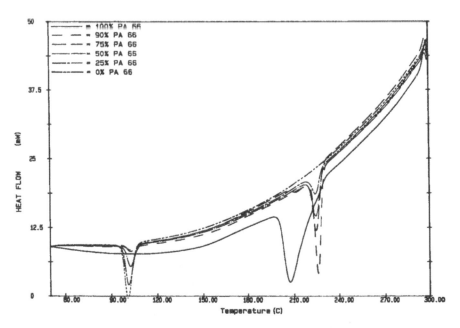

Figure 7 Crystallization temperature of nylon in nylon/Santoprene blends.

and the blends crystallize at higher temperatures, i.e., 224 and 226°C. The crystallization temperature of Santoprene was not affected as a result of blending it with nylon.

The increase in crystallization temperature in the blend should not be confused with blend miscibility. It simply shows that good mixing of both components did occur during blending even though they are immiscible. On the other hand, the increased crystallization temperature may indicate a partial miscibility, which can be attributed to the EPDM component in Santoprene. Two things may have occurred in the blend when cooling from the melt:

1. The EPDM may contain a component that is slightly soluble in nylon, thereby plasticizing it. The plasticizing effect may result in high chain mobility of nylon.
2. The Santoprene, via the EPDM component, may have acted as a nucleating agent. Both effects lead to an early crystallization of nylon upon cooling. The EPDM component, which provides nuclei for heterogeneous crystallization, accelerates the formation of crystallized regions at high temperature. The increased crystallization temperature, which is independent of blend composition, attests that once the Santoprene is melt-mixed with nylon, the crystallization rate is increased due to plasticizing and diluent effects of blending. This raises an important phenomenon, i.e., supercooling or undercooling temperature. The extent of interaction between the two components can be explained based on supercooling temperature. It is defined as the temperature difference between the melting and crystallization temperatures, i.e.,

$$T_{sc} = T_m - T_c \qquad (4)$$

Supercooling temperature depends on the cooling rate and nucleation mechanism. Three mechanisms have been identified by Utracki (70), i.e.,

1. Spontaneous homogeneous nucleation, which rarely occurs in a supercooled homogeneous melt
2. Orientation induced nucleation caused by alignment of macromolecules and spontaneous crystallization
3. Heterogeneous nucleation on the surface of a foreign phase

However, mechanisms (2) and (3) prevail due to the presence of EPDM in the Santoprene component. As explained above, the plasticizing effect of Santoprene enhances nylon chain mobility, and it provides nuclei for heterogeneous crystallization. The concept of supercooling in relation to blend miscibility requires further investigation, as it has not been clearly addressed. Some investigators (62,69,71,72) have used the concept to explain temperature depression in polymer blends. Since the depression of the melting temperature of a crystallizable component is evidence of partial miscibility, one may conclude that low supercooling

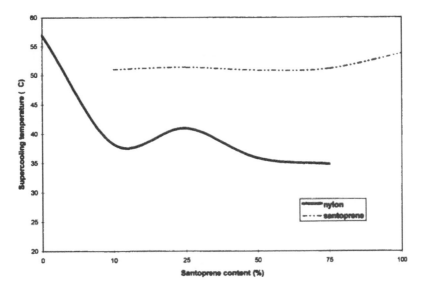

Figure 8   Supercooling temperatures of nylon and Santoprene in nylon/Santoprene blends.

favors miscibility. It is seen in Fig. 8 that lower supercooling in the Santoprene region (above 50/50 composition) can be taken as an evidence of partial miscibility of the two components.

III.   MECHANICAL PROPERTIES

One of the main reasons for blending is to create a new material with improved mechanical properties. This is achieved by combining the good properties of the individual components while diminishing their inferior properties. However, the mechanical properties of the blend will depend on whether the components form a single phase (miscible) or separate phases (immiscible) after blending. Also, the mechanical properties rely on the degree of adhesion at the interface of the two components. Thus incompatibility, which is due to the absence of specific interaction between the two phases, often leads to inferior mechanical properties. In the case of some TPEs, improved adhesion at the interface may not necessarily lead to improved mechanical properties (73,74). However, in most cases, a significant improvement in mechanical properties results from good adhesion between the two components.

The mechanical properties will be represented by discussing some recent results on tensile and impact properties of nylon/Santoprene blends.

A. Tensile Properties

This is unarguably the most widely reported of mechanical properties. The load–elongation curve of thermoplastic elastomer is characterized by low modulus, tensile strength, and high elongation at break. It is well documented that blending a rigid polymer with a thermoplastic elastomer will improve the tensile properties of the thermoplastic elastomer component at the expense of the rigid component. Therefore addition of a soft segment block copolymer to a rigid thermoplastic dramatically increases the ductility and gives some reduction in yield strength and modulus (75). It can be seen in Fig. 9 that the stress–strains of nylon/Santoprene blends show different deformation characteristics. The deformation in this case ranges from moderately ductile behavior in the nylon-rich region to highly ductile in the Santoprene-rich region. Nylon on its own displays a ductile behavior, which is reduced on addition of Santoprene. The reduction of strain in the nylon-rich region suggests that the presence of rubbery segment in Santoprene hinders cold-drawing of nylon, which eventually leads to a premature failure. A similar observation was reported on PA6/EPM-g-SA (76). The authors attributed the reduction of strain to the presence of EPM-g-SA, i.e., it makes cold-drawing of nylon matrix more difficult. On the other hand, Santoprene, which is more elastomeric, exhibits a high deformation, and an addition of 25% nylon reduced the strain. This may indicate a partial miscibility of the two components. Owing to the addition of Santoprene, all stress levels of nylon were reduced with increasing Santoprene content (Fig. 9). Figure 10 shows the tensile strength of binary blends of nylon/Santoprene and ternary blends of nylon/Santoprene/SAN as functions of Santoprene content. 5% of SAN was added to the binary blend in order to study the influence it might have on the tensile properties of the blends. As expected, in the binary blend, the tensile strength reduces with increasing Santoprene content up to the 50/50 level. Above the 50% level, i.e., Santoprene-rich compositions, the reduction was minimal. In this case, it is certain that even though both components are immiscible, addition of Santoprene causes a significant reduction of tensile strength of nylon with increasing Santoprene content within the nylon-rich region. Thus the EPDM component in the Santoprene is likely to affect the spherulitic growth of nylon crystals, which eventually leads to premature failure. This fact is supported by the reduction of the crystallinity of nylon with increasing Santoprene content. The results for the ternary blends (Fig. 10) are almost identical with those of binary blends except for a little improvement with addition of SAN. This is an expected trend in most thermoplastic elastomers, where the addition of the hard phase reinforces the matrix (77). It can be seen in Fig. 10 that irrespective of blend compositions, ternary blends had higher tensile strength than binary blends. The results for tensile modulus shown in Fig. 11 show a similar trend in both cases, with the ternary blend showing a higher modulus. It is seen in the figure that the tensile modulus of neat

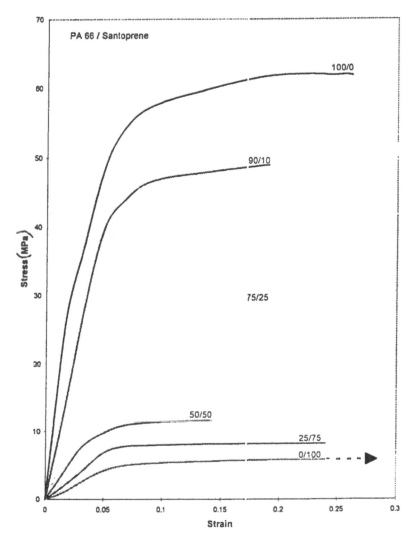

Figure 9 Stress–strain curves of nylon/Santoprene blends.

Santoprene was about 0.2 GPa, which is eight times lower than that of neat nylon (about 1.6 GPa). However, addition of 5% SAN increases the modulus of neat nylon from 1.6 to around 1.8 GPa. The modulus of neat nylon was hardly affected until the Santoprene content was more than 25% in the ternary blend, whereas a significant drop in the modulus occurred in the binary blend with as little as 10% Santoprene content.

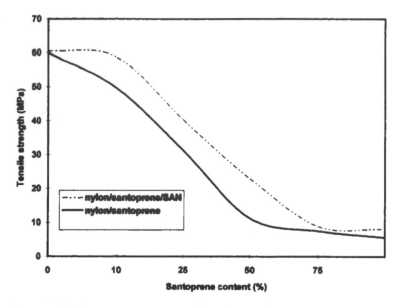

Figure 10   Tensile strength as a function of Santoprene content.

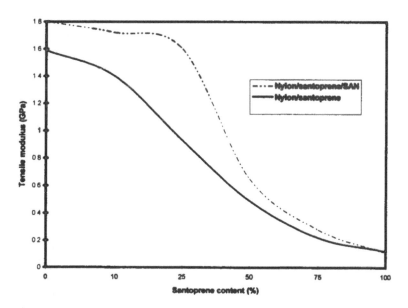

Figure 11   Tensile modulus as a function of Santoprene content.

B. Fracture Toughness

Fracture toughness is the resistance of a material to the propagation of an existing crack. In a simple way, it is the ability of a material to resist fracture by absorption of energy by plastic deformation. Thus ductile materials will have high toughness due to a large deformation before failure whilst brittle materials will not, because fracture occurs at lower deformation level. The fracture toughness of the single-edge notched dumbbell sample can be obtained from the well-known equation

$$K_{IC} = \sigma_c Y \sqrt{\pi a} \tag{5}$$

where K_{IC} is the mode I fracture toughness (applicable when the sample shows a brittle behavior), σ_c is the fracture initiation stress, a is the crack length, and Y is the correction factor for a single-edge tensile specimen given as (78)

$$Y = 1.12 - 0.231 \left(\frac{a}{W}\right) + 10.55 \left(\frac{a}{W}\right)^2 - 21.72 \left(\frac{a}{W}\right)^3 + 30.39 \left(\frac{a}{W}\right)^4$$

where W is the specimen width. Within the framework of linear elastic fracture mechanics (LEFM), K_{IC} is converted to the critical energy release rate (G_{IC}) using the relationship

$$G_{IC} = \frac{K_{IC}^2}{E} \tag{6}$$

where E is the Young's modulus. In the case of ductile material, the J-integral (J_c) method is applicable. In this case, J_c is obtained from the load–displacement curves on the assumption that the material behaves plastically at fracture initiation.

$$J_c = \frac{2U_c}{hb} \tag{7}$$

where U_c is the total work area under the load–displacement curve at fracture initiation, h is the specimen thickness, and b is the ligament length.

Table 3 shows the fracture toughness of nylon, Santoprene, and their blends. The mode of failure is either brittle or ductile, depending on the blend composition. The values of the fracture energy are fairly large, probably because of the elastomeric nature of the samples, especially 100% Santoprene. The fracture toughness of nylon based on Eq. (5) (obtained from an average of six specimens), which is extremely high and which reduces with increasing Santoprene content within the nylon-rich region, is very interesting. The behavior is unanticipated because the EPDM component is expected to enhance the toughness of nylon. A drop of about 10% between the neat nylon and the 75/25 blend is attributed to the polypropylene content in Santoprene. Polypropylene is known to have low tough-

Table 3 Fracture Toughness Data

Composition	G_{ic} (kJ/m^2)	J_c (kJ/m^2)	Fracture behavior
100/0	82.4	—	brittle
90/10	74.9	—	brittle
75/25	73.4	—	brittle
50/50	89.8	—	brittle
25/75	—	56.8	ductile
0/100	—	350.5	ductile

ness, and a thorough mixing, which occurred during blending with nylon, reduces the toughness of nylon. The increased fracture toughness of a 50/50 blend sample confirms the region where the interaction between the two components is highest. This is in agreement with the results of the tensile test (Fig. 10) where the improvement in tensile strength of Santoprene commenced in the region of 50/50 composition. It is evident from the above results that immiscible blends can generate improved properties without compatibilization due to good mixing (79). Although phase separation may occur at a micro level, careful blending may result in one component partially dissolved in the other. The fracture toughness approach is just another method that can be used to study the blend miscibility.

C. Impact Strength

The impact resistance is the ability of a material to withstand a sudden shock loading. The impact resistance of a polymer is a complex function of geometry, mode of loading, load application rate, material properties, and environment (80). There are several published articles in the literature on blending of rigid thermoplastics with elastomeric materials. All investigations have one main objective, to improve the impact resistance of rigid thermoplastics.

It is well known that the impact strength of a polymer blend will depend on the amorphous or rubbery content in the blend, as well as the interfacial adhesion between the dispersed phase and the matrix (81). In Fig. 12, the impact test results of neat nylon, Santoprene, and their blends are shown. The figure shows that nylon has a very low impact strength whilst the impact strength of Santoprene is about six times higher. It can be seen in the figure that the low impact strength of nylon remains the same in the nylon-rich region. However, with 50/50 blend composition, the impact strength of nylon was slightly improved, and it increased for the Santoprene-rich compositions. Thus if the two components are miscible, with the addition of as little as 10% Santoprene, one would expect an improvement in impact strength of nylon in the nylon-rich compositions. It has been shown (82)

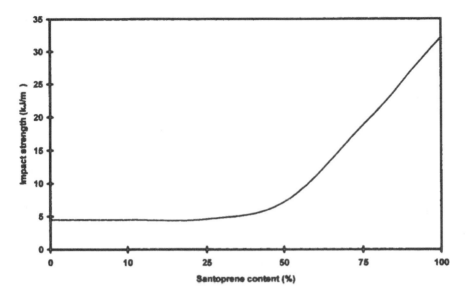

Figure 12   Impact strength vs. Santoprene content.

that achieving a high impact strength requires a high rubber content, and the integrity of the rubber domain is maintained by cross-linking. In view of this, low impact strength in the nylon-rich region suggests that the EPDM content in the nylon-rich compositions in insufficient to cause any dramatic effect on nylon. The mode of failure (not shown) ranges from brittle failure in the nylon-rich region characterized by rapid crack propagation to a ductile mode for Santoprene-rich compositions with much greater plastic deformation. Poor impact strength for the nylon-rich compositions is simply due to lack of adhesion at the interface of the two components resulting from the large size of the dispersed phase (83). Santoprene, being a flexible polymer, has a very high impact strength, which is reduced on blending with nylon, and as is observed in Fig. 12, a dramatic reduction of impact strength of Santoprene occurred with increasing nylon content. The same trend is observed in impact energy as a function of Santoprene content, are shown in Fig. 13. Once again, the low impact energy of nylon remains unchanged until 50/50 composition. The EPDM in Santoprene, which would be expected to enhance the impact energy of nylon, is ineffective until around 50/50 composition. As mentioned earlier, the presence of low toughness polypropylene in the Santoprene may be responsible for inactivity of Santoprene in the nylon-rich compositions.

The results of mechanical tests indicate that nylon is not particularly affected by Santoprene in the nylon-rich region.

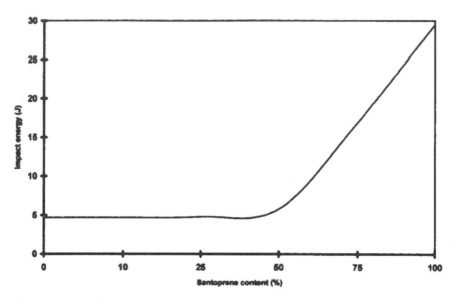

Figure 13   Impact energy vs. Santoprene content.

IV. MORPHOLOGY

Studies on morphology of the blends can provide some useful information on miscibility of the components. Scanning electron microscopical (SEM) observation of fractured surfaces of nylon 66, Santoprene, and their blends are shown in Fig. 14. Neat nylon 66 shown in Fig. 14a contained a spherical aggregate of spherulites, which is a typical characteristic of semicrystalline polymers. The EPDM and polypropylene, which are the main Santoprene components, show one phase that is an evidence of miscibility of the two components (Fig. 14b). There is no phase separation, and the morphology, which is homogeneous, displays a typical ductile failure. As for the blends, the behavior is quite different, i.e., the morphology depends on blend composition. As can be seen in Fig. 14c, a significant phase change occurred with as little as 10% Santoprene added to nylon (90/10). Thus the occurrence of two phases of the components is an evidence of immiscibility. In this case, the 90/10 blends contain large dimples embedded in a nylon matrix. The dimples of irregular shape, which are loosely placed, indicate poor adhesion at the interface of the two components. It is obvious from the micrograph that the Santoprene phase, which is partially detached from the nylon phase, leads to phase segregation of the components, although it is possible that total phase separation did not occur during blending, as is confirmed by thermal

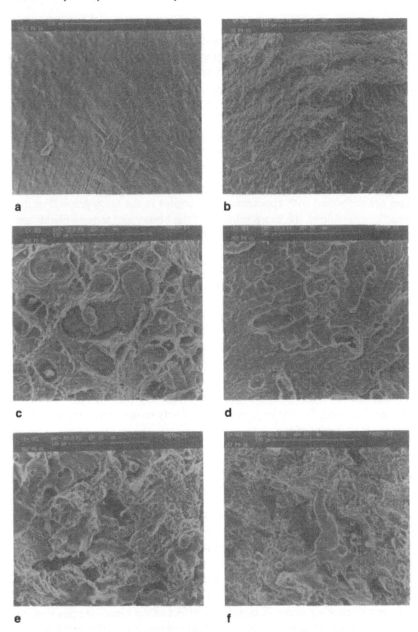

Figure 14 Morphology of nylon 66/Santoprene blends: (a) 100/0, (b) 0/100, (c) 90/10, (d) 75/25, (e) 50/50, and (f) 25/75.

and mechanical properties. In the 75/25 blend shown in Fig. 14d, it is seen that the shape of the dimples is reduced and oriented. This orientation indicates a more ductile behavior as a result of increased santoprene content. Both 50/50 and 25/75 blends on the other hand displayed different behavior, i.e., the two components are well mixed with good interaction. The shape of the dimples is now smaller than in either 90/10 or 75/25 blends and is a sign of little adhesion between the components.

The morphology of a ternary blend containing nylon 66, Santoprene, and 5% stryrene acrylonitrile (SAN) copolymer is shown in Fig. 15. The platelike spherulitic structure observed in neat nylon (Fig. 14a) has disappeared because of the presence of 5% SAN in the neat nylon resin. As can be seen in Fig. 15a, the spherulites are deformed with appearance of granules in the micrograph. For the 90/10 blend compositions, the morphologies of the binary and ternary blends are similar, except that the size of the particles is reduced in the ternary blend. Addition of 5% SAN caused major morphological changes in the rest of the blends, i.e., 75/25, 50/50, and 25/75. In the 75/25 blend (Fig. 15d), both the size and the quantity of the particles are reduced and oriented. The 50/50 blend suggests a partial dissolution of Santoprene into nylon and at 25/75 a single phase morphology appeared. Based on the above SEM observation of both binary and ternary blends, it seems that SAN compatibilizes the nylon/Santoprene blends to a certain extent.

The phase change as a result of various blend compositions is reflected in the mechanical properties, where good interfacial adhesion leads to improved impact and tensile properties. However, the improvement occurred in above 50/50 blend compositions, and since the blend is uncompatibilized, it is assumed that if a good compatibilizer is added, further improvement is likely to occur as a result of improved adhesion at the interface.

V. CLOSING REMARKS

The research on blending of nylon 66 with Santoprene has revealed that both components are immiscible but good interaction does occur during blending, based on the results of thermal and mechanical properties. The blends showed double melting temperatures corresponding to those of the polypropylene component in Santoprene and nylon 66 irrespective of blend compositions. The increase in crystallization temperature of nylon was due to the EPDM component in Santoprene, which acts as a nucleating agent. A significant effect of the blend composition on mechanical properties occurred at around 50/50 composition. The SEM micrographs of the blends revealed a two-phase structure confirming phase segregation of the two components especially for nylon-rich compositions. However, it may be concluded that the addition of as little as 5% SAN compatibilizes the blend, and the interfacial adhesion of the components is improved above the 75% composition as was revealed by SEM micrographs.

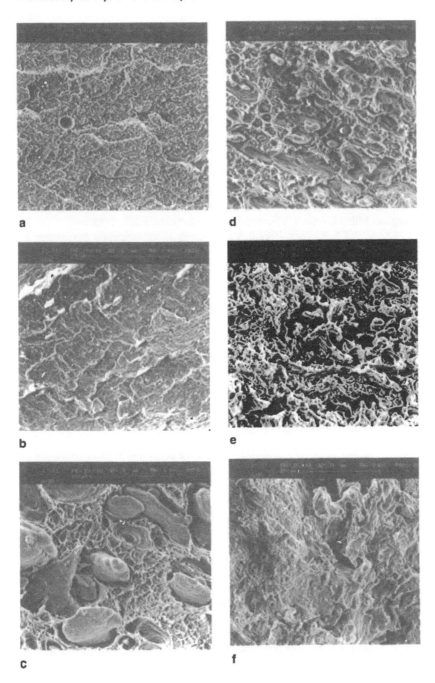

Figure 15 Morphology of nylon 66/Santoprene blends containing 5% SAN:
(a) 100/0, (b) 0/100, (c) 90/10, (d) 75/25, (e) 50/50, and (f) 25/75.

ACKNOWLEDGMENTS

The author gratefully acknowledges the support of Universiti Sains Malaysia for a research grant to carry out this work. I am also grateful to my students Teh How Kiat and Heng Siew Choo for carrying out most of the work.

REFERENCES

1. C. P. Rader and B. M. Walker. Thermoplastic elastomers and the rubber and the plastics industries. In: Handbook of Thermoplastic Elastomers (B. M. Walker and C. P. Rader, eds.). Van Nostrand Reinhold, 1987, Chap. 1.
2. H. Witich and K. Friedrich. Interlaminar fracture of laminates made of thermoplastic impregnated fiber bundle. J. Thermoplas. Comps. Mater. 1:221, 1988.
3. I. Y. Chang and J. K. Lees. Recent developments in thermoplastic composites. J. Thermoplas. Comp. Mater. 1:227, 1988.
4. T. Matsuo. Prospect of thermoplastic composites. J. Jap. Soc. Polym. Proc. 5:225, 1993.
5. W. J. Cantwell, P. Davies, and H. H. Kausch. The effect of cooling rate on deformation and fracture in IM6/PEEK composites. Compos. Structure 14:151, 1990.
6. D. C. Leach. Continuous fiber reinforced thermoplastic composites. In: Advanced Composites (I. Partridge, ed.). Elsevier, 1989, Chap. 2.
7. J. B. Cattanach and S. N. Cogswell. Processing of aromatic polymer composites. In: Development of Reinforced Plastics-5 (G. Pritchard, ed.). Elsevier, 1987, Chap. 1.
8. G. O. Shonaike and T. Matsuo. An experimental study of impregnation conditions on glass fiber reinforced TPE composites. J. Reinf. Plast. Compos. 15:16, 1996.
9. N. K. Dutta, A. K. Bhomick, and N. M. Choudhury. Thermoplastic elastomers. In: Standard Handbook of Thermoplastics (O. Olabisi, ed.). Marcel Dekker, 1997, Chap. 15.
10. N. Ogata. Studies on polycondensation reactions of nylon salt: 1. The equilibrium in the systems of polyhexamethylene adipamide and water. Makromol. Chem. 43:117, 1961.
11. E. I. DuPont de Nemours, Inc. Zytel Design Handbook. Wilmington, 1965.
12. M. I. Kohan. Nylon Plastics. Wiley Interscience, 1973.
13. D. B. Jacobs and J. Zimmerman. Polymerization Process (C. E. Schildknecht and I. Skeist, eds.). John Wiley, 1977, Chap. 12.
14. R. B. Seymour and G. S. Kirshenbawn. High Performance Polymers: Their Origin and Development. Elsevier, 1986.
15. J. R. Fried. Polymer Science and Technology. Prentice-Hall, 1995, Chap. 10.
16. M. I. Kohan. Polyamides. In: Engineering Materials Handbook: Engineering Plastics 2:124, 1995.
17. Monsanto. General Handbook of Santoprene: Processing and Physical Properties.
18. A. F. M. Barton. Handbook of Solubility Parameters and Other Cohesion Parameters. Boca Raton: CRC Press, 1983, p. 1.
19. A. Y. Coran and R. P. Patel. Rubber thermoplastic compositions, Part I. EPDM-polypropylene thermoplastic vulcanizates. Rubber. Chem. Technol. 53:141, 1980.

20. A. Y. Coran and R. P. Patel. Rubber–thermoplastic compositions, Part II. NBR-nylon thermoplastic elastomeric compositions. Rubber. Chem. Technol. 53:781, 1980.
21. A. Y. Coran and R. P. Patel. Rubber–thermoplastic compositions, Part III. Predicting elastic moduli of melt-mixed rubber–plastic combinations. Rubber Plast. Technol. 54:91, 1981.
22. A. Y. Coran and R. P. Patel. Thermoplastic compositions, Part IV. Thermoplastic vulcanizates from various rubber–plastic combinations. Rubber Chem. Technol. 54:892, 1981.
23. A. Y. Coran and R. P. Patel. Rubber–thermoplastic compositions, Part V. Selecting polymers for thermoplastic vulcanizates. Rubber Chem. Technol. 55:116, 1982.
24. A. Y. Coran, R. P. Patel, and D. Williams. Rubber–thermoplastic compositions, Part VI. The swelling of vulcanized rubber–plastic compositions in fluids. Rubber Chem. Technol. 55:1063, 1982.
25. A. Y. Coran and R. P. Patel. Rubber–thermoplastic compositions, Part VII. Chlorinated polyethylene rubber–nylon compositions. Rubber Chem. Technol. 56:210, 1983.
26. A. Y. Coran and R. P. Patel. Rubber–thermoplastic compositions, Part VIII. Nitrile rubber polyolefin blends with technological compatibilization. Rubber Chem. Technol. 56:1045, 1983.
27. A. Y. Coran, R. P. Patel, and D. Williams-Headd. Rubber–thermoplastic compositions, Part IX. Blends of dissimilar rubbers and plastics with technological compatibilization. Rubber Chem. Technol. 58:1014, 1985.
28. A. Y. Coran. Thermoplastic elastomers based on elastomer–thermoplastic blends dynamically vulcanized. In: Thermoplastic Elastomers (N. R. Legge, G. Holden, and H. E. Schroeder, eds.). Hanser, 1987, Chap. 7.
29. M. Akiba and A. S. Hashim. Vulcanization and crosslinking in elastomers. Prog. Polym. Sci. 22:475, 1997.
30. A. Y. Coran. Thermoplastic elastomers. In: Science and Technology of Rubber. 2d ed. (J. E. Mark, B. Erman, and F. R. Eirich, eds.). Academic Press, 1994, p. 377.
31. P. Rader. Elastomeric alloy thermoplastic vulcanizates. In: Handbook of Thermoplastic Elastomers. 2d ed. (B. M. Walker and C. P. Rader, eds.). Van Nostrand Reinhold, 1988, Chap. 4.
32. G. O. Shonaike, T. Hayase, and Y. Takenaka. Miscibility of PET/PBT blends. Omron Technics 32:195, 1992.
33. J. W. Barlow and D. R. Paul. Mechanical compatibilization of immiscible blends. Polym. Eng. Sci. 24:525, 1985.
34. M. Bank, J. Leffingwell, and C. Thies. The influence of solvent upon the compatibility of polystyrene and poly(vinyl methyl ether). Macromolecules 4:43, 1971.
35. D. Heikens and W. Barentsen. Particles dimmension in polystyrene/polyethylene blends as a function of their melt viscosity of the concentration of added graft copolymer. Polymer 18:69, 1977.
36. R. Gelles and C. W. Frank. Energy migration in the aromatic vinyl polymers. 2. Miscible blends of polystyrene with poly(vinyl methyl ether). Macromolecules 15:741, 1982.
37. M. Xanthos, M. W. Young, and J. A. Biesenberger. Polypropylene/polyethylene

terephthalate blends compatibilized through functionalization. Polym. Eng. Sci. 30:355, 1990.

38. R. L. Markham. Introduction to compatibilization of polymer blends. Adv. Polym. Tech. 10:231, 1990.

39. M. Xanthos and S. S. Dagli. Compatibilization of polymer blends by reactive processing. Polym. Eng. Sci. 31:929, 1991.

40. Y. Y. Wang and S. Chen. Polymer compatibility: ternary blends of poly(vinylidene chloride-co-vinyl chloride), poly(vinyl chloride) and poly(acrylonitirle-co-butadiene). Polym. Eng. Sci. 21:47, 1981.

41. N. C. Liu and W. E. Baker. Reactive polymers for blend compatibilization. Adv. Polym. Tech. 10:231, 1990.

42. J. Simitzis, C. Paitontzis, and N. Economides. Influence of compatibilizer and additives on mechanical properties of polyethylene–polystyrene blends. Polym. Polym. Compos. 3:427, 1995.

43. S. W. Lee, C. S. Ha, and W. J. Cho. Miscibility of nylon 6 with poly(maleic anhydride-co-vinyl acetate) and hydroxylated poly(maleic anhydride-co-vinyl acetate) blends. Polymer 37:3347, 1996.

44. DuPont. Nylon-ST Super Tough Nylon resin, Bulletin E29250, 1975.

45. D. Newray and K. H. Ott. New rubber-modified thermoplastics. Angew. Makromol. Chem. 98:213, 1981.

45. S. Y. Hobbs, R. C. Bopp, and V. H. Watkins. Toughened nylon resin. Polym. Eng. Sci. 23:380, 1983.

46. S. Wu. Phase structure and adhesion in polymer blends: a criterion for rubber toughening. Polymer 26:1855, 1985.

47. L. Z. Pillon and L. A. Utracki. Compatibilization of polyester/polyamide blends via catalytic ester-amide interchange reaction. Polym. Eng. Sci. 24:1300, 1984.

48. S. Wu. A generalized criterion for rubber toughening: the critical matrix ligament thickness. J. Appl. Polym. Sci. 35:549, 1988.

49. A. J. Oshinski, H. Keskkula, and D. R. Paul. Rubber toughening of polyamides with functionalized block copolymers: 2. Nylon 66. Polymer 33:284, 1992.

50. Y. Takeda, H. Keskkula, and D. R. Paul. Effect of polyamide functionality on the morphology and toughness of blends with a functionalized block copolymer. Polymer 33:3173, 1992.

51. B. Majumdar, H. Keskkula, and D. R. Paul. Mechanical behavior and morphology of toughened aliphatic polyamides. Polymer 35:1399, 1993.

52. Q. Guo, J. Huang, T. Chen, and Z. Feng. Miscibility of poly(hydroxyether phenolphthalen) with aromatic polyester and polycarbonate. Polymer Comm. 31:240, 1990.

53. G. O. Shonaike. Studies on miscibility of glass fiber reinforced blends of polyethylene terephthalate with polybutylene terephthalate. Europ. Polym. J. 28:777, 1992.

54. W. J. MacKnight, F. E. Karasz, and J. R. Fried. Solid state transition behavior of blends. In: Polymer Blends, Vol. 1 (D. R. Paul and S. Newman, eds.). New York: Academic Press, 1978, Chap. 5.

55. O. Olabisi, L. M. Robeson, and M. T. Shaw. Polymer–Polymer Miscibility. New York: Academic Press, 1979, Chap. 3.

56. T. Nishi and T. T. Wang. Melting point depression and kinetic effects of cooling on

crystallization in poly(vinylidene fluoride)–poly(methyl methacrylate) mixtures. Macromolecules 8:909, 1975.

57. R. L. Imken, D. R. Paul, and J. W. Barlow. Transition behavior of poly(vinylidene fluoride)/poly(ethyl methacrylate) blends. Polym. Eng. Sci. 16:593, 1976.

58. T. K. Kwei, G. D. Patterson, and T. T. Wang. Compatibility in mixtures of poly(vinylidene fluoride) and poly(ethyl methacrylate). Macromolecules 9:780, 1976.

59. E. Martuscelli, C. Silvestre, and G. Abbate. Morphology, crystalization and melting behavior of films of isotactic polypropylene blended with ethylene-propylene copolymers and polyisobutylene. Polymer 23:229, 1982.

60. T. S. Ellis. Influence of structure on phase behavior of polyamide blends. Macromolecules 24:3845, 1991.

61. T. S. Ellis. Miscibility and immiscibility of polyamide blends. Macromolecules 22:742, 1991.

62. E. Martuscelli, M. Pracella, and W. P. Yue. Influence of composition and molecular mass on the morphology, crystallization and melting behavior of poly(ethylene oxide)/poly(methyl methacrylate) blends. Polymer 25:1097, 1984.

63. I. C. Sanchez and R. K. Eby. Thermodynamics and crystallization of random copolymers. Macromolecules 8:638, 1975.

64. S. W. Lee, C. S. Ha, and W. J. Cho. Miscibility of nylon 6 with poly(maleic anhydride-co-vinyl acetate) and hydroxylated poly(maleic anhydride-co-vinyl acetate) blends. Polymer 37:3347, 1996.

65. E. Martuscelli. Influence of composition, crystallization conditions and melt phase structure on solid morphology, kinetics of crystallization and thermal behavior of binary polymer/polymer blends. Polym. Eng. Sci. 24:563, 1984.

66. R. Greco, C. Mancarella, E. Martuscelli, G. Ragosta, and Y. Jinghua. Polyolefin blends, 2: Effect of EPR composition on structure, morphology and mechanical properties of iPP/EPR alloys. Polymer 28:1929, 1987.

67. R. Greco, M. Malinconico, E. Martuscelli, G. Ragosta, and G. Scarinzi. Role of degree of grafting of functionalized ethylene-propylene rubber on the properties of rubber-modified polyamide-6. Polymer 28:1185, 1987.

68. B. K. Kim, S. Y. Park, and S. J. Park. Morphological, thermal and rheological properties of blends: polyethylene/nylon-6, polyethylene/nylon-6/maleic anhydride-g-polyethylene) and maleic anhydride-g-polyethylene/nylon-6, Europ. Polym. J. 27:349, 1991.

69. A. Eshuis, E. Roerdink, and G. Challa. Multiple melting in blends of poly(vinylidene fluoride) with isotactic poly(ethyl methacrylate). Polymer 23:735, 1982.

70. L. A. Utracki. Polymer Alloys and Blends. Munich: Hanser, 1989, Chap. 2.

71. M. R. Kamal, M. A. Sato, and L. A. Utracki. Some solid state properties of blends of polyethylene terephthalate and polyamide-66. Polym. Eng. Sci. 22:1127, 1982.

72. K. Yoshikai, K. Nakayama, and M. Kyotani. Thermal behavior, morphology, and mechanical properties of blend strands consisting of poly(ethylene terephthalate) and semiaromatic liquid crystalline polymer. J. Appl. Polym. Sci. 62:1331, 1996.

73. V. W. Srichatrapimuk and S. L. Cooper. Infra-red thermal analysis of polyurethane block polymers. J. Macromol. Sci. Phys. B15:267, 1978.

74. T. A. Speckhard and S. L. Cooper. Ultimate tensile strength of segmented polyurethane elastomers: factors leading to reduced properties of polyurethane based on non-polar soft segments. Rubber Chem. Technol. 59:405, 1986.
75. C. R. Lindsey, D. R. Paul, and J. W. Barlow. Mechanical properties of HDPE-PS-SEBS blends. J. Appl. Polym. Sci. 26:1, 1981.
76. S. Cimmino, L. D'Orazio, R. Greco, G. Maglio, M. Malinconico, C. Mancarella, E. Martuscelli, R. Palumbo, and G. Ragosta. Morphology-properties relationships in binary polyamide 6/rubber blends: influence of the addition of functionalized rubber. Polym. Eng. Sci. 24:48, 1984.
77. C. S. S. Namboodiri, S. Thomas, S. K. De, and D. Khastgir. Thermoplastic elastomers from epoxidized natural rubber (ENR)-styrene-acrylonitrile copolymer (SAN) blends. Kautsch. Gum. Kunststoffe 42:1004, 1989.
78. K. Takahashi, G. O. Shonaike, J. Kusumoto, Y. Sakurada, and S. Yamauchi. Application of the shadow-optical method of caustic to toughness evaluation of polymer alloy. Optics Lasers Eng. 14:101, 1991.
79. S. Aktar and J. L. White. Characteristics of binary and ternary blends of poly(phenylene sulfide) with poly(bis-phenol A) sulfone and polyetherimide. Polym. Eng. Sci. 31:84, 1991.
80. A. F. Yee. Impact resistance. In: Encyclopaedia of Polymer Science and Engineering, Vol. 8, 1987, p. 36.
81. J. Harrison, S. Cartasegna, and P. K. Agarwal. Morphology, thermal and mechanical properties of polyacetal/ionomer blends. Polym. Eng. Sci. 36:2061, 1996.
82. G. O. Shonaike and H. K. Teh. Miscibility of uncompatibilized nylon/santoprene blends. J. Appl. Polym. Sci. (in press).
83. C. B. Bucknal. Toughened Plastics. London: Applied Science, 1977.

10

High-Performance Polymer Blends and Alloys: Structure and Properties

Martin Weber
BASF AG, Ludwigshafen, Germany

SUMMARY

Blends or alloys of high-performance polymers with engineering thermoplastics offer high potential as new materials with optimized cost/performance balance. By blending polyethersulfone with polycarbonate, materials with improved toughness and processing behavior can be obtained. Due to the amorphous nature of polycarbonate, these blends show no improvements in stress crack resistance. This property can be improved by blending polyethersulfone with polyamides. Functionalized polyethersulfones were used to compatibilize polyethersulfone and polyamide during the extrusion process. Poyethersulfone/polyamide alloys offer improved toughness and a good stress crack resistance.

I. INTRODUCTION

Blending of polymers has become a popular route to develop materials with new properties (1–5). Especially in the area of engineering thermoplastic materials this approach has led to a significant number of large-volume products such as PPE/HIPS blends (Noryl®), PC/ABS blends (Bayblend®), PC/PBT blends (Xenoy®), and PA/PPE blends (Noryl GTX®) (6–9). Because of their broad range of properties, the sales volumes of these polymer blends have increased with higher growth rates than the sales of engineering thermoplastic materials in recent years (10). The main reason for the success of these materials is their broad range of properties, which can be tailored to fulfill the needs of a multitude of applications.

The high end of the engineering thermoplastic materials, the so-called high-performance polymers (Table 1), offer exceptional heat resistance (short term and long term), good dimensional stability, and resistance to various chemical environments (11,12). Owing to their chemical structures, these materials have very high melt viscosity, which causes severe problems during subsequent processing steps.

The amorphous high-performance polymers like polyarylates, polysulfones, polyethersulfones, and polyetherimides furthermore suffer from a low stress crack resistance, especially during exposure to organic solvents. But the major drawback of these materials is of course the very high sales price compared to engineering thermoplastics like polyamides, polyesters, and polycarbonates. For this

Table 1   Structure and Thermal Properties of High-Performance Polymers

Structure	Name	T_g/T_m (°C)
	Polyarylate	189
	Polysulfone	185
	Polyetherimide	221
	Polyethersulfone	225
	Polyphenylensulfide	89/284
	Polyetherketone	145/335

reason the applications of high-performance polymers are rather limited and restricted to highly sophisticated small-volume areas in the electronics and aerospace industries and the medical area.

Blending of high-performance polymers with engineering thermoplastics might be an opportunity to improve the processibility as well as the chemical resistance of the base materials. Depending on the amount of incorporated engineering thermoplastic, a significant price reduction of the high-performance polymer might as well be achieved. These improvements could accelerate the penetration of high-performance polymers into large-volume applications in the near future. The following sections will deal with the possibilities and limits of this approach.

II. HIGH-PERFORMANCE POLYMERS

The structures and the thermal properties of the most familiar high-performance polymers are summarized in Table 1. The synthesis and properties of these materials are described in detail elsewhere (11–14). The focus of this paper is directed to polymer blends and alloys based on polyarylethers (polyethersulfone, polysulfone). As amorphous polymers, polyarylethers offer high heat resistance, dimensional stability, and good mechanical properties. Like many other amorphous thermoplastic materials, these products have low stress crack resistance and high melt viscosity.

III. HIGH-PERFORMANCE POLYMER BLENDS

A. Polyethersulfone/Polycarbonate Blends

Although blends of polysulfone (PSU) and polycarbonate (PC) are well known and already commercialized (15–20), little is known about blends of polyethersulfone (PES) and polycarbonate (21–22).

Compared to polyethersulfone, blends of polyethersulfone and polycarbonate could offer significant improvements regarding processibility as well as raw materials costs. Therefore melt blended samples were prepared and characterized. Due to the significant difference in heat stability between the two polymers, attention has to be paid to the heat stability of these blends.

1. Experimental

The properties and the sources of the raw materials used are given in Table 2. The blends were prepared using a ZSK 30 (Werner & Pfleiderer) twin screw extruder, operating at a barrel temperature of 350°C with a screw speed of 250

Table 2 Properties and Source of the Polyethersulfones and
Polycarbonates Used

Polymer	Trade name	T_g (°C)	Viscosity number (ml/g)
PES 1	Ultrason E 1010[a]	225	48.2[c]
PES 2	Ultrason E 2010[a]	225	55.9[c]
PC 1	Makrolon 2800[b]	149	61.4[d]
PC 2	Makrolon 3118[b]	150	66.5[d]

[a] BASF.
[b] Bayer.
[c] Solvent: N-Methyl-pyrrolidone, 0.005g/ml.
[d] Solvent: Dichloromethane, 0.005 g/ml.

rpm. The throughput was 10 kg/h. For mechanical testing the samples were injection molded at 350°C, and the mold temperature was 120°C. The mechanical
testing of the materials was performed according to ISO procedures (Table 3).
DSC measurements were done using a Perkin-Elmer DSC-7, with a heating rate
of 20 K/min; TGA measurements were performed on a DuPont instrument using a heating rate of 10 K/min (air atmosphere). The melt viscosity was measured in a capillary rheometer (Goettfert Rheograph 2003) at a constant shear
rate of 55 Hz.

2. Results and Discussion

In order to characterize the phase behavior of PES/PC blends, melt mixed samples
were investigated by DSC. As can be seen from the DSC traces (Fig. 1), all blends
have two T_gs (Table 4), exactly at the same positions as those of the blend com-

Table 3 ISO Procedures Used for Mechanical Testing

Property	Notation	Method
E-modulus	E	ISO 527
Tensile strength	σ	ISO 527
Tensile elongation	ε_R	ISO 527
Charpy impact	a_n	ISO 179 1eU
Charpy notched impact	a_K	ISO 179 1eA
Puncture test	W_T	ISO 6603
Melt flow index	MFI	ISO 1133
Vicat softening temperature B	VST B	ISO 306

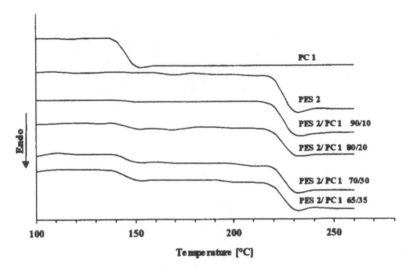

Figure 1 DSC traces of PC 1, PES 2, and PES 2/PC 1 blends with different compositions; heating rate 20 K/min.

ponents. This result clearly indicates the immiscibility of polyethersulfone and polycarbonate.

This result was further confirmed by transmission electron microscopy (Figs. 2a and 2b). The TEM images were obtained using the natural contrast between polyethersulfone and polycarbonate, so polyethersulfone appears dark and polycarbonate bright. The TEM image of the blend with a polyethersulfone/polycarbonate 80/20 composition shows a polyethersulfone matrix phase with dispersed

Table 4 T_gs of the PES 2/PC 1 Blends

Blend composition (wt%) PES 2/PC 1	T_{g1} (°C)	T_{g2} (°C)
100/0	—	225
90/10	149	225
75/25	149	225
50/50	148	225
25/75	148	225
0/100	149	—

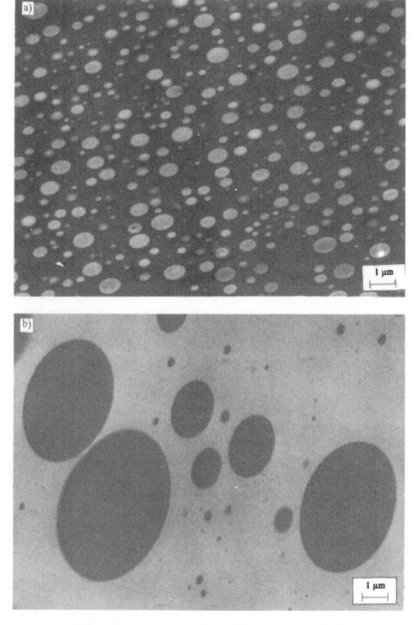

Figure 2 TEM micrographs of PES 2/PC1 blends. (a) 80/20 composition; (b) 50/50 composition.

polycarbonate particles of spherical shape having diameters ranging from 0.2 to 1.0 μm, whereas the blend with the 50/50 composition has a polycarbonate matrix phase with dispersed polyethersulfone particles that have diameters up to 5 μm.

Since this work focused on the development of materials with high heat distortion temperatures, it is obvious that the blends should have a polyethersulfone matrix phase. The influence of the composition on the nature of the matrix phase can be studied using the Vicat B temperature. As depicted in Fig. 3, the addition of approximately 15 wt% of polycarbonate to polyethersulfone does not influence the Vicat B temperatures significantly, but the incorporation of higher amounts of polycarbonate causes a dramatic reduction of the Vicat B temperature, indicating the beginning of the phase inversion from a continuous polyethersulfone phase to a polycarbonate matrix phase.

As is already known from the literature (23), the composition at which the phase inversion occurs can be predicted by Eq. (1):

$$\frac{\eta_1}{\eta_2} = \frac{\phi_1}{\phi_2} \tag{1}$$

where

η_1, η_2 = viscosity of the components 1 and 2

ϕ_1, ϕ_2 = volume fractions of the components 1 and 2

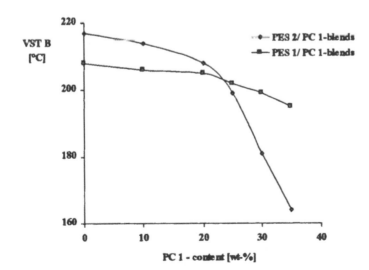

Figure 3 Vicat softening temperature as a function of the PC 1 content of the PES/PC blends.

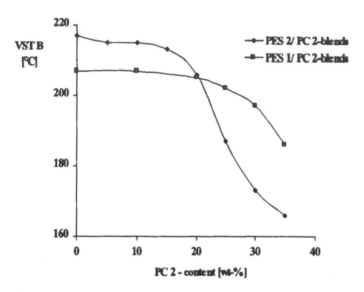

Figure 4 Vicat softening temperature as a function of the PC 2 content of the PES/PC blends.

By changing the viscosity ratio of the blend components it should be possible to shift the onset of the phase inversion in the blends to higher amounts of polycarbonate. As can be seen from the curves in Fig. 3 and Fig. 4, this also holds for PES/PC blends. Both pictures reveal the possibility of shifting the phase inversion in PES/PC blends to a higher PC content either by lowering the molecular weight of the polyethersulfone used or by increasing the molecular weight of the polycarbonate component.

The influence of the composition on the mechanical properties of PES 2/PC 1 blends is given in Table 5. The E-modulus and the tensile strength of the blends decrease with increasing amount of polycarbonate in the blends (Fig. 5), whereas the elongation at break shows an increase with increasing amount of polycarbonate (Fig. 5).

The notched impact strength and the absorbed energy in the puncture test increases with the polycarbonate content of the blends (Table 5). These results reveal the good interfacial adhesion between the immiscible polymers in polyethersulfone/polycarbonate blends. The flow behavior of the blends also shows the beneficial influence of polycarbonate (Table 5). The addition of 20 wt% of polycarbonate caused a 100% increase of the MFI value compared to pure polyethersulfone. In the area of the phase inversion (above 20 wt% PC content) the further addition of polycarbonate gave rise to an even higher increase of the MFI values. This correlates with the change of the matrix phase in the blends from the highly viscous polyethersulfone to the low-viscosity material polycarbonate.

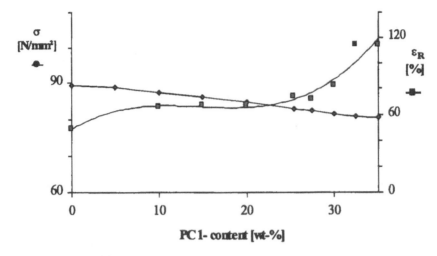

Figure 5 Tensile strength and tensile elongation at break as a function of the PC1 content of the PES 2/PC 1 blends.

In order to evaluate the blend system with the best performance in terms of flow, mechanical properties, and heat distortion temperature, the blends with an 80/20 composition based on different components were compared (Table 6). Using the low-molecular-weight PES 1 improves the flow properties of the blends, but unfortunately these blends have a higher notch sensitivity and lower energy of fracture in the puncture test than the blends based on PES 2.

In the blends with high-molecular-weight PES 2, the influence of the polycarbonate type is only moderate, so the blend based on PES 2 and PC 1 shows the most promising performance.

ible 5 Mechanical Properties of the PES 2/PC 1 Blends

S 2 (wt%)	100	95	90	85	80	75	70	65
1 (wt%)	—	5	10	15	20	25	30	35
T B (°C)	217	214	215	213	206	187	173	166
(N/mm²)	2760	2749	2714	2695	2667	2626	2615	2594
(N/mm²)	90	89	88	86	85	83	82	81
(%)	48	36	66	67	67	74	83	114
(kJ/m²)	4.6	4.4	4.8	4.7	5.6	5.4	5.9	6.3
T (Nm)	71	98	82	106	98	105	90	109
Fl (g/10')	35	44	49	62	74	102	136	148

Table 6 Mechanical Properties of PES/PC 80/20 Blends Based on Different
PES and PC Types

	PES 1/PC 1	PES 1/PC 2	PES 2/PC 1	PES 2/PC 2
VST B (°C)	205	205	206	208
E (N/mm^2)	2797	2792	2667	2741
σ (N/mm^2)	88	80	85	85
ε_R (%)	55	14	66	40
a_K (kJ/m^2)	1.0	1.0	5.6	4.2
W_T (Nm)	73	27	98	118
MFI (g/10')	227	214	74	78

The processing of polyethersulfone takes place at melt temperatures of up to
380°C, so attention has to be paid to the thermal stability of the added component.
To judge the thermal stability, TGA measurements were used. As can be seen
from Fig. 6, polyethersulfone/polycarbonate blends have almost the same thermal
stability as pure polyethersulfone. By the addition of 20 wt% polycarbonate the
temperature at which a 1% weight loss occurs decreases only moderately from
445 to 439°C.

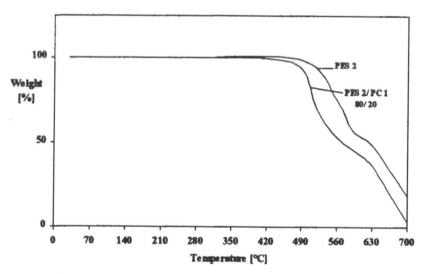

Figure 6 TGA traces of PES 2 and a PES 2/PC 1 blend with an 80/20 composition, heating rate 10 K/min; air atmosphere.

During injection molding or extrusion processes the material is also exposed to shear stress. For this reason it is also necessary to study the influence of shear on the melt viscosity at high temperatures. These measurements were performed at 340, 360, and 400°C at a constant shear rate of 55 Hz over a period of 30 minutes. At 340 and 360°C the melt viscosity is almost constant for 30 minutes. At 400°C a small decrease of the viscosity occurs during the first 10 minutes of the measurement, but after this time the viscosity stays quite constant. Figure 7 reveals the material's excellent melt stability even at 400°C. Due to their lower melt viscosity, PES/PC blends have a wider processing window than pure PES, which is of high relevance especially for the preparation of large parts or parts with low wall thickness.

Since polyethersulfone/polycarbonate blends contain no compatibilizers, the question arises whether the morphologies of these physical mixtures are stable during the common processing steps. To answer this question TEM micrographs were taken from samples after melt blending (Fig. 8a) and after subsequent injection molding (Fig. 8b). The small increase of the particle size of the dispersed polycarbonate during injection molding can be correlated to the limited shear sensitivity of these blends. Nevertheless attention has to be paid to this problem during processing of complicated parts from a noncompatibilized polymer blend.

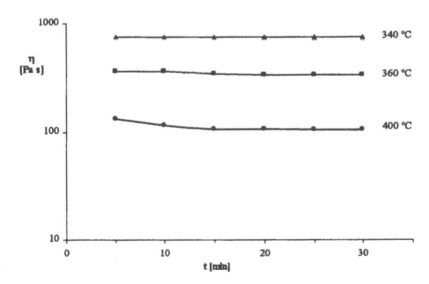

Figure 7 Melt viscosity of a PES 2/PC 1 blend with an 80/20 composition as a function of time at 340, 360, and 400°C; shear rate 55 Hz.

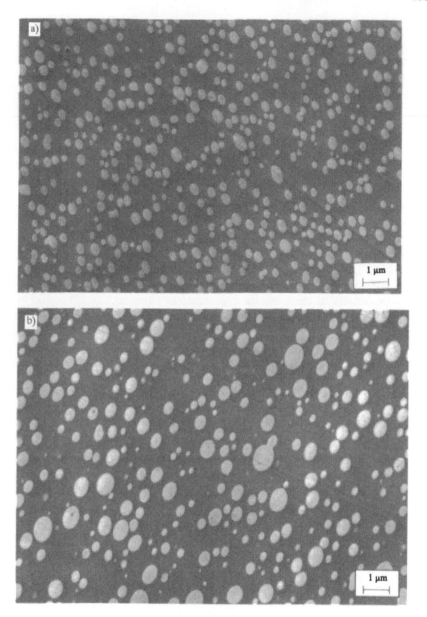

Figure 8   TEM micrographs of a PES 2/PC 1 blend with an 80/20 composition, (a) after melt blending, and (b) after subsequent injection molding.

IV. HIGH-PERFORMANCE POLYMER ALLOYS

As a consequence of the amorphous nature of polycarbonate, blends of polyethersulfone and polycarbonate show no significant improvements regarding environmental stress crack resistance, which is one of the main drawbacks of amorphous high-performance polymers. Another drawback of noncompatibilized polymer blends might be the shear sensitivity of the material during processing, as discussed previously.

In order to improve the stress crack resistance of polysulfone or polyethersulfone significantly, blends with partially crystalline materials like polyesters or polyamides offer much higher potential. Due to the limited thermal resistance of common polyesters like polyethylene terephthalate or polybutylene terephthalate (24), blends of these polymers with polysulfone may suffer from a low thermal stability. The main focus of our work is therefore related to polyarylether/ polyamide blends. Nevertheless it should be mentioned that fiber-reinforced polysulfone/polyethylene terephalate blends are available (tradename Mindel B) (25–27) and are used for applications like print cartridges and electronic component potting cups (28–29).

Since polyarylethers and polyamides are completely immiscible, the mechanical properties of the binary blends are very poor (30–32). As can be seen from Fig. 9, the tensile elongation of binary polyethersulfone/polyamide blends is extremely low compared to that of the blend components. On the other hand, the E-modulus of the mixtures increases with increasing amount of polyamide in the blends (Fig. 9).

Two different ways to compatibilize polyarylether/polyamide blends are known from the literature. Several authors use "pre-made" polyarylether–polyamide copolymers as compatibilizers. Polyarylether– polyamide block copolymers can be prepared by anionic polymerization techniques starting from polysulfone or amino-terminated polysulfones and caprolactam (32,33–36). The influence of the added PSU-b-PA species was judged by using TEM and DSC measurements.

Jeong reported an increase in the T_g of the amorphous PA 6-phase of about 5 to 6 K after the addition of a PA 6–PES-PA 6 block copolymers to PES/PA 6 blends (33). Polysulfone–PA 6 blockcopolymers and alloys were also investigated by McGrath and coworkers, who reported a tremendous increase in chemical and stress crack resistance compared to pure polysulfone (31).

A polyhydroxy polymer can also be used as compatibilizer for polysulfone/polyamide blends (37).

Reactive blending techniques have also been elaborated to compatibilize polyarylether/polyamide blends. Reactions between polysulfones functionalized with acid groups and polyamides led to polymer alloys with improved mechanical performance (38–40).

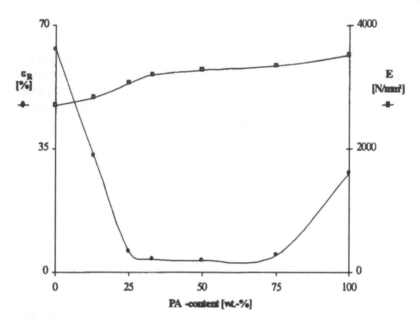

Figure 9 Elongation at break and E-modulus of PES 1/Ultramid T blends as a function of the blend composition.

Another approach to compatibilize polyarylether/polyamide blends is to use polyarylethers with anhydride functionality (40–44). The synthesis of anhydride-terminated polysulfones was described by Myers, starting from amino-terminated polysulfone and a subsequent reaction with a dianhydride. The use of such a type of anhydride-terminated polysulfone in polysulfone/polyamide blends causes a significant increase in toughness. These alloys also offer improved chemical resistance (41). Polyarylether/polyamide blends can also be compatibilized by polyarylethers grafted by maleic anhydride units (PSU-g-MA). The grafting of maleic anhydride onto polysulfone is possible in solution and in the melt (42). Another method to attach anhydride functionalities to polysulfone chains is the conversion of terminal hydroxyl groups with fluorophthalic anhydride (43).

In a previous paper, we reported on the strength of polysulfone/polyamide interfaces that were modified by polysulfones with different functional groups. Polysulfone grafted with maleic anhydride (PSU-g-MA) gave a significant increase in interfacial strength between polysulfone and polyamide (40). This effect was explained by an in-situ copolymer formation during the annealing steps.

Strong evidence for the in situ formation of copolymers was also found in model experiments where polysulfones with various functional groups (carboxyl-, different anhydride-) were melt blended with polyamide (44).

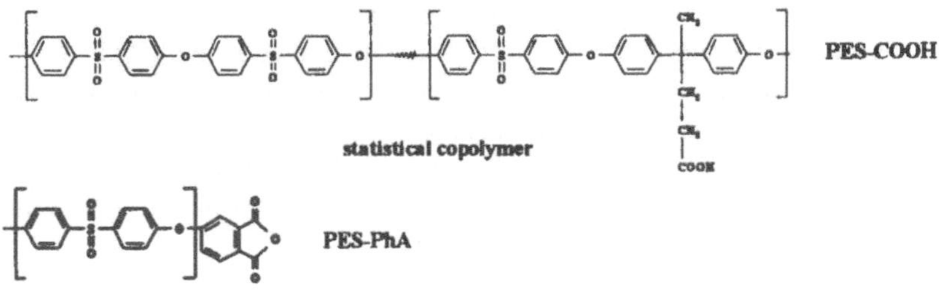

Figure 10   Chemical structures of the functionalized polyethersulfones PES-COOH and PES-PhA.

Since the reaction times used in model experiments are higher than the residence times during extrusion in a twin screw extruder, it was investigated whether the already discussed possibilities of compatibilization are feasible for a real extrusion process. Furthermore, the mechanical performance of the different polyethersulfone/polyamide alloys will be discussed.

A. Experimental Part

1. Functionalized Polyethersulfones

For the study of polymer alloys two different functionalized polyethersulfones were prepared in 10 kg quantities. The structures of these polymers are presented in Fig. 10. The procedures for the preparation and characterization of these polymers are given in the literature (43,45). The properties of the components used are summarized in Table 7. The morphologies of the alloys were characterized by TEM. The ultrathin sections were stained with RuO_4, which predominantly reacts with the polyethersulfone phase and/or with polytungsten acid, which contrasts with the polyamide. TEM was performed in the bright field mode on a Hitachi H 7100 transmission electron microscope operating at 125 kV.

2. Preparation of the Polymer Alloys

The polymer alloys were prepared on a ZSK 30 twin screw extruder (Werner & Pfleiderer) at a barrel temperature of about 340°C with a screw speed of 250 rpm

Table 7   Properties of the Functionalized Polyethersulfones

Product	Intrinsic viscosity (ml/g)	T_g (°C)	Equiv. funct. groups (μmol/g)
PES-COOH	35.1	217	88
PES-PhA	52.5	224	45

and 10 kg/h throughput. The samples for mechanical testing were injection molded at a melt temperature of 340°C and a mold temperature of 100°C.

For the blending studies a commercial copolyamide consisting of units derived from caprolactam, hexamethylene diamine, and terephthalic acid (PA 6.6T, T_g = 100°C, T_m = 285°C, Ultramid®T) was used. To increase the impact strength of the polyarylether/polyamide alloys a maleic anhydride-grafted ethylene-propylene-rubber was incorporated (Exxelor®VA 1803).

B. Results and Discussion

1. PES/PA Alloys Compatibilized with PES-COOH

In order to check the influence of PES-COOH on the morphology and properties of PES/PA blends, the amount of PES-COOH was gradually increased, at a constant PES/PA ratio (Table 8).

The MFI values of the alloys are lower than the MFI value of the pure mixture. Since the molecular weight of the PES-COOH component was quite low compared to the polyethersulfone, no conclusions regarding the copolymer formation during blending can be drawn from the MFI results. The extraction studies clearly indicate that only a very small amount of copolymer was formed during the extrusion procedure.

The TEM images of samples 1 and 3 also show no influence of the reactive component on the morphology of the alloys (Figs. 11a and b). In both cases the particle shape of the dispersed PA phase is quite heterogeneous. The diameters of the dispersed particles range from 0.5 up to 2 μm.

Nevertheless, a significant improvement of the mechanical properties could be obtained by the addition of PES-COOH (Figs. 12 and 13). Both the impact strength and the work of fracture in the puncture test show a nice increase with increasing amount of PES-COOH in the alloys. These results suggest that the

Table 8 Formulations of the Melt Blended PES/PES-COOH/PA/EP Rubber Alloys

PES (wt%)	66	60	54
PES-COOH (wt%)	—	6	12
PA (wt%)	29	29	29
EP rubber (wt%)	5	5	5
MFI (g/10′)	287	>300	>300
Extraction with DMF			
Soluble part (wt%)	66.4	66	66.1

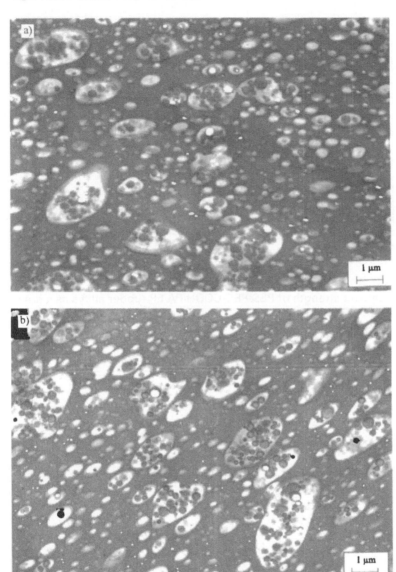

Figure 11 TEM micrographs of (a) PES/PA/EP rubber 66/29/5 blend, (b) PES/PES-COOH/PA/EP rubber 54/12/29/5 alloy (stained with PTA and RuO$_4$).

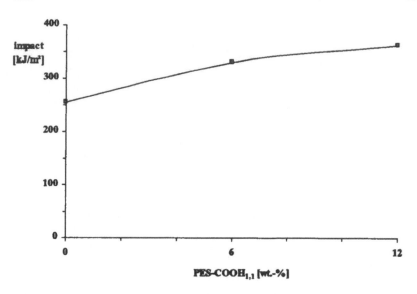

Figure 12 Impact strength of PES/PES-COOH/PA/EP rubber alloys as a function of the PES-COOH content.

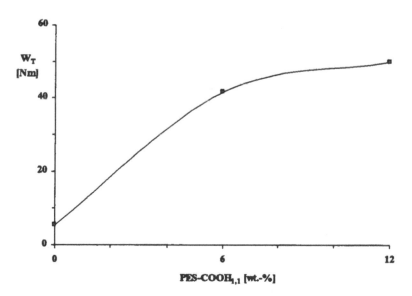

Figure 13 Work of fracture of PES/PES-COOH/PA/EP rubber alloys as a function of the PES-COOH content.

addition of PES-COOH significantly improves the phase adhesion between PES and PA.

As already mentioned, by using the fracture test, a substantial increase of the fracture toughness was obtained in annealed bilayer specimens of PSU-COOH and Ultramid®T (40). At that time the increase in fracture toughness was attributed to the probable chemical reactions between PSU-COOH and polyamide. Due to the high annealing times in those experiments, this might be the right explanation, but since extraction studies show almost no evidence for chemical reactions during extrusion, another interaction should be responsible for the improvement of the phase adhesion in the polymer alloys. IR studies should reveal whether hydrogen bonding between PES-COOH and PA occurs.

2. PES/PA Alloys Compatibilized by PES-PhA

To study reactions between PES-PhA and PA during the extrusion process, PES was partially substituted by PES-PhA (Table 9). Since the molecular weight and the melt viscosity of PES and PES-PhA were almost identical in this system, the decrease of the MFI values of the alloys with increasing amount of PES-PhA clearly indicates the formation of highly viscous copolymers during the extrusion procedure. This conclusion is further supported by extraction studies, which show a tremendous decrease of soluble matrix material (PES, PES-PhA) with increasing amount of PES-PhA in the system (Table 9).

The insoluble material of the sample with the highest amount of PES-PhA was further extracted with 1,1,1,3,3,3-hexafluoro-isopropanole (HFIP). With this solvent the pure polyamide as well as the in-situ formed copolymer can be separated from the rubber phase, which remains insoluble. After evaporation of the HFIP the remaining material was further extracted with formic acid. By this procedure the polyamide can be separated from the copolymer. The morphology of

Table 9 Formulations of the Melt Blended PES/PES-PhA/PA/EP Rubber Alloys

PES (wt%)	70	63	56	35
PES-PhA (wt%)	—	7	14	35
PA (wt%)	25.5	25.5	25.5	25.5
EP rubber (wt%)	4.5	4.5	4.5	4.5
MFI (g/10')	201	169	122	66
	Extraction with DMF			
Soluble part (wt%)	70.4	65.6	59.1	30.4

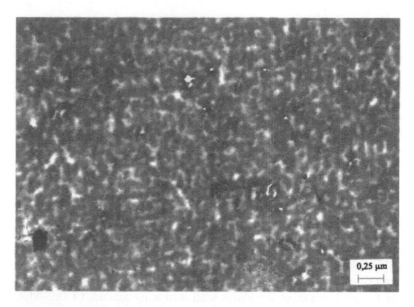

Figure 14 TEM micrograph of the in-situ formed copolymer extracted from a PES/PES-PhA/PA/EP rubber alloy (stained with RuO₄).

the remaining material was investigated by TEM (Fig. 14). The sample was stained with RuO_4, so the PES phase appears dark whereas the PA phase appears bright. Although this material has no clear microstructure like anionic polymerized block copolymers, the domain spacing as well as the size of the domains confirms the existence of direct links between the different chemical components.

Also in this system, the addition of PES-PhA significantly increases the toughness of the resulting PES/PA alloys (Figs. 15 and 16). The impact strength shows a plateau after the addition of 14 wt% of PES-PhA (Fig. 15). The notched impact strength reaches a maximum at a PES-PhA content of 14 wt%, and a further addition of PES-PhA cause no further increase (Fig. 16).

The influence of the addition of PES-PhA on the morphology of the resulting alloys was studied by TEM (Figs. 17a–c). By the addition of 14 wt% of PES-PhA a tremendous size reduction of the dispersed particles occurs. The TEM picture (Fig. 17b) furthermore reveals almost a bimodal particle size distribution composed of very small PA particles (particle diameter from 0.05 to 0.15 μm), which contain almost no rubber, and bigger particles (particle size 0.5 to 1 μm, long axis), which have high rubber content. The rubber-containing particles have an elongated shape, whereas the PA particles are almost spherical. The material with a PES-PhA-content of 35 wt% (Fig. 17c) has almost the same morphology as the sample discussed before. The higher amount of in-situ–created compatibilizer

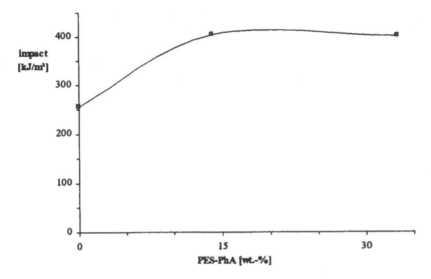

Figure 15 Impact strength of PES/PES-PhA/PA/EP rubber alloys as a function of the PES-PhA content.

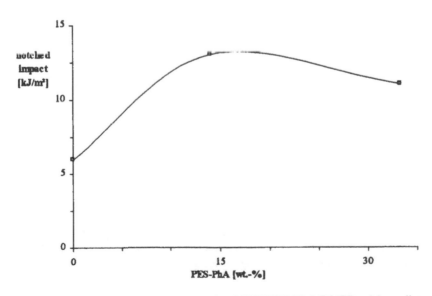

Figure 16 Notched impact strength of PES/PES-PhA/PA/EP rubber alloys as a function of the PES-PhA content.

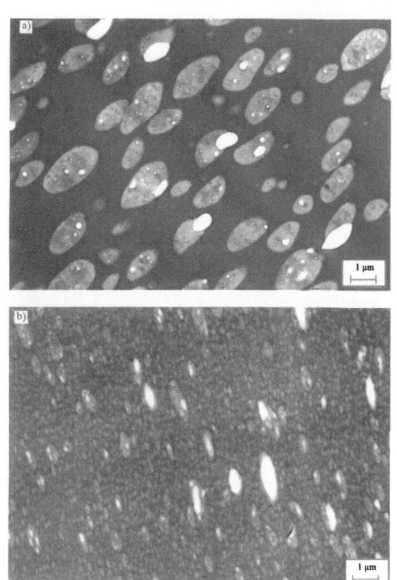

Figure 17 TEM micrographs of (a) PES/PA/EP rubber 70/25,5/4,5 blend, (b) PES/PES-PhA/PA/EP rubber 56/14/25,5/4,5 alloy, and (c) PES/PES-PhA/PA/EP rubber 35/35/25,5/4,5 alloy (stained with RuO₄).

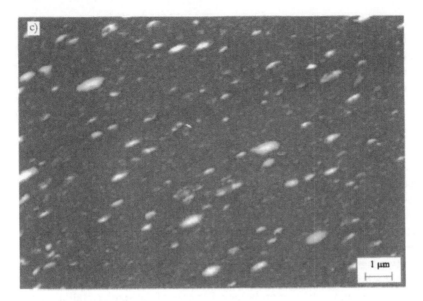

Figure 17 Continued

causes even smaller PA particles and also smaller particles with high rubber content. The rubber particles are just coated with a thin PA layer. In this system, the increase in toughness correlates with the decrease in particle size due to the formation of copolymers.

The mechanical properties of polyethersulfone and the polymer alloys discussed are summarized in Table 10. The addition of PA leads to a reduction of the Vicat B temperature, but both PES/PA alloys reach at least a value of 200°C. The

Table 10 Comparison of the Basic Mechanical Properties of PES and PES/PA Alloys

	PES 1	PES1/ Ultramid T/ EP rubber/ PES-COOH (54/29/5/12)	PES 1/ Ultramid T/ EP rubber/ PES-PhA/ (56/25.5/4.5/14)
Vicat B (°C)	216	202	200
E-modulus (N/mm^2)	2700	2650	2750
Charpy notched impact (kJ/m^2)	7	7.4	13
MFI (g/10′)	110	>300	122
Chemical resistance	+/−	+	+

alloys have the same stiffness as the pure polyethersulfone, whereas both alloys have a higher impact strength. Furthermore, the alloys offer higher MFI values. Preliminary results regarding the chemical resistance of the materials show significant advantages for the PES/PA alloys, especially against organic solvents like toluene, isopropanole, and methylethylketone.

V. CONCLUSIONS

The possibilities of improving the deficiencies of the high-performance polymer polyethersulfone have been discussed. The combination of polyethersulfone with polycarbonate leads to materials with good flow properties that have almost the same mechanical performance as pure polyethersulfone. The processing window of polyethersulfone can be broadened by the addition of polycarbonate. Owing to the amorphous nature of polycarbonate, polyethersulfone/polycarbonate blends have no advantages regarding to chemical resistance over polyethersulfone.

Polymer alloys of polyethersulfone and polyamide show improved flow, improved toughness, and a stable morphology. Because of the semicrystalline component polyamide, the chemical resistance of these alloys is significantly higher than the resistance of pure polyethersulfone.

These results clearly show the possibilities of improving the properties of high-performance polymers by the addition of engineering thermoplastics. On the other hand, it is also evident that the incorporation of engineering thermoplastics with lower temperature resistance will reduce the short-term (Vicat B temperature) as well as the long-term heat stability of high-performance polymers.

ACKNOWLEDGMENT

A part of this work was sponsored by the Ministerium für Bildung, Wissenschaft, Forschung und Technologie (BMBF), project number 03N3028 3. The author would like to thank W. Heckmann, W. Schrepp. V. Andre, I. Hennig, H. -M. Laun, and J. Hofmann for continuous support and B. Cunningham and D. Garau for technical assistance.

REFERENCES

1. L. A. Utracki. Polymer Alloys and Blends. Munich, Vienna, New York: Hanser, 1989, pp 1–27.
2. L. A. Utracki. Encyclopaedic Dictionary of Commercial Polymer Blends. Toronto: ChemTec, 1994, pp 1–26.
3. D. R. Paul, S. Newman. Polymer Blends. New York: Academic Press, 1978.

4. O. Olabisi, L. M. Robeson, M. T. Shaw. Polymer–Polymer Miscibility. New York: Academic Press, 1979.
5. C. Bailey, W. L. Sederel. Product design for productivity and innovation with engineering thermoplastics and their blends in the nineties. Makromol Chem, Macromol Symp 75:55–66, 1993.
6. L. Bottenbruch. Technische Thermoplaste, Technische Polymer Blends. Munich, Vienna: Hanser, 1993, pp 26–97.
7. L. Bottenbruch. Technische Thermoplaste. Technische Polymer Blends. Munich. Vienna: Hanser, 1993, pp 149–214.
8. L. Bottenbruch. Technische Thermoplaste, Technische Polymer Blends. Munich, Vienna: Hanser, 1993, pp 215–258.
9. L. Bottenbruch. Technische Thermoplaste, Technische Polymer Blends. Munich, Vienna: Hanser, 1993, pp 97–105.
10. B. S. Kaith, A. S. Singha. Polymer blends and alloys. In: N.P. Cheremisinoff, ed. Handbook of Engineering. New York: Marcel Dekker, 1997, pp 649–665.
11. J. P. Critchley, G. J. Knight, W. W. Wright. Heat-Resistant Polymers. New York: Plenum Press, 1983.
12. R. B. Seymour, G. S. Kirshenbaum. High Performance Polymers: Their Origin and Development. New York, London: Elsevier, 1986.
13. P. E. Cassidy. Thermally Stable Polymers. New York, Basel: Marcel Dekker, 1980.
14. M. J. M. Abadie, B. Silion. Polyimides and Other High-Temperature Polymers. Amsterdam, New York, Tokyo: Elsevier, 1991.
15. R. J. Petersen, R. D. Cornelinssen, L. T. Rozelle. Recent advances in polymer compatibility. ACS, Polym Prepr 10:385–390, 1969.
16. F. S. Myers, J. O. Brittain. Mechanical relaxation in polycarbonate-polysulfone blends. J Appl Polym Sci 17:2715–2724, 1973.
17. J. E. McGrath, T. C. Ward, E. Shchori, A. Wnuk. Polycarbonate-polysulfone block copolymers. Polym Prepr 18:346–351, 1977.
18. P. Sanchez, P. M. Remiro, J. Nazabal. Influence of reprocessing on the mechanical properties of a commercial polysulfone/polycarbonate blend. Polym Eng Sci 32:861–867, 1992.
19. R. J. Kumpf, R. Archey, W. Kaufhold, D. Meltzner, H. Pielartzik. BPA-polycarbonate/BPA-polyethersulfone block copolymers by reactive processing. Polym Prepr 34/2:580–581, 1993.
20. W. Kaufhold, R. Kumpf, A. D. Meltzner, R. Cohen, C. Dancy, M. Hutnick. Block copolymers of polycarbonate/polyethersulfone as compatibilizer in PC/PSU blends. Polym Prepr 34/2:793–794, 1993.
21. B. P. Barth. U.S. Pat. 3,365,517,23.01.1968, Union Carbide Corporation.
22. M. Weber. High temperature resistant polymer blends. Proceedings of the Polymer Processing Society Regional Meeting, Tokyo, 1993, pp 117–118.
23. D. R. Paul, J. W. Barlow. Polymer blends (or alloys). J Macromol Sci, Rev Macromol Chem C18:109–168, 1980.
24. I. Goodmann. Polyesters. In: H. F. Mark, N. M. Bikales, C. G. Overberger, G. Menges, eds. Encyclopedia of Polymer Science and Engineering. Vol. 12. New York: John Wiley, 1988, pp 25–26.

25. L. A. Utracki. Encyclopaedic Dictionary of Commercial Polymer Blends. Toronto: Chem Tec, 1994, pp 314–315.
26. R. H. Snedeker. EP 36 959, 04.03.81, Union Carbide Corporation.
27. E. Nield. U.S. 3,742,087, 26.06.1973, Imperial Chemical Industries.
28. Xerox Thermal Ink Jet Printer Uses Amoco's Mindel® Resin. Polymer News 19:348, 1996.
29. Electronic Component Potting Cups in Mindel® Resin. Polymer News 15:379, 1992.
30. E. Nield. U.S. 3,729,527, 28.04.1971, Imperial Chemical Industries.
31. J. E. McGrath, L. M. Robeson, M. Matzner. Polysulfone-nylon 6 block copolymers and alloys. Polym Prepr 14:1032–1037, 1973.
32. W. Bu, J. He. The effect of mixing time on the morphology of immiscible polymer blends. J Appl Polym Sci 62:1445–1456, 1996.
33. T. O. Ahn, S. C. Hong, H. M. Jeong, J. H. Kim. Nylon 6-polyethersulfone-nylon 6 block copolymer: synthesis and application as compatibilizer for polyethersulfone/nylon 6-blend. Polymer 38:207–215, 1997.
34. C. E. Koning, R. Fayt, W. Bruls, L. V. D. Vondervoort, T. Rauch, P. Teyssie. From incompatible poly(aryl ether sulfone)/polyamide 4.6 blends to new impact resistant alloys by a synergistic combination of a block copolymer emulsifier and an impact modifier. Macromol Symp 75:159–166, 1993.
35. W. Kaufhold, H. Schnablegger, R. T. Kumpf, H. Pielarzik, R. E. Cohen. Morphological and thermal investigations of nylon-6-poly(sulfone ether)-nylon-6 triblock copolymers. Acta Polym 46:307–311, 1995.
36. H. Kye, J. L. White. Continuous polymerization of caprolactam-polyether sulfone solutions in a twin screw extruder to form reactive polyamide-6/polyether sulfone blends and their melt spun fibers. Polym Process 11:310–319, 1996.
37. D. Lausberg, G. Blinne, P. Ittemann, G. Heinz, E. Seiler, M. Knoll. DE 36 17 501, 24.05.86, BASF AG.
38. H. H. Hub, G. Blinne, H. Reimann, P. Neumann, G. Schaefer. EP 185 237, 30.11.1985, BASF AG.
39. P. Maréchal, T. Chiba, T. Inoue, M. Weber, E. Koch. Melt reactivity of carboxylic acid functional polysulfone in polyamide/polysulfone blends: phase morphology and mechanical properties aspects. Polymer, 39:5655–5662, 1998.
40. D. Maeder, J. Kressler, M. Weber. Studies on the compatibilization between polyamide and poly(arylether sulfone). Macromol Symp. 112:123–130, 1996.
41. C. L. Myers. Compatibilization of poly(arylether sulfone)/polyamide blends. SPE Conference Proceedings ANTEC 1:1420–1423, 1992.
42. M. Weber, K. Muehlbach. EP 513 488, 30.03.1991, BASF AG.
43. M. Weber, C. Fischer. WO 97/04018, 20.07.1996, BASF AG.
44. M. Weber, W. Heckmann. Compatibilization of polysulfone/polyamide-blends by reactive polysulfone—evidence for copolymer formation. Polym Bull 40:227–234, 1998.
45. T. Koch, H. Ritter. Functionalized polysulfones from 4,4-bis(4-hydroxyphenyl)-pentanoic acid, 2,2-isopropylidenediphenol and bis(4-chlorophenyl)sulfone: synthesis, behaviour and polymer analogous amidation of the carboxylic groups. Macromol Chem Phys 195:1709–1717, 1994.

11

"Natural" Polymer Alloys: PC/ABS Systems

Roberto Greco
Institute of Research and Technology of Plastic Materials, National Research Council of Italy, Arco Felice, Naples, Italy

I. INTRODUCTION

Polycarbonate (PC) and acrylonitrile-butadiene-styrene (ABS) alloys are well known commercial products, having received great attention in the past in patents and technical applications (1–3). The combination of component properties in such systems is particularly good, giving rise to materials with satisfactory thermal, mechanical, and impact performances. Their limitations, as well as their positive contributions, are shown in Table 1.

PC/ABS alloys are complex systems consisting of four polymeric species, PC, polystyrene (PS), polyacrylonitrile (PAN) and polybutadiene (PB), the last three in the form of copolymers, S-co-AN (SAN) and PAN grafted onto PB. The nature of their interactions as a function of numerous chemical, physical, and processing variables is not clearly understood from a scientific point of view. A certain number of papers in the literature deal with particular aspects of their behavior, but a unifying view is still lacking. The aim of this chapter is, therefore, to review the properties of these alloys as described in the literature and to try to understand how their good qualities can be obtained in a "natural" way, that is, without any specific compatibilization technique. Their case can be considered rather an exception, in fact, in comparison with other incompatible systems, for which more complex toughening procedures are generally needed, such as the addition of suitable interfacial agents, often achieved by in-situ reactive blendings.

Table 1 Comparison Between Technological Properties of PC and ABS

Behavior	PC properties	ABS properties
Positives	High heat distortion temp.	Economical
	Mechanical resistance	Processibility
	Low-T toughness	Impact strength
	Transparency	Notch sensitivity
	Dimensional stability	
	Electric properties	
Negatives	Chemical resistance	
	Processibility	Low heat distortion
	Notch sensitivity	temperature
	Stress cracking	

For a systematic study, addressed to the attempt to highlight the intimate reasons of the alloys' performance characteristics, it may be convenient to provide a preliminary brief description of the component (PC, SAN, and ABS) characteristics as well as of those of the PC/SAN alloys, as they can be considered as precursors of these systems. Finally, the properties of the more complex PC/ABS alloys will be analyzed and discussed. The final knowledge acquired for such alloys could be useful for developing similar compatibilization techniques to be applied to other multicomponent systems.

With respect to nomenclature the terms "blend" and "alloy" used in this paper follow Utracki's (3) usage: the term blend means two or more polymers and/or copolymers simply melt mixed, in the molten state or in the solid state, with no reference to their overall properties; an alloy is a compatibilized immiscible blend, for which the adhesion among the components is strong enough to impart a satisfactory mechanical performance to the system. As mentioned above this can be achieved in a natural way, as in PC/ABS systems, or by the addition of suitable additives.

II. ALLOY COMPONENTS

A. Polycarbonates

PCs are linear thermoplastic polyesters of carbonic acid with aliphatic or aromatic dihydroxy compounds (4–6). Among the members of this large family only bisphenol A polycarbonate (PC-BPA), with the structure below will be taken into account in this paper and simply indicated as PC from now on:

$$\left[\underset{CH_3}{\overset{CH_3}{-O-\underset{|}{\overset{|}{C}}-O-\overset{O}{\overset{\|}{C}}-O-}} \right]_n$$

In the solid state, PC exhibits a good thermal stability up to its glass transition temperature (about 150°C). In the dry, molten state, PC may be kept at 310°C for hours, since its thermal degradation starts above 400°C.

Its injection molding grades have an average molecular weight (MW) generally between 22,000 and 32,000. Tensile, unnotched impact and flexural strength steadily increase with increased MW up to about 22,000 and level off beyond this value, whereas the melt viscosity continues to increase in this range. A compromise is therefore necessary for providing a MW sufficiently high for obtaining satisfactory mechanical properties but sufficiently low for obtaining melt viscosities capable of imparting flow characteristics suitable for filling complex molds.

PC in the solid state (particularly in stress–strain and in unnotched impact tests) is tougher than other amorphous glassy polymers such as polystyrene and polymethylmethacrylate. This has been attributed to a broad loss modulus G'' maximum, the so-called γ-relaxation, occurring at about −100°C (6).

A brittle-to-ductile transition occurs at temperatures below room temperature and depends on several factors: (1) variables causing internal physical changes prior to testing (i.e., free volume content) such as MW, additives, thermal treatments (physical aging and annealing); (2) test variables such as temperature, loading rate, state of stress, external media; (3) sample geometry, such as thickness and notch sharpness, to which PC is particularly sensitive.

B. Styrene–Acrylonitrile Copolymers

SAN copolymers are low-cost materials of increasing commercial importance, made by copolymerization of acrylonitrile and styrene monomers, usually in their azeotropic ratio (76 wt% of styrene), in order better to control the copolymer composition (7–9).

The copolymerization of AN with S improves the bad processibility of PAN, making it easily moldable so that it can be shaped by conventional equipment. Such a process moreover improves the stress cracking resistance of general-purpose PS in several environments.

SAN copolymers are strong, rigid, and transparent materials, showing a high dimensional and thermal stability, craze resistance, low creep behavior, excellent tensile and flexural strength, surface hardness, and good resistance to weather agents and external media such as water, acid and basic aqueous solutions, and detergents and bleaches, as well as to solvents such as oils, gasolines, and kerosenes.

Most of the SAN characteristics can be improved by increasing the AN amount in the copolymers, showing best results for AN contents ranging from 20 wt% up to 35 wt%. Beyond these values a yellowing effect increases too, requiring the addition of suitable antioxidants. SAN materials are utilized for buildings, automobiles, major and minor domestic appliances, packaging, home furnishings, and so on.

C. Acrylonitrile-Butadiene-Styrene Copolymers

ABS copolymers are a large family of thermoplastic materials, containing as elastomeric component a polybutadiene-based copolymer, in the form of a dispersed phase in a thermoplastic matrix of SAN (10–13). The SAN is partially grafted onto the elastomer in order to reduce the domain size and obtain a good dispersion of rubber into the matrix.

ABS can be manufactured by several procedures: (1) mechanical blending of SAN with a SAN-co-B copolymer in common mixing equipment; (2) several polymerization processes which include the following.

1. Emulsion polymerization is a double-step process. First, one produces an elastomeric substrate latex, in a stirred batch reactor. The rubber can be PB, styrene-co-B (SBR), or AN-co-B (NBR) random copolymers (containing an S or AN amount less than 35 wt%, in order to keep the T_g of the rubber sufficiently low). Careful control must be exerted on the process with respect to average particle size (in the range of 0.05–0.5 μm) and cross-linking degree, since both variables greatly affect the grafting efficiency in the second step. Second, one copolymerizes S and AN, and simultaneously causes a grafting reaction of SAN with the formed rubbery substrate latex. High rubber contents can be obtained by this process (up to 50%). These master batches can be successively blended with SAN or ABS of different composition, in order to vary the rubber concentration and the particle size distribution. Advantages are low temperatures and pressures used in the process and a great versatility in the product range. Disadvantages are high energy requirements.

2. Suspension polymerization also involves a double-step process. First is prepolymerization. The rubber, dissolved in a styrene–acrylonitrile mixture together with a chain-transfer agent and an initiator, is charged to a prepolymerization reactor. The average particle size, larger than in the previous case (generally in the range of 0.5–5 μm), is partially controlled by adjusting the MW of the rubber and the stirring intensity. The phase inversion sets the maximum rubber amount that can be incorporated (about 15 wt%). Higher rubber contents, in fact, increase the viscosity too much and consequently the average size of the rubber particles. Second is polymerization. The prepolymer is charged to a suspension reactor together with water, a suspending agent, an initiator, and a chain-transfer agent. Water and energy consumption are lower, but the waste water is more concentrated than for emulsion process.

3. Bulk polymerization is a continuous two-step process. First is prepolymerization. A butadiene rubber is dissolved in a styrene–acrylonitrile mixture in a prepolymerization vessel, together with a transfer agent, an initiator, and a diluent, for controlling the viscosity. Discrete rubber particles including SAN and monomer are formed, whose size (0.5–10 μm) is controlled by a high shear stirring. Next is polymerization. The prepolymerized material is continuously charged to a polymerization reactor, where the rubber particles are cross-linked, retaining their shape. The rubber percentage is limited to 15–18 wt%, since the viscosity increases so much beyond this concentration that the processing becomes very difficult. Advantages of this method are low energy requirements and low waste water amounts and hence low costs of production. Disadvantages are the high cost of equipment, minor product flexibility (due to the difficulty of processing highly viscous polymer melts), and less complete conversion from monomer to polymer. This last effect requires, for most ABSs, a devolatilization process in order to free the products from residual monomer prior to compounding the final product.

All of these methods of preparation yield a large ABS family of great flexibility, depending on composition, MW, degree of grafting, rubber particle size, and morphology, allowing the tailoring of properties suitable to meet specific end uses.

ABSs are engineering thermoplastics exhibiting good processibility, excellent toughness, and sufficient thermal stability. They have found application in many fields, such as appliances, building and construction, business machines, telephones, transportation, automotive industries, recreation, and electronics.

III. ALLOYS

A. PC/SAN Alloys

A number of papers have analyzed the behavior of PC/SAN alloys (14–46); such analysis is a useful precursor to the analysis of the more complex multicomponent PC/ABS systems.

All the literature results can be briefly summarized as follows, whilst referring the reader elsewhere (39) for a more detailed analysis.

1. PC and SAN blends are completely immiscible in the melt (21) as well as in the solid state (14,33).

2. There is a narrow interval of SAN composition (around 25 wt% of AN in SAN) where a variety of overall alloy properties such as lap shear stress, tensile modulus, elongation at break, notched impact Izod strength, and inward T_g shifts of alloy components (with respect to homopolymer values) exhibit maximum values (14,29,30,33–36,38–40). In Fig. 1 the lap shear stress of adhesion, between sheets of PC and SAN, is shown, as an example, as a function of the AN percentage in SAN.

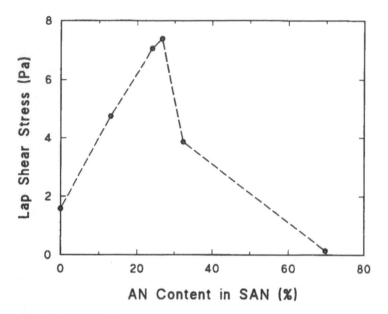

Figure 1 Average lap shear stress versus SAN internal composition. [Adapted from Keitz et al. (14).]

3.   The interfacial tension between PC and SAN shows a minimum in the same AN composition range in SAN, indirectly confirming the existence of a miscibility window (14,35–40). In Fig. 2, particle average diameters versus AN wt% in SAN are shown (35): a minimum is observed at about 25% of AN, which is consistent with morphological observations made by other authors (29).

All these findings were interpreted as due to a miscibility window existing in the same SAN internal composition range where the PC/SAN alloys exhibit maximum adhesion values.

A simple binary interaction model was proposed for explaining such behavior (14,15). Its central hypothesis is that the exothermic mixing responsible for the miscibility window is due not only to intermolecular interactions of the components (PC and SAN) but mainly to the combination of these with intramolecular ones (S and AN in SAN copolymers).

The overall interaction energy density B of the alloy, made by a copolymer (of internal monomer components 1 and 2) and a polymer 3 can be written, in fact, as a function of the three binary interaction parameters:

$$B = B_{13} \, \phi'_1 + B_{23} \, \phi'_2 - B_{12} \, \phi'_1 \, \phi'_2 \qquad (1)$$

where ϕ'_1 and ϕ'_2 are the monomer volume fractions in the copolymer.

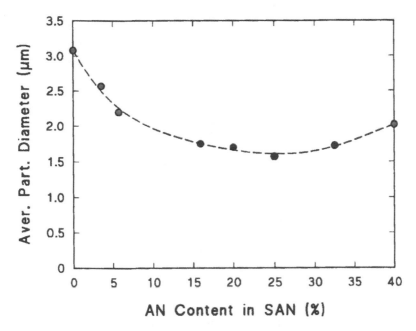

Figure 2 Average diameter of SAN dispersed particles as a function of AN content in SAN for PC/SAN alloys. [Adapted from Callaghan et al. (35).]

The first two terms' interaction parameters of monomers 1 and 2 with polymer 3 are linearly additive. The third term (intramolecular interaction parameter between the two monomers) is of a quadratic form, as shown in Fig. 3. In the case of endothermic mixings, B_{ij} in Eq. (1) are all positive, and B, as a function of ϕ'_1 tends to exhibit a minimum ($d^2B/d\phi'^2_1 = 2B_{12} > 0$).
Furthermore (14), for

$$B_{12} > (\sqrt{B_{13}} + \sqrt{B_{23}})^2 \qquad (2)$$

B becomes negative, and therefore a miscibility window exists between ϕ'_{1a} and ϕ'_{1b}, as shown in Fig. 3.

In the case of SAN being the copolymer and PC the polymer, the lap shear stress maximum can be considered to represent an indirect evidence of the existence of a miscibility window.

It is interesting to note, for later discussion, that the range of AN contents (25–27 wt%), at which the adhesion between SAN and PC is maximum, is very close to the azeotropic composition of SAN, often used as a matrix phase for ABS materials, as already mentioned above.

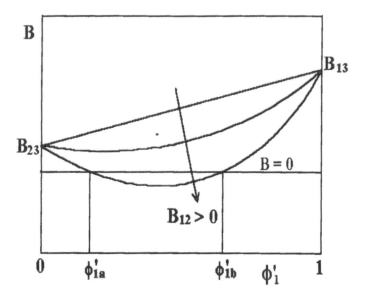

Figure 3 Overall interaction parameter B of a blend, made of a polymer and a copolymer, as a function of the copolymer composition (ϕ'_1) for positive B_{12} values increasing in the arrow direction. [Adapted from Keitz et al. (14).]

4. The PC/SAN alloys show inward shifts of the component T_g when compared to homopolymer values (14,16,21,22,25,35,37,38), as shown in Fig. 4.

These T_g shifts have often been interpreted by several authors as due to a partial miscibility (14,16,21,25,37) of the components. The real cause is due to migrations of low-MW species of SAN towards PC domains during the melt mixing (22,35,38). This species, diluting the PC, decreases its T_g and on the other hand, leaving the SAN domains, it tends to enhance the SAN average molecular weight, increasing SAN T_g.

5. The migrating MW SAN species, presumably located in interzones close to the PC/SAN interface boundaries, tend to dilute the local macromolecular concentration. The result is a net decrease in the number of molecular entanglements between PC and SAN in those regions. This lowered interconnectivity in turn decreases the adhesion between the PC and SAN phases (38). This effect is in evidence in Fig. 5, where the fracture toughness of adhesion between PC and SAN is presented as a function of the amount of benzonitrile (simulating the SAN low-MW species) for two alloys containing SAN with two different AN contents (24 and 29 wt%). Both curves undergo a monotonic decrease with increasing the benzonitrile content, starting from two different initial values that depend on the AN content in SAN.

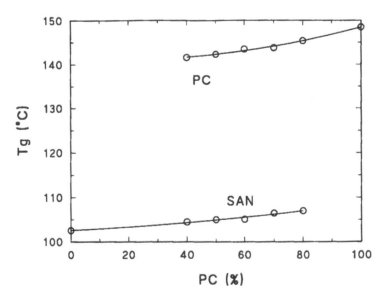

Figure 4 Glass transition temperature of PC and of SAN for PC/SAN blends as a function of the alloy composition. [Adapted from Kim and Burns (21).]

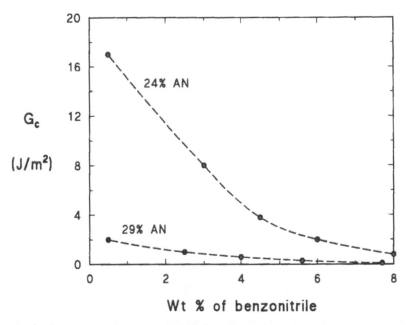

Figure 5 Fracture toughness of PC/SAN interface G_c versus benzonitrile content, at two different values of AN% in SAN, as indicated. [Adapted from Janarthanan et al. (38).]

297

Thermal and dynamic-mechanical behaviors of PC/SAN alloys reinforced with Kevlar and carbon short fibers were comparatively analyzed by Kodama (41). The Kevlar fibers increase the T_g inward shifts of the alloys, indicating strong interactions with PC and SAN domains and a consequent compatibilization of the components, better than in the case where no fibers are added. The carbon fibers did not alter this situation, indicating that they have no functional groups interacting with PC and SAN.

6.   Compression and injection molding can yield different morphologies, which can be investigated by dynamic-mechanical tests, as shown by several authors (17,18,23,24,26–29,38–40). The storage modulus G' versus T shows measurable changes (new peaks or shoulders) between the T_gs of the two components. This suggests that PC and SAN changes of T_g in alloys depend not only on the thermodynamics of the interface but also on the final morphology of the systems (for instance, on the contact surface-to-volume ratio). Therefore any external cause, such as thermal treatments, that produce variations in the morphological features of the systems, can be monitored by such a technique.

7.   Special composites, made of alternating coextruded layers of PC and SAN, were observed under an optical microscope while subjected to tensile tests (17,18,31,42–46). The specimen thickness was kept constant, whereas the PC and SAN layers were changed by varying the PC/SAN volume ratio as well as the total number of layers from 49 up to 1857. The crazing mechanisms were found to depend only on the PC thickness δ and not on the SAN one, according to three conditions: (a) for δ > 6 μm, only single crazes, randomly distributed in SAN layers, were observed; (b) for 1.3 μm < δ < 6 μm, a transition from single crazes to doublets took place; (c) for δ < 1.3 μm, craze arrays were developed.

B.   PC/ABS Alloys

1.   Rheological Behavior and Processibility

The melt viscosity of PC/ABS alloys as a function of composition was measured by Dobrescu and Cobzaru (47) at low (γ = 1 s^{-1}) as well as at high shear (γ = 10^3 s^{-1}) rates, as shown in Fig. 6. In both cases a minimum is present, whose depth is the larger the lower the shear rate. Moreover its position with respect to the composition axis seems to depend on the shear rate value. The trend of the curve is due to the immiscibility in the melt of PC and ABS. In other words, the two components exert a reciprocal lubricating effect during the mixing, lowering the overall internal friction of the material.

In the same figure, two analogous curves of PC/PMMA alloys are reported as well. In this case the minimum is not observed. This could be an indirect indication of miscibility between PC and PMMA. Caution should be taken, however, in drawing such conclusions only from the shape of viscosity curves, unless independent evidence is available from other techniques, as suggested by Utracki (3).

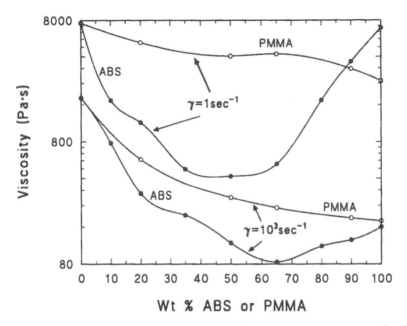

Figure 6 Viscosity as a function of ABS or PMMA percentage for PC/ABS (full circles) and PC/PMMA (open circles) blends at two different shear rates, as indicated. [Adapted from Dobrescu and Cobzaru (47).]

Kim and Burns (21) found that the viscosity of PC/ABS alloys, measured in a capillary rheometer, went through a minimum at an alloy weight ratio of 50/50, as found by Dobrescu and Cobzaru (47).

In Fig. 7, the extrudate swell ratio is reported as a function of SAN, ABS, and Kodar contents. The last one is a copolyester, miscible with PC, formed by 1.4-cyclohexanedimethanol and a mixture of terephthalic and isophthalic acids, which does not show the pronounced maximum exhibited by the other two blends.

The viscosity–composition relationship of PC/ABS systems was analyzed by Kumar et al. (48) by using the Lecyar model, a simplified version of a more accurate one proposed by McAllister (49). The experimental data showed a strong negative deviation from the linear trend with respect to the homopolymer values, as found by other authors (21,47,51,52). A good fit of the data with the model was obtained.

The processibility of PC/ABS alloys was analyzed by Greco et al. (50) by using an ABS(M), whose composition is given in Table 2.

The ratio of a Brabender-like torque τ, per unit volume, of the alloy mixing, to roller speed n, as a function of ABS content, decreases from the PC with the

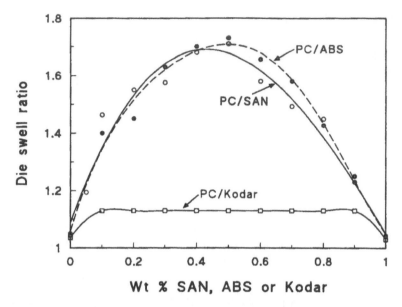

Figure 7 Die swell ratio as a function of SAN, ABS, or Kodar percentage for PC/SAN, PC/ABS, and PC/Kodar blends. [Adapted from Kim and Burns (21).]

ABS addition (see Fig. 8). At low n values the curves exhibit only a slight decrease with a series of intermediate peaks. At high roller speeds (n = 32 and 64 rpm), the curves, after an initial sharp decrease, level off to the final ABS value at an alloy composition of about 40% of ABS, confirming the ability of ABS to improve the PC processibility.

Babbar and Mathur (51) confirmed, once more, the PC processibility improvement by ABS by using an ABS of an internal composition (AN/B/S = 24/16.5/59.5) similar to the one utilized by Greco et al (50). Curves at different

Table 2 Some Characteristics of Three ABSs Used in PC/ABS Blends

Trade name	Code	AN wt%	B wt%	S wt%	MFI (g/10 m) (ASTM D-1238)
Sinkral B32	ABS(B32) or B32	27	22	51	4
Sinkral M122	ABS(M) or M	23	14	63	18
Sinkral A12	ABS(A) or A	27	11	62	26

Source: Ref. 40.

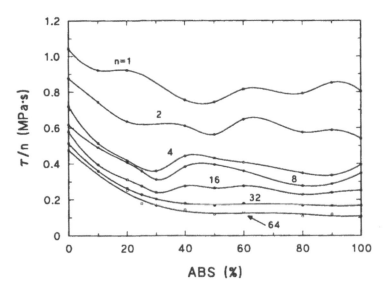

Figure 8 Ratio τ/n of mixing torque per unit volume τ to roller rotation speed n as a function of ABS% at different n values for PC/ABS(M) blends, as indicated. [Adapted from Greco et al. (50).] ABS(M) is defined in Table 2.

mixing temperatures (210,220, and 230°C), with n values ranging between 40 and 70 rpm, were obtained. In Fig. 9 only the data relative to a roller speed rotation of 40 rpm have been reported. The torque per unit volume and roller speed exhibits a decreasing trend with an increase in ABS and temperature, and the curves obtained at different temperatures are parallel to each other. These features of the curves are similar at all other n values, not reported in Fig. 9.

Chiang and Hwung (52) analyzed alloys containing ABS of two different compositions (AN/B/S): I (22/19/59) and II (20/32/48). The Brabender torque exhibited a monotonic decrease from the value of pure PC with ABS addition, the improvement in PC processibility being larger for II than for I. This effect was attributed to a diverse morphology existing in the two systems. In other words the outer shell of the ABS particles was supposed to be made by AN in I and by B in II. Since B exhibits a worse affinity with PC than AN, a larger lubricating effect is induced in II and hence the result is a lower torque.

The effect of the ABS composition on the PC processibility was studied by Greco (40) as well, by using the three different commercial ABSs described in Table 2. Extended trade name, code, internal composition, and melt flow index (MFI) calculated according to ASTM D-1238, of the ABS, manufactured by Enichem Inc., with the common trade name of Sinkral, are also shown in Table 2.

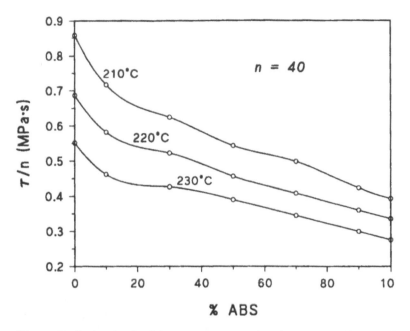

Figure 9   Ratio τ/n of mixing torque per unit volume τ to roller rotation speed n as a function of ABS% at a value of 40 rpm and at different mixing temperatures, as indicated. [Adapted from Babbar and Mathur (51).]

The B content decreases for the three ABSs: from 22% for B32, to 14% for M, and to 11% for A, whereas the MFI increases in the same order. The M and A compositions, are rather similar with minor differences. The AN value is very close to the one for which a maximum of adhesion has been observed (25%) between PC and SAN in PC/SAN alloys (14,31,33–36,37,38–40).

A rheological characterization of the PC and ABS polymers used is shown in Fig. 10, where the ratio of the Brabender torque per unit volume to the roller rotation speed τ/n, having the dimensions of an apparent viscosity (MPas), is reported.

The PC exhibits a less pseudoplastic trend than the three ABSs. ABS(B32) shows τ/n values higher than those of the two other ABSs. This effect, due to its higher B content or MW, develops a higher internal friction during the treatment in the Brabender-like mixer, and its curve intersects with that of PC at an n value of 16 rpm. Therefore at the rotation speed at which the alloys were prepared (32 rpm), its "viscosity" is lower than that of PC. The other two ABS show values much lower than the PC one and very close to each other.

These rheological features can have a great influence on alloy processibility, and of course on their final morphology and overall properties.

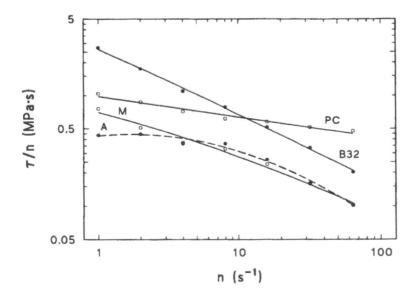

Figure 10 Ratio τ/n of mixing torque per unit volume τ to roller rotation speed n as a function of n values, for PC and ABS polymers. ABS(M), ABS(B32), and ABS(A) are defined in Table 2. [Adapted from Greco et al. (40,56).]

As a reference to real conditions of the alloy mixing, the same parameter τ/n has been reported in Fig. 11 as a function of alloy composition. As it is possible to see, all three ABSs reduce the PC internal friction, greatly improving the PC processibility when this is the system matrix. The best effect is provided by the ABS(B32), the most viscous among the three ABSs. Its greater B amount probably enhances the reciprocal lubricating effect necessary to lower the internal friction of the system, confirming the results of Chiang and Hwung (52).

With increasing the ABS content a phase inversion occurs and the properties of the ABS matrix become predominant for all the alloys. The internal friction of PC/ABS(M) and PC/ABS(A) at 32 rpm (mixing roller speed) is lower than the PC one at high ABS contents, and τ/n levels off on the ABS value. When the internal friction of ABS and PC is comparable, as in the case of ABS(B32), the alloy exhibits a pronounced minimum, already observed by other authors (21,47).

Lee et al. (53) analyzed the influence of the injection molding process on the formation of the solid-state morphology of an PC/ABS (90/10) alloy. A bead-and-string structure was observed in the skin regions of the plaques, whereas in the middle of the body the ABS phase was isotropically dispersed in the PC matrix in the form of round particles. In a subsequent paper (54) the analysis was extended

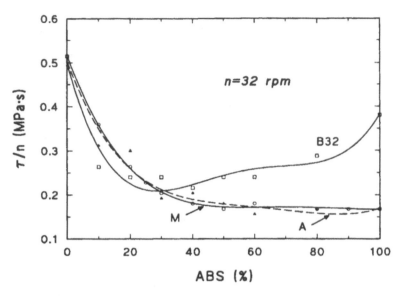

Figure 11 Ratio τ/n of mixing torque per unit volume τ to roller rotation speed n as a function of ABS% at an n value of 32 rpm, for three PC/ABS alloys, containing three ABSs of different internal composition, coded M, B32, and A, as indicated in Table 2. [Adapted from Greco et al. (40,56).]

to the entire alloy composition. Three composition ranges were identified: (1) a PC-rich alloy, where the situation was as described above; (2) a midrange (between PC/ABS 70/30 and 60/40), where the previous structure evolved at the edges to a coalesced, stratified morphology and in the center to a coarse, dispersed ABS phase, with some regions of cocontinuity with PC; (3) an ABS-rich alloy with dispersed PC domains. Qualitatively the above results were explained as due to the melt flow patterns during the mold filling and to relaxations and coalescence acting during the cooling of the materials prior to the complete solidification.

An appearance of flow marks on the surface of thin PC/ABS items made by injection molding was studied by Hamada and Tsunasawa (55), by analyzing the internal structure of the alloy obtained from diverse processing conditions. The conclusion was that the gate position was crucial in order to avoid superficial defects such as flow mark formations.

2. Thermal Properties

Inward T_g shifts are observed for PC/ABS alloys, as in the case of the previously mentioned PC/SAN ones. In Fig. 12 PC/SAN data, already reported in Fig. 4, have

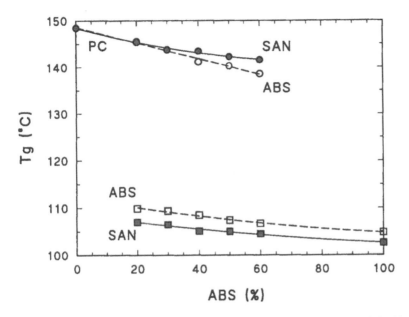

Figure 12 Comparison between T_g of PC (circles) and T_g of SAN or ABS (squares) for PC/SAN (filled symbols) and PC/ABS (open symbols) alloys as a function of alloy composition (SAN or ABS%) [Adapted from Kim and Burns (21).]

been compared with those relative to PC/ABS systems, as found by Kim and Burns (21). It is interesting to note that the T_g of PC decreases for PC/ABS alloys more than for PC/SAN ones. In the case of PC/SAN alloys the T_g shifts were attributed to low-MW SAN species migrating toward PC domain boundaries, as discussed. This further increase in T_g shift for PC/ABS(B32) alloys with respect to PC/SAN ones can be ascribed to low-MW species of B. These migrate (together with SAN species) toward PC domains and induce a higher plasticization of PC than in the case of PC/SAN alloys.

An effect of the internal composition of the ABS used in the alloy on the inward T_g shift of the PC, observed by Greco and Dang (40,56), is shown in Fig. 13. The B32-coded alloy, containing the greatest concentration of butadiene among all, exhibits the largest T_g shift for PC, whereas the other two (M and A), both containing lower and similar butadiene contents, show higher PC T_g values. This seems to be due to a more marked contribution to PC plasticization, specifically due to the low-MW species of butadiene, contained in a larger amount in ABS(B32). The total content of both AN and S is lower, in fact, in ABS(B32) than in the other two ABS materials; therefore if the shift effect were attributable only to the AN and S low-MW

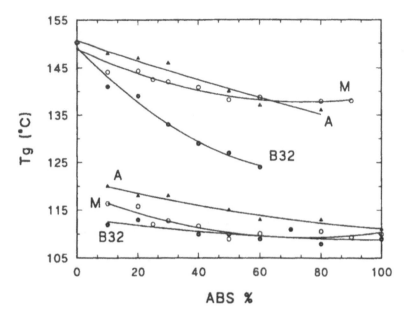

Figure 13 T_g of PC and ABS versus ABS% for PC/ABS alloys, containing three ABSs of different internal composition, coded M, B32, and A, as indicated in Table 2. [Adapted from Greco et al. (40,56).]

species, it should be lower for PC/ABS(B32) alloys, but this is not the case. This finding is in agreement with the increase of the T_g shifts passing from PC/SAN (where no B is present) to PC/ABS alloys, as shown in Fig. 12.

Morbitzer et al. (57) analyzed thermal and dynamic-mechanical behaviors of two PC/ABS alloys, made according the following two procedures.

1.   A SAN-g-B terpolymer and a SAN copolymer were blended in a 60/40 ratio. Various amounts of PC were successively added to this blend.

2.   PC and SAN were premixed in a 50/50 ratio and different amounts of a SAN-g-B terpolymer added to the alloy.

In both cases the T_g of PC and ABS, detected by DSC and by dynamic-mechanical tests as a function of the alloy composition, varied in an analogous manner to that found for PC/SAN alloys by other authors (14,16,21, 24,35,39,40). The T_g of B decreased with decreasing the ABS content, and consequently the B (butadiene) amount in the alloy. This effect was attributed to thermal stresses built up, during the cooling of the material, by the different thermal expansion coefficients of the grafted rubber particles and the surrounding SAN matrix (58). The stress fields in the matrix, around the rubber particles, tended to overlap dependent on the rubber content and the final morphology.

Since these stresses were lower than the interfacial forces, the rubber particles underwent a negative hydrostatic pressure. This induced an increase in free volume and a subsequent decrease in T_g.

3. Morphology

The morphology of PC/ABS alloys has been investigated by several authors generally by means of scanning (SEM) and transmission electron microscopy (TEM) (14,39,40,50,52–59,65,66,69,73,74,76–83,85–87,90,91,93).

Dong et al. (59) treated specimen surfaces, smoothed by a glass knife, with two different and complementary etching solutions: (1) an NaOH aqueous solution in order to hydrolyze the PC, leaving the ABS unaltered; (2) a strong oxidant aqueous solution to etch the ABS, leaving the PC phase unaltered. In Fig. 14 a few photomicrographs of a 75/25 PC/ABS alloy are shown. In Fig. 14a the surface of this alloy is treated with the first solution and shows remaining ABS particles, after the PC has been hydrolyzed with a kind of wispy coating of PC attached to the particles. Only a successive strong washing with hot water can get rid of the PC, revealing underneath very neat ABS particle surfaces (see Fig. 14b). The PC material attached to the ABS particles even after the etching, as well as the features on the ABS surfaces observable after washing, can be considered the first qualitative evidence of the good adhesion existing between PC and ABS phases. In the third photograph (Fig. 14c), the specimen surface has been treated with the second oxidant solution, giving rise to a reverse type of micrograph, where the PC matrix is visible and the ABS particles have been etched, leaving holes where they previously existed.

4. Mechanical Behavior

Several papers have reported and discussed the mechanical properties of PC/ABS alloys (39,40,50,52,56,57,60,64,80,82,83,85). The general features of tensile behavior with respect to the alloy composition, starting from PC and progressively increasing the ABS amount, are as follows:

1. The alloy modulus E_b is a linear combination of those of the two components.
2. The stress at yield, as well as the stress at break, decrease with increasing ABS content.
3. The elongation at break decreases with increasing ABS content, often reaching a minimum value in the middle of the alloy composition range.

Typical stress–strain curves of PC, ABS, and PC/ABS alloys at different compositions are shown in Fig. 15. They were obtained by testing unnotched specimens at RT and a low deformation rate (0.1 mm/min). Going from pure PC and increasing ABS content up to about 50 wt%, the yield peak decreases and continuously broadens. The effect is probably due to the local action of the ABS particles

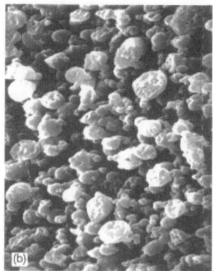

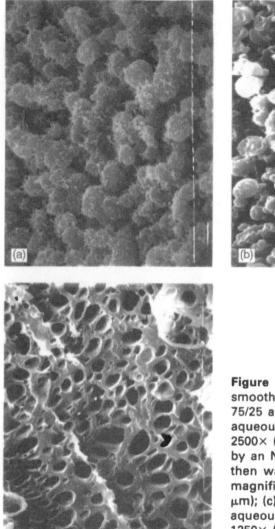

Figure 14 Photomicrographs of smoothed surfaces of a PC/ABS(M) 75/25 alloy (a) etched by an NaOH aqueous solution, magnification 2500× (10 mm = 4 μm); (b) etched by an NaOH aqueous solution and then washed by hot water (90°C), magnification 2000× (10 mm = 5 μm); (c) etched by a strong oxidant aqueous solution, magnification 1250× (10 mm = 8 μm). ABS(M) is defined in Table 2.

on the PC surroundings, which experience very high stress concentrations. The total mechanical energy is dissipated through a largely diffused PC shear yielding, with less concentrated neck formation and lower and broader yield peaks. At alloy concentrations higher than about 50% ABS, a phase inversion occurs; the ABS becomes the system matrix and the yield assumes the typical ABS yield peak with no successive cold-drawing.

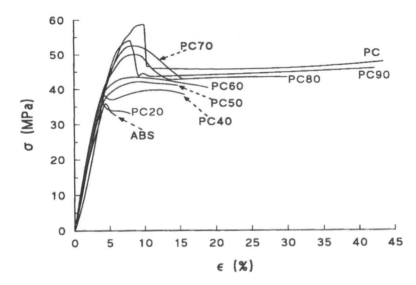

Figure 15 Stress–strain curves (nominal stress σ as a function of percentage of elongation ε) of PC, ABS(M), and of their PC/ABS(M) alloys, at different compositions (PC wt%) as indicated. [Adapted from Greco et al. (40,50).] ABS(M) is defined in Table 2.

The lowering of the elongation at break with increasing ABS content, observed in Fig. 15, represents an increasing difficulty of the PC matrix to undergo to cold-drawing because it is hampered by the ABS particles. In the middle of the composition, cocontinuous morphologies of PC and ABS coexist with low or minimum elongation values. When the phase inversion takes place, the elongation at break of the alloy approaches that of pure ABS.

Kunori and Geil (61) applied various mechanical theories to a number of alloys (PC/PS, PC/HDPE, and PC/LDPE) in order to obtain suitable qualitative tests of adhesion for two-component systems. They showed, by the comparison of experimental data with model predictions, that for PC/PS alloys the adhesion between the PC matrix and the dispersed PS phase was rather strong. When PS was substituted by the two polyethylenes, the adhesion was, instead, very poor, owing to the nonpolar nature of the polyolefin interface as well as to the marked volume contractions due to their crystallization. This effect, occurring after PC solidification, determined the detachment of the dispersed particles from the matrix.

Following this way of thinking, predictions of Kerner's models (61–63), relative to conditions of perfect and of no adhesion between alloy components, were compared by Greco et al. (50,56) with experimental data relative to a PC/ABS(M) alloy. The aim was a way of providing an indirect determination of the level of adhesion between PC and ABS.

For the case of perfect adhesion, the Young's modulus of the alloy, E, as a function of composition in Kerner's model is expressed by

$$E = E_t \frac{15(1 - m)\phi_2 E_2 + [(7 - 5m)E_1 + (8 + 10m)E_2]\phi_1}{15(1 - m)\phi_2 E_1 + [(7 - 5m)E_1 + (8 + 10m)E_2]\phi_1} \qquad (3)$$

where E_1, ϕ_1, m are the modulus, the volume fraction, and the Poisson's ratio of the matrix and E_2, ϕ_2 the corresponding values of the dispersed phase. The model assumes that the system is isotropic, that the dispersed phase particles are spherical, and that the adhesion between the two phases is perfect.

The above expression can be modified for particular systems such as foams, alloys containing rubber particles in a rigid matrix, or alloys in which inclusions are sitting loosely in holes of the matrix and cannot give any contribution to the modulus (since for very small deformations no stress can be transmitted to the particles across the interface).

For all these systems the modulus of the dispersed phase, E_2, can be assumed to be zero, and Kerner's equation (50,62) simplifies to

$$E = E_1 \frac{(7 - 5m)\phi_1}{15(1 - m)\phi_2 + (7 - 5m)\phi_1} \qquad (4)$$

Kerner's equation, originally derived for shear modulus G, can be applied to the tensile modulus E as well, by using the well-known equation

$$E = 2G(1 + m) \qquad (5)$$

assuming that the Poisson's ratio for matrix and inclusions are the same. This is a first approximation for the case of PC and ABS whose m values are 0.39 and 0.46, respectively. Furthermore it was shown by Kunori and Geil (61) and by Dickie (63) that the Kerner's expressions are scarcely sensitive to m changes.

From the work of Dong et al. (59) it has been possible to verify that the ABS(M) and the PC dispersed phases were in the form of spherical particles in PC and in ABS matrices, respectively, up to a concentration of 20–30% in weight. Therefore Eqs. (3) and (4) are applicable with confidence to the examined PC/ABS alloys, at least within 20–30% of the dispersed phase, in both cases. Beyond these values the assumptions of the theory cannot hold.

A further assumption for an appropriate application of Kerner's equations is that the two alloy components have the same mechanical behavior of the corresponding homopolymers. This seems to be the case for PC/ABS alloys, since the component T_gs have similar values to those of the homopolymer, except for a very thin interlayer responsible for the observed T_g inward shifts.

The results of Eq. (3) and Eq. (4) are shown (solid lines) in Fig. 16, for both cases where either PC or ABS is the matrix, together with the experimental data (open circles). The theoretical curves of perfect adhesion follow an additivity behavior. Those relative to conditions of no adhesion show, on the contrary, a strong

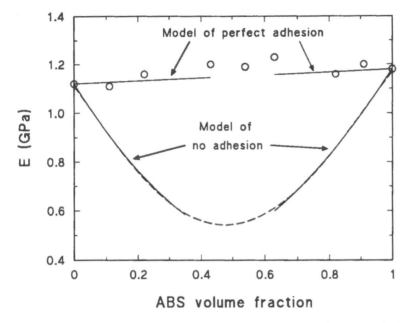

Figure 16 Young's modulus of PC/ABS(M) alloys as a function of alloy composition (ABS volume fraction): circles, experimental data; solid lines, Kerner's models for perfect and no adhesion, as indicated (40,50,56,61,62). ABS(M) is defined in Table 2.

negative deviation from the linear additivity of pure PC and ABS moduli. The experimental data are almost coincident with the former, indicating a very good adhesion at the interface between PC and ABS phases. The physical reason for this strong adhesion is the chemical affinity existing between PC and the SAN component of ABS. As already seen in PC/SAN alloys (14,30,32,34–40), an AN content of about 25% in weight, generally contained in all ABS materials, provides the maximum level of affinity with PC.

In Fig. 17 a comparison is made between the yield stress values relative to two ABS materials of different composition, as described in Table 2. The behavior versus the alloy composition seems to be rather similar in spite of the different B amounts of the two ABSs.

The elongation at break of the alloys of Fig. 17 is plotted in Fig. 18 as a function of composition. This parameter decreases strongly with increasing ABS content for both systems; but the B32 alloy, with a larger B amount, exhibits a lower value in the middle of the composition range. Similar results were found by Suarez et al. (60) for extruded sheets and injection-molded bars, whereas Chiang and

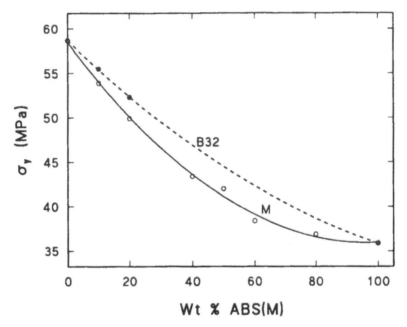

Figure 17 Yield stress as a function of ABS alloy content for two PC/ABS alloys, containing ABS(B32) and ABS(M) of different internal composition. ABS(B32) and ABS(M) are defined in Table 2. (From Refs. 40,56.)

Hwung (52) found higher elongation values for alloys containing larger B contents. The diverse findings can be attributed to the range of processing techniques and conditions used in the different cases. This determines the morphological features of the alloy, which influence in a different way the cold-drawing process following the yield of the specimens.

A model able to predict the yield strength of binary alloys was applied by Kolarik (64) to several polymeric systems; it showed that an additivity trend with respect to the values of the alloy components indicates good adhesion between them. A few authors (50,56,60) also confirmed such behavior for ABS/PC systems.

5. Impact Behavior

A number of papers have reported and discussed the impact behavior of PC/ABS alloys (25,39,40,50,52,56,57,60,65,66,67–73,76,77,80–83,85,87–92). Their impact properties are of utmost importance, since PC is brittle, particularly when scratches are present on the specimen surfaces (notch sensitivity), and this short-

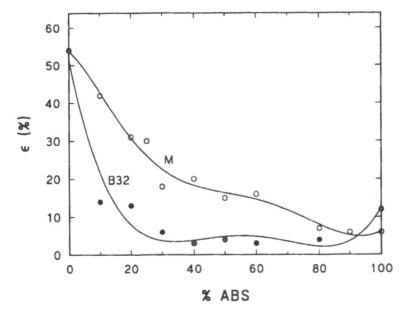

Figure 18 Elongation at break as a function of ABS content for two PC/ABS alloys, containing ABS(B32) and ABS(M). ABS(B32) and ABS(M) are defined in Table 2. (From Refs. 40,56.)

coming is one the main reasons for blending PC with ABS. This is clearly illustrated in Fig. 19, where a number of curves of Charpy impact tests are reported for the previously mentioned PC/ABS(M) alloy (39,40,50,56). Pure PC and PC90 alloy exhibit a classic, brittle behavior with a very fast fracture propagation mechanism of fracture. A further 10% addition of ABS to the PC determines a brittle-to-ductile transition, characterized by a slow ductile propagation mechanism of the crack opening. This is followed, with further ABS additions, by a ductile-to-brittle transition at 50% of ABS in the alloy and finally a return to semiductile behavior for alloys containing more than 80% of ABS.

This change in the fracture mechanism is better illustrated in Fig. 20, where the maximum impact stress and the impact strength, relative to the curves of Fig. 19, are reported as a function of alloy composition.

In Fig. 20 composition zones relative to brittle, ductile, and semiductile behaviors are indicated by the letters B, D, and SD respectively.

Photographs of the fracture surfaces of the alloys representing the four composition regions illustrated in Fig. 20 are shown in Fig. 21. In Fig. 21a the PC fracture starts with an initiation center close to the notch, and then the crack follows

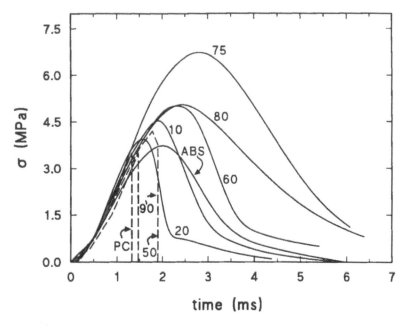

Figure 19 Curves of impact stress versus time of deformation in a flexural Charpy test, on notched specimens of PC, ABS(M), and PC/ABS(M) alloys of different alloy composition (PC wt%), as indicated. ABS(M) is defined in Table 2. [Adapted from Greco et al. (56).]

at a fast propagation rate, as evidenced by the smooth surface of the rest of the specimen. The addition of 10 wt% ABS does not change the mechanism, even though the surface appears less smooth due to the presence of the ABS particles (see Fig. 21b). Furthermore in both cases the lateral dimensions of the specimens appear to be unaltered by the fracture, indicating a condition of plain strain.

By adding 10 wt% more ABS the mechanism changes sharply and the fracture surface reveals marked flow patterns, accompanied by a strong lateral contraction of the specimen (see Fig. 21c). This suggests that very large deformations have occurred prior to fracture initiation, during the crack opening and propagation, and hence the process occurs in a plain-stress condition, confirming the kind of curve shown in Fig. 19 for the alloy of 80 wt% PC. The situation does not change up to 40 wt% of ABS addition to the PC matrix (Figs. 21d and 21e). The next brittle zone starts at 50 wt% PC down to 20 wt% PC with low impact strengths (cf. Fig. 20), and the fracture surface appears rough but homogeneous, suggesting again a plain-strain condition. The mechanisms may be a little different from those of the previous brittle zone, as we will demonstrate shortly (Figs. 21f and 21g). Finally, at low

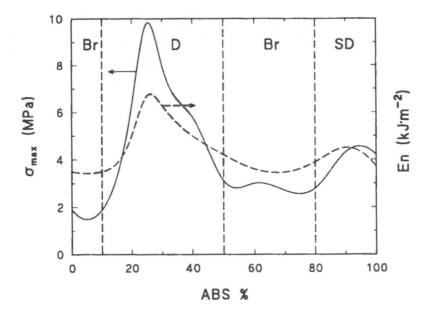

Figure 20 Maximum impact stress (on the left-hand axis) and energy for crack initiation (on the right-hand axis) versus ABS content for PC/ABS(M). Impact behavior zones: Br, brittle; D, ductile; SD, semiductile. [Adapted from Greco et al. (56).] ABS(M) is defined in Table 2.

compositions, semiductile behavior is restored, and the specimens acquire an impact mechanism similar to that of pure ABS (Fig. 21h).

From a microscopic point of view, the mechanism may be similar to the one proposed by Wu and Margolina (65–71) for the case of rubber toughened nylons. The crucial parameter is the average interparticle distance: at 10 wt% of ABS in the PC matrix this variable is rather large, due to the relatively small amount of the dispersed phase; a plain-strain condition is acting at the microscopic level, inducing a brittle catastrophic failure; at a concentration of 20 wt% of the ABS the average particle diameter increases but the number of particles increases too; therefore the interparticle distance decreases below its critical value, creating microscopic plain-stress conditions, which induce deformations in large regions of the specimen surrounding the notched area, whose patterns are clearly visible on the fracture surface. A lateral overall striction is also visible indicating that all the specimen is involved in the phenomenon along the fracture surface, with a ductile crack opening.

In the midcomposition range (50–80 wt% of PC), cocontinuous morphologies with thick layers of both the components are formed, local and large-scale plain-

Figure 21 Fracture surface photomicrographs of notched specimens relative to PC, ABS(M), and PC/ABS(M) alloys at different composition (PC wt%): (a) PC; (b) 90; (c) 80; (d) 75; (e) 60; (f) 40; (g) 20; (h) ABS. ABS(M) is defined in Table 2. Magnification 20× (10 mm = 500 μm).

Figure 21 Continued

strain conditions are established, inducing once again a brittle behavior, with no lateral contraction of the specimen.

Finally, when ABS becomes the matrix and PC the dispersed phase, the behavior is semiductile, characteristic of pure ABS. This indicates the inability of hard particles such as those of PC to provide improvements to the impact performances of the ABS matrix, the reason being the high T_g value of the PC-dispersed particles.

The notched Izod impact strength, investigated by Suarez et al. (60), was much lower for ABS than for PC. It leveled off at the ABS value up to 50% of PC content and then increased almost linearly up to the PC value.

Chun et al. (72) found that a synergistic effect in notched Izod impact strength extended over a broad composition range (PC/ABS from 80/20 up to 10/90), much wider than in other authors' findings (21,39,40,50,56). This was attributed to the PC/SAN miscibility, to a suitable ABS composition (low B content), and to an improved mixing device.

Weber and Schoeps (73) studied the morphology of PC/ABS alloys in relation to the PC/SAN composition (keeping constant the amount of the SAN-g-B copolymer) and to the melt processing temperature. The Vicat temperature slightly depended on the melt temperature and increased with increasing PC content. The morphology and the impact resistance were strongly influenced not only by the alloy composition but also by the melt temperature. An enhancement of the latter parameter caused a demixing of PC and SAN and a worsening of the impact performances. However, when PC was the matrix, this effect became negligible (72).

The toughening mechanisms of these alloys were analyzed by Ishikawa and Chiba (74) by three-point bending of round notched bars. The addition of small ABS amounts (2%) decreased the PC toughness, whereas contents of ABS from 5 wt% up to 20 wt% yielded much greater improvements. A stress whitening effect was attributed to the formation of voids at the interphase existing between PC and ABS.

Radusch et al. (75) analyzed the dynamic-mechanical and dielectric properties, suggesting that a partial compatibility exists between PC and ABS in boundary layers having the order of magnitude of 1 nm. The macroscopic properties were, therefore, determined by the interactions among the phases in these layers. An optimal concentration for toughened behavior was about 70% PC.

Real time small-angle x-ray scattering was performed by Bubeck et al. (76) on a series of rubber-modified thermoplastics in order to investigate the modes of deformation in tensile impact tests in such materials. For PC/ABS alloys the predominant mechanisms were shear yielding in the PC and associated rubber particle cavitation in the ABS. The cavitation mode seemed to provide a direct relationship between rubber content and impact toughness. The mechanism did not change whether the tensile impact direction was perpendicular or parallel to that of the injection-molding direction. Finally, crazing, although a precursor of the final fracture, occurred after the prevailing non-crazing mechanisms, contributing only a few percent to the total plastic deformation.

Suyun (77) found that the fracture morphology of PC/ABS alloys with high notched impact strength was a synergistic combination of the rupture characteristics of PC (smooth surfaces and striations with branches) and those of ABS (microcavities and parabolic markings). The fracture morphology of alloys with low impact strengths was very different from that of the two components. Both fracture morphology and impact strength were dependent on PC and ABS characteristics, as well as on alloy composition.

An accurate fractographic study was performed by Lee et al. (78) on PC/ABS single notched specimens obtained by injection molding and fractured in tensile mode in an Instron machine at a relatively low strain rate (48 mm/s). Stress whitening was observed on the surfaces of ABS and of all the alloys but PC, indicating voiding formation during the deformation and fracture. Plane-stress flow lines were observed for PC and for PC/ABS 90/10 and 70/30, accompanied by a lateral contraction of the overall sample. Also the 50/50 alloy exhibited some features of plain stress at the specimen's edges. In the center, however, a valley on one surface and a corresponding ridge on the opposite surface was evidence of a plain-strain region the resulting mechanism being therefore of mixed fracture mode. No overall contraction was observed on 30/70 and 10/90 PC/ABS alloys. A characteristic feature (called herringbone or chevron fracture) was however still visible in the center of the specimen with narrow shear lips, and this kind of failure mechanism was considered to be of a ductile nature. The ABS surface showed no contraction and no shear lips at the edge, indicating a macroscopic condition of plain strain. The fracture mode as a function of composition was reported (77). There was, in fact, a gradual change from a shear fracture of PC under plain-stress conditions to failure by crazing of ABS under plain-strain conditions. The main influence of the ABS on a PC matrix was the cavitational mechanism of the rubber particles. The plain strain observed in the central region of the specimens in intermediate compositions was due to the increase of the ABS content, while the shear lips at the edges characteristic of the plain-stress conditions became narrower and narrower, completely disappearing in pure ABS specimens. An S-shaped curve was observed between the ductile-to-brittle transition temperature and composition. The most ductile compositions were PC/ABS 70/30 and 60/40, whereas the most brittle were 30/70 and 10/90.

In a subsequent paper Lee et al. (79) performed a fractographic and a morphological analysis on an injection-molded specimen of PC/ABS composition 70/30. Fracture occurring perpendicular to the injection direction yielded a herringbone feature in the surface. Fracture occurring parallel to it yielded an inverse herringbone feature, and the alloy toughness decreased in the second instance. The herringbone was determined by interactions of the main crack with secondary

cracks started along the center line. The inverse herringbone had the same origin,
but the secondary cracks were initiated near the edges. Such differences in behavior between perpendicular and parallel directions were attributed to the PC orientation induced by the shear flow during the processing. The findings described
above seem to be somewhat contradictory, since they may depend on several variables such as the internal composition of the ABS blended with PC and the average particle size and size distribution.

In Fig. 22 the impact Charpy strength of Fig. 20 as a function of composition
of a PC/M alloy is compared with those of PC/B32 and PC/A (alloys containing
the different ABSs described in Table 2), obtained by Greco et al. (40,56). The
general trend of the curves is similar for the first and the second alloy with pronounced peaks and an alternation of brittle and ductile behaviors, whereas the
third one exhibits low values all over the composition field. However, minor but
significant differences are present even between the first and the second alloy:
comparing the PC90 alloy of both, the PC/M is brittle whereas the PC/B is ductile. The reason seems to be the lower rubber amount of ABS(M) with respect to
ABS(B). The difference in behavior of the PC/ABS(M) and the PC/ABS(B) is
clearly confirmed by the inspection of Fig. 23: the first alloy shows complete brittle fracture surface under a plain-strain condition (Fig. 23a), whereas the second

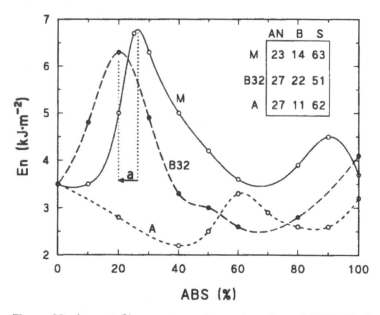

Figure 22 Impact Charpy strength as a function of PC/ABS alloys, containing different ABSs, codes, and compositions as indicated in the upper corner
box as well as in Table 2. [Adapted from Greco et al. (40,56).]

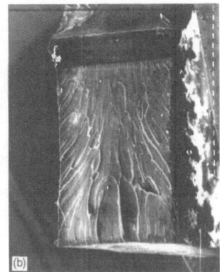

Figure 23 Photomicrographs of fracture surfaces of PC/ABS 90/10 notched specimens for different alloys, containing two ABSs, PC/ABS(M), and PC/ABS(B32), as defined in Table 2. Magnification 20× (10 mm = 500 μm).

one, containing a higher rubber amount, is largely deformed under a plain-stress condition (Fig. 23b). The behavior of the third alloy (A), for which almost no synergistic effect is present, can be tentatively attributed to a lower B content, as well as to a lower amount of SAN, which could diminish the ABS-PC adhesion with respect to the other two alloys. The two effects lead to the inability of ABS(A) to toughen the PC matrix.

Other authors too have analyzed the influence of the ABS composition on PC/ABS alloy impact properties (79–83). Kurauchi and Ohta (80) analyzed two alloys, made of a commercial PC and two commercial ABSs, of different (AN/B/S) compositions: 22/29/49 and 20/37/43. Only the first alloy showed synergistic effects in tensile stress–strain parameters (energy absorption and elongation at break) as well as in unnotched impact Charpy strength, with maxima at high PC contents of composition. In the second alloy, which had a larger rubber amount than the first, these parameters instead showed an almost linear trend at high PC contents and minima at low PC amounts. This effect was explained by the different B contents of the two ABSs (29%, 37%): in the second case, the B encapsulates most of the SAN, reducing the PC/SAN contact surface. Moreover it has a lower AN content than α, with the result that the PC affinity with SAN is lowered. Herpels and Mascia (81) used two different ABSs with 20% and 30% of

B in PC/ABS alloys. The two kinds of alloys were compared at equal rubber level in the final product. Small changes of T_g of the three phases were interpreted in terms of a partial compatibility between AN and PC. No correlation of the rubber alloy content was found with fracture, obtained at low and high propagation speed. However a toughness increase was obtained under plain-strain conditions. The B was found to be encapsulated in the SAN matrix.

Lombardo et al. (82), analyzed PC/ABS alloys based on the following types of ABS:

1. SAN (with no B) containing 25% AN (SAN 25)
2. An emulsion-made ABS, with 50% rubber content and very small uniform particle size (about 0.1 μm)
3. A bulk-polymerized ABS, with 16% rubber content and large rubber particle sizes (0.5–1 μm)

Sample 2 exhibited lower modulus and tensile strength but excellent performance in standard and sharp notched tests at RT as well as at low temperatures close to the T_g of the rubber. These results were attributed to the high rubber content, but it was stressed that composition and rubber particle size and distribution could have also been partially responsible for the impact behavior.

The influence of the type of rubber particles was, therefore, separated from that of the amount of rubber. This was achieved by preparing alloys of SAN 25 with different amounts of ABS, in order to obtain PC/ABS alloys with same rubber concentrations (5%, 10%, and 15 wt%).

Small and uniform rubber particles, obtained by emulsion polymerization, toughened PC/ABS materials at rubber concentrations and temperatures lower than those of large rubber particles produced by bulk polymerization.

Wu et al. (83) analyzed the effect of the rubber content in ABS and the phase distribution of the extra added rubber on the toughening of PC/ABS alloys. They found that increasing rubber content resulted in toughness improvement. Additional core-shell rubber particles such as those of MBS (methylmethacrylate-butadiene-styrene copolymer) enhanced the alloy toughness further. Both effects were negatively compensated by an increase in melt viscosity, so a balance is always needed when selecting starting materials. Other factors such as morphology, MW, degree of grafting, and agglomeration of particles affect the alloy behavior.

Cooney (84) tested the photostability of films of a commercial PC/ABS alloy that was rather sensitive to UV and visible radiation. In particular the weak alloy component was the polybutadiene, which was easily oxidized and cross-linked on the specimen surface. This effect embrittled the surface, causing superficial cracks on bending and thus lowered the alloy impact strength.

Eguiazàbal and Nazàbal (85) investigated the effect of recycling on the alloy properties. They observed that just one reprocessing by injection molding changed only slightly the mechanical behavior of PC/ABS alloys. More than two reprocessing steps produced deep changes in the rubbery phase of the ABS, which was oxidized and cross-linked. This effect decreased the failure properties and the impact performance by the resulting rubber hardening, whereas the small deformation properties, such as modulus and yield strength, remained substantially unaffected. Density and flow behavior were altered as well.

Stefan and Williams (86) analyzed a 50/50 commercial PC/ABS alloy by means of nuclear magnetic resonance (NMR), varying the temperature between 100°K and 500°K. SEM, TEM, and impact properties were studied as well. At temperatures close to the T_g of PC, the chain motions of the zones surrounding the PC were strongly influenced by the PC itself. Therefore the apparent volume (interaction zone) of the PC increased by about 50% in comparison with its actual value. At temperatures close to the T_g of PB, the rubber influenced the surrounding regions, so that the PB apparent volume was effectively twice its actual value. The larger interaction of the rubber was attributed to the diverse morphology (small round particles) as well as to the higher flexibility of the PB chains. The improved impact performance of the alloy with respect to that of pure PC was related to the apparent volume of the rubber phase. These findings suggested the existence of interacting regions between the PC and ABS phases.

Lu et al. in a series of papers compared the effect of specimen thickness, side grooves (87,88), and temperature (89) on the fracture toughness of PC/ABS alloys determined by different methods, such as crack tip opening displacement, J-integral, and hysteresis energy. Seidler and Grellmann (90) determined fracture toughness of such alloys under dynamic loading, based on J-integral, using a special block method. The toughness values were very sensitive to changes in temperature as well as to alloy composition. Scanning and electron microscopy investigations allowed quantitative toughness–morphology correlations to be made, which aids toughness optimization of the materials.

Wu et al. (91) studied the influence of MW of PC on PC/ABS alloy properties, in the range of PC MW going from 1.8×10^4 up to 3.6×10^5. They observed higher impact strengths, higher critical strain–energy release rate, and lower brittle–ductile transition temperatures with increasing MW. However, as one would expect, a continuous MW increase yielded a high melt viscosity, which renders alloy processability more difficult. Therefore a compromise must be reached in choosing the MW of PC, in order to take into account both effects. With the highest MW used (3.6×10^5), the best compromise between impact properties and processibility, relatively to the alloy composition, was found for the PC/ABS (65/35) alloy.

Guest and Van Daele (92) studied the influence of thermal aging on impact properties of PC/ABS alloys containing 25, 60, and 75 wt% of PC, at temperatures of 90, 110, and 130°C for times up to 1500 h. PC and ABS underwent physical aging, becoming more brittle, when aged at 90°C and 130°C, temperatures below their T_g (about 155°C and 110°C, respectively). The ABS aged at 110°C exhibited a measurable reduction of toughness beyond 10 hours of aging, because of degradative oxidation of the B component on the specimen surface (see also Refs. 84 and 85). More severe effects were revealed by the alloys for long times of exposure. However, alloys containing high PC amounts maintain most of their properties at high temperature and long times. Some additional factors need to be taken into account such as the redistribution of stabilizers during the alloy mixing or during the thermal aging.

IV. CONCLUSIONS

From the above discussion it is possible to draw the following concluding remarks.

1.   The lowering of viscosity of PC (and of torque in the mixing apparatus) with SAN and ABS addition, up to about 30–40% of such components, yields an improved PC processibility. This effect seems to be due to a reciprocal lubricating effect that lowers the internal overall friction of the alloys.

2.   PC and ABS are completely immiscible in both melt and solid state. The observed inward shifts of T_g detected for PC/ABS and PC/SAN systems (relative to pure PC and ABS values), which are a function of the alloy composition, are due to low MW SAN (and B) species migration towards PC boundary domains and not to partial miscibility of PC and SAN, as proposed by some authors.

It is interesting to note that these shifts are larger for PC/ABS than for PC/SAN alloys, probably because of an additional contribution of low MW B species to PC plasticization. As a confirmation of this hypothesis, in PC/ABS alloys containing ABS of different compositions, the higher the B amount the larger the shifts.

3.   Some authors propose the existence of interzone layers located at the PC/SAN or PC/ABS boundaries in which PC and SAN (or PC and ABS) chains can form mixed entanglements, responsible for the adhesion between the phases. The low-MW SAN and B species, migrating from SAN and ABS bulk domains toward the PC, tend to dilute the local entanglement density, reducing the interphase adhesion. This hypothesis can only be inferred for PC/ABS alloys, since it has only been verified for PC/SAN alloys using benzonitrile as a model for low-MW species; see Janarthanan et al. (38).

4.   PC exhibits in general a very ductile behavior in tensile tests on unnotched specimens. The addition of ABS lowers and broadens the yield stress peaks and reduces the cold drawing ability of PC and the consequent elongation at break of the alloys. After the phase inversion, the alloy tensile curves resemble that of pure ABS.

The main deformation mechanism is shear yielding for pure PC and alloys where PC is the matrix. With increasing ABS content the mechanism changes smoothly to crazing. The microstructural sequence of the deformation is initially craze initiation and propagation in the ABS, then arrest of crazes in the PC, and finally extensive matrix and rubber particle deformation. In contrast with these conclusions, other authors found, by flexural (73) and tensile tests (75), that the formation of voids around rubber particles and cavitation, together with shear yielding in the PC matrix, are the primary mechanisms acting during stress relaxation and toughening. Crazing is the precursor to final fracture, but it comes after noncrazing mechanisms and gives only a small contribution to the overall plastic deformation.

5. Synergistic effects in impact performances are the general features of alloys where PC is the matrix.

An alternation of brittle-to-ductile mechanisms with changing alloy composition was observed from both impact curves and fractography. In fact, moving from PC to PC/ABS alloys with increasing ABS content, plain-strain to plain-stress transitions were observed. These overall effects were likely due at a microscopic level to a kind of percolation mechanism; a critical interparticle distance of the PC matrix plays the crucial role. If the interparticle distance is lower than its critical value, a local plain-stress condition is established, and the alloys become ductile; when it is larger than the critical value, a plain-strain condition determines the alloy brittleness.

6. ABS internal composition is very important in determining the alloy properties. The S/AN ratio of the SAN copolymers is crucial for the adhesion between PC and SAN, but for most ABS used it is generally very close to the azeotropic S/AN ratio for SAN (about 25 wt% of AN). Of the three components the most important is the B content. However the literature results are somewhat contradictory: an increase of PB yields synergistic effects in impact behavior in several cases, only minor improvements in others, and even a worsening in a few cases.

It should be stressed that parameters other than ABS composition can play a role of great importance in the impact performance. These include ABS and PC molecular characteristics, molecular orientation, different kinds of processing, processing variables, previous thermal history of the materials, ABS rubber particle type, ABS particle size and size distribution, test variables such as temperature, strain rate, specimen thickness, notch radius, etc., interfacial adhesion, low-MW contents of ABS species, and so on.

7. Thermal aging, ending in physical (increase in density) and chemical (thermo-oxidation of butadiene) change, is of great importance in determining the impact performance of these alloys.

REFERENCES

1. D. R. Paul, J. W. Barlow. Polymer blends (or alloys). J Macromol Sci-Rev Macromol Chem C18(1):109–168, 1980.

2. D. R. Paul, J. W. Barlow, H. Keskkula. Polymer blends. In: H. F. Mark, N. M. Bikales, C. G. Overberger, G. Menges, eds. Encyclopedia of Polymer Science and Engineering. New York: John Wiley, 1985, Vol. 12, pp 399–461.
3. L. A. Utracki. Polymer Alloys and Blends: Thermodynamics and Rheology. Munich: Hanser, 1989, Appendix II.B and pp 1–27.
4. L. Bottenbruch. Polycarbonates. In: H. F. Mark, N. G. Gaylord, N. M. Bikales, eds. Encyclopedia of Polymer Science and Technology. New York: John Wiley, 1969, Vol. 10, pp 710–764.
5. D. Freitag, U. Grigo, P. R. Muller, W. Nouvertné. Polycarbonates. In: H. F. Mark, C. G. Overberger, G. Menges, eds. Encyclopedia of Polymer Science and Engineering. New York: John Wiley, 1985, Vol. 11, pp 648–718.
6. N. G. McCrum, B. E. Read, G. Williams. Anelastic and Dielectric Effects in Polymeric Solids. New York: John Wiley, 1967, Ch. 13, pp 534–537.
7. G. A. Morneau, W. A. Pavelich, L. G. Roettger. Survey and styrene-acrylonitrile polymers. In: H. F. Mark, D. F. Othmer, C. G. Overberger, G. T. Seaborg, eds. Kirk Othmer Encyclopedia of Chemical Technology, 3d ed. New York: John Wiley, 1978, pp 427–442.
8. G. P. Ziemba. Acrylonitrile-styrene polymers. In: H. F. Mark, N. G. Gaylord, N. M. Bikales, eds. Encyclopedia of Polymer Science and Technology. New York: John Wiley, 1969, Vol. 1, pp 425–435.
9. F. M. Peng. SAN copolymers. In: H. F. Mark, N. M. Bikales, C. G. Overberger, G. Menges, eds. Encyclopedia of Polymer Science and Engineering. New York: John Wiley, 1985, Vol. 1, pp 452–470.
10. C. B. Bucknall. Toughened Plastics. Essex (England): Applied Science, 1977, pp 90–106.
11. G. A. Morneau, W. A. Pavelich, L. G. Roettger. ABS Resins. In: H. F. Mark, D. F. Othmer, C. G. Overberger, G. T. Seaborg, eds. Kirk Othmer Encyclopedia of Chemical Technology. 3d ed. New York: John Wiley, 1978, pp 442–456.
12. A. Lebovits. Acrylonitrile-butadiene-styrene copolymers. In: H. F. Mark, N. G. Gaylord, N. M. Bikales, eds. Encyclopedia of Polymer Science and Technology. New York: John Wiley, 1969, Vol. 1, pp 436–444.
13. D. M. Kulich, P. D. Kelley, J. E. Pace. Acrylonitrile-butadiene-styrene polymers. In: H. F. Mark, N. M. Bikales, C. G. Overberger, G. Menges, eds. Encyclopedia of Polymer Science and Engineering. New York: John Wiley, 1985, Vol. 1, pp 388–426.
14. J. D. Keitz, J. W. Barlow, D. R. Paul. Polycarbonate blends with styrene/acrylonitrile copolymers. J Appl Polym Sci 29:3131–3145, 1984.
15. D. R. Paul, J. W. Barlow. A binary interaction model for miscibility of copolymers in blends. Polymer 25:487–494, 1984.
16. R. A. Mendelson. Miscibility and deformation behavior in some thermoplastic polymer blends containing poly(styrene-co-acrylonitrile). J Polym Sci, Polym Phys Ed 23:1976–1995, 1985.
17. B. Gregory, A. Hiltner, E. Baer. Dynamic mechanical behavior of continuous multilayer composites. Polym. Eng Sci 27:568–572, 1987.
18. B. Gregory, A. Siegmann, J. Im, A. Hiltner, E. Baer. Deformation behavior of coextruded multilayer composites with polycarbonate and poly(styrene-acrylonitrile). J Mat Sci 22:532–538, 1987.

19. R. E. Skochdopole, C. R. Finch, J. Marshall. Properties and morphology of some injection-molded polycarbonate-styrene acrylonitrile copolymer blends. Polym Eng Sci 27:627–631, 1987.

20. L. L. Berger, E. J. Kramer. Microdeformation in partially compatible blends of poly(styrene-acrylonitrile) and polycarbonate. J Mat Sci 22:2739–2750, 1987.

21. W. N. Kim, C. M. Burns. Thermal behavior, morphology, and some melt-properties of blends of polycarbonate with poly(styrene-acrylonitrile) and poly(acrylonitrile-butadiene-styrene). Polym Eng Sci 28:1115–1125, 1988.

22. M. J. Guest, J. H. Daily. The use of glass transition data for characterizing polycarbonate-based blends. Eur Polym J 25:985–988, 1989.

23. M. J. Guest, J. H. Daily. Dynamic mechanical spectroscopy of blends of bisphenol-A-polycarbonate with styrene-acrylonitrile copolymers. Eur Polym J 26:603–620, 1989.

24. K. W. Mclaughin. The influence of microstructure on the dynamic mechanical behavior of polycarbonate/poly(styrene-acrylonitrile) blends. Polym Eng Sci 29: 1560–1568, 1989.

25. J. C. Huang, M. S. Wang. Recent advances in ABS/PC blends. Adv Polym Tech 9:293–299, 1989.

26. D. Quintens, G. Groeninckx, M. Guest, L. Aerts. Mechanical behavior related to the phase morphology of PC/SAN polymer blends. Polym Eng Sci 30:1474–1483, 1990.

27. D. Quintens, G. Groeninckx, M. Guest, L. Aerts. Phase morphology coarsening and quantitative morphological characterization of a 60/40 blend of polycarbonate of bisphenol A (PC) and poly(styrene-co-acrylonitrile) (SAN). Polym Eng Sci 30:1484–1490, 1990.

28. D. Quintens, G. Groeninckx, M. Guest, L. Aerts. Viscoelastic properties related to the phase morphology of 60/40 PC/SAN blend. Polym Eng Sci 30:1207–1214, 1991.

29. D. Quintens, G. Groeninckx, M. Guest, L. Aerts. Phase morphology characterization and ultimate mechanical properties of 60/40 PC/SAN blend: influence of the acrylonitrile content of SAN. Polym Eng Sci 31:1215–1221, 1991.

30. V. S. Shah, J. D. Keitz, D. R. Paul, J. W. Barlow. Miscible ternary blends containing polycarbonate, SAN, and aliphatic polyesters. J Appl Polym Sci 32:3863–3879, 1986.

31. J. Im, E. Baer, A. Hiltner. Microlayer composites. In: E. Baer, A. Moet, eds. High Performance Polymers. Munich: Hanser, 1990, pp 175–198.

32. C. B. Arends. Percolation in injection molded polymer blends. Polym Eng Sci 32:841–844, 1992.

33. T. W. Cheng, H. Keskkula, D. R. Paul. Property and morphology relationships for ternary blends of polycarbonate, brittle polymers and an impact modifier. Polymer 33:1606–1619, 1992.

34. Y. Takeda, D. R. Paul. Morphology of nylon-6 blends with styrenic polymers. J Polym Sci Part-B, Polym Phys 30:1273–1284, 1992.

35. T. A. Callaghan, K. Takakuwa, D. R. Paul, A. R. Padwa. Polycarbonate-SAN copolymer interaction. Polymer 34:3796–3808, 1993.

36. V. H. Watkins, S. Y. Hobbs. Determination of interfacial tensions between BPA polycarbonate and styrene-acrylonitrile copolymers from capillary thread instability measurements. Polymer 34:3955–3959, 1993.

37. J. Kolarik, F. Lednicky, M. Pegoraro. Blends of polycarbonate with poly(styrene-acrylonitrile): miscibility, interfacial adhesion, tensile properties, and phase structure. Polym Networks Blends 3(3):147–154, 1993.

38. V. Janarthanan, R. S. Stein, P. D. Garrett. Effect of oligomers and acrylonitrile content on the interfacial adhesion between PC and SAN. J Polym Sci Part-B, Polym Phys 31:1995–2001, 1993.

39. R. Greco, A. Sorrentino. Polycarbonate/ABS blends: a literature review. Adv Polym Techn 13:249–258, 1994.

40. R. Greco. Polycarbonate toughening by ABS. In: E. Martuscelli, P. Musto, G. Ragosta, eds. Advanced routes for polymer toughening. Amsterdam: Elsevier, 1996, pp 469–526.

41. M. Kodama. Surface induced compatibilization of immiscible poly(styrene-co-acrylonitrile) and polycarbonate blend by short Kevlar fiber. Polym J 25:95–98, 1993.

42. D. Haderski, K. Sung, J. Imm, A. Hiltner, E. Baer. Crazing phenomena in PC/SAN microlayer composites. J Appl Polym Sci 52:121–133, 1994.

43. K. Sung, D. Haderski, A. Hiltner, E. Baer. Mechanisms of interactive crazing in PC/SAN microlayer composites. J Appl Polym Sci 52:147–162, 1994.

44. K. Sung, A. Hiltner, E. Baer. Three-dimensional interaction crazes and micro-shear bands in PC-SAN microlayer composites. J Mat Sci 29:5559–5568, 1994.

45. E. Baer. Micro and nano layered polymers. Macrom Symp 104:31–32, 1996.

46. E. Baer, A. Hiltner. Hierarchical structures in natural and synthetic polymer science. Vysokomol Soedin Seriya A 38:549–563, 1996.

47. V. Dobrescu, V. Cobzaru. Some rheological properties of polycarbonate blends. J Polym Sci Polym Symp 64:27–42, 1978.

48. G. Kumar, R. Shyam, N. Sriram, N. R. Neelakantan, N. Subramanian. Estimated blend viscosities by a simple and workable model. Polymer 34:3120–3122, 1993.

49. R. A. McAllister. AIChE J6:427, 1960.

50. R. Greco, M. F. Astarita, L. S. Dong, A. Sorrentino. Polycarbonate/ABS blends: processability, thermal properties and mechanical and impact behavior. Adv Polym Techn 13:259–274, 1994.

51. I. Babbar, G. N. Mathur. Rheological properties of blends of PC and ABS. Polymer 35:2631–2635, 1994.

52. W. Y. Chiang, D. S. Hwung. Properties of polycarbonate/acrylonitrile-butadiene-styrene blends. Polym Eng Sci 27:632–639, 1987.

53. M. P. Lee, A. Hiltner, E. Baer. Formation and break-up of a bead-string structure during injection moulding of a polycarbonate/acrylonitrile-butadiene-styrene blend. Polymer 33:675–684, 1992.

54. M. P. Lee, A. Hiltner, E. Baer. Phase morphology of injection moulded polycarbonate/acrylonitrile-butadiene-styrene blends. Polymer 33:685–697, 1992.

55. H. Hamada, H. Tsunasawa. Correlation between flow marks and internal structure of thin PC/ABS blend injection moldings. J Appl Polym Sci 60:353–362, 1996.

56. R. Greco, L. S. Dong. PC/ABS blends: compatibilization and mechanical behavior. In R. Greco, E. Martuscelli, eds. Polymer blends. Macromol Symp 78:141–153, 1994.

57. L. Morbitzer, H. J. Kress, C. Lindner, K. H. Ott. Struktur und eigenschaften von mehrphasenkunststoffen. Angew Makromol Chemie 132:19–42, 1976.

58. L. Morbitzer, D. Kranz, G. Humme, K. H. Ott. Structure and properties of ABS polymers. X. Influence of particle size and graft structure on loss modulus temperature dependence and deformation behavior. J Appl Polym Sci 20:2691–2704, 1985.

59. L. S. Dong, R. Greco, G. Orsello. Polycarbonate/acrylonitrile-butadiene-styrene blends: 1. Complementary etching techniques for morphology observations. Polymer 34:1375–1382, 1993.

60. H. Suarez, J. W. Barlow, D. R. Paul. Mechanical properties of ABS/polycarbonate blends. J Appl Polym Sci 29:3253–3259, 1984.

61. T. Kunori, P. H. Geil. Morphology–property relationships in polycarbonate based blends. I. Modulus. J Macrom, Sci-Phys, B18(1):93–134, 1980.

62. E. H. Kerner. The elastic and thermoelastic properties of composite media. Proc Phys Soc 69B:808, 1956.

63. R. A. Dickie. Heterogeneous polymer–polymer composites. I. Theory of viscoelastic properties and equivalent mechanical models. J Appl Polym Sci 17:45–63, 1973.

64. J. Kolarik. A model for the yield strength of binary blends of thermoplastics. Polymer 35:3631–3637, 1994.

65. S. Wu. Impact fracture mechanisms in polymer blends: rubber-toughened nylons. J Polym Sci-Phys Ed 21:699–716, 1983.

66. S. Wu. Phase structure and adhesion in polymer blends: a criterion for rubber toughening. Polymer 26:1855–1863, 1985.

67. S. Wu. Formation of dispersed phase in incompatible polymer blends: interfacial and rheological effects. Polym Eng Sci 27:335–343, 1987.

68. A. Margolina, S. Wu. Percolation model for brittle-tough transition in nylon/rubber blends. Polymer 29:2170–2173, 1988.

69. S. Wu. A generalized criterion for rubber toughening: the critical matrix ligament thickness. J Appl Polym Sci 35:549–561, 1988.

70. S. Wu, A. Margolina. Reply to comments. Polymer 31:972–974, 1990.

71. S. Wu. Chain structure, phase morphology, and toughness relationships in polymer and blends. Polym Eng Sci 30:753–761, 1990.

72. J. H. Chun, K. S. Maeng, K. S. Suh. Miscibility and synergistic effect of impact strength in polycarbonate/ABS blends. J Mat Sci 26:5347–5352, 1991.

73. G. Weber, J. Schoeps. Morphologie und technologische eigenschaften von polycarbonat/ABS-mischungen. Angew Makromol Chemie 136:45–64, 1985.

74. M. Ishikawa, I. Chiba. Toughening mechanisms of blends of poly(acrylonitrile-butadiene-styrene) copolymer and BPA polycarbonate. Polymer 31:1232–1238, 1990.

75. H. J. Radusch, G. W. Schreyer, H. Kausche. Investigation of dynamic-mechanical relaxation on ABS-copolymer/PA and ABS-copolymer/PC blends. Preprints part 1: Lectures and Posters of 3d Dresden Polymer Discussion on "High Performance Materials", Dresden, 16–19 April, 1991, Section L4, pp 27–55.

76. R. A. Bubeck, D. J. Buckley Jr, E. J. Kramer, H. R. Brown. Modes of deformation in rubber-modified thermoplastics during tensile impact. J Mat Sci 26:6249–6259, 1991.

77. L. Suyun. Fracture morphology and strength of polycarbonate/ABS blends. Gaofenzi Cailiao Kexue Yu Gongcheng 7(6):62–67, 1991.

78. M.-P. Lee, A. Hiltner, E. Baer. Fractography of injection molded polycarbonate acrylonitrile-butadiene-styrene blends. Polym Eng Sci 32:909–919, 1992.

79. M.-P. Lee, A. Hiltner, E. Baer, C.-I. Cao. Herringbone fracture of a polycarbonate/ABS blend. J Mat Sci 28:1491–1502, 1993.
80. T. Kurauchi, T. Ohta. Energy absorption in blends of polycarbonate with ABS and SAN. J Mat Sci 19:1699–1709, 1984.
81. J. J. Herpels, L. Mascia. Effects of styrene-acrylonitrile/butadiene ratio on the toughness of polycarbonate/ABS blends. Eur Polym J 26:997–1003, 1990.
82. B. S. Lombardo, H. Keskkula, D. R. Paul. Influence of ABS type on morphology and mechanical properties of PC/ABS blends. J Appl Polym Sci 54:1697–1720, 1994.
83. S. J. Wu, S. C. Shen, F. C. Chang. Effect of rubber content in acrylonitrile-butadiene-styrene and additional rubber on polymer blends of polycarbonate and acrylonitrile-butadiene-styrene. Polym J 26:33–42, 1994.
84. J. D. Cooney. The weathering of engineering thermoplastics. Polym Eng Sci 22:492–498, 1982.
85. J. I. Eguiazàbal, J. Nazàbal. Reprocessing polycarbonate/acrylonitrile-butadiene-styrene blends: influence on physical properties. Polym Eng Sci 30:527–531, 1990.
86. D. Stefan, H. L. Williams. Molecular motions in bisphenol-A polycarbonates as measured by pulsed NMR techniques. II. Blends, block copolymers, and composites. J Appl Polym Sci 18:1451–1476, 1974.
87. M. L. Lu, F. C. Chang. Fracture toughness of a PC/ABS blend by the ASTM E813 and hysteresis energy J integral methods: effect of specimen thickness and side groove. Polymer 36:2541–2552, 1995.
88. M. L. Lu, K. C. Chiou, F. C. Chang. Elastic-plastic fracture toughness of PC/ABS blend based on CTOD and J-integral methods. Polymer 37(19):4289–4297, 1996.
89. M. L. Lu, K. C. Chiou, F. C. Chang. Effect of temperature on fracture toughness of PC/ABS based on J-integral and hysteresis energy methods. J Appl Polym Sci 62(6):863–874, 1996.
90. S. Seidler, W. Grellmann. Fracture behavior and morphology of PC/ABS blends. J Mat Sci 28:4078–4084, 1993.
91. J.-S. Wu, S.-C. Shen, F.-C. Chang. Effect of molecular weight on polymer blends of polycarbonate and ABS. J Appl Polym Sci 50:1379–1389, 1993.
92. M. J. Guest, R. Vandaele. Thermal aging of bisphenol-A polycarbonate and acrylonitrile-butadiene-styrene blends. J Appl Polym Sci 55:1417–1429, 1995.

12

Properties of Thermotropic Liquid Crystalline Polymer Blends

Tsung-Tang Hsieh, Carlos Tiu, and George P. Simon
Monash University, Clayton, Victoria, Australia

Stuart R. Andrews and Graham Williams
University of Swansea, Singleton Park, Swansea, Wales

Kuo-Huang Hsieh and Chao-Hsun Chen
National Taiwan University, Taipei, Taiwan, Republic of China

I. INTRODUCTION AND BRIEF LITERATURE REVIEW

Thermotropic liquid crystalline polymers (TLCPs) form a special class of material that incorporates the anisotropic behavior of the liquid crystalline state with the characteristics of polymers, leading to unique chemical and physical properties. TLCP melts are mesomorphic with low viscosity and excellent mechanical properties in the solid state and can be molded with equipment used for conventional random coil, thermoplastic (TP) materials. When blended with TPs, the melt-processible TLCP phase often functions as a processing aid to reduce the melt viscosity of the host TP matrix, and the blend tends to develop a morphology of TLCP fibrils under appropriate shear and extensional flow fields that, when solidified, act to reinforce the TP matrix and lead to improved mechanical performance. A considerable number of studies have been conducted to elucidate the rheology and thermodynamics of binary TP + TLCP blends as well as the processing variables controlling the morphology and mechanical properties (1–5).

Compared to extensive researches on TP + TLCP blends, much less attention has been directed towards blends consisting of two TLCPs. Jin et al. (6) reported on early work regarding miscibility of two semiflexible TLCPs. They concluded

331

that TLCP blends that possess the same mesophase are miscible and that those with different mesophases (nematic and cholesteric or nematic and smectic) are immiscible, as characterized by DSC and optical microscopy. Darragas et al. (7) showed that liquid crystalline semiflexible polyesters that differed only in the length of the spacer unit were also completely miscible, whereas blends with a structurally similar thermoplastic, polybutylene terephthalate (PBT), were partially miscible.

DeMeuse and coworkers (8–11) have also investigated a number of systems of TLCP blends usually based around rigid Hoechst-Celanese Vectra-like copolyesters consisting of units of p-hydroxybenzoic acid (HBA), 6-hydroxy-napthoic acid (HNA), terephthalic acid (TA), and hydroquinone (HQ). Blends of TLCPs comprising copolymers of HBA and HNA of different copolymer ratios were examined (8), and torque vs. mixing time experiments and DSC results demonstrated that chemical reaction is not a predominant factor in the miscibility; this was supported by x-ray analysis. The behavior in the molten state is rather more complex, blends of materials of very different copolymer ratios being immiscible, and may indicate the need in these systems for chemical similarity to yield miscibility. In blends of HBA/HNA copolymers with HBA/HNA/TA/HQ (9), the blends were once again miscible in the solid state but showed viscosity behavior that did not indicate a single phase in the melt. Since TLCPs have a long melt relaxation time, it seems unlikely that materials will become molecularly mixed on cooling. DeMeuse and Jaffe therefore hypothesized (9) that the degree of mixing in the melt may have been on the scale of the domain structure, rather than between different molecules. Further work in these systems (10) emphasized the nonequilibrium nature of the miscible blend, with long annealing times altering the apparent miscibility.

Blends of an HBA/HNA copolymer with a TLCP consisting of HNA, TA, and acetoxy-acid aniline were found to show the synergistic properties that have characterized TCLP + TLCP blends to date (11). Viscosities of some of the blend concentrations were lower than both of the homopolymer values, and modulus of the blends was likewise synergistic (higher). A single blend glass transition indicated their miscibility. The viscosity and modulus synergy was explained by the greater orientation developed in the melt and maintained in the solid state due to mutual, intimate interaction and lubrication of molecules. Lin and Winter (12) also reported synergy in that the shear viscosity of their phase-separated TLCP blends was far lower than that of either TLCP homopolymers, and they suggested that slippage at the interface of the different components may cause this reduced viscosity. The scale of the miscibility required for such good properties is not yet understood and is part of the driving force for the current research.

Akhtar and Isayev (13) examined blends of HBA/HNA copolymers with HBA/TA/HQ materials and found them to be phase separated to some degree, with small changes in the location of distinct glass transitions. Though these blends appear largely immiscible and the viscosities appeared intermediate between the homopolymer extremes, the TLCP blends showed better mechanical properties than the homopolymers in tensile modulus and strength. It appears that either improved viscosity or tensile properties (or both) is characteristic of many such rigid chain polyester blends, which has led to patents in this area; see for example Refs. 14 and 15.

Miscibility between low-molecular-weight liquid crystals (LMWLCs) is often used to identify the type of the liquid crystals formed (e.g., nematic, cholesteric, smectic); molecular liquid crystals of the same class tend to be miscible and form one phase. Lack of miscibility between LMWLCs forming the same mesophases has also been observed, but this is more the exception than the rule (16). For polymeric liquid crystals, theory (17) predicts that a single anisotropic solution should be obtained if two lyotropic liquid crystalline polymers (LLCPs) that possess the same kind of mesophase dissolve in a solvent at sufficient concentration, and experiments have been carried out whose results are in agreement with this theory (18). However, immiscibility of LLCPs of the same mesogenic group with different chemical structures (16) or similar chemical structures (19) has also been observed. In TLCP + TLCP blends containing the same type of mesophase, usually nematic, it is not yet known whether the same situation also applies.

Ballauff modeled the various entropic and enthalpic forces involved in miscibility in such systems (20). It is widely realized that TP blends tend to be immiscible owing to low entropy of mixing because of the high molecular weight of polymers and endothermic heats of mixing, and it is often the presence of favorable intermolecular interactions (and thus negative enthalpy of mixing) that leads to miscibility. In the case of TP + TLCP blends, the entropic factor is even more unfavorable due to the different molecular conformations of each component. For blends of rigid TLCPs with other rigid TLCPs, the entropy term is similar to that of TP + TP blends, and the possibility of enthalpic interactions between TLCPs once again becomes an important driving force. An innovative use of such properties was made by Lee and DiBenedetto (21), who used a TLCP to compatibilize a TLCP and thermoplastic (poly(ethylene terephthalate), PET) blend. The TLCP used as the compatibilizer was immiscible with the other TLCP used in the blend but strongly adhered to it in the solid state.

In TP blends there is a close interrelationship between the packing of polymer chains, miscibility, and molecular motion, and this is likely to be particularly true in rigid chain TLCP systems that are close packed. Whilst miscibility and molecular motion can be investigated by relaxational techniques such as dynamic mechanical and dielectric relaxation, a measure of packing (and hence free or excluded volume) has been measured by the positron annihilation lifetime spectroscopy (PAL). This technique involves placing a radioactive source such as ^{22}Na between two small coupons of the sample. This source ejects positrons (the antiparticle of the electron) which, when they enter the sample, release a characteristic energy in the form of a detectable burst of γ-radiation. The positron decays within the sample or forms other subatomic species by combination with an electron. One of the particles is the orthopositronium particle (o-Ps), which represents a bonded species of an electron and a positron. This seeks and becomes trapped in regions of low electron density (spaces between chains) and eventually annihilates with electrons on the surface of the free volume. When these positrons and other species decay, they emit a different, detectable γ-radiation energy signature. Both bursts of radiation are detected and a lifetime spectrum results. This spectrum can be deconvoluted into three lifetimes, the longest of which relates to o-Ps annihilation and is given by the lifetime parameter τ_3, which is of the order of 1 to 3 ns. The size of the voids in which the positronium species reside corresponds to molecular dimensions in the range of a few angstroms, and the larger the size, the longer (the greater the value of) the τ_3. Also obtained from the deconvolution is the intensity I_3, which is related to the probability of o-Ps formation and the relative number of free volume sites. Further details about PAL and polymers can be found elsewhere (22).

Of particular interest in the context of this work is the application of PAL to free volume in TLCPs. Only a moderate amount of work has been done on TP blends and PAL to date (23–28). It has been shown (28) that in miscible blends there often is a negative deviation of the size of the free volume radii upon blending of miscible polymers, owing to the squeezing out of free volume between chains and conformational adjustments that accompany chains experiencing even weak attractive forces. In the case of partially miscible or immiscible blends, the size of the free volume holes appears to fall on a line that is a weighted average of the homopolymer values (27,29). In the limited work to date on TLCP and PAL, the values that have been measured for rigid chain (29) and semiflexible TLCPs (30) tend to be small in size and concentration, due to the close-packed structure of the rigid chains. Miscible blends of TLCP copolyesters have been found to contain greater concentrations of free volume (due to both an increase in the size of the cavities and their concentration) than the corresponding compositionally equivalent copolymer (29), which was ascribed to poorer packing in the blends, a point supported by the fact that the blends have a lower T_m than the copolymer. Few PAL studies have been performed on blends of TLCPs and thermoplastics (31), and there has been one mention of a PAL of a TLCP + TLCP blend (29).

In this paper, we seek to examine the relationship between rheological properties, miscibility, and free volume of a TLCP blend between two rigid chain copolyesters, Vectra® A950 and HIQ45, both of which contain a common p-hydroxybenzoic acid unit. Vectra® A950 (referred to as Vectra hereafter) consists of 73 mol% p-hydroxybenzoic acid (HBA) and 27 mol% 6-hydroxy-2-naphthoic acid (HNA). The HBA homopolymer is capable of a forming a liquid crystalline phase with a high melting point greater than its degradation point, whilst the HNA homopolymer melts to a smectic mesophase at 440°C. Random copolymerization of

HBA and HNA results in a large melting point reduction that makes melt processing possible (32,33). HIQ45 has a copolymer composition based on 45 mol% HBA, 27.5 mol% isophthalic acid (IA), and 27.5 mol% hydroquinone (HQ). The IA/HQ copolymer does not form a mesophasic melt but is a fusible crystalline polymer that forms an isotropic melt. When HBA is compolymerized with IA and HQ, the resultant polymer is thermotropic within a certain composition range (34). As a result, the HIQ polymer exhibits biphasic morphology, coexistence of liquid crystalline and crystalline phases in the solid state. In the molten state it is thought to have a broad biphasic region where there exists an anisotropic phase and an isotropic phase (from the crystalline phase). This can lead to a positive temperature dependence of viscosity of HIQ polymer, as the viscosity of the isotropic phase is higher than that of the anisotropic phase, and the quantity of the isotropic phase increases during heating (35,36). The biphasic nature of this material means that it is difficult to obtain reproducible synthesis and rheological behavior (36), and work on HIQ materials has previously involved in situ blends of HIQ within a fully polymerized TLCP (36). Alternately, blending of the HIQ with a thermoplastic has made use of the different temperature dependencies of HIQ and a polyetherimide thermoplastic, ensuring good processing behavior over a wide temperature range (37).

In this work we seek likewise to produce materials with improved properties by melt blending HIQ45 with the widely available, rheologically well-behaved TLCP Vectra. In addition, because it is a TLCP + TLCP blend, there remains the possibility that synergistic properties may result, as discussed above. Because HIQ materials do not contain the expensive HNA unit, and have shown other interesting annealing and permeability properties (38), they are TLCPs of much interest.

II. EXPERIMENTAL

A. Materials

The materials chosen for this study were two nematic thermotropic copolyesters, Vectra and HIQ45, both supplied by Celanese. They have the following chemical structures.

HBA   HNA

Vectra® A950

HIQ45

Blends with different weight percentages of the two TLCPs (25 wt%, 50 wt%, and 75 wt% of Vectra) were prepared in the molten state using a Haake Rheocord internal mixer for 5 minutes at 320°C after all materials used were vacuum dried at 100°C for 24 hours. 0.2 wt% of a transesterification inhibitor, Irganox 1098 from Ciba-Geigy, was added to limit chemical reactions taking place during melt blending. In order to experience the same thermal and shear history, pure TLCPs were processed under the identical conditions as their blends.

B. Rheological Measurement

Rheological measurements were performed on a Rheometrics Dynamic Analyzer II (RDAII) in an environmental chamber at 320°C under nitrogen atmosphere and performed within the linear viscoelastic region obtained by strain sweep test using 25 mm gap parallel-plate fixtures with a gap of 0.9 mm. Higher shear rate viscosities were obtained on a Galaxy III capillary rheometer with capillary die of diameter 0.76 mm and length 25.4 mm ($L/D = 33$). The Rabinowitch correction for nonparabolic velocity profiles was applied to the capillary rheometer data.

C. Thermal Characterization

Thermal properties of the blends and TLCPs were measured on a Perkin-Elmer DSC-7 in a nitrogen atmosphere at the heating rate of 10°C/min over a temperature range from 50°C to 380°C. Two-point temperature calibration was performed using high-purity zinc and indium. In order to erase the effect of previous thermal history, the results reported are second heating runs after the samples were scanned at the cooling rate of 10°C/min from 380°C to 50°C.

D. Dynamic Mechanical Thermal Analysis

Dynamic mechanical thermal analysis (DMTA) was carried out on a Perkin-Elmer DMA-7 at a frequency of 1 Hz over a temperature range from −80 to 180°C using a three-point bending mode. Two-point temperature calibration was performed by high-purity n-octane and indium. The heating rate was 2°C/min and the specimen dimensions were 20×5×2 mm^3. The dynamic torsion testing was performed by Rheometrics RDAII equipped with rectangular torsion fixtures.

Specimen dimensions were $50 \times 5 \times 2$ mm^3 and oscillated every 2°C with a soak time of 1 minute to obtain thermal equilibrium. The loss tangent and storage modulus of each sample were determined at 1, 10, and 100 Hz over a temperature range from 45°C to 180°C.

E. Dielectric Relaxation Spectroscopy

Dielectric relaxation spectroscopy (DRS) was obtained using a NovoControl dielectric spectrometer based on a Solatron 1260 Frequency Analyzer in conjunction with a Chelsea Dielectric Interface device and controlled by a computer and WINDETA software capable of measurements between 10^{-1} to 10^5 Hz. Temperature control was possible in a purpose-built cell that incorporates heating and liquid nitrogen cooling, with the lowest temperature measured for this work being -140°C. The data were obtained using three terminal guarded cells, and silver foil was adhered to the sample with vacuum grease to improve electrical contact with the electrodes. The loss parameter determined was the imaginary component of the dielectric permittivity ε''. This is determined under the assumption that the sample is equivalent to a capacitor and resistor in parallel, as calculated by the equation

$$\varepsilon'' = \frac{G}{C_0 \cdot \omega} \tag{1}$$

where G is the conductance, ω is angular frequency ($=2\pi f$), and C_0 is the empty capacitance of the sample. In addition, the dielectric constant, the real component of the dielectric permittivity ε', was also determined according to the equation

$$\varepsilon' = \frac{C_a}{C_0} \tag{2}$$

where C_a is the capacitance of the sample. The loss tangent, also used as a measure of dielectric loss, is then determined by

$$\tan \delta = \frac{\varepsilon''}{\varepsilon'} \tag{3}$$

F. Positron Annihilation Lifetime Spectroscopy

The positron annihilation lifetime spectroscopy (PAL) measurements were made on an automated EG&G Ortec fast–fast coincidence system using a ^{22}Na spot source approximately 2 mm in diameter and instrument resolution of 250 ps. The source was contained between 0.0001 inch thick titanium foils. At least 10 spectra were obtained with 30,000 peak counts and were analyzed using the PFPOS-FIT program. Further experimental details can be found elsewhere (25). The magnitude of the free volume radius can be estimated using the following semiempirical equation (39):

$$\tau_3 = \frac{1}{2}\left[1 - \frac{R}{R_0} + \frac{1}{2\pi}\sin\left(\frac{2\pi R}{R_0}\right)\right]^{-1} \tag{4}$$

where τ_3 (ns) and R (angstroms) are the o-Ps lifetime and radius, respectively. $R_0 = R + \Delta R$, where ΔR is the fitted empirical electron layer (and equals 1.66 angstroms). The average free volume of a hole can then be calculated as

$$\langle V_f \rangle = \frac{4}{3}\pi R^3 \tag{5}$$

The total fractional free volume h expressed as a percentage can be determined (40) from the equation

$$h = CV_f I_3 \tag{6}$$

where V_f is in angstroms cubed calculated from Eq. (3), I_3 is in %, and C is empirically determined from comparison with pressure-volume-temperature data and is found to be approximately 1.8×10^{-3} in glassy polymer systems (22,41–43).

G. Solid Density Measurements

The density measurements of the blends were determined by a Micromeritics® Accupyc 1330 gas pycnometer on crushed samples at ambient temperature under helium atmosphere. The tested samples were purged by nitrogen ten times to ensure a consistent dry atmosphere on testing.

III. RESULTS AND DISCUSSION

A. Rheology

The storage moduli (G') and loss moduli (G'') of Vectra and HIQ45 from dynamic shear flow measurement are shown in Fig. 1a and Fig. 1b for measurements at 320°C. In the case of Vectra, the elastic characteristic is dominant in the low-frequency region, with G' being greater than G''. As frequency increases, the viscous component G'' rises faster than G', eventually becoming greater than G'. By comparison, HIQ45 shows that the viscous nature of the TLCP is dominant at low frequencies, and G' and G'' are comparable within the high-frequency region. There has been a model proposed to describe the frequency dependence of G' and G'' of viscoelastic liquids as determined by dynamic rheology (44) as in Fig. 2. According to the relative magnitudes of G' and G'', a logarithmic plot of G' and G'' as a function of frequency consists of three zones: terminal, plateau, and transition, due to the influence of molecular weight and entanglements in flexible polymers. The appearance of the plot is interpreted in terms of an entanglement network and

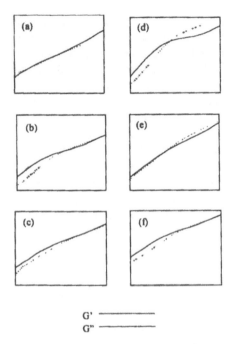

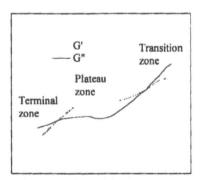

Figure 1 Dynamic moduli vs. frequency of various blend compositions at 320°C: (a) Vectra, (b) HIQ45, (c) 90 wt% Vectra, (d) 75 wt% Vectra, (e) 50 wt% Vectra, (f) 25 wt% Vectra.

Figure 2 Generalized frequency dependence of dynamic moduli of a viscoelastic polymer melt.

two sets of relaxation times relating to short-range (between entanglements) and long-range (beyond entanglements) configurational rearrangements. The long-range relaxation occurs at the time scale of the terminal zone, and the short-range relaxation at that of the transition zone. At frequencies related to the plateau to transition zone, and within the transition zone itself, the glass transition comes into play. In the event of very-low-molecular-weight polymers, there is no plateau zone, and the terminal zone merges with the transition zone owing to the absence of entanglements. The characteristic frequency of the progression of the terminal zone to the plateau zone at a given temperature occurs at lower frequencies with increasing molecular weight, because of an increase in entanglements. Decreasing the melt temperature would lead to the same effect for a given molecular weight, the changeover from terminal to plateau zone occurring at a lower frequency because of a stronger entanglement response (affinity between polymer chains).

However, for TLCPs it may be expected that entanglements amongst rigid, somewhat linear polymer chains are not as important as they are in more flexible polymers. In TLCPs the frequency crossover seen between the terminal and plateau zones, behavior is more likely to result from effects of chain rigidity, packing, and intermolecular association between different molecules (12). In Fig. 2 it can be seen that Vectra behaves at low frequencies as a material in its plateau zone, changing to the transition zone at higher frequencies. Rodlike macromolecules are believed to exhibit longer relaxation times than flexible polymers owing to the difficulty of molecular reorientation arising from close packing. The dynamic behavior of Vectra supports this, as its terminal zone seems to occur at very low frequencies (lower than can be measured with our equipment), indicating both that the chains are rigid and that strong interaction exists between chains at the measurement temperature. By contrast, HIQ45 behaves more as a thermoplastic, moving from terminal zone to plateau zone with increasing frequency in the frequency window of the experiment; such behavior has been seen in other semiflexible TLCPs (45,46). It seems that HIQ45 has less intermolecular interaction and rigidity and is more loosely packed than Vectra. This may be due to the metalinkage of its isophthaloyl moiety in HIQ45 and its consisting of three different ester units rather than the two in Vectra. This could result in a greater diversity of molecular environments in HIQ45, possibly frustrating tight packing.

Dynamic rheological behavior of TLCP + TLCP blends with 90 wt%, 75 wt%, 50 wt%, and 25 wt% Vectra content is shown in Fig. 1c–f. All four blends display behaviors similar to that of pure HIQ45, even with addition of 90 wt% Vectra. That is, all the blends showed a terminal-to-plateau-zone behavior in the frequency range studied, as did HIQ45, even when the HIQ45 concentration was only 10 wt%. The fact that HIQ45 is so influential in determining the dynamic rheological behavior of all the blends (rather than the blend rheology being some weighted average as occurs in immiscible blends) seems to indicate a degree of molecular miscibility.

In order to depict the composition dependence of dynamic behavior, the magnitude of G' and G'' at 1 Hz and the crossover frequency between terminal and plateau zone are shown in Fig. 3. The crossover frequencies of the blends are roughly in the same range as that of HIQ45, with that of Vectra apparently being far lower than others and not measurable in this work. Unusually, there is a slightly higher value of the crossover frequency for 90 wt% Vectra, but this may be only a small deviation compared with the much lower, unmeasurable value of Vectra. Likewise G' and G'' also seem dominated by the HIQ45 values, only varying towards the Vectra values at very high Vectra content.

Figure 4 shows the results of complex viscosity η^* versus frequency from oscillatory measurement for the blend system at 320°C. Vectra shows a steady shear thinning behavior over the whole frequency range studied. However, in the low-frequency region, HIQ45 exhibits a rather moderate shear thinning, showing more pseudoplastic flow at high frequencies. HIQ45 has a higher viscosity than Vectra and HIQ45 + Vectra blends. This can be attributed in part to the biphasic nature of HIQ45—the coexistence of a more viscous isotropic phase and a less viscous liquid crystalline phase, rather than a single anisotropic polydomain structure (35). The complex viscosities of blends are clearly also dominated by HIQ45, as was the viscoelastic behavior described above. Even though they lie between the values of the two TLCP homopolymers, they lie closer to the data of the HIQ45 homopolymer. The change in viscosity with concentration as viewed in Fig. 4 is not simple nor monotonic due to differing shear thinning behaviors of the components and is very much dependent on frequency of measurement. For example, even though Vectra has the lower complex viscosity at low frequencies, the 25 wt% Vectra blend has a lower viscosity than 75 wt% Vectra blends.

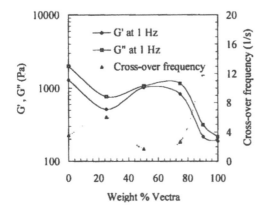

Figure 3 Composition dependence of G' and G'' at 1 Hz and crossover frequency from terminal zone to plateau zone of Vectra/HIQ45 blends.

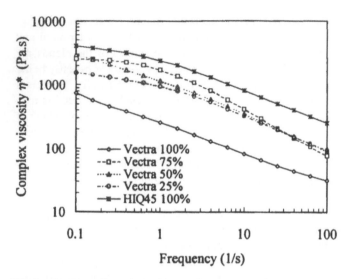

Figure 4   Complex viscosity vs. frequency of Vectra/HIQ45 blends at 320°C.

The steady shear viscosities of the two TLCPs and their blends measured by capillary at 320°C are depicted in Fig. 5. Both of the pure components and their blends show strong non-Newtonian behavior in the shear rate range investigated. At the shear rates obtainable using the capillary rheometer, the viscosities of the blends are much closer to that of neat Vectra, and the 50 wt% Vectra blend is less

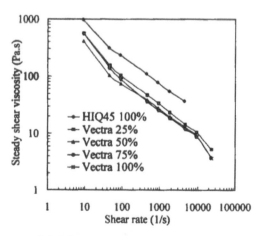

Figure 5   Capillary shear viscosity vs. shear rate of Vectra/HIQ45 blends at 320°C.

viscous than Vectra between 10 s^{-1} to 200 s^{-1}. This is illustrated in Fig. 6, which shows the composition dependence of the blend viscosity at a shear rate of 50 s^{-1}, as interpolated from the capillary rheometer measurements and from dynamic oscillatory measurements. Most isotropic polymers follow the Cox–Merz rule, which predicts that the magnitude of complex viscosity should be equivalent to that of the steady shear viscosity at equal value of frequency and shear rate, and this has proven useful in predicting steady shear viscosity when only linear viscoelastic data are available. Figure 6 reveals that the complex viscosity is much higher than the steady shear viscosity at the same composition and frequency (shear rate). The steady shear may more effectively align the polydomain structure of the mesophasic melts of the TLCPs and their blends and may even align the isotropic phase (if any exists) into an anisotropic phase of the melt in the flow direction (12), resulting in a lower viscosity. In addition, since the continuous shear rate data are obtained from capillary rheometry, greater alignment is possible due to the extensional forces that deform the melt as it converges at the entrance of the capillary.

The relationship between viscosity and blend composition has been proposed to follow either the logarithmic rule of mixture for miscible blends,

$$\ln \eta_{BL} = w_1 \ln \eta_1 + w_2 \ln \eta_2 \tag{7}$$

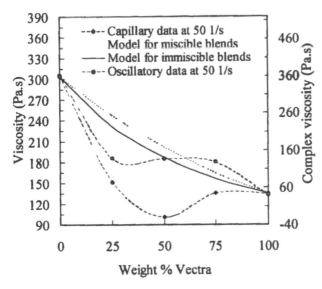

Figure 6 Comparison of two models and experiment results interpolated at 50 s^{-1}.

or the serial model for complete immiscible blends,

$$\frac{1}{\eta_{BL}} = \frac{w_1}{\eta_1} + \frac{w_2}{\eta_2} \tag{8}$$

where η_{BL} is the viscosity of the blend, w_i represents the weight fraction of the components in the blend, and η_i is the viscosity of the pure components (8). The viscosities of the blends do not follow either miscible or immiscible models in this frequency regime and capillary rheometry, showing a very strong negative deviation, with the blend viscosities being very similar to those of the less viscous Vectra. This is in contrast to the very low shear rate, dynamic properties which were shown earlier to be dominated throughout all blend compositions by HIQ45. It appears that the Vectra component is more readily aligned under high shear rates and that this influence persists for all blend concentrations. Figure 7 shows a higher shear rate (1000 s^{-1}) in which blend viscosities are almost equivalent to those of the Vectra homopolymer. This is further evidence that in the molten state there must be some level of miscibility (at the very least a breakdown of the domain structure) for one component to dominate the flow across almost the entire composition range.

As mentioned in the introduction, it has been reported that in TLCP + TLCP blends, some blend compositions often show a lower viscosity than either component homopolymer and a higher modulus and/or tensile strength than either of

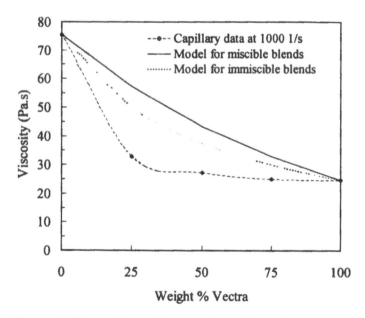

Figure 7 Comparison of two models and experiment results interpolated at 1000 s^{-1}.

the pure components (8–11,14,15). In this case, a synergistic minimum is seen in the higher shear rate capillary data of Fig. 6, and the other data in Figs. 6 and 7 are clearly much lower than would be expected. One explanation for why blending two TLCPs can cause such synergistic properties, and low viscosity generally, is that one TLCP "lubricates" the other on a molecular level, giving rise to a greater degree of alignment of the molecular order that results in high stiffness and lower viscosity, although it is not readily apparent on what level lubrication and enhanced alignment occurs or whether the lubrication is on a truly molecular level or on a larger domain scale. It is possible that at the higher shear rates the flow of the HIQ45 chains is lubricated by Vectra molecules, as more rigid Vectra chains would further align the HIQ45 chains in the liquid crystalline state and, if HIQ45 is biphasic, encourage the nonmesogenic phase to become anisotropic. At lower shear rates, behavior such as the values of G', G'' and the crossover frequency are dominated by HIQ45.

Owing to their long relaxation times compared to TP, TLCP melts can be expected to retain in the solid state features such as packing and alignment found in the molten material, especially since dynamic measurements are performed within the linear viscoelastic limits where the molecular arrangements are unperturbed by the applied strain. As a consequence, it is reasonable to seek to correlate the molten state behavior and the solid state behavior, and this will be discussed subsequently.

B. DSC

The DSC thermograms of the TLCPs and their blends are shown in Fig. 8. A weak crystalline-to-nematic melting transition of Vectra occurs at about 280°C, whilst HIQ45 has two larger endothermic peaks at about 310°C and 350°C. It is tempting to consider the first as a crystalline-to-nematic melting transition, whilst the second is either another crystalline-to-nematic melting transition or a nematic-to-isotropic transition (known as a clearing point). However, examination of HIQ45 by cross-polarized optical microscopy showed a birefringent liquid crystalline behavior up to 370°C (the limit of the equipment), well above the second peak. This is, the nature of the melt in this temperature range is not readily concluded from optical microscopy. No melting transition temperatures of Vectra or HIQ45 can be observed by DSC in the blends. This is quite different from other TLCP + TLCP blends (6–13) and may result from the miscibility of the blends preventing segregation of pure components and thus suppressing crystalline formation of each TLCP. Alternatively, the kinetics of cooling may be such as to freeze in an amorphous structure in the TLCPs where the crystallinity is usually of a nonperiodic layer (NPL) type (47) and total degree of crystallinity is normally very low. Although it is clear that the heat of fusion of Vectra is low, the peaks in homopolymer HIQ45 are quite large, and their nonappearance in blends may thus be

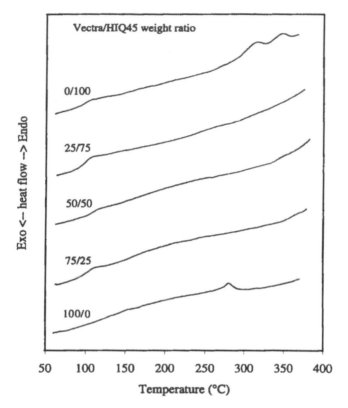

Figure 8 DSC thermograms of Vectra/HIQ45 blends.

further indication of intimate molecular mixing. A blend of 75/25 and 30/70 copolymers made up of HBA and HNA units, respectively, was concluded to be a miscible blend, as it exhibits only one melting transition well below those of component TLCPs or a TLCP of the same composition (29). Lin and Winter (12) reported that immiscibility of their blends is evidenced by two independent crystallization peaks in a DSC run attributed to the exotherms of the pure components. DeMeuse and Jaffe (8) concluded that their blends are miscible judging from the single shifting melting points observed, which suggests that molecular mixing has occurred. It therefore appears that melting points from DSC have been a useful guide in assessing miscibility in these types of blends.

The glass transition temperature T_g of polymer materials can be detected according to their DSC thermograms and seen as the slope change in the specific heat curve. According to this method, HIQ45 exhibits a T_g change between 120°C and 140°C indicative of a T_g at 127°C. The T_g of Vectra is undetectable by DSC owing to the small change in heat capacity during its glass–rubber transition. However, even with the addition of only 25 wt% HIQ45, the magnitude of the blend T_g behavior is significantly enhanced. The 75 wt% Vectra and 25 wt% Vectra blends also show a much greater heat capacity change than the 50 wt% Vectra blend and HIQ45 itself. This suggests that the molecular motions of these two blends through glass–rubber transition is somewhat stronger than that of the other blends and indeed the homopolymers themselves. Referring to Table 1, the temperature locations of T_gs of blends measured by DSC increase as the HIQ45 content increases, also implying miscibility of the blends in the solid state.

C. DMTA

Figure 9 shows the tan δ ($=E''/E'$) as a function of temperature for the pure TLCPs and their blends using DMTA. Vectra displays three transitions in the temperature range of interest, −80 to 180°C. There is a weak γ-transition ranging from −70 to −40°C attributed to the motion of HBA units. The β-transition results from the local reorientation of the HNA moieties and occurs between −10 and 50°C. The γ-transition temperature or T_g associated with the onset of the cooperative micro-Brownian motion along the main chain is at about 106°C (47,48). By comparison, HIQ45 shows the T_g at about 132°C and a stronger sub-T_g transition, attributed to the motion of its phenylene units both from HBA and HQ, as the oxygen-aromatic bonding of HQ is able to rotate more freely than the HBA unit, which is bonded by carbonyl groups to the main chain. The kinked molecular structure of HIQ45, due to the meta configuration of IA, causes sharp deviation of the para chain axis of HBA and HQ by about 60° at each IA locus along the main chain and results in stronger damping behavior at T_g than that of Vectra, whose

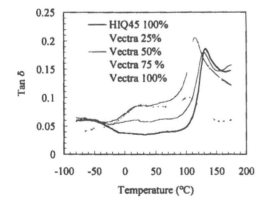

Figure 9   tan δ vs. temperature of Vectra/HIQ45 blends by DMTA (three-point bend mode at 1 Hz).

parallel offset HNA units do not upset the chain linearity sufficiently, and which thus has a weak α-relaxation. Since the glass transition temperatures of the two TLCPs are 30°C apart, it is possible to determine blend miscibility from them.

In miscible TP blends a particular blend tends to demonstrate a single tan δ peak that varies smoothly in temperature location from one homopolymer to the other. By contrast, immiscible blends show distinct T_gs of the homopolymer components. In these blends the intensities of the tan δ relaxations are all comparable in height to that of HIQ45, the T_gs of the blends shifting from 130°C (HIQ45-rich blend) to 114°C (Vectra-rich blend). In terms of height, the relaxation peaks of T_gs of the blends do not follow simple additivity, and the tan δ peaks (especially 75% and 25% Vectra) are not sharp, suggesting a partially miscible morphology. The intensified damping characteristics of the α-relaxation of 25 wt% and 75 wt% Vectra blends may result from the lack of constraint on molecular chain motion by the crystallites which, although found in HIQ45 and Vectra homopolymers, are less prevalent in the blends (according to DSC). The glass transition behavior demonstrated by the DSC technique (Fig. 8) also revealed an unusual compositional dependence of the strength of α-transition. To confirm this glass transition behavior, torsion mode dynamic mechanical measurements of the solid samples with the RDAII were performed and the α-relaxations of the blend system studied. Figure 10 depicts the results for 1 Hz, the same frequency employed by DMTA in Fig. 9, and the results, along with T_g from DSC and DMTA, are shown in Table 1. Note that temperature dependencies as a function of composition are similar for the different techniques and frequencies, the glass transition temperature increasing with increasing measurement frequency. The height of the tan δ peak from the torsion measurement of

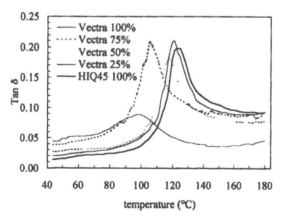

Figure 10   tan δ vs. temperature of Vectra/HIQ45 blends by RDAII (torsion mode at 1 Hz).

Table 1 Glass Transition Temperatures T_g from Different Techniques and the Apparent Activation Energies E_a of the Blends

nd 6 nposition (wt%)	T_g (°C) (DSC)	T_g (°C) (DMTA 1 Hz)	T_g (°C) (RDAII 1 Hz)	T_g (°C) (RDAII 10 Hz)	T_g (°C) (RDAII 100 Hz)	E_a (RDAII) (kJ/mol)
100% HIQ45	127	132	123	127	131	332 ± 1.2
25% Vectra	118	130	121	125	129	328 ± 1.9
)% Vectra	107	128	117	121	125	322 ± 1.9
i% Vectra	104	114	105	111	115	240 ± 29.5
)0% Vectra	—ᵃ	106	99	101	103	581 ± 1.8

T_g not observable by DSC.

the neat TLCPs shows a similar magnitude to that of DMTA but is different between techniques for the blends. However, as in the case of the DMTA data of Fig. 9, the torsion data of Fig. 10 show that there is a change in temperature location of the peak as a function of temperature, although the heights of the tan δ peaks appear random. In addition, the torsion spectra of 25 wt% and 75 wt% Vectra blends have very weak shoulders on the low-temperature side of the main peaks, indicating the possibility of a slight two-phase morphology. Because of the possible biphasic nature of HIQ45, the Vectra molecules may not be as miscible with the small amount of isotropic phase, which could lead to multiple phases. Alternately, it may be that there is a limited solubility of one polymer in the other and more than one phase thus exists.

The secondary or β-relaxations of the blends can be seen in the DMTA data at about 25°C in Fig. 9 and relates to motion of the rigid HNA group. It is quite weak when there is greater than 25 wt% HIQ45. All blends exhibit a similar magnitude of γ-transition comparable to that of the HIQ45 homopolymer.

Although the α-relaxation of polymers usually follows a free volume WLF type temperature dependence, it is found that the limited frequency range measurable by the RDAII means that the data often follows a linear Arrhenius-like relationship, and this is used to allow comparison between the activation energies of various blends. The apparent activation energies E_a of torsion mode dynamic mechanical measurements derived from log f_{max} versus $1/T$ plots are shown in Table 1. The high E_a of Vectra indicates that molecular chain motion is strongly hindered by its liquid crystalline chain alignment and the bulky HNA moiety. HIQ45 exhibits a lower E_a than Vectra owing to its less linear molecular configurations, which give more degrees of freedom for molecular chain motions at the glass transition. All the blends (other than the 75 wt% Vectra blend) show a similar E_a to HIQ45, which indicates that mobility is dominated by the HIQ45 component, which perhaps is not surprising given the similarity in appearance of HIQ45 and all the blends in the dynamic melt rheology data presented earlier. Figure 11 illustrates the temperature dependence of storage moduli at 1 Hz by RDAII torsion measurements, and this further illustrates the difference in the nature of the glass transition behavior between the two homopolymers. HIQ45 exhibits a clear glass transition region where the shear modulus G' decreases abruptly by three orders of magnitude in a 20–30°C range, whereas Vectra shows a gradual reduction of G' with temperature over a wider temperature span, leveling off with highest rubbery modulus among the blend systems. With the addition of Vectra up to 50%, the blends behave the same way as HIQ45 with clear glass transition regions. The blend with 75% Vectra content resembles Vectra at low-temperature range with G' gradually decreasing with temperature. However, as temperature approaches T_g, its behavior has a similarity to that of HIQ45 and other blends, with a strong drop in modulus attributed to glass transition. Normally, the value of G' ($>T_g$) is dependent on molecular weight between entanglements for linear flexible polymers, but this may not be relevant for rigid TLCPs and is more likely a measure of chain rigidity, and, as with E_a values from the α-relaxation, it shows that main-chain motion and interaction is dominated by HIQ45.

D. DRS

The tan δ loss spectra of the range of blend samples determined by DRS at 1 kHz is shown in Fig. 12. The temperature location of the maxima in the loss spectra compared to DSC and DMTA data clearly identifies it with the α-relaxation

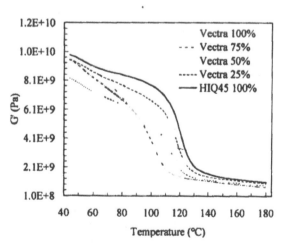

Figure 11   Storage moduli vs. temperature of Vectra/HIQ45 blends by RDAII (torsion mode at 1 Hz).

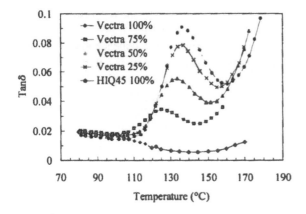

Figure 12   tan δ vs. temperature of Vectra/HIQ45 blends at 1 kHz by DRS.

peak of the HIQ45 homopolymer. The strength of the α-relaxation seen for HIQ45 is not normally seen so distinctly in rigid, main-chain TLCPs, as is well illustrated by the broad, ill-defined dielectric spectra of Vectra homopolymer, also shown in Fig. 12. The lack of a clear loss maximum for the α-relaxation in Vectra by dielectric relaxation spectroscopy has been reported and summarized elsewhere (49–52). This may be because most of the dipoles are relaxed prior to the T_g region by secondary γ- and β-relaxations (due to HBA and HNA motions) and thus there is little dipolar strength remaining at the glass transition itself. The strength and pronounced shape of α-relaxation for the HIQ45 material indicate that the nature of the motion of the dipoles on the rigid chain is somewhat different from that for Vectra, not least because of the absence of the rigid HNA unit, and most of the dipolar relaxation strength is relaxed by the primary relaxation.

The relaxation peaks of the blends occur at intermediate temperatures and heights (contrasting with mechanical spectroscopy) and thus are miscible as judged by this technique, the relaxation appearing to get progressively broader with higher Vectra content. The reason for the contrast between the classical dielectric relaxation data for a miscible blend, compared to the variation of relaxation strength in the mechanical spectroscopy techniques, is not fully understood, but possibly it demonstrates that the alignment of chains may influence mechanical relaxation strengths. It may be that dielectric relaxation that probes the mobilities of local chain dipoles is not so sensitive to such larger scale order and alignment, or alternatively the thinner dielectric samples (films) may be more evenly aligned because of the higher shear rate squeezing flow required to press the thin films required in DRS.

The temperature–frequency locus of HIQ45 and the blend α-relaxation peaks as seen by DRS (the broad Vectra peak is not observed) are shown in Fig. 13. All blends and HIQ45 show a curved frequency–temperature dependence characteristic of the primary α-relaxation of polymers, in which there is a decrease in free volume as the glass transition is approached from above, leading to a rapid increase in the relaxation time of chain motion (53). It appears that the wider frequency range of the dielectric technique is sufficient to capture this behavior, as opposed to the narrower range possible using DMTA. The dependency observed in Fig. 13 can be modeled using the Vogel–Fulcher (VF) equation (54,55).

$$f_m = A \cdot exp \left(- \frac{B}{R \cdot (T - T_0)} \right) \tag{9}$$

where A is the pre-exponential factor, B is a measure of the energy of the rotational energy barrier that bonds must overcome in the motion, T is the temperature, and T_0 is the temperature at which all motion is supposed to cease and is often found experimentally to be some 50°C below the glass transition temperature determined by DMTA (56). The data were fitted to Eq. (9) and the results for these samples are given in Table 2.

It can be seen that the rotational energy barrier B in Table 2 is greatest for the Vectra-rich materials and lowest for the HIQ45 homopolymer, and this is similar to the trend in activation energy due to mechanical spectroscopy seen in Table 1. Comparison is difficult because a Vectra homopolymer value of B could not be determined, but it appears as though for blends of 50 wt% or more Vectra, the behavior is dominated by the more rigid Vectra component, although an alternative

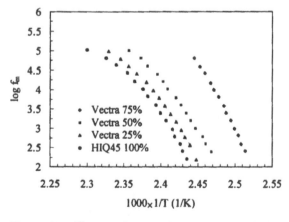

Figure 13 The maximum of α-transition of HIQ45 and the blends at various frequencies as a function of temperature.

Table 2 Vogel–Fulcher Fitted Parameters [Eq. (9)] from DRS Measurements

Blend composition (wt%)	A	B (kJ/mol)	T_0 (°C)
75% Vectra	12.10±1.10	6.19±1.50	91.4±5.1
50% Vectra	10.05±0.70	5.69±1.23	94.7±4.8
25% Vectra	8.08±0.13	2.57±0.16	112.6±0.9
100% HIQ45	7.36±0.18	1.89±0.18	118.1±1.2

Relaxation peak for Vectra not observable by DRS.

interpretation could be that B is determined by whichever is the majority component. Although the rotational energy barrier of Vectra could not be determined, it appears that the motions of polymer chains in the solid state are strongly affected by HIQ45 molecules and point to a likely scenario that the addition of HIQ45 creates an environment that can only be achieved by intimate molecular mixing. The value of T_0 is approximately 15–25°C below the glass transition measured by DMTA, and the deviation decreases as Vectra content decreases.

The low temperature, sub-T_g relaxations as measured by DRS are shown in Fig. 14 for a frequency of 1 kHz. The secondary relaxation spectra of the two homopolymers look quite different. The spectrum of Vectra is well known, its β-relaxation (25 kJ/mol, 20°C at 1 Hz) and γ-relaxation (12 kJ/mol, −50°C at

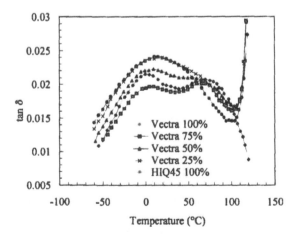

Figure 14 tan δ vs. temperature of Vectra/HIQ45 blends at 1 kHz by DRS for sub-T_g relaxation.

1 Hz) having been often reported with DRS, for example by Alhaj-Mohammed et al. (50) for a variety of frequencies. As mentioned previously, the β-relaxation has been assigned to the motion of carbonyl groups associated with the HNA moiety (48), and the assignation as the β-relaxation has been confirmed by its absence in related systems with no HNA groups (52). The lowest temperature motion, the γ-relaxation, has thus been ascribed to relaxation of carbonyl dipoles associated with motion of the HBA moiety. It is able to rotate without disturbing the motion of adjacent groups and thus is a local motion (51). The low-temperature relaxation spectrum of HIQ45 has not been reported before. It can be seen that it is much broader than that of Vectra, although it appears to be centered around the similar position as the γ-relaxation in Vectra, whilst no higher temperature β-relaxation is observed, which is reasonable given the absence of an HNA unit. The similar location of the γ-relaxation peak of the HIQ45 sample is also reasonable, since it possesses, like Vectra, an HBA group whilst also containing an isophthalic unit. Both these units have carbonyl groups and are thus polar and clearly relax at the same temperature, leading to the broader HIQ45 relaxation. The HBA γ-relaxation of Vectra is much narrower, although it is soon followed at higher temperatures by the local HNA relaxation. The hydroquinone moiety in HIQ45 is not as dielectrically active, as it has no dipoles perpendicular to the chain.

The addition of a little HIQ45 (25 wt%) results in a slightly suppressed motion of both secondary relaxation peaks of Vectra (a slight decrease in height, with little change in peak position), and this could be due to miscibility resulting in a slight restriction of the secondary relaxations. Although miscibility predominantly affects primary dielectric and mechanical relaxations, in some polymer systems either secondary relaxations can be suppressed or their transition temperatures can be changed when molecularly mixed in a miscible blend (57). Further addition of HIQ45 results in a broadening of the relaxation peak until, with greater than 50% HIQ45, the spectrum looks very similar to that of the HIQ45 homopolymer. The γ-transition detected by DMTA shows a similar trend, except that it is enhanced by the addition of HIQ45, and it is much weaker for the Vectra homopolymer (Fig. 9).

In addition to being presented as a temperature scan, the γ-relaxation is particularly strong in both HIQ45 and Vectra and can be viewed in the frequency domain, as is shown for $-20°C$ measurement temperature in Fig. 15 of ε'' as a function of frequency. It can be seen clearly that the broadness of the relaxation peak is greatest for the HIQ45 material (as discussed above) and intermediate for all other blend compositions. It can be seen from this figure that there is only a very small change to the γ-relaxation temperature with blending, as can be expected for such a local motion. The β- and α-relaxations cannot be similarly viewed in the frequency domain in these polymers, the relaxations being too broad and weak.

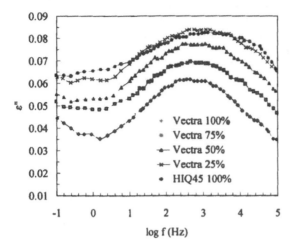

Figure 15 Dielectric loss ε'' vs. frequency of Vectra/HIQ45 blends at $-20°C$.

E. PAL

Figure 16 shows the dependence on Vectra composition of τ_3 and I_3 by PAL. The size of the free volume sites, as indicated by values of τ_3, appears generally to show an average between those of the homopolymers with a slight sigmoidal fluctuation around the mean. This contrasts with the negative deviation (smaller-than-average size) observed in miscible amorphous blends (23–26), where it was found that the value of τ_3 invariably showed a negative deviation from linear additivity, ascribed to closer proximity of chains in a miscible system. Rather, the data in this TLCP + TLCP blend appear to be closer to those of a two-phase partially miscible thermoplastic blend of polycarbonate and poly(methyl methacrylate) (27). In that system, the additivity of size may be because limited solubility of one phase in the other meant that the average size of the free volume sites in the blend (which PAL measures) is a weighted sum of the components. The implication is therefore also that because the system is partially miscible, interfacial adhesion is good and no excess free volume is created at the interfaces. There is some evidence that in amorphous, immiscible blends a positive deviation above average may occur (58) if the interface is poor.

However, the intensity data indicative of the concentration of free volume sites does shows a negative deviation with blends of greater than 25 wt% Vectra, being much lower than an average of the homopolymer data. It appears that in the blend there exist fewer free volume sites, possibly indicating close packing. These issues are further illustrated by the magnitude of the free volume holes V_f and the percentage fractional free volume h calculated using Eqs. (4), (5), and (6) shown

Hsieh et al.

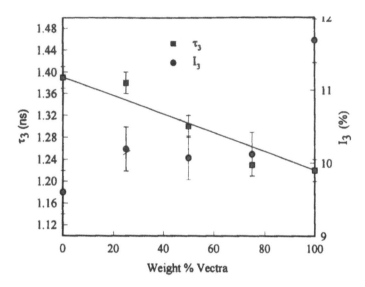

Figure 16 τ_3 and I_3 vs. Vectra content of Vectra/HIQ45 blends by PAL measurements.

in Fig. 17. It can be seen here that whilst the size of the holes in the blends is a weighted average of the homopolymers, the total free volume fraction h shows a negative deviation from linearity, which may be expected in close-packed systems. This differs from that of miscible thermoplastic blends, in which a negative deviation of the *size* of the free volume sites V_f, rather than free volume fraction h (58), often lies on the rule of mixtures line. This difference in behavior between miscible thermoplastics blends and this TLCP + TLCP blend may be indicative more of improved close packing than true molecular miscibility. Nonetheless, it is important to recognize that blends of these TLCPs do seem to pack better in the solid state than their homopolymers, according to the PAL technique, and the synergy in the 75 wt% Vectra blend having a lower free volume fraction than the homopolymers correlates well with the ideas of synergistic low viscosity and high modulus in such systems owing to improved molecular alignment. As with thermoplastic materials, the free volume appears to get squeezed out in this system, especially at high Vectra compositions.

There is scientific precedent for the comparison we make between free volume and mechanical/rheological properties. For example, a negative deviation in free volume (close packing) often seems to lead to a strong change in properties that rely on molecular mobility, such as impact strength. It is known, for example, that miscible thermoplastic polyester blends often have much lower impact strength than the homopolymer components (23,59), since the motion of the

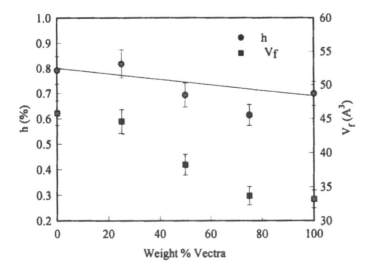

Figure 17 Average free volume fraction *h* and average volume of free volume sites V_f vs. Vectra content of Vectra/HIQ45 blends using the PAL technique.

chains in the glassy region at room temperature is restricted by the interactions that lead to miscibility.

F. Solid Density

The solid density of TLCPs and their blends is shown in Fig. 18. The trend of density versus Vectra content shows very much the same trend as average free volume fraction *h* versus Vectra content from the PAL, i.e., a high free volume fraction leads to a low value of density. It seems that in this system the variation of average free volume fraction sampled by PAL is directly related to a measurable, macroscopic density change. This is not necessarily the case for other non-liquid-crystalline thermoplastic blend systems examined by PAL (25), where density did not change with composition, although the PAL parameters did. This may indicate that the rigidity of polymer chains means that macroscopic order and local molecular packing are in some sense self-similar and reflective of each other. Taking this comparison between the melt and solid states further, the concentration behavior of solid density appears to correlate with the abnormal results of complex viscosity. For example, the 75 wt% Vectra blend has the highest density among the blends because of closer molecular packing and also has a higher complex viscosity. By contrast, the low-density 25 wt% Vectra blend shows looser packing,

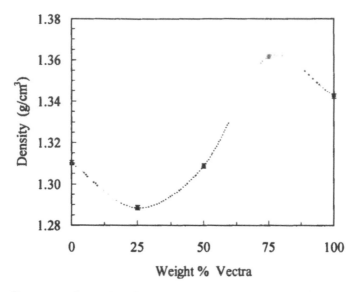

Figure 18 Solid density vs. Vectra content of Vectra/HIQ45 blends.

and lower complex viscosity results. By making such comparison between solid density and melt properties, we are making use of the concept expressed earlier that packing in the melt and solid states in these systems will be similar owing to the long relaxation times of the TLCPs and that the structure of the molten blends becomes readily frozen in in these systems as they solidify.

IV. CONCLUSIONS

A blend system of two TLCPs has been prepared with inhibitor to limit chemical reaction (transesterification), and it blend properties were studied. A model of frequency dependence of G' and G'', consisting of terminal zone, plateau zone, and transition zone, was used to interpret the dynamic rheological results. Vectra shows plateau-to-transition zone behavior that is different from the terminal-to-plateau zone behavior of HIQ45 and blends and indicates that Vectra has a greater relaxation time than that of HIQ45 and blends. The blend behavior at lower frequency oscillatory rheometry is strongly influenced by HIQ45 across the whole composition range (even at quite low HIQ45 concentrations), which indicates that they may be intimately miscible for such behavior to occur. The steady shear viscosities of the blends under capillary flow do not obey the models for either miscible or immiscible blends and exhibit a very strong negative deviation from rule of additivity. Rigid Vectra has the lowest viscosity, and this is reflected across the whole blend range, further indicating that the blends may be miscible. In non-liquid-crystalline blends it is often found that blends that are miscible in the molten state have a positive deviation of their blend viscosities owing to lack of free volume in the molten state (60). It seems that in this TLCP blend system the low viscosity occurs despite close packing, and indeed profits from it by "self-lubrication" on a molecular scale of the two TLCPs.

Further evidence for some level of miscibility exists. The disappearance of the endothermic peak of the blends in the DSC thermograms indicates that the crystallization of either pure component is suppressed and may be due to close packing interfering with either the kinetics or the enthalpy of crystallization. The tan δ peaks by DMTA and torsional mechanical measurements are intermediate between those of the homopolymer. They are however broadened, and small shoulder peaks can be observed, revealing that although the blends are largely miscible, they may not be fully molecularly mixed. By contrast, from DRS, a clear, primary α-relaxation is seen for HIQ45 and blends that is intermediate in position and height between HIQ45 and Vectra. The temperature and frequency locations of the secondary relaxations are not found to change much with blending, commensurate with the local nature of their motion and the lack of strength of the intermolecular interactions driving the degree of miscibility involved. PAL parameters indicate that close packing in the blend generally leads to a negative deviation in the free volume fraction, although the size appears to be something of an average. This may be indicative of a reduction in free volume due to the propensity for the chains of HIQ45 and Vectra to lie close to one another on something close to a molecular level, which is consistent with notions of the intimate mixing that occurs in miscible blends. Thus even though free volume is less than additive in these blends (measured by PAL and density at room temperature), viscosity in the molten state remains low owing to the self-lubricating nature of different TLCPs that are closely packed. Whilst the results of solid density measurements support the free volume fraction variation determined by PAL, they can also be tentatively used to explain the abnormal behavior of the complex viscosity of the blends. Pressure-volume-temperature curves across a wide temperature range are a better way of making such free volume/viscosity comparisons, and such work is currently underway on this system.

ACKNOWLEDGMENTS

Thanks to Linda Sawyer and Mike Jaffe at Hoechst Celanese for providing the HIQ45 sample. TTH would like to acknowledge a scholarship grant from the Targeted Institution Link (TIL) program of DEET (Australian Government). SRA acknowledges support from the EPSRC (UK) for a Post-doctoral Research Assistantship, and we thank the EPSRC (UK) for a grant to purchase the Novocontrol dielectric equipment.

REFERENCES

1. D Beery, S Kenig, A Siegmann. Structure and properties of molded polyblends. Polym Eng Sci 31:459–465, 1991.
2. W Brostow, TS Dziemianowicz, M Hess, R Kosfeld. Blending of polymer liquid crystals with engineering polymers. In: RA Weiss, CK Ober, eds. Liquid-Crystalline Polymers. Washington DC: ACS Symposium Series 435, 1990, pp 402–413.
3. SM Hong, BC Kim, SS Hwang, KU Kim. Rheological and physical properties of polyarylate/LCP blend systems. Polym Eng Sci 33:630–639, 1993.
4. SH Jang, BS Kim. Copolyester-amide and amorphous polyamide blends. Polym Eng Sci 34:847–856, 1994.
5. O Roetting, G Hinichen. Blends of thermotropic liquid crystalline and thermoplastic polymers. Adv Polym Technol 13:57–64, 1994.
6. JI Jin, EJ Choi, KY Lee. Miscibility of main-chain thermotropic polyester. Polym J 18:99–101, 1986.
7. K Darragas, G Groeninckx, H Reynaers, C Samyn. Miscibility of blends of LC polyesters and blends of polybutylene terephthalate with LC polyester. Eur Polym J 30:1165–1171, 1994.
8. MT DeMeuse, M Jaffe. Model system for liquid crystal polymer blends. Mol Cryst Liq Cryst Inc Nonlin Opt 157:535–566, 1988.
9. MT DeMeuse, M Jaffe. Investigations into the structure of liquid-crystalline polymer–liquid-crystalline polymer blends. In: RA Weiss, CK Ober, eds. Liquid-Crystalline Polymers. Washington DC: ACS Symposium Series 435, 1990, pp 439–457.
10. MT DeMeuse, M Jaffe. Studies of the structure of blends containing two liquid crystal polymers. Polym Adv Technol 1:81–92, 1990.
11. S Kenig, MT DeMeuse, M Jaffe. Properties of blends containing two liquid crystalline polymers. Polym Adv Technol 2:25–30, 1991.
12. YG Lin, HH Winter. Rheology of phase separated blends of two thermotropic liquid crystalline copolyesters. Polym Eng Sci 32:773–776, 1992.
13. S Akhtar, AI Isayev. Self-reinforced composites of two thermotropic liquid crystalline polymers. Polym Eng Sci 33:32–42, 1993.
14. M Matsumoto, K Teruhisa. US Patent 4,837,268 (1989).
15. AI Isayev, PR Subramanian. US Patent 5,070,157 (1991).
16. E Marsano, E Bianchi, A Ciferri. Mesophase formation and polymer compatibility. 2. Cellulose acetate/(hydroxypropyl)cellulose/diluent system. Macromolecules 17:2886–2889, 1984.
17. PJ Ferry, RS Frost. Statistical thermodynamics of mixtures of rodlike particles. 3. The most probable distribution. Macromolecules 11:1126–1133, 1978.
18. SM Aharoni. Rigid backbone polymers: 6. Ternary phase relationships of polyisocyanates. Polymer 21:21–30, 1980.
19. J Lin, H Wu, S Li. Compatibility of rodlike polymers. Eur Polym J 30:231–234, 1994.

20. M Ballauff. Polymer blends containing liquid crystal polymers. Polym Adv Technol 1:109–116, 1990.
21. WC Lee, AT DiBenedetto. The processing of ternary LCP/LCP/thermoplastic blends. Polymer 34:684–690, 1993.
22. YC Jean. Positron annihilation spectroscopy for chemical analysis: a novel probe for microstructural analysis of polymers. Microchem J 42:72–102, 1990.
23. MD Zipper, GP Simon, GM Stack, MR Tant, AJ Hill. A free volume study of miscible polyester blends. Polymer International 36:127–136, 1995.
24. GP Simon, MD Zipper, AJ Hill. On the analysis of positron annihilation lifetime spectroscopy data in semicrystalline miscible polymer blend systems. J Appl Polym Sci 52:1191–1202, 1994.
25. MD Zipper, GP Simon, P Cherry, AJ Hill. The effect of crystallinity on chain mobility and free volume in the amorphous regions of a miscible polycarbonate polyester blend. J Polym Sci Polym Phys Ed 52:1237–1247, 1994.
26. J Liu, YC Jean, H Yang. Free-volume hole properties of polymer blends probed by positron annihilation spectroscopy. Macromolecules 28:5774–5779, 1995.
27. MD Zipper, GP Simon, V Flaris, JA Campbell, AJ Hill. Interfacial phenomena in polymer blends probed by positron annihilation lifetime spectroscopy. Materials Science Forum 189–190:167–172, 1995.
28. JA Campbell, AA Goodwin, M Safari Ardi, GP Simon, CJT Landry-Coltrane. Free volume studies in miscible polymer blend systems. Macromol. Chem 118:383–388, 1997.
29. CM McGullagh, Z Yu, AM Jamieson, J Blackwell, JD McGervey. Positron annihilation lifetime measurements of free volume in wholly aromatic copolyesters and blends. Macromolecules 28:6100–6107, 1995.
30. A Uedono, R Sadamoto, T Kawano, S Tanigawa, T Uryu. Free volumes in liquid-crystalline polymer probed by positron annihilation. J Polym Sci Phys Ed 33:891–897, 1995.
31. RA Naslund, PL Jones. Characterization of thermotropic liquid crystalline polymer blends by positron annihilation lifetime spectroscopy. Materials Research Society Proceedings, Submicron Multiphase Materials 274, 1991.
32. AM Donald, AH Windle. Liquid Crystalline Polymers. Cambridge: Cambridge Univ. Press, 1992.
33. P Magagnini. Molecular design of thermotropic main-chain liquid crystalline polymers. In: FP La Mantia, ed. Thermotropic Liquid Crystal Polymer Blends. Lancaster: Technomic, 1993, pp 1–42.
34. AB Erdemir, DJ Johnson, JG Tomka. Thermotropic polyesters: synthesis, structure and thermal transitions of poly(p-oxybenzoate-co-p-phenylene isophthalate)s. Polymer 27:441–447, 1986.
35. G Kiss. Anomalous temperature dependence of viscosity of thermotropic polyesters. J Rheol 30:585–598, 1986.
36. B Gupta, G Calundann, LF Charbonneau, HC Linstid, JP Shepherd, LC Sawyer. Development of in-situ blends of poly(p-oxybenzoate-co-p-phenyleneisophthalate) with other thermotropic liquid crystalline polymers. J Appl Polym Sci 53:575–586, 1994.
37. KH Wei, G Kiss. Liquid crystalline polymer blends with stabilized viscosity. Polym Eng Sci 36:713–720, 1996.

38. JY Park, DR Paul, I Haider, M Jaffe. Effect of thermal annealing on the gas permeability of HIQ-40 films. J Polym Sci Polym Phys Ed 34:1741–1745, 1996.
39. H Nakanishi, SJ Wang, YC Jean. Microscopic surface tension studied by positron annihilation. In: SC Sharma, ed. Proceedings of International Conference on Positron Annihilation in Fluids. Singapore; World Scientific, 1988, pp 292–298.
40. YY Wang, H Nakanishi, YC Jean, TC Sandreczki. Positron annihilation in amine-cured epoxy polymer-pressure dependence. J Polym Sci Polym Phys Ed 28:1431–1441, 1990.
41. H Higuchi, Z Yu, AM Jamieson, R Simha, JD McGervey. Thermal history and temperature dependence of viscoelastic properties of polymer glasses: relation to free volume quantities. J Polym Sci Polym Phys Ed 33:2295–2306, 1995.
42. Y Kobayashi, W Zheng, EF Meyer, JD McGervey, AM Jamieson, R Simha. Free volume and aging of poly(vinyl acetate) studied by positron annihilation. Macromolecules 22:2302–2306, 1989.
43. Z Yu, U Kobayashi, JD McGervey, AM Jamieson, R Simha. Can positron annihilation lifetime spectroscopy measure the free-volume hole size distribution in amorphous polymers. Macromolecules 28:6268–6272, 1994.
44. JD Ferry. Viscoelastic Properties of Polymers. 3d ed. New York: John Wiley, 1989, pp 366–369.
45. BY Shin, IJ Chung, BS Kim. Rheological and mechanical properties of thermotropic polyesters with long flexible spacers in the main chain. Polym Eng Sci 34:949–957, 1994.
46. DJ Alt, SD Hudson, RO Garay, K Fujishiro. Oscillatory shear alignment of a liquid crystalline polymer. Macromolecules 28:1575–1579, 1995.
47. S Hanna, AH Windle. Geometrical limits to order in liquid crystalline random copolyesters. Polymer 29:207–223, 1986.
48. DJ Blundell, KA Buckingham. The β-loss process in liquid crystal polyesters containing 2,6-naphthyl groups. Polymer 26:1623–1630, 1985.
49. MH Alhaj-Mohammed, GR Davies, S Abdul Jawad, IM Ward. Dielectric properties of liquid-crystal random copolyesters of 4-hydroxybenzoic acid and 2-hydroxy-6-naphthoic acid. J Polym Sci Polym Phys Ed 26:1751–1760, 1988.
50. GR Davies, IM Ward. Structure and properties of oriented thermotropic liquid crystalline polymers in the solid state. In: AE Zachariades, RS Porter, eds. High Modulus Polymers—Approaches to Design and Development. New York: Marcel Dekker, 1988, pp 37–70.
51. DI Green, GR Davies, IM Ward, MH Alhaj-Mohammed, S Abdul Jawad. Mechanical and dielectric relaxations in liquid crystalline copolyesters. Polym Advanced Technol 1:41–47, 1990.
52. GP Simon. Dielectric properties of polymeric liquid crystals. In: JP Runt and JJ Fitzgerald, eds. Dielectric Spectroscopy of Polymeric Materials. Washington D.C.: ACS Publications, 1997.
53. NG McCrum, BE Read, G Williams. Anelastic and Dielectric Effect in Polymeric Solids. London and New York: John Wiley, 1967 (rereleased by Dover Publications, New York, 1991).
54. H Vogel. The law of the relation between the viscosity of liquids and temperature. Phys Z 22:645–646, 1921.

55. GS Fulcher. Analysis of recent measurements of the viscosity of glasses. J Amer Ceram Soc 8:339–355, 1925.
56. HW Starkweather. Frequency temperature relationships for the glass transition. Macromolecules 26:4805–4808, 1993.
57. GP Simon. Dielectric relaxation spectroscopy of thermoplastic polymer and blends. Materials Forum 18:235–264, 1994.
58. GP Simon. The use of positron annihilation lifetime spectroscopy in probing free volume of multicomponent polymeric systems. TRIP 5:394–400, 1997.
59. AF Yee, MA Maxwell. Mechanical properties of polymer mixtures effect of compatibility. J Macromol Sci Phys B17:543–564, 1979.
60. LA Utracki. Polymer Alloys and Blends. Munich Hanser, 1990.

13

Polymer Liquid Crystals in High-Performance Blends

Witold Brostow

University of North Texas, Denton, Texas

I. INTRODUCTION

As argued by Travinska and her coworkers (1), there exist two main directions of modern polymer technology. One is the synthesis of new compounds. The other—the topic of this book—is the creation of new compositions by blending.

The blending route is particularly important for polymer liquid crystals (PLCs). Mechanical and thermophysical properties of PLCs are much better than those of engineering polymers (EPs) (2). EPs are typically flexible, with relatively low mechanical strength, low upper service temperature, and large thermal expansivity. A well-established procedure for improving the properties of EPs consists of inserting rigid reinforcement (carbon fibers, boron fibers, glass spheres, etc.); the resulting materials have been called heterogeneous composites (HCs) (3). While HCs necessarily have better mechanical properties than EPs, the improvement of thermophysical properties along the EP → HC route is not significant. The presence of the reinforcement does not prevent the softening of the EP matrix. Even in the mechanical properties, that improvement manifests itself only in certain directions. When an HC is loaded in an unfavorable direction, we observe fiber pullout. Similarly, delamination takes place in layer HCs (4–6).

Every PLC molecule contains also rigid reinforcement—but at the molecular rather than macroscopic level. Thus the problems associated with HCs are eliminated from the start. The thermophysical properties are also better than those of the HCs: the service temperature range is large and the isobaric expansivity α is small. Sometimes even $\alpha < 0$ is observed. We conclude that mechanical and thermophysical properties of PLCs are better than those of HCs and much better than those of EPs.

This is not yet a complete list of the advantages of PLCs. Their processing is much easier than that of HCs since ordinary equipment for EP processing can be used. Moreover, PLCs have typically lower viscosities than EPs, a consequence of the presence of rigid LC sequences in the chains. The viscosities of EP + PLC blends can also be lower than those of pure EPs, sometimes by two orders of magnitude.

With all these advantages, PLCs are more expensive than the massively produced EPs. This situation is expected to change when the demand for PLCs will increase. The PLC + EP blending is a way of achieving good properties at reasonable prices. In the remainder of this chapter we shall review the mechanical, thermophysical, and rheological properties of such blends. The present volume contains also a chapter on PLC + PLC blends by Hsieh et al. (7).

PLCs have to be distinguished from monomer liquid crystals (MLCs), the latter typically used in display devices. It cannot be stressed enough that the mere formation of an EP + PLC blend by no means assures the properties desired. PLCs are multiphase materials and they have hierarchical structures (8,9). A blend containing 50% of each of the components can have properties close to the pure EP, or close to the pure PLC, or in between (nearly additive), or far outside of the additivity curve. We shall discuss factors determining which of these various types of behavior prevails.

II. MECHANICAL PROPERTIES OF EP + PLC BLENDS

It will be convenient in discussing properties of such blends to consider consecutively two distinct factors. The first factor is related to the inherent properties that the PLC component contributes to the blend. The second factor is related to the interactions between the components.

To analyze the inherent nature of PLCs, consider the process of drawing a specimen in the flow direction. LC sequences (or MLC molecules) are oriented approximately parallel to a preferred axis in space called the director. This is related to packing, and the driving force is the usual thermodynamic one: the tendency to minimize the Gibbs function G. Melt processing operations such as extrusion or injection molding enhance the orientation further. However, we can also improve the orientation by the solid-state operation of cold working. Given the proclivity for orientation prevailing in PLCs, the degrees of alignment achieved here are higher than can be produced by the cold working of metals.

To demonstrate the effects of cold drawing, we shall use some of the results reported in (10). We have investigated PET/0.3PHB, in which PET is poly(ethylene terephthalate) and PHB is the *p*-hydroxybezoic acid; the latter is LC, and 0.3 is the mole fraction of PHB in the copolymers, a relatively low value. The specimens were drawn at 25°C up to the elongation of 300% and then the tensile properties determined. The elastic modulus E as a function of elongation ε is shown in Fig. 1. We see that drawing has resulted in a fourfold increase of the modulus. The tensile strength σ_{max} changes to a comparable extent, as shown in Fig. 2. For definitions of basic mechanical properties see a textbook of materials science and engineering such as (11).

Before going to blending, or other forms of solid or melt processing, we can vary the concentration of LC sequences during synthesis. In Fig. 3 we show the modulus E as a function of x in PET/xPHB copolymers (10) in the direction parallel to flow; x is the mole fraction of the liquid crystalline PHB component. We observe that at first an increase in the concentration of the LC sequences x enhances E. However, after reaching a maximum, E begins to decrease. This fact can be explained by another inherent feature of semirigid PLCs: their brittleness. The tensile strength σ_{max} behaves in a similar way; see Fig. 4. However, the results shown in Figs. 3 and 4 pertain to along-the-flow processing direction. In the directions perpendicular (transverse) to the flow the behavior is quite different. Here increasing x leads to decreases in both E and σ_{max}, as shown in Figs. 5 and 6. Thus the natural tendency for PLCs to become oriented is a two-edged sword.

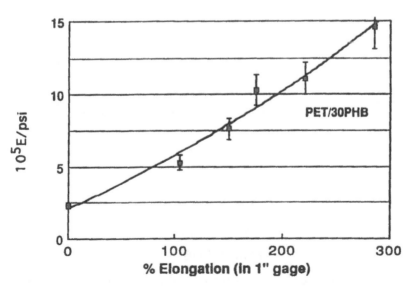

Figure 1   Elastic modulus vs. percentage elongation for PET/0.3PHB. (After Ref. 10.)

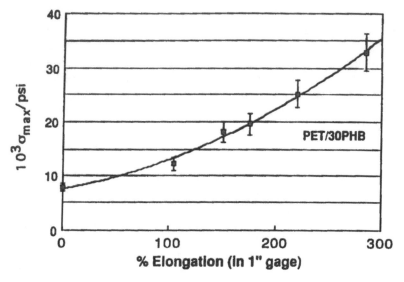

Figure 2   Tensile strength vs. percentage elongation for PET/0.3PHB. (After Ref. 10.)

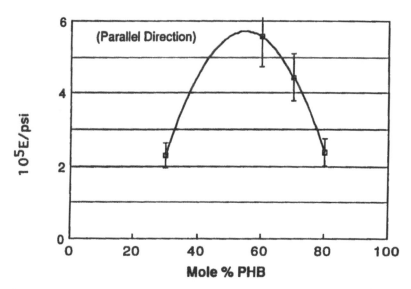

Figure 3   Elastic modulus in the direction parallel to flow vs. PHB concentration. (After Ref. 10.)

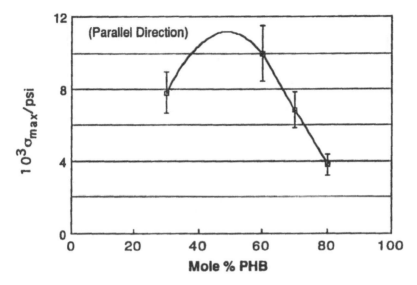

Figure 4   Tensile strength in the direction parallel to flow vs. PHB concentration. (After Ref. 10.)

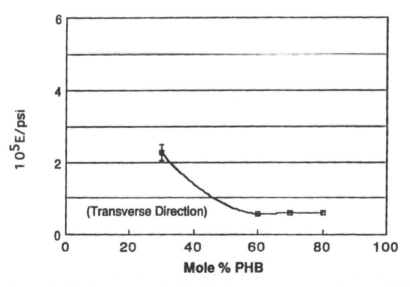

Figure 5   Elastic modulus in the direction transverse to flow vs. PHB concentration. (After Ref. 10.)

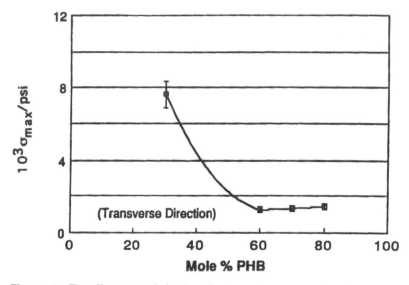

Figure 6 Tensile strength in the direction transverse to flow vs. PHB concentration. (After Ref. 10.)

We have not mentioned before that there exists a variety of molecular structures of PLC dependent on their shapes (approximately a rectangle, a disc, a cone) of the LC units, their orientation (such as perpendicular with respect to the chain backbone), and their locations. A classification of PLCs on this basis has been developed in Ref. 4 and amplified in Ref. 5. It will not be repeated here, since various aspects of it have been discussed earlier (4–6). However, we need to note that each class has different properties. The behavior shown in Figs. 1–6 pertains to longitudinal polymers in which the LC sequences are present in the main chain and are oriented approximately parallel to the backbone. To acquire a wider picture, let us now turn briefly to PLC networks. Finkelmann and coworkers (12,13) have developed a two-step procedure for creating such networks. First they mechanically align a PLC which is cross-linked only weakly. In the second step there is a cross-linking reaction of the remaining free reactive groups that freezes the alignment achieved. This step can be performed at different temperatures, which produce various extents of the alignment. Necessarily, the alignment is higher if the second stage is performed in a LC state and lower if it is performed above the isotropization (clearing) temperature.

We shall now consider the second factor that affects the properties of the blends, namely the EP + PLC interactions. We have argued before how important these interactions are (14), as did Singer et al. (15), Sęk and Kaczmarczyk (16), and Chiou et al. (17).

One option here is to use blends in which there is some compatibility between the constituents. One should realize that a typical EP + PLC system has three constituents: EP, the flexible sequences in the PLC, and the LC sequences in the PLC. In a study of blends of polycarbonate (PCarb) with PET/0.6PHB we have found that there is some miscibility between PCarb and PHB, but not between the other pairs. This is sufficient for improvement of the mechanical performance of PCarb by adding only some 15–20 wt% of PET/0.6PHB (14). We have argued before the importance of the phase diagrams for mechanical and other properties (10,14). We have also noted that in constructing phase diagrams of PLC-containing systems one needs to include the long-lived nonequilibrium phases that are characteristic for such systems (10). To give a well-known example, glassy phases tend towards thermodynamic equilibrium by slow crystallization, but this fact does not prevent us from using glassy materials. Since in PLCs we have also nonequilibrium but long-lived phases, their inclusion in the phase diagrams provides additional and useful information.

However, there is also another factor involved; Fig. 7 shows the effect of annealing time on the bending modulus in the same PCarb + PET/0.6PHB system. We show two modulus vs. annealing temperature curves for two different annealing times at each T. Clearly longer annealing time enhances the modulus, particularly so at lower temperatures. This can be explained by the progressive crystal-

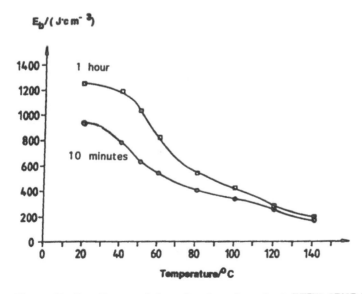

Figure 7 Bending modulus of polycarbonate + PET/0.6PHB blends (50 wt% each) as a function of temperature and annealing time. (After Ref. 14.)

lization of PCarb. For higher annealing temperatures, just bringing the sample to the desired temperature is by itself tantamount to annealing.

Since miscibility promotes better mechanical properties, compatibility helps in this regard as well. The miscibility is defined in thermodynamic terms of lowering the Gibbs function. By contrast, there is no generally accepted definition of compatibility. Practical definitions that we find useful are that miscible components "like" one another while compatible ones "tolerate" one another. In the paper by Singer and coworkers already mentioned (15) there is a discussion of the effect of compatibilization on properties of a specific EP + PLC blend. There is no question that the use of the compatibilizers to achieve improved properties of such blends is going to increase. Another option involves taking advantage of reactions taking place in the molten state such as esterification. This problem has been reviewed by Porter and Wang (18) and applied by Amendola and coworkers (19). The latter have achieved better fiber formation and also improved adhesion. Somewhat similarly, Chiou et al. (17) took advantage of a chemical reaction to achieve better compatibility. This has lowered the tensile modulus but increased the elongation and also the impact strength.

III. MECHANICAL PROPERTIES OF HP + PLC BLENDS

An HC consists of an EP and a rigid reinforcement (RR). This opens another option: systems of the kind EP + RR + PLC, in other words HC + PLC. This results in mixed composites or "supercomposites" in which there are reinforcements of the EP at both macroscopic and molecular levels. We are currently working on such systems (20,21). Our HC involves three improvements over typical such composites. First, the RR consists of thin and therefore flexible glass fibers (GFs). Second, the GFs are comingled with an EP using a coupling agent. Third, the GFs we use are not only unidirectional but also bidirectional, in the form of mats. These are the advantages we have even before adding a PLC.

We create laminate type structures from two kinds of layers. The first kind is the glass mat, which already contains the comingled EP. Ordinarily, one makes HC laminate structures using such layers only. We, however, intercalate such layers with a second kind: EP + PLC blend layers at various locations (in the middle only, close to the surfaces, every second layer, etc.). Our previously accumulated experience on EP + PLC blends "alone" is used to advantage. Together with quasi-static and dynamic mechanical testing, we investigate also thermophysical properties. We find that a relatively low fraction ($\geq$15 wt%) of the PLC in EP + PLC layers increases the damping energy of the composite and also increases the glassy plateau modulus. A further increase in the PLC concentration actually worsens the performance (as characterized for instance in tensile tests), since interplay slip becomes more probable at low temperatures. We explain these effects by the formation of the LC-rich islands above the PLC concentration of 15 wt%. Apparently the PLC in the EP + PLC layer reinforces the former more as long as it does not form a separate phase. At 15% PLC, however, by substituting partly the glass fiber mat by the EP + PLC blend, we achieve comparable mechanical properties combined with significant weight savings. In other words, replacing in the laminate several glass mat + PP layers by EP + PLC layers does not change the performance significantly, while the total weight of the composite layers is distinctly lower.

IV.   THERMOPHYSICAL PROPERTIES OF EP + PLC BLENDS

At the end of Sect. II we have noted the importance of phase diagrams, miscibility, and compatibility for the service performance of the blends.

Before discussing the blends, we need to consider pure LC materials. Even here multiphasicity appears a rule rather than an exception. Thus the phase diagram of copolymers with the formula PET/xPHB was determined (10). The diagram contains 12 phase regions. Long-lived nonequilibrium phases—typical in LC systems—are included, and thus each region contains up to four phases, such as PET crystals, PHB-rich islands, isotropic PET-rich glass, and PHB-rich glass.

A new phase called *quasi-liquid* (q-l) was defined in the course of our work on phase diagrams (10). It originates from the amorphous state that existed below the glass transition temperature T_g, but q-l appears between T_g and the melting transition T_m. The q-l phase has the following characteristic features:

1.   A q-l is not cross-linked, but it does *not* exhibit the ordinary liquid mobility. The presence of another (LC) component below its glass transition and/or of crystallites prevents this phase from flowing as a liquid does. The hindrance to the flow is caused by the LC sequences, which in this respect act somewhat similarly to the junctions in polymer networks. This situation is in contrast to non-LC polymers between T_g and T_m, in which the formerly amorphous phase flows around the crystalline regions, and the liquid viscosity depends on the temperature only. The viscosity of a q-l depends also on the *concentration* of LC sequences.

2.   An ordinary liquid upon heating can only undergo vaporization or, if it is an isotropic polymer melt, no further phase transition at all. By contrast, a q-l has to undergo at least two more transitions: melting and isotropization at the *clearing point*. If more than one LC phase is formed, then even more transitions will be observed; for instance, PET/xPHB forms a smectic E and a smectic B phase (10).

3.   The cold crystallization occurs in the q-l phase.

4.   The q-l phase shows an analogy with the leathery state in the elastomers. Both types of systems are immediately above their glass transition regions, and both exhibit *retarded* responses to application of external forces.

In Sect. I we have noted that PLCs exhibit lower thermal expansivity than EPs. Two quantities are used to characterize that behavior. One is the isobaric expansivity, or volumetric isobaric expansivity

$$\alpha = V^{-1}\left(\frac{\partial V}{\partial T}\right)_{P} \tag{1}$$

where V is the volume, T the temperature, and P the pressure. The term coefficient of thermal expansion (abbreviated as CTE) is also in use. The other is the linear isobaric expansivity

$$\alpha_{L} = L^{-1}\left(\frac{\partial L}{\partial T}\right)_{P} \tag{2}$$

where L is the specimen length. If a material is isotropic, then the two quantities in question can be related by simple algebra. If a material is anisotropic, that is, oriented during melt flow, as PLCs are, we have the linear expansivity parallel to the flow $\alpha_{\parallel}$ and linear expansivity $\alpha_{\perp}$ perpendicular to the flow.

One can measure α_{L} in a thermomechanical analysis (TMA) apparatus and then calculate α from it. Or one can use an apparatus that produces full P-V-T data, that is, specific volume V as a function of temperature T, $V(T)$, plus $\alpha(T)$ plus also isothermal compressibility $\kappa_{T}(T)$, where

$$\kappa_{T} = V^{-1}\left(\frac{\partial V}{\partial P}\right)_{T} \tag{3}$$

one such apparatus for polymer solids and melts is called the Gnomix (22). Our own results obtained with this apparatus (23) provide a comparison between EPs and PLCs. The expansivity of polyethylene α(PE, 140°C) is $7.20 \cdot 10^{-4}$ K^{-1}. At the same temperature we have α(PET/0.6PHB, 140°C) = $1.97 \cdot 10^{-4}$K^{-1}.

This chapter is concerned with EP + PLC blends, but in the present section we have so far discussed pure PLCs only—where the only variable was the concentration of the LC sequences in the copolymers. One can easily imagine that blends have even more complicated phase diagrams, as found for blends of PET/0.6PHB with EPs, for instance with polycarbonate (14). The diagram has 18 regions, and in each region we have 2–4 phases, including the quasi-liquid discussed above, PET glass, PCarb glass, PCarb crystals, PHB-rich islands in the solid state, smectic E, smectic B, PHB-rich isotropic liquid, etc. The diagram shows partial miscibility of PCarb with PHB but not with PET—which has consequences for the mechanical properties of the blends. We have already discussed in the preceding section the connection between phase diagrams and mechanical performance. Another important reason for the study of these diagrams is using them as a basis of *processing optimization*. In other words, in contrast to the usual establishment of processing parameters by trial and error, the knowledge of a phase diagram makes possible *intelligent processing*.

V. RHEOLOGICAL PROPERTIES OF EP + PLC BLENDS

We have mentioned already in Sect. 1 that addition of a PLC to an EP can lower the viscosity of the latter. Let us look at an example. In Fig. 8 we show logarithmic viscosity, log η, of polypropylene (PP), to which up to 20 wt% of PET/0.6PHB is added (24). The addition of only 5% of PET/0.6PHB to PP causes a pronounced log η lowering, while further addition of the PLC has little effect.

The conclusion from Fig. 8 that the presence of LC sequences in a flexible matrix lowers viscosity to a certain degree is *not* universally valid. Jackson and Kuhfuss (25) determined viscosity of PET/xPHB copolymers as a function of x at 275°C. They found that, starting from pure PET ($x = 0$), log η first increases, then goes through a maximum, then falls, then goes through a minimum, and then for $x > 0.7$ increases rapidly. In contrast to PET/0.6PHB + PP blends discussed above, Jackson and Kuhfuss investigated pure copolymers. Their results can be explained in terms of the orientation and LC-rich island formation. One has to agree with Roetting and Hinrichsen (26) that PLC rheology "is a rather complex subject." The findings of Jackson and Kuhfuss reflect a combination of effects of molecular structure, hierarchical structures, and results of varying composition.

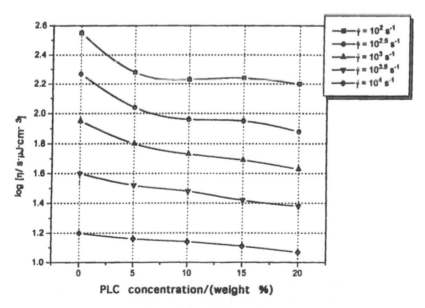

Figure 8 Logarithmic viscosity as a function of shearing rate and PLC concentration in polypropylene + PET/0.6PHB blends. (After Ref. 24.)

Certain generalizations are possible, however. In Fig. 9 we show a representation of normalized viscosities for a number of EP + PLC systems (24) such that the PLC is always the same, namely the PET/0.6PHB already well-known to us. It is clear that, the usual limited accuracy of the experiments notwithstanding, all the results can be represented by the following equation (24):

$$\ln\left(\frac{\eta_{blend}}{\eta_{matrix}}\right) = a_0 + a_1 \ln\left(\frac{\eta_{matrix}}{\eta_{PLC}}\right) + a_2\ln^2\left(\frac{\eta_{matrix}}{\eta_{PLC}}\right) \tag{4}$$

where a_0, a_1, and a_2 are parameters for a given class of blends with a common PLC.

The viscosity lowering such as is shown in Fig. 8 is a consequence of the ease of orientation of the rigid sequences in the PLC. This is why thermotropic PLCs (those in which the LC properties show in a certain temperature range) are often processible with conventional processing equipment for thermoplastics—in fact more easily than thermoplastics. This is the reason that PLCs are also known under the name of self-reinforcing plastics.

PLC rheology is by itself a fairly broad subject, and there exists an entire book on it (27).

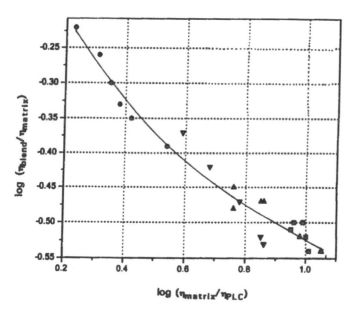

Figure 9 Normalized viscosities of blends containing 20 wt% PET/0.6PHB; experimental (points) and calculated from Eq. (4) (continuous line). The major 80% weight components are ■ polycarbonate, ● polypropylene, ▲ poly(vinylidene fluoride), ▼ poly(butylene terephthalate). (After Ref. 24.)

VI. THEORY OF PLCS

There are several theories of PLCs that help us to connect the macroscopic behavior to molecular structures and interactions. The most complete is the statistical mechanical theory originally formulated by Flory in 1956 (28) and then developed further by him, his students, and his collaborators (29–33). The Flory theory involves using a lattice. A chain segment can be placed on a lattice site; the solvent molecule has the same dimensions. One assumes that the combinatorial contribution to the partition function is independent of the orientational contribution. The latter arises from the possible orientations of rigid LC sequences and from the anisotropic interactions between the segments.

The Flory theory, together with its extentions, makes possible a number of predictions. We know that there are several types of LC phases, classified as nematic, cholesteric, and smectic (11). The theory predicts conditions necessary for the formation of each of these types of phases in PLCs (32,33). Moreover, for a ternary system of the kind flexible EP + PLC + solvent one obtains predictions of miscibility gaps in good agreement with experiment (34). That diagram can be viewed on the Gibbs triangle as an isothermal cross section. We have homogenous regions close to the triangle tops where one of the components dominates. In a large part of the diagram we have two phases, one PLC-rich and one EP-rich. The theory predicts *channeling*, namely that the LC sequences in PLC chains cause orientation of flexible EP chains between them. This prediction is confirmed by subsequent experimental data, including the P-V-T results for a series of PET/xPHB copolymers with varying x, these in the solid as well as in molten states (23).

VII. PERSPECTIVES FOR THE FUTURE

We are far from having fully explored the advantages of PLCs. Some of these advantages have been pointed out, but there are more. For instance, Kawagoe and coworkers (35) have shown how the addition of a PLC to an EP (in their case polypropylene) has extended considerably the fatigue lifetime. There is no doubt that the advantages that PLCs have, those already known and probably also still new ones, will lead to much wider applications of these materials in the future.

ACKNOWLEDGMENTS

This work has benefitted from discussions with Cosimo Carfagna, Hans-Eckart Carius, Nandika Anne D'Souza, Lew Faitelson, Michael Hess, Georg Hinrichsen, Robert Maksimov, Monika Plass, Andreas Schönhals, George P. Simon, Jürgen

Springer, Tomasz Sterzynski and Janusz Walasek. Partial financial support was provided by the Robert A. Welch Foundation, Houston (Grant # B-1203). Constructive input from two referees is also appreciated. Dr. Brostow can be contacted at The University of North Texas, Denton, Texas 76203-5310, USA, brostow@ unt.edu.

REFERENCES

1. T. V. Travinska, Y. S. Lipatov, Y. V. Maslak, and V. F. Rosovitski. Polymer Eng. Sci. 38: to be published, 1998.
2. W. Brostow, ed. Polymer Liquid Crystals. 3. Mechanical and Thermophysical Properties. London: Chapman and Hall, 1998.
3. W. Brostow, T. S. Dziemianowicz, J. Romanski, and W. Werber. Polymer Eng. Sci. 28:785–795, 1988.
4. W. Brostow. Kunststoffe 78:411–419, 1988.
5. W. Brostow. Polymer 31:979–995, 1990.
6. W. Brostow. Chapter 33 in: J. E. Mark, ed. Physical Properties of Polymers Handbook. Woodbury, NY: American Institute of Physics Press, 1996.
7. T.-T. Hsieh, C. Tiu, G. P. Simon, S. R. Andrews, G. Williams, K.-H. Hsieh, and C.-H. Chen. Chapter 12 in this volume.
8. L. C. Sawyer and M. J. Jaffe. J. Mater. Sci. 21:1897, 1986.
9. W. Brostow and M. Hess. Mater. Res. Soc. Symp. 255:57–73, 1992.
10. W. Brostow, M. Hess, and B. L. López. Macromolecules 27:2262–2269, 1994.
11. W. Brostow. Science of Materials. New York: John Wiley, 1979; Malabar, FL: Krieger, 1985. W. Brostow. Introducción a la ciencia de los materiales. México, D.F.: Editorial Limusa, 1981. W. Brostow. Einstieg in die moderne Werkstoffwissenschaft. Leipzig: VEB Deutscher Verlag für Grundstoffindustrie, 1985; München–Wien: Hanser, 1985.
12. J. Küpfer and H. Finkelmann. Makromol. Chem. Rapid Commun. 12:717, 1991.
13. J. Küpfer, E. Nishikawa, and H. Finkelmann. Polymers Adv. Tech. 5:110, 1994.
14. W. Brostow, M. Hess, B. L. López, and T. Sterzynski. Polymer 37:1551–1560, 1996.
15. M. Singer, G. P. Simon, R. Varley, and M. R. Nobile. Polymer Eng. Sci. 36:1038–1046, 1996.
16. D. Sęk and B. Kaczmarczyk. Polymer 37:2925–2931, 1997.
17. Y.-P. Chiou, K.-C. Chiou, and F.-C. Chang. Polymer 37:4099–4106, 1997.
18. R. S. Porter and L.-H. Wang. Polymer 33:2019, 1992.
19. E. Amendola, C. Carfagna, P. Netti, L. Nicolais, and S. Saiello. J. Appl. Polymer Sci. 50:83, 1993.
20. C. S. Own, D. Seader, N. A. D'Souza, and W. Brostow. Polymer Comp. 19:107, 1998.
21. W. Brostow, N. A. D'Souza, S. Maswood, and P. Punchaipetch. Work in progress, 1998.
22. P. Zoller, P. Bolli, V. Pahud, and H. Ackermann. Rev. Sci. Instrum. 47:948, 1976.

23. J. M. Berry, W. Brostow, M. Hess, and E. G. Jacobs. Polymer 39:4081, 1998.
24. W. Brostow, T. Sterzynski, and S. Triouleyre. Polymer 37:1561–1574, 1996.
25. W. J. Jackson, Jr., and H. F. Kuhfuss. J. Polymer Sci. Phys. 14:2043, 1976.
26. O. Roetting and G. Hinrichsen. Adv. Polym. Tech. 13:57, 1994.
27. D. Acierno and A. A. Collyer, eds. Polymer Liquid Crystals. 2. Rheology and Processing. London: Chapman and Hall, 1996.
28. P. J. Flory. Proc. Royal Soc. A 234:60–72, 73–85, 1956.
29. P. J. Flory and A. Abe. Macromolecules 11:1119–1122, 1978.
30. R. R. Matheson, Jr., and P. J. Flory. Macromolecules 14:954, 1981.
31. R. R. Matheson, Jr. Macromolecules 19:1286, 1986.
32. W. Brostow and J. Walasek. J. Chem. Phys. 105:4367–4376, 1996.
33. W. Brostow and J. Walasek. J. Chem. Phys. 108:6484, 1998.
34. S. Blonski, W. Brostow, D. A. Jonah, and M. Hess. Macromolecules 26:84–88, 1993.
35. M. Kawagoe, M. Nomiya, J. Qiu, M. Morita, and W. Mizuno. Polymer 38:113–118, 1997.

14

Structure–Property Relationships in Poly(aryl Ether Ketone) Blends

Andy A. Goodwin,* George P. Simon, and Marcus D. Zipper†
Monash University, Clayton, Victoria, Australia

I. INTRODUCTION

Poly(ether ether ketone) (PEEK) is a commercial semicrystalline thermoplastic with a glass transition temperature of 143°C, a melting temperature of 340°C, and outstanding mechanical properties that are utilized in a wide range of engineering applications in both unfilled and reinforced form (1).

PEEK

The polymerization chemistry, history, and development of PEEK and related polyaryls have been extensively reviewed (2,3), and since 1993 PEEK has been produced by Victrex PLC. The impressive properties of PEEK have made it a candidate for many blending studies, in particular in blends with other high-temperature engineering polymers. The aim of this chapter is to review the structure–property relationships of PEEK blends, with the greater emphasis being placed on the more commonly studied PEEK blends.

Current affiliations:

* Boral Plasterboard, Port Melbourne, Victoria, Australia.

† Huntsman Chemical Company Australia Pty Ltd., West Footscray, Victoria, Australia.

II. BLENDS WITH POLYIMIDES

One of the most commonly studied blends containing PEEK has been that with
poly(ether imide) (PEI), probably because these polymers form a miscible blend
across the whole composition range (4,5) in which the morphology can be varied
according to thermal history. Harris and Robeson (5) reported on the blending of the
pure polymers in an extruder and processing the blend by injection and compression
molding; miscibility was confirmed by differential scanning calorimetry (DSC)
scans on amorphous samples. A maximum in toughness was found in both as-
molded and annealed specimens at intermediate compositions that, for the semicrystalline samples, corresponded to a maximum in PEEK crystallinity. This may be due
to the greater extent of interlamellar slip occurring as a result of the presence of PEI
(6). The crystalline nature of PEEK was reported to increase significantly the chemical resistance of the amorphous PEI, particularly in the presence of aqueous bases.
It was also noted that the crystallization kinetics of PEEK were modified by the
presence of the noncrystallizing PEI. The yield stress and tensile modulus of amorphous PEEK/PEI were later reported to follow simple additive behavior (7).

The issue of the morphology of PEEK/PEI blends was later studied in more
detail in a number of articles. DSC, dynamic mechanical analysis (DMA), and
small-angle x-ray scattering (SAXS) have been used to examine the miscibility
and crystallization behavior (8). Upon cold crystallization at 250°C, blends rich in
PEI showed two peaks in the dynamic mechanical spectrum, one due to interlamellar amorphous PEEK and the other a higher temperature transition corresponding to phase-separated PEI, which appeared close to the T_g of pure PEI. As
the SAXS data indicated that the long period remained unaltered with blend composition, the authors concluded that significant interlamellar segregation did not
occur. The rate of PEEK crystallization in the blends was found to be substantially
lowered, compared with pure PEEK, but it was thought that PEEK crystallized in
much the same manner as in the pure state, with little change in the degree of crystallinity with blend composition. Optical and electron microscopy studies on melt
and cold crystallized samples added to these findings and confirmed that blend
morphology was primarily dependent upon the crystallization temperature (9). At
high crystallization temperatures (around 320°C), the PEI was mainly excluded to
the interspherulitic regions. At intermediate temperatures the PEI was segregated
between bundles of lamellae, while at low temperatures (35°C above T_g) a single-
crystal type dendritic morphology resulted. Figure 1 is a schematic diagram representing the possible size scales of segregation in a crystallized PEEK/PEI blend.
A recent report using SAXS (10) has noted that in high-PEI-content blends PEI is
excluded from the interlamellar regions while, in PEEK-rich blends, the opposite
was found. Additionally, it was found that the formation of secondary lamellae
was impeded by the PEI. An interesting method for investigating the phase morphology is the use of a solvent to extract the PEI in cystalline samples (11) with

(A)

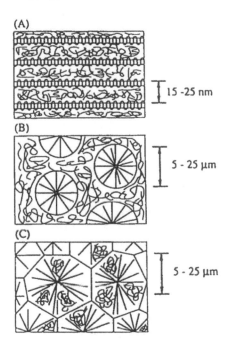

15 -25 nm

(B)

5 - 25 μm

(C)

5 - 25 μm

Figure 1   Three types of noncrystalline component segregation morphology in semicrystalline/amorphous blend: (A) interlamellar, (B) interspherulitic, and (C) interfibrillar [or inter (lamellar-bundle)]. (From Ref. 19, B. B. Sauer and B. S. Hsiao. Broadening of the Glass Transition in Blends of Poly(Aryl Ether Ketones) and a Poly(Ether Imide) as Studied by Thermally Stimulated Currents. J. Polym. Sci. B. 31:917, 1993. © 1993 John Wiley and Sons, Inc. Reprinted by permission of John Wiley and Sons, Inc.)

subsequent examination by DMA and electron microscopy. Dichloromethane was found to remove the PEI contained between fibrillar bundles and in interspherulitic regions, while SEM revealed a bimodal pore size scale. The presence of residual PEI in interlamellar regions was confirmed by DMA studies. The immersion of an amorphous 50/50 PEEK/PEI blend in acetone was found to induce crystallization of the PEEK, although the initial rate of uptake was slower than that of pure PEEK (12), most likely because interactions between the blend components increase the blend density (13). Another suggestion was that chemical exposure produces a thin crystalline surface film that would make subsequent ingress of solvent more difficult (6). Yield stress and tensile strength of a blend saturated with acetone were dominated by PEEK, while Young's modulus was dependent on the extent of plasticization by the solvent (14).

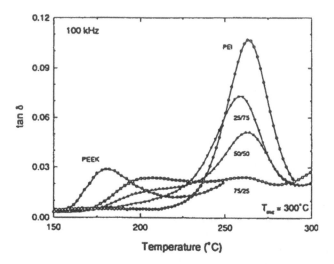

Figure 2   Dielectric tan δ (100 kHz) vs. temperature (°C) for isothermally melt crystallized PEEK/PEI blends of varying composition (T_{mc} = 300°C). (From Ref. 16, J. F. Bristow and D. S. Kalika. Investigation of semicrystalline morphology in poly(ether ether ketone)/poly(ether imide) blends by dielectric relaxation spectroscopy. Polymer 38:287, 1997, © 1996 Elsevier Science Ltd. Reprinted with permission from Elsevier Science Ltd, The Boulevard, Langford Lane, Kidlington OX5 1GB, U.K.)

Dielectric relaxation is well suited to probing phase composition and has been used to give further insights into the morphology of PEEK/PEI blends (15,16). Figure 2 shows the dielectric response of a series of crystalline PEEK/PEI blends, which is similar to the DMA data of earlier reports in that the observed loss processes appear to correspond to the PEEK and segregated PEI phases; the high-temperature relaxation is due to an almost pure PEI phase that is contained within interfibrillar regions, while the lower temperature relaxation arises from motions of a mixed interlamellar phase. The increase in temperature location of this process with increasing PEI is due to a relative increase in the concentration of PEI in the mixed phase. For the 50/50 blend the weight fraction of PEI in the mixed phase was estimated to be about 50%. In semicrystalline PEEK/PEI blends the constraints imposed by the PEEK crystal lamellae on the amorphous PEEK chains alter the nature of the interactions between the PEEK and the PEI so that a significant extent of PEI segregation usually occurs outside of the interlamellar regions. In both DMA and DRS studies no PEEK relaxation is observed in crystalline blends, which confirms the absence of a pure PEEK crystal–amorphous interface region. This is consistent with studies on miscible blends containing a flexible, crystallizing component where a crystal–amorphous interface region is observed at the lamellar surface (17), while for blends containing a more rigid crystallizing polymer such an interface is not detected, indicating the presence of the noncrystallizing component within this interface region (18). These dielectric studies also suggest that the nature of the phase separation is kinetically controlled by the crystallization temperature, while the segregation of PEI leads to a wider range of local relaxation environments and a broadening of the relaxation process.

Dielectric relaxation, dynamic mechanical analysis, and thermally stimulated current techniques have been used to investigate the molecular dynamics of PEEK/PEI blends in both the crystalline and the amorphous state (19–21). It is generally found that the glass transition is significantly broader in the blends and that the relative rate of relaxation is slower, both effects arising from the presence of concentration fluctuations and perhaps a change in the cooperative nature of segmental motions, as is commonly found in miscible blends. In crystalline samples the T_g was found to broaden by as much as 60°C, although no relaxation associated with the crystal–amorphous interface was detected (19).

The interactions between PEEK and PEI have been shown by equilibrium melting studies (22) and analysis of T_g data (20) to be small, which is consistent with the slight undershoot in T_g with composition commonly reported, and only a slight increase in density of the blends (22). A detailed investigation of the intermolecular interactions between PEEK and PEI in solution cast blends has been carried out using FT-IR (23). Based on an examination of the PEI carbonyl stretching modes, the authors concluded that for favorable interactions to occur between the oxygen lone-pair electrons of the PEEK ether group and the electron-deficient imide rings of PEI, the imide rings rotate about the phenylene axis to an angle of about 45°. This explanation was proposed to account for the observed minimum in the intensity ratio of the two carbonyl bands in the 70% blend.

Another study focused on the polyimide isopropylidene linkages to analyze the underlying reasons for miscibility between PEEK and polyimides (6). The heats of mixing between several model compounds were determined using adiabatic calorimetry and were used in a thermodynamic model to calculate interaction parameters for a range of mer units. The results suggested that polyimides containing isopropylidene linkages would be miscible with PEEK, as is often found in practice. The authors (6) attributed this to unfavorable interactions between the aliphatic phenyl isopropylidene and the polar phenyl imide mers in the polyimide. However, this finding is not universal, since a commercial polyimide (XU-218) that contains a significant aliphatic character is immiscible with PEEK.

A number of articles in the patent and journal literature have referred to blends of PEEK with polyimides of varying chemical structure, different from that of PEI (24–26). A semicrystalline thermoplastic polyimide (trade name Aurum) and two LARC polyimides have been melt blended with PEEK and investigated by DSC (27,28). In all cases these materials have been found to be completely immiscible with PEEK, which contrasts with the behavior of other poly(aryl ether ketone)s such as poly(ether ketone) (PEK), poly(ether ketone ether ketone ketone) (PEKEKK) and poly(ether ketone ketone) (PEKK). These higher ketone-containing polymers showed varying degrees of miscibility with the various polyimides, which was discussed in terms of chain flexibility and the nature of interactions between the blend components. A commercial PEEK has also been blended with poly(amide imide) and found to be totally immiscible, whereas blending with sulphonated PEEK produced a miscible system with strong positive T_g deviations. The blends were examined by UV-Vis and FT-IR techniques, and the change in phase behavior with sulphonation was explained by the formation of electron donor–acceptor complexes between the sulphonated PEEK phenylene rings and the phenylene units of the polyimides (29,30).

III. BLENDS WITH PEK-C AND PES-C

The polymers phenolphthalein poly(ether ketone) (PEK-C) and phenolphthalein poly(ether sulphone) (PES-C) are thermoplastic polyethers to which has been added a phenolphthalein pendant group in order to increase T_g beyond 200°C and eliminate crystallinity so that they may be considered for use as high-temperature matrices for polymer composites (31).

PEK-C

They have been blended with PEEK in an attempt to achieve materials with high operating temperatures and good mechanical properties (32–34). Both polyethers are partially miscible with PEEK as determined by DSC and DMA studies, as shown in Fig. 3. Although the extent of miscibility differs from that of PEEK/PEI blends, the effects of PEK-C and PES-C are essentially the same. Both polymers dilute the PEEK nuclei and cause a slowing of crystallization rate as well as a lowering of melting temperature and PEEK crystallinity. PES-C was also shown to be trapped in the interfibrillar and interlamellar regions of the crystalline PEEK when a relatively low crystallization temperature was used. The addition of PES-C to PEEK has been found to increase the ease with which PEEK crystals nucleate on the surface of untreated carbon fibers (35), and this was believed to be due to changes in the relative rates of bulk and fiber nucleation of PEEK induced by PES-C. This effect is likely to have important implications for the mechanical behavior of PEEK composites.

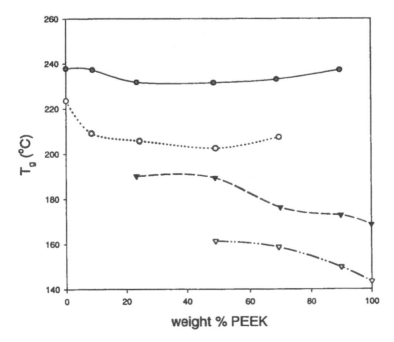

Figure 3 The glass transition temperatures of PEK–C/PEEK and PES-C/PEEK blends: circles are from dynamic mechanical; triangles are from DSC; filled symbols are PES-C/PEEK blends; and open symbols are PEK-C/PEEK blends. (From Ref. 33.)

IV. BLENDS WITH FLUORINATED POLYMERS

The primary motivation for the blending of PEEK with fluoropolymers (and most commonly poly(tetrafluoroethylene) (PTFE)) has been to produce a material with low friction and wear rate properties. Both materials have their own strengths in these areas, and their combination in a such a composite has proven advantageous. The nature of PTFE is such that it is impossible to melt process; most components are formed into their final shape by sintering at high temperature and pressure. Therefore in multicomponent blends it is most often used in the form of powder and combined with another thermoplastic, often with a further rigid, included-component to optimize mechanical properties such as modulus. In this sense it is more of a dispersion of PTFE in PEEK, rather than a traditional polymer melt blend.

PTFE is a semicrystalline polymer with a melting point of 327°C and a high density (2.1–2.3 g cm^{-3}), the low amorphous content meaning that its glass transition is difficult to measure. Due to its smooth molecular profile, it has a very low coefficient of friction, especially at low velocities of sliding, due to its low surface tension and ready ability smoothly to lay down an anisotropic transfer film onto many surfaces. As such it is widely used as the material of choice in moving parts such as bearings. It suffers from poor wear properties, particularly at higher velocities, precisely because of its ability to transfer polymer layers. By contrast, PEEK has good abrasion resistance, and in concert with its high fatigue, solvent, and creep resistance, this means that it has potential for use in moving parts in under-the-bonnet applications. However, it has a reasonably high coefficient of friction μ between 0.4 to 0.6, depending on conditions, compared with 0.1 to 0.2 for PTFE, and it leaves very patchy transfer films (36) and appears to wear better in continuous, rather than reciprocating, motion conditions (37). Whilst PEEK's high friction coefficient is undesirable at low velocities, it also shows a failure known as scuffing at high speeds and loads, limiting its use (38).

The first primary study on blends of PEEK and PTFE has been carried out by Briscoe et al. (38), who combined PEEK powder or pellets with 25 μm PTFE powder, forming it into the appropriate rods or cylinders for testing by compacting and using a heating cycle that finishes with a 380°C anneal for two hours. It was found that the coefficient of friction of the blend reduced quite rapidly from about 0.27 for PEEK to about half its value with the first 10 wt% addition of PTFE, thereafter remaining constant. The difference in the mechanism of friction with composition can be seen by the presence of transfer "flakes" (rather than transfer films) in low PTFE additions to PEEK, whereas local heating and melting is observed for PEEK alone. The favorable wear properties of the PEEK were not compromised significantly with increasing PTFE composition: at 10 wt% PTFE the wear rate had increased by less than half, far less than expected by a rule of mixtures, since the wear rate of PTFE under comparable conditions is some six orders of magnitude higher. The decrease in friction coefficient of PEEK means that it can be used at higher loads with little reduction in wear properties. Mechanical properties such as modulus and hardness decrease with the inclusion of PTFE, but not to any great extent, and 10 wt% PTFE seems to be an optimum composition. Also of interest is that the PTFE-rich end (with about 20 wt% PEEK) shows a rapid decrease in wear (about five orders of magnitude) with a low value of μ, although processing of such material makes fabrication difficult. Similar results have been seen in compression molded disks (39), where the behavior was much the same except that the 15 wt% PTFE wear rate was synergistically lower than that of PEEK and the frictional coefficient almost as low as that of PTFE homopolymer (Fig. 4).

The very slight differences between the studies are probably due to the complexity of the competing mechanisms, which are dependent on morphology and test conditions. Whilst the addition of PTFE reduces the frictional coefficient of PEEK (and thus reduces wear), the increased concentration of PTFE, which itself wears poorly, may serve to increase the wear rate, and it is the balance of these that lead to the final properties. Overlaid on this is that the morphology of the

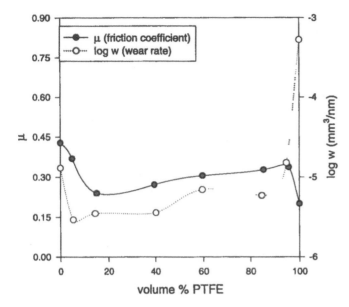

Figure 4 Friction coefficient μ and specific log wear rate w shown as functions of composition for a blend of PEEK and PTFE particles. (From Ref. 39.)

PEEK, such as the degree of crystallinity and the size and nature of the spherulitic structure, can play a role. The amorphous phase would be expected to wear to a greater extent than the crystalline fraction, and it has been found that lower molecular weight PEEK, with fewer entanglements in the amorphous region, has poorer wear properties (40). The rate of cooling and the presence of foreign particles such as PTFE can affect the degree of crystallinity of PEEK and be an additional reason for property modification. Stuart and Briscoe (41) used Raman spectroscopy to determine an increase in intensity of carbonyl stretches from PEEK's crystalline region, indicating an increase in crystallinity with the biggest change occurring at additions of about 10 wt% PTFE. This supports the conclusion that the decrease of friction (and only a modest increase in wear) in the 10–15 wt% range is due in part to the presence of the PTFE phase but also to the increased crystallinity induced in PEEK. Work by Damman et al. (42) on PEEK films laid on to aligned PTFE films showed that PEEK crystallization may be particularly encouraged in the presence of PTFE due to epitaxial growth of the chain axes of the PEEK along those of the PTFE, which is thought to be particularly important in this polymer pair because of a fortuitous close matching of chain axis periodicities and interchain distances. In relation to the blends discussed above, this may encourage crystallization of PEEK at the surfaces of internal particles but also on any aligned PTFE transfer layer produced.

In reality, commercial blends often involve other additives such as fibers and solid lubricants to achieve a range of properties, and these can often make up for the slight reduction in mechanical properties such as the blending of PTFE into PEEK may cause. Lu and Friedrich (39) found that the addition of graphite can also reduce wear rate and friction and proposed that an optimum combination for a low-friction wear-resistant material that performs well at a high pv (pressure–velocity) level (i.e., extreme conditions) may involve 10% PTFE, 10% graphite flakes, and 10% carbon fiber in a PEEK matrix. By comparison, CuS is one of a number of particulate fillers that can reduce wear rates but often increases friction. Inclusion of 20 wt% CuS in PEEK decreases wear rates by three times but increases friction by 25%, and this can be halved if only 10 wt% PTFE is added (43). In this instance it is the ease of formation of a smooth transfer film that reduces wear. Care must always be taken to test materials at the appropriate conditions, since it has been shown that PEEK with PTFE, graphite, and polybenzimidazole (PBI) additives appears good at moderate conditions but appears to wear rapidly at high pv. As in all tribological measurements, the countersurface used is crucial, as is shown in a reciprocating wear study by Braza and Furst (44) where the wear rate of a composite of PEEK and PTFE fibers was some 90% less against titanium nitride than against stainless steel, possibly owing to differences in the ease of transfer film formation.

Many of the mechanical properties of such blends are still to be measured, especially impact strength and fracture properties generally. This may be particularly important because there is unlikely to be much strong bonding between the nonpolar PTFE and polar PEEK. Some efforts have been made to address the quality of the interface. Briscoe et al. (38) used γ-irradiation to improve wear properties for small doses, although this got worse with greater dosage. It is the PTFE phase that shows significantly reduced wear (two orders of magnitude) due to irradiation in this way, and thus the effect is dependent on the concentration of PTFE being studied. Nishi et al. (45) used excimer-laser radiation to improve the PEEK/PTFE interface and showed almost an order of magnitude increase in peel strength due to the formation of functional and polar groups (including carbonyl) on the PTFE surface that were able to interact with PEEK. Interestingly, such groups only occurred when PTFE was in contact with PEEK and other polyesters, not on its own.

It should be noted that some fluorinated polymers can be melt processible, and fluorinated copolymers, usually elastomeric, may be used. These offer new possibilities to improve properties of PEEK, bringing with them additional properties of solvent and heat resistance and self-extinguishing properties, as well as modifying mechanical properties. Senior (46) gives an example of a PEEK blend with a flexible propylene-tetrafuoroethylene copolymer decreasing the PEEK modulus and allowing it to be used as a valve seat in a demanding underwater oil pipeline ball valve. The flexibility introduced allows a good sealing ability, along with the good heat and solvent resistance of the PEEK, neither of which is compromised much by the addition of the fluoropolymer phase. Such melt blends of PEEK with fluorocopolymers can also lead to materials with other improvements, such as dramatically improved dielectric breakdown, thereby making them suitable for wire coatings (47).

V. BLENDS WITH LIQUID CRYSTALLINE POLYMERS

Among the engineering thermoplastics that made their entrance into the commercial marketplace in the mid- to late 1980s were thermotropic liquid crystalline polymers (TLCPs). These were the melt processible versions of well-known lyotropic LCPs such as Kevlar (poly(p-phenylene terepthalamide). Lyotropic LCPs such as Kevlar have such great chain rigidity and intermolecular bonding that melt processing is not possible (they degrade prior to the crystals melting) and thus they must be spun from a solution such as of sulphuric acid. To allow melt processing of TLCPs, rigid molecules with units that disrupted chain linearity and packing slightly (and thus reduced melting points) were developed. These were often copolyester materials, and comonomers would use strategies such as increased chain flexibility, introduction of lateral substituents, and kinked monomers. The resultant material can be heated above its melting point into an isotropic liquid crystalline mesophase. The advantage of this is that the chain alignment and lack of entanglement lead to low melt viscosities whilst also yielding high moduli in the direction of flow. Combined with good properties in the solid state, such as low permeability and low thermal expansion, these materials showed much promise. This has been tempered over the years by issues such as poor weldline strength, anisotropy of molded parts, and a moderately high expense per kg. Nonetheless, they flourish in markets such as the electronics industry where they are prized for their ability to be used in the production of small, accurate parts and in other industries also as high-modulus melt spinnable fibers. A good summary about types of TLCPs and their use in processing and blends can be found in recent books (48,49). Much research has also gone into their use in thermoplastic blends, where they often tend to be immiscible but have a range of useful properties. The seminal paper in the area of TLCPs and thermoplastics (TPs) is the oft-referenced one of Kiss (50), who blended TLCPs with a range of engineering thermoplastics, including PEEK. As in other systems, it was found that the low viscosity of the TLCP improved the mechanical properties of the blend owing to the propensity for the TLCP to form fibrils (small, elongated included phases of approximate diameter 1 μm). Kiss (50) used one of the most widely used of TLCPs, the Vectra A950 material from Hoechst-Celanese, which is a copolymer of p-hydroxybenzoic acid (HBA), 6-hydroxynapthoic acid (HNA), which has the structure

$$\left[\begin{array}{c} \overset{O}{\underset{\parallel}{C}} - \bigcirc - O \end{array} \right]_{0.73} \quad \left[\begin{array}{c} \overset{O}{\underset{\parallel}{C}} - \bigcirc\bigcirc - O \end{array} \right]_{0.27}$$

HBA   HNA

Although the PEEK modulus and tensile strength increased by some 15–20% on addition of 30 wt% TLCP, the strain to failure decreased markedly. Mehta and Isayev (51) examined the miscibility properties of PEEK and Vectra A950 in some detail. Although both are polymers with a high degree of aromaticity, the two are immiscible, as evidenced by a glass transition of PEEK that does not change significantly, remaining at about 150°C. (Vectra A950 has a much weaker glass transition at about 100°C and cannot be easily observed in blends due to dilution). Similarly, the melting temperature of the PEEK crystals remains constant with blend composition and does not change significantly from about 338°C. As with its glass transition, the melting point T_m of Vectra A950 is weak. This is because the ordered nature of the Vectra A950 material means that the entropy change from crystal to mesophase is low, and since enthalpy $\Delta H_m = \Delta S_m \cdot T_m$, it is a weak transition (usually enthalpy of melting about 1 J/g).

Such immiscibility between TLCPs and TPs is relatively common, and it arises in part from the mixing of rigid rods and flexible coils, which is entropically unfavorable, quite apart from a general lack of strong intermolecular interactions required to drive miscibility. In terms of crystallinity of the PEEK phase, however, TLCPs can play a role, as has been shown in many TLCP + TP systems (52). Mehta and Isayev (51) showed that small amounts of TLCP increased the energy of crystallization with a maximum at 2.5 wt% Vectra A950 because of some heterogeneous nucleation at low values. At levels greater than this optimum, the heat of crystallization stayed constant at the same value as that of pure PEEK.

As mentioned above, one of the motivations in blending TLCPs is that their intrinsic anisotropy can be reduced. In the case of Vectra A950 in this work, it was approximately a factor of three in terms of storage modulus along and transverse to the flow path, but this reduces to about two at 50 wt% PEEK. The mechanical and rheological properties and the morphologies that result in this blend have been studied in some detail (53), since they represent an alternative to glass-fiber reinforced PEEK. Glass-filled materials have very good mechanical properties such as high modulus, but they suffer from difficulty in processing owing to their high viscosities, and this can lead to degradation of the matrix polymer and wear on processing machinery. Viscosity of the Vectra A950 is at least an order of magnitude lower than that of PEEK and shows a linear decrease with wt% Vectra A950

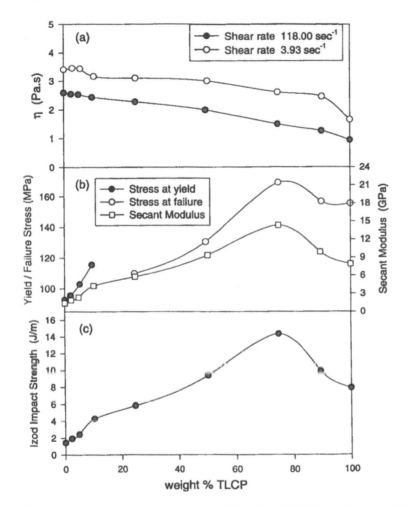

Figure 5 Mechanical properties of blends of PEEK and Vectra A950 TLCP (copolymer of HBA and HNA). (From Ref. 53.)

(Fig. 5a). The yield observed in PEEK did not occur above 10 wt%, although the resultant failure stress continues to increase to approximately double that of the PEEK yield value. Some synergy is observed at about 80 wt% PEEK with a slight maximum in failure stress (Fig. 5b). Surprisingly, this is also seen very strongly in the modulus of the blends, since modulus is a low-strain property. However, this may be partly because it is a secant modulus being reported, which includes a small amount of deformation.

In terms of impact strength, Vectra A950 is greater than PEEK owing to its fibrous nature and the dissipation of the crack (and increase in fracture surface) over a wider area. This favorable aspect is preserved at greater than 50 wt% Vectra A950, with a maximum near 90 wt% Vectra A950 (Fig. 5c). The blend morphology is the classic fibrillar morphology of TLCP + TP blends, the precise nature of which depends on concentration. Fibrils are observed for Vectra A950 concentrations above 10 wt%; at lower TLCP compositions they are predominantly ellipsoidal. There is a skin-core effect as well, with the stretching, fountain flows that occur at the skin causing greater alignment and a higher length-to-diameter ratio of the fibrils. It is perhaps not surprising that the failure properties are not good at medium Vectra A950 concentrations, since the adhesion between phases is moderately poor and shows little evidence of fiber adhesion.

Further improvement in fibril orientation and aspect ratio (and thus mechanical properties) can often be obtained by drawing the blends into fibers. However, in a PEEK blend with a TLCP known as Ekonol E6000 from Sumitomo Chemicals [HBA, terephthalic acid (TA), biphenol (BP)], an increased draw ratio did not correspondingly increase the modulus of the blend (54) and may be due to the poor mechanical interface between components.

Similar blend structure–property morphology results as described above were found in a PEEK blend with a different rigid copolyester TLCP, Xydar (from Amoco, originally Dartco), which is a copolymer of TA, BP with a majority of HBA units (55), where another problem has been reported in TLCP blends, that of degassing during processing. Degradation studies showed an increasingly lower level of degradation temperature with addition of Vectra A950, as well as a greater weight loss at a given temperature. This notwithstanding, there have been a number of patents in which PEEK was one of the semicrystalline matrices with which rigid chain TLCPs were to be blended to obtain improved properties such as easier processibility, improved mechanical properties, and (particularly relevant for PEEK) better solvent resistance (56–58).

Much more detailed studies of the effects of TLCP addition on crystallization kinetics have been carried out by De Carvalho and Bretas (159) and Zhong et al. (60,61). Blends of PEEK with a DuPont TLCP consisting of terephthalic acid (TA), phenylhydroquinone (PHQ), and hydroquinone (HQ) units and known as HX4000 have been studied by a number of workers (59,62). It was found that the TLCP reduces the crystallization process across the composition range. This seems to indicate that there must be some miscibility in the molten state for the HX4000 to have the broad effect that it does. Heterogeneous nucleation of the PEEK would be expected if they were simply immiscible, but rather a retardation occurs with less perfect crystals due to melt miscibility of the components. This is an interesting system to study because, unlike most TLCPs, the HX4000 also has a strong crystallization peak, and growth of this phase is found to be promoted by blending with PEEK, which gives an increase in HX4000 crystallization and melting temperatures. In the solid state, HX4000 and PEEK are miscible up to 50 wt% HX4000 but partially miscible afterwards (62).

A few TLCPs have been reported that are actually quite miscible with certain thermoplastics, particularly if cast from solution. One of these has been described by Zhong et al. (61), in which PEEK is solution blended with a TLCP consisting of HBA, TA, and resorcinol (RE) units in p-chlorophenol. A single T_g is observed on heating, with phase separation on heating, partly due to the release of kinetically entrapped miscibility and partly due to crystallization. The addition of the TLCP leads to a minimum in the crystallization half-life (inversely proportional to the rate of crystallization) at 20 wt% TLCP as seen in Fig. 6, and this is thought to be the result of a combination of dilution of PEEK by the TLCP phase (reducing crystallization rate due to hindrance of the crystallization process) and an increase in rate due to enhanced nucleation. Overall the total amount of crystallinity increases with the addition of TLCP.

Further work by the same group with another TLCP consisting of BP, HBA, TA, and m-phthalic acid (MPA) units and PEEK used optical microscopy and growth kinetics to determine whether it is an increase in spherulite growth rate (Fig. 7) or nuclei density (Table 1) that leads to such a higher crystallinity in

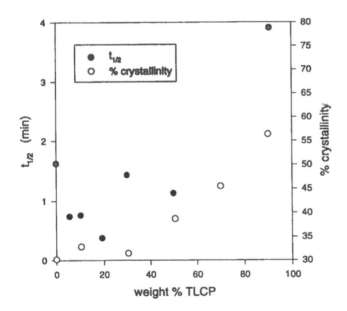

Figure 6 Crystallization properties of a blend of PEEK and TLCP (HBA, TA, and RE) indicative of rate of crystallization ($t_{1/2}$, half-time of crystallization) and % crystallinity from cooling runs of the DSC (scan rate = 20°C/minute). (From Ref. 60.)

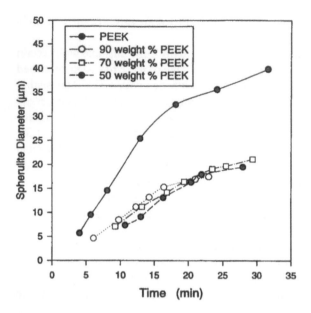

Figure 7 Increase in spherulitic diameter with time at 300°C for a blend of PEEK and TLCP (HBA, TA, MPA, BP) for different compositions. (From Ref. 61.)

PEEK. It can be seen that the linear growth rate was slower in the blends than PEEK homopolymer but that the nucleation density in the blends is some two orders of magnitude higher in PEEK homopolymer. Interestingly, the nucleation density is not related simply to the amount of foreign TLCP phase and thus does not rise monotonically with TLCP content. Rather, the *size* of the inclusions affects crystallization efficacy, and above 10 wt% TLCP the number of nuclei de-

Table 1 Nuclei Density N for Blends of PEEK and TLCP (Copolymer of HBA, TA, MPA, BP) Crystallized at 300°C

PEEK (wt%)	N (cm^3)
100	8.83×10^9
90	2.53×10^{11}
70	1.22×10^{11}
50	1.61×10^{11}

creases again with larger size particles, although it remains two orders of magnitude higher than for PEEK alone.

Some TLCPs have been found to be miscible to different degrees with PEEK, even if melt blended. Aceirno and Naddeo (63) showed that a widely used semiflexible TLCP, the copolymer of HBA and ethylene terephthalate (ET), was miscible with PEEK. This TLCP was commercialized by Eastman Kodak and is now produced by Unitika in Japan under the Rodrun tradename. It is essentially a copolymer that is a two-phase material with both mesogenic HBA-rich and flexible PET-rich phases. Most commonly used is HBA 60/PET 40, a copolymer with 60 mole % HBA material. The PET phase has a T_g of around 55–75°C and a melting point around 250°, in the PHB phase a broad T_g of around 140–175°C and a possibly a clearing point/degradation temperature of some 450°C (52). Aciemo and Naddeo (63) showed that for concentrations of up to 30 wt% HBA 60/PET 40 the glass transition decreases in a manner that suggests that PEEK is miscible with the flexible PET phase. There is a slight decrease in melting point (possible miscibility), and interfacial adhesion between them appears strong, as shown by electron microscopy of a blend failure surface.

VI. TERNARY AND COMPATIBILIZED BLENDS

There have been two primary motivations behind production of ternary blends, including PEEK. The first has already been mentioned in the section on blends of PEEK with PTFE. A ternary blend improves the already good friction and wear properties that a binary blend affords by addition of reinforcer or additional components that lubricate dry-sliding properties or reduce wear and may include materials such as graphite powder or carbon fibers. Most descriptions of these more complicated systems reside in the patent literature. An example of this is a Japanese patent in which PEEK is blended with PTFE, polyimides, and graphite for good abrasion and heat resistance (64). In addition, there are other examples of ternary blends in which the addition of other polymeric components leads to good properties. These may be for traditional uses such as reduced friction and wear, such as PEEK with isophthalic acid-m-phenyldiamine copolymers and PTFE for improved abrasion resistance (65).

Many of the ternary PEEK polymer blends arise due to mixing with polyimides, with which they are either miscible or compatible, and the polyimide is used to obtain higher glass transition temperatures (heat distortion). A further component is added to improve crystallization, which may be hindered by the higher T_g caused by the addition of the polyimide. An example is the addition of a small amount (about 1 wt%) of poly(phenylene sulphide) (PPS), which improves the crystallization rate of the blend and is even more effective than a filler such as talc (66). Higher crystallinity is attractive because of the greater chemical resistance and mechanical and thermal properties it affords. Ternary blends with PEEK are used in thermosetting resins to obtain materials that adhere well to metal and can resist water, often even to 95°C. Examples are PEEK with PES and thermosetting resins (67) or with poly(arylethersulfone) and an imido thermoset (68).

In some cases, the PEEK may be the minority phase, such as in a 1993 patent in which 5 wt% PEEK is mixed with PTFE and poly(arylene sulphide) to produce a very low coefficient of friction material for metals (69).

The second aspect of research with ternary blends of PEEK has been the use of a third component as a compatibilizer, preferably sitting at the interface of PEEK and the other component. The purpose of such compatibilizers in polymer blends generally is usually twofold: to increase phase dispersion through decreased interfacial tension in the melt of an immiscible blend system, and to improve solid state adhesion, particularly to improve failure properties such as impact strength, fracture toughness, and other ultimate properties. For a compatibilizer to locate preferentially at an interface, it usually must possess characteristics of both phases (such as a diblock or graft material) in which each end is either chemically similar to or miscible/compatible with one of the phases. Details of the mechanism of compatibilizers and the range available in many polymer systems have recently been summarized (70). The use of such materials in PEEK systems has been somewhat more limited than in the first class of PEEK ternary blends discussed above, and the classical use of copolymers as compatibilizing agents has been used even less. Yazaki et al. (71) used oligomers of PEEK to compatibilize a TLCP blend (Xydar, as mentioned previously). The oligomers had the structure

n=3, 9, 15

Whilst modulus increased with the addition of TLCP to PEEK, the flexural strength was reduced by about 30% and the tensile strength similarly, especially at TLCP concentrations of between 30 and 60 wt%. The addition of oligomers served to increase the flexural strength and tensile strength, even at quite high additions of oligomer (30 wt%). There did not seem to be a significant effect due to the length of the oligomer itself. Electron microscopy of blend failure surfaces showed improved dispersion and fibrillation of the TLCP phase and better bonding between it and the PEEK matrix. The explanation for this effect is not the simple copolymer model described above but the more reactive compatibilization. ESCA analysis showed that the carbonyls of the TLCP reacted with the PEEK oligomer. In this work it is proposed that this is with the terminal aromatic unit of the oligomer on which the fluorine atom resides.

Qipeng and coworkers (72) looked at blends of PEEK and PEK-C and noted that when this material was blended with another moderately high temperature polymer, poly(phenylene oxide) (PPO), the resultant blends are immiscible and the properties quite poor, with a dramatic decrease in ultimate tensile strength in the middle composition range (Fig. 8), smooth cavities of the order of 1 to 8 μm remaining on the fracture surface with little adhesion upon pullout. A copolymer of polystyrene (PS) and poly(ethylene oxide) (PEO) was used to compatibilize the materials, because PS and PEO are known to be miscible with PPO and partially miscible with PEK-C, respectively. The morphology of the compatibilized blend (9 wt% of the block copolymer included) was much finer, and the tensile strength increased in the low middle regions although the properties at the extremes were not improved.

Copolymers are not always the only materials that can be used to compatibilize blends. It is easier, cheaper, and more convenient if another thermoplastic can be used. This may be a plastic that is either fully or partially miscible with either or both of the blend components. An example in the case of PEEK relates to the

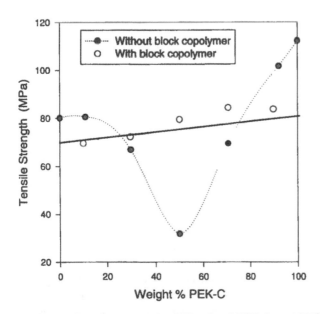

Figure 8 Tensile strength of blends of PEK-C and PPO, with and without the addition of a PEO-PS-PEO triblock copolymer. (From Ref. 72.)

partial miscibility of the DuPont polymer HX4000 mentioned previously, made up of TA, PHQ, and HQ units and found to be partially miscible with PEEK. However, as also mentioned previously, PEEK is fully miscible with PEI, and it has been found that the HX4000 liquid crystalline polymer is partially miscible with PEI, showing synergistic properties in tensile strength and modulus (73). The possibility of further synergies therefore arises if these miscible and partially miscible materials are mixed together (62), and it was found that high moduli are possible at high HX4000 concentrations (i.e., 80 wt% HX4000 with 10 wt% each of PEEK and PEI) and good tensile strengths achievable at high loadings of the PEEK or PEI component (80 wt% PEI with 10 wt% of both PEEK and HX4000). This synergy is carried into the rheological properties: ternary blends with low HX4000 concentrations had lower viscosities than the component materials, coinciding with a fibrillar morphology of the TLCP phase (74). The ternary nature of the material makes for a complex system, and Morales and Bretas have further investigated crystallinity (75) and phase behavior in these materials (76). As reported in most of the papers concerning this system, precise phase behavior depends on annealing conditions, and the thermal history of either the HX4000 or the PEI phase has a strong influence on crystallinity and spherulitic morphology of the PEEK. The miscibilities of the ternary system gave good high-temperature friction properties (77). It was found that both the miscible PEEK/PEI blends and the HX4000 showed high friction near their glass transition temperatures; addition of HX4000 to the binary blend resulted in properties better than those of the constituents, which was explained by the presence of the reinforcing HX4000 phase and the inducement of cold crystallization during sliding, which improves the strength and integrity of the boundary layer.

VII. BLENDS WITH POLY(ETHER SULPHONE)

The miscibility behavior of PEEK/PES blends has been studied by a number of workers (78–84). Melt blended systems have been reported to exhibit immiscible behavior, as shown by two glass transition temperatures close in value to those of the homopolymers over the entire composition range (79,80,82,84). Malik (81) reported that the blends were partially miscible and showed good adhesion between the PEEK and the PES as indicated from scanning electron micrograph results. Arzak et al. (80,84) found that amorphous immiscible blends also exhibit good phase adhesion.

Blends prepared by solution blending have been shown to exhibit a single T_g and are believed to be compatible (78,79). When the solution blended materials were processed at a molding temperature below the T_m of PEEK, the blends remained compatible, but when they were molded above the PEEK melting point, the blends exhibited phase separation behavior (79). It was believed that processing the materials at temperatures above the PEEK T_m caused an increase in the motion of the PEEK chains that disturbed the uniform mixing of the PEEK and PES chains, once again indicating kinetically entrapped miscibility. Ni (83) prepared blends by mixing solutions of PEEK and PES in concentrated H_2SO_4. These samples were found to be only partially miscible, and an increase in percentage PES led to an increase in the compatibility of the two phases (the two T_gs came closer together). Clearly, miscibility arising from solution blending depends on the solvent used.

Melt blended immiscible and partially miscible blends have been reported to show good mechanical compatibility (modulus, tensile strength, yield stress, and impact strength are all close to linearity with composition) (80,81,84). Even though the blends are phase separated, the excellent mechanical performance is expected to be a result of good adhesion between the phases. Quenched amorphous blends were found to be ductile, while slow cooled or annealed semicrystalline blends were more brittle (80,84). These trends are well illustrated by the impact strength measurements presented in Fig. 9. The decrease in ductility of annealed samples is believed to be due to a number of factors including their semicrystalline nature, a decrease in free volume of the amorphous region, and the weakening of the secondary transition (84). As-moulded samples can be produced in an amorphous state by quenching from the melt (80,84), while others have found that room temperature cooling of samples from the melt leads to relatively high degrees of crystallinity for samples with more than 30% PEEK, and samples

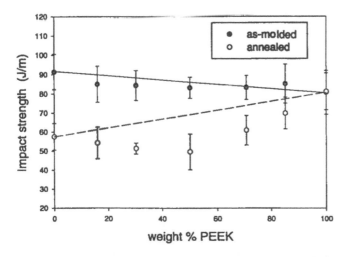

Figure 9 Impact strength–composition relationship for PEEK/PES as molded (black circles) and annealed (white circles) blends. (From Ref. 84.)

with less than 30% PEEK were amorphous (81). This observation of Malik (81) indicates that PES influences the crystallinity of PEEK. Ni (83) also observed that PES reduces the crystallinity of PEEK as a result of the compatibility between the two components when prepared by mixing in H_2SO_4. This indicates perhaps some limited solubility or very fine dispersion of the PES phase. Others workers have concluded, however, that the presence of PES has only a small effect on the crystallization and melting behavior of PEEK (80,84) owing to the phase separation of the PEEK and PES when melt blended. Wu et al. (78) observed that the crystallization behavior was dependent on the conditions of forming the blend (either formed above or below the T_m of PEEK). Nakayama and Malek (82) found that drawing PEEK/PES sheets affected the crystallinity and increased the molecular orientation and modulus.

It is interesting to note that when PEK is reacted with PES in the presence of potassium carbonate at 300°C, transetherification reactions occur, resulting in a block copolymer (85). Further transetherification changes the block copolymer to a random copolymer. Reactions between PEEK and PES were however not reported.

PES is an amorphous engineering thermoplastic which has a higher T_g than PEEK, is cheaper than PEEK, and has high-temperature performance, good hydrolytic and thermo-oxidative stability, and low flammability and smoke emission. As described above, it is found that PEEK/PES blends tend to be immiscible. These properties of PES can be of benefit to PEEK. For example, PES is added in order to produce oriented PEEK films to improve the material's heat resistance and the fluidity in extrusion processing (86). This patent also claims that other heat-resistant polymers such as polyarylate, polyesters, nylon, or polycarbonate could also be added to PEEK to improve these properties. However, because PES is amorphous and has low chemical resistance, blending of PEEK with PES is a potential route to produce an improved material. Several patents describe PEEK/PES compositions (some systems include glass fibers and other inorganic fillers) that exhibit excellent solvent and chemical resistance as well as injection-molding stability (87–89).

Making use of the immiscibility and fine dispersion, PEEK/PES blends have also been described in the patent literature as being used to form membranes (90–92). One patent described extruded blends as immiscible, interpenetrating networks. A solvent for PES was then used to leach out the PES phase, leaving a PEEK membrane with uniform porosity. The patent by Damrow et al. (92) is also based on removing the amorphous component of the blend to form a membrane and involves the addition of a plasticizer. Microporous membranes of this sort are useful for a variety of applications including liquid separations, ultrafiltration, and membrane distillation. Another patent describes the addition of PES to PEEK to decrease the warpage and shrinkage of the crystalline PEEK when molded (93). It is claimed that PES is a good polymer to add to PEEK to decrease the warpage and shrinkage, as it is amorphous and thus will not add to the warpage and shrinkage, and its high T_g will improve the stiffness at temperatures between 150 to 200°C.

VIII. BLENDS WITH OTHER POLY(ARYL ETHER KETONES)

Blending PEEK with a range of poly(aryl ether ketones), such as PEK, PEEEK, PEEKK, and random PEK copolymers, has shown that these blends are miscible and isomorphic, while PEEK/PEKK blends are immiscible and not isomorphic (94–97). The ether and ketone unit cells of poly(aryl ether ketones) are nearly identical and interchangeable in the crystalline state (98), and if a blend is miscible in the melt, the two types of chains can be in close proximity during crystallization and be isomorphic (94,97). Harris and Robeson (95) concluded that PEEK and other poly(aryl ether ketones) are miscible and display isomorphic behavior if the difference in ketone units between the two polymers is less than 25 wt%. Sham et al. (96) found that for PEEK/PEK blends, the polymers cocrystallize only when quenched from the melt and do not show this behavior when the system is subjected to other thermal histories, such as being cooled from the melt or isothermally crystallized. Blends of PEEK with poly(ether diphenyl ether ketone) (PEDEK) have shown that for blends where the PEDEK content is greater than 20 wt% or less than 75 wt%, the PEEK and PEDEK components form independent crystalline states and are immiscible (99). When the PEDEK content was ≤ 20 wt% or ≥ 75 wt%, the blends were found to be miscible.

A patent by Harris and Robeson (94) found that most miscible, isomorphic blends of PEEK with other poly(aryl ether ketones) exhibit a melting point higher than that observed for PEEK. The patent also claims that the blends show excellent mechanical compatibility and provide a broader range of use temperatures and processibility than the constituent polymers.

Patents by Harris and Winslow (100,101) describe additions of 2 wt% or less of a poly(aryl ether ketone), having a melting point higher than 400°C, as an immiscible nucleating agent to PEEK. The blends were found to exhibit faster crystallization rates than PEEK and retained high level of crystallinity even after prolonged treatment in the melt, resulting in excellent mechanical properties and chemical and heat resistance. Crystallization rates for miscible PEEK/PEI blends were also found to improve on addition of the other poly(aryl ether ketone). Harris and Winslow (101) claim that the blends can be fabricated into any desired shape and are particularly useful as electrical insulation for electrical conductors.

IX. POLY(ARYLENE SULFIDE) BLENDS

Poly(aryl ether ketones) such as PEEK require very high processing temperatures, and many materials that could act as plasticizers or processing aids are not stable

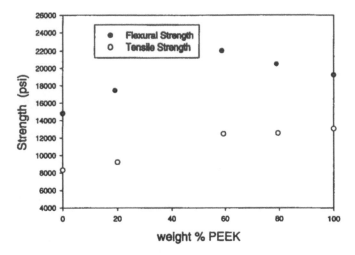

Figure 10 Weight percent poly(ether ether ketone) in blends with poly(phenylene sulfide) composites. (From Ref. 102.)

at these temperatures. To improve the processing of PEEK, blends with poly(phenylene sulphide) (PPS) (which is immiscible with PEEK) have been studied over the entire composition range, as PPS has excellent temperature resistance (102). The addition of PPS results in improved melt properties of PEEK in that it can be processed at lower temperatures. The blends also displayed excellent mechanical compatibility as shown by the synergistic flexural strength behavior (Fig. 10). PEEK/PPS blend films (103) have also shown mechanical properties, such a modulus, that are better than PEEK. PEEK/PPS blends, prepared as bearing materials, have been shown to exhibit high fatigue strength and do not suffer excessively from overheating (104).

PEEK/PPS blends exhibit a higher heat distortion temperature (HDT) than PEEK (102) due to the much higher HDT of the PPS. Adding PPS to PEEK is also economically desirable due to the lower cost of PPS. Blends have shown two crystallization temperatures and two melting temperatures that correspond to the constituent polymers (103). Heat-resistant PEEK/PPS films have been prepared by biaxially stretching the material in a temperature range at least equal to the crystallization temperature of PPS, but not higher than the crystallization temperature of PEEK (103). The resulting films exhibit excellent solder heat resistance and dimensional stability, and superb electrical characteristics, transparency, and surface smoothness.

The desired properties of poly(arylene sulfides), such as excellent temperature and solvent resistance and outstanding strength, have been shown to be affected by the crystallinity behavior of the polymer. PEEK has been shown to act as an excellent nucleating agent for poly(arylene sulfides), especially for low-concentration additions of PEEK (105,106). The addition of PEEK results in an increase in the melt crystallization temperature and provides sites for crystallization. This leads to an effective modification of the crystalline morphology of the poly(arylene sulfide). The addition of a small amount of PEEK to PPS, in forming carbon-fiber-reinforced prepreg tapes by a pultrusion process, has also been shown to be beneficial in improving the properties of composite samples (105).

X. POLYARYLATE BLENDS

Polyarylates are high-temperature, high-performance thermoplastics with good thermal and mechanical properties and excellent melt stability at high temperatures. Polyarylates have a glass transition temperature above that of PEEK, and so it is of interest to study the blends of the two components. The miscibility, crystallization, morphology, and phase behavior of blends of an amorphous polyarylate, bisphenol-A polyarylate (PAr), with PEEK has been studied by Krishnaswamy et al. (107,108). These workers observed that on quenching the material to an amorphous state the blends exhibit two glass transitions corresponding to a nearly pure PEEK phase and a PAr-rich mixed phase (Fig. 11). Robeson and Har-

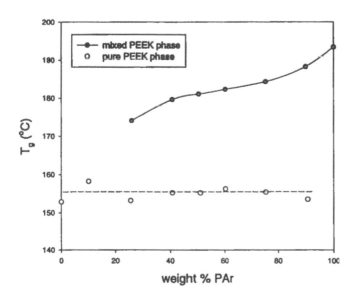

Figure 11 Glass transition temperatures (DSC, 120°C/min) as a function of blend composition. (From Ref. 107.)

ris (109) have shown similar results. Crystallized samples have also been shown to display two relaxations, corresponding to the mobilization of interlamellar amorphous PEEK in a PEEK-rich phase and to the mobilization of mixed amorphous material in a PAr-rich phase (107,108). The addition of PAr to PEEK also retarded the rate of crystallization of PEEK in the blends. Microscopy revealed that amorphous blends with low to intermediate levels of PEEK exhibited a PEEK-rich phase (as inclusions) in a continuous PAr-rich matrix with good adhesion between the phases (108).

The mechanical and physical properties of PEEK/polyarylate blends have also been reported (109). Notched Izod impact strength and tensile impact strength values were found to be higher for blends than for the individual components (i.e., synergism). Krishnaswamy and Kalika (108) believe that this improvement in impact properties is a result of the morphology that they describe: the PEEK acts as an impact modifier for the PAr. Robeson and Harris (109) also found that the addition of PEEK to polyarylate results in an improvement in the environmental stress rupture resistance and that the blend HDT was not found to increase significantly with the addition of polyarylate (as compared to pure PEEK). The patent also describes a blend of PEEK with a liquid crystalline polyarylate. This blend displays synergistic behavior in tensile strength, and the melt viscosity of the material increases with the addition of the liquid crystalline polyarylate.

Ternary blends of PEEK/PEI/polyarylate have been studied (110). Most of the ternary blends were found to be miscible (Fig. 12) and displayed enhanced

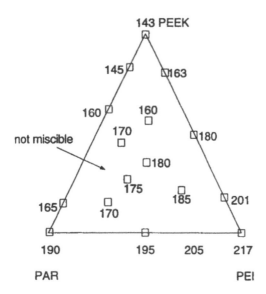

Figure 12 Glass transition temperatures in °C of ternary PEEK-PEI-PAr blends as determined by DSC. (From Ref. 110.)

ease of processing compared to pure PEEK or miscible binary PEEK/PEI blends. It is interesting to note that in this study binary blends of PEEK/polyarylate showed single glass transition temperatures, which differs from what others have reported (107–109).

XI. MISCELLANEOUS PEEK BLENDS

There have been a number of other polymers that have been blended with PEEK and studied. These include poly(vinylidene fluoride) (111), poly (phenylene ether) (95), polybenzoxazole (112), polybenzimidazole (113), poly(p-phenylenebenzo-bisthiazole) (114), poly(arylene ether phosphine oxide) (115), poly(phenylene oxide) (116), Noryl (116), poly(caprolactone) (117), poly(aryl ether ketone co-carbonate (118), poly(ether ether ketone)–poly(dimethylsiloxane) block copolymers (119), aliphatic polyesters (120), fluorinated elastomers (121), epoxy resin (as a particulate filler) (122), Thermid resin (a curable aromatic polymer) (123), and polyindane resin (124). Functionalized poly(aryl ether ketones) have been blended with bismaleimide resin as thermoplastic modifiers (125). Another interesting paper discusses synthesized poly(ether ketones) containing ester groups that are blended with and transesterify with aromatic polycarbonate to form block copolymers (126). It is beyond the scope of this review to discuss each of these blend systems in detail, and the interested reader is referred to the corresponding references.

XII. CONCLUDING REMARKS

Mixing of PEEK with other polymers produces a range of blend morphologies; blends with PEI are miscible and exhibit a composition-dependent glass transition temperature, while blends with fluoropolymers, liquid crystal polymers, PES, PPS, and poly(arylates) are generally immiscible. Intermediate to these are blends with PEK-C and PES-C, which are partially miscible. In miscible and partially miscible systems the crystallisation rate of PEEK is significantly reduced by the presence of a noncrystallising component with a higher glass transition temperature. PEEK has been used to improve the wear and solvent resistance of fluoropolymers, such as PTFE, by combining the advantages of both in an immiscible blend. Although these blends show some promise, their fracture properties have yet to be investigated. In PEEK and thermotropic liquid crystal polymer blends the liquid crystal phase acts as a nucleating agent for PEEK crystallization and also aids processing by lowering melt viscosity below that of pure PEEK. The mechanical properties of these blends show synergistic behavior. There have also been reports of PEEK/liquid crystal blends which have a degree of miscibility, and in these cases the crystallization of PEEK is strongly affected. Strong phase adhesion is the feature of blends with PES and poly(arylates), which leads to useful mechanical properties, while PPS improves the melt properties of PEEK and is itself nucleated for crystallization by additions of small amounts of PEEK.

ACKNOWLEDGMENTS

The authors are grateful to Victrex PLC for providing financial assistance for the writing of this article.

REFERENCES

1. G. Pritchard. Anti-corrosion polymers: PEEK, PEKK and other polyaryls. Rapra Review Report 7(8), 1995.
2. P. A. Staniland. In: G. Allen and J. C. Bevington, eds. Comprehensive Polymer Science. Pergamon Press, 1989.
3. J. B. Rose. In: R. B. Seymour and G. S. Kirshenbaum, eds. High Performance Polymers: Their Origin and Development. Elsevier, 1986.
4. J. P. Gavula, J. E. Harris, and L. M. Robeson. Blends of a poly(aryl ketone) and a polyetherimide. International Patent No. WO 85/01509, 1985.
5. J. E. Harris and L. M. Robeson. Miscible blends of poly(aryl ether ketone)s and polyetherimides. J. Appl. Polym. Sci. 35:1877, 1988.
6. J. E. Harris and L. M. Robeson. Blends of poly(aryl ether ketone). In: B. M. Culbertson, ed. Contemporary Topics in Polymer Science. Vol. 6. New York: Plenum Press, 1989, p. 519.
7. A. A. Goodwin, J. N. Hay, G. A. C. Mouledous, and F. Biddlestone. A compatible blend of poly(ether ether ketone) (PEEK) and poly(ether imide) (Ultem 1000). Int. Fund. Sci. Tech. 5:44, 1980.
8. G. Crevecour and G. Groeninckx. Binary blends of poly(ether ether ketone) and poly(ether imide): miscibility, crystallization behavior, and semicrystalline morphology. Macromolecules 24:1190, 1991.
9. S. D. Hudson, D. D. Davis, and A. J. Lovinger. Semicrystalline morphology of poly(aryl ether ether ketone)/poly(ether imide) blends. Macromolecules 25:1759, 1992.
10. C. H. Lee, T. Okada, H. Saito, and T. Inoue. Exclusion of non-crystalline polymer from the interlamellar region in polymer blends—poly(ether ether ketone) poly(ether imide) blend by small-angle X-ray scattering. Polymer 38:31, 1997.
11. R. H. Mehta, D. A. Madsen, and D. S. Kalika. Microporous membranes based on poly(ether ether ketone) via thermally-induced phase separation. J. Memb. Sci. 107:93, 1995.
12. M. M. Browne, M. Forsyth, and A. A. Goodwin. Solvent diffusion in poly(ether ether ketone)/poly(ether imide) blends. Polymer 36:4359, 1995.

13. H. L. Chen and R. S. Porter. Phase and crystallization behaviour of solution-blended poly(ether ether ketone) and poly(ether imide). Polym. Eng. Sci. 32:1870, 1992.

14. M. M. Browne, M. Forsyth, and A. A. Goodwin. The effect of solvent uptake on the relaxation behaviour, morphology and mechanical properties of a poly(ether ether ketone)/poly(ether imide) blend. Polymer 38:1285, 1997.

15. A. M. Jonas, T. P. Russell, and D. Y. Yoon. Dielectric spectroscopy and x-ray scattering studies on semicrystalline blends of poly(ether-ether-ketone) with poly(ether imide). PMSE 70:394, 1994.

16. J. F. Bristow and D. S. Kalika. Investigation of semicrystalline morphology in poly(ether ether ketone)/poly(ether imide) blends by dielectric relaxation spectroscopy. Polymer 38:287, 1997.

17. Y. Ando and D. Y. Yoon. Phase separation in quenched non-crystalline poly(vinylidene fluoride)/poly(methyl methacrylate) blends. Polym. J. 24:1329, 1994.

18. J. P. Runt, C. A. Barron, X. Zhang, and S. Kumar. Melting point depression in PBT/Par blends. Macromolecules 24:3466, 1991.

19. B. B. Sauer and B. S. Hsiao. Broadening of the glass transition in blends of poly(aryl ether ketones) and a poly(ether imide) as studied by thermally stimulated currents. J. Polym. Sci. B. 31:917, 1993.

20. A. A. Goodwin and G. P. Simon. Glass transition behaviour of poly(ether ether ketone)/poly(ether imide) blends. Polymer 37:991, 1996.

21. A. A. Goodwin and G. P. Simon. Dynamic mechanical relaxation behaviour of poly(ether ether ketone)/poly(etherimide) blends. Polymer 38:2363, 1997.

22. H. L. Chen and R. S. Porter. Melting behavior of poly(ether ether ketone) in its blends with poly(ether imide). J. Polym. Sci. B. 31:1845, 1993.

23. H. L. Chen, J. W. You, and R. S. Porter. Intermolecular interaction and conformation in poly(ether ether ketone)/poly(ether imide) blends—an infrared spectroscopic investigation. J. Polym. Res. 3:151, 1996.

24. T. Tsutsumi, S. Morikawa, T. Nakakura, K. Shimamura, T. Takahashi, N. Koga, M. Ohta, A. Morita, and A. Yamaguchi. Polyimide based resin composition. European Patent No. 0 430 640 A1, 1990.

25. Y. Okawa, N. Koga, H. Oikawa, T. Asanuma, and A. Yamaguchi. Polyimide composition. European Patent No. 0 564 299 A2, 1993.

26. T. Tsutsumi, S. Morikawa, T. Nakakura, T. Takahashi, A. Morita, and Y. Gotoh. Polyimide based resin composition. U.S. Patent No. 5,312,866, 1994.

27. B. B. Sauer and B. S. Hsiao. Miscibility of three different poly(aryl ether ketones) with a high melting thermoplastic polyimide. Polymer 35:3315, 1993.

28. B. B. Sauer, B. S. Hsiao, and K. L. Faron. Miscibility and phase properties of poly(aryl ether ketone)s with three high temperature all-aromatic thermoplastic polyimides. Polymer 37:445, 1996.

29. R. J. Karcha and R. S. Porter. Miscible blends of a sulfonated poly(aryl ether ketone) and aromatic polyimides. J. Pol. Sci. B. 27:2153, 1989.

30. R. J. Karcha and R. S. Porter. Miscible blends of modified poly(aryl ether ketones) with aromatic polyimides. J. Pol. Sci. B. 31:821, 1993.

31. H. Zhang and T. Chen. Chinese Patent No. 85108751, 1985.

32. G. C. Alfonzo, V. Chiappa, J. Lui, and E. R. Sadiku. Crystallization of poly(aryl-ether-ether-ketone)s blends. Eur. Polym. J. 27:795, 1991.

33. B. Li, G. Li, Y. Zhang, T. He, T. L. Chen, and G. C. Alfonzo. In: Polymers and Bio-
    materials. 29. Elsevier, 1991.
34. Z. Zhang and H. Zeng. Morphology, crystallization kinetics and melting behaviour
    of the blends of poly(ether ether ketone) with poly(ether sulfone with cardo side
    group). Polymer 34:4032, 1993.
35. Z. Zhang and H. Zeng. Interfacial crystallisation of PEEK/PES-C blends on carbon
    fibre. Eur. Polym. J. 29:1647, 1993.
36. J. Vande Voort and S. Bahadur. The growth and bonding of transfer film and the role
    of CuS and PTFE in the tribological behaviour of PEEK. Wear 181–183:212–221,
    1995.
37. A. Schelling and H. H. Kausch. The influence of long term reciprocating dry fric-
    tion on the wear behaviour of short fibre reinforced composite materials. In P. K.
    Rohtagi, P. J. Blau, and C. S. Yust, eds. Tribology of Composite Materials. New
    York: ASM International, 1990, pp. 227–238.
38. B. J. Briscoe, L. H. Yao, and T. A. Stolarski. The friction and wear of poly(tetraflu-
    oroethylene)-poly(etheretherketone) composites: an initial appraisal of the optimum
    composition. Wear 108:357–374, 1986.
39. Z. P. Lu and K. Friedrich. On sliding friction and wear of PEEK and its composites.
    Wear 181–183:624–631, 1995.
40. J.-N. Chu and J. M. Schultz. The influence of microstructure on the failure be-
    haviour of PEEK. J. Mater. Sci. 25:3746–3752, 1990.
41. B. H. Stuart and B. J. Briscoe. A fourier transform Raman spectroscopy study of
    poly(ether ether ketone)/polytetrafluoroethylene (PEEK/PTFE) blends. Spec-
    trochimica Acta 50A(11):2005–2009, 1994.
42. P. Damman, C. Fougnies, M. Dosieré, and J. C. Wittmann. Liquid–liquid phase
    separation and oriented growth of poly(aryl ether ether ketone) on friction-trans-
    ferred poly(tetraflouroethylene) substrates. Macromolecules 28:8272–8276,
    1995.
43. J. Vande Voort and S. Bahadur. The growth and bonding of transfer film and the role
    of CuS and PTFE in the tribological behaviour of PEEK. Wear 181–183:212–221,
    1995.
44. J. F. Braza and R. E. Furst. Reciprocating sliding wear evaluation of a
    polymeric/coating tribological system. Wear 162–164:748–756, 1993.
45. M. Nishi, S. Sugimoto, Y. Shimuzu, N. Suzuki, S. Kawanishi, T. Nagase, M. Endo,
    and Y. Eguchi. Surface modification of polytetrafluoroethylene by excimer-radia-
    tion. Polym. Mater. Sci. Eng. 68:165–166, 1993.
46. J. M. Senior. Proceedings of Fluoropolymers '92, RAPRA Technology, Shrews-
    bury, U.K., Paper 19, pp. 1–4.
47. M. Petersen, Raychem Corp., US Patent No. 4,777,214, 1988.
48. A. M. Donald and A. H. Windle. Liquid Crystalline Polymers. Cambridge: Cam-
    bridge Univ. Press, 1992.
49. F. P. La Mantia, ed. Thermotropic Liquid Crystal Polymer Blends. Lancaster: Tech-
    nomic, 1993.
50. G. Kiss. In situ composites: blends of isotropic polymers and thermotropic liquid
    crystalline polymers. Polym. Eng. Sci. 27(6):410, 1987.

51. A. Mehta and A. I. Isayev. The dynamic properties, temperature transitions and thermal stability of poly(etherether ketone)–thermotropic liquid crystalline polymer blends. Polym. Eng. Sci. 31(13):963–970, 1991.

52. G. P. Simon. Longitudinal PLC + EP blends: miscibility and crystallization phenomenon. In: W. Brostow, ed. Polymer Liquid Crystals, Volume 3. London: Thomson Scientific, 1997.

53. A. Mehta and A. I. Isayev. Rheology, morphology and mechanical characteristics of poly(ether ether ketone)–liquid crystal polymer blends. Polym. Eng. Sci. 31(13):971–980, 1991.

54. G. Crevecoeur and G. Groeninckx. Melt spinning of in-situ composites of a thermotropic liquid crystalline polyester in a miscible matrix of poly(ether ether ketone) and poly(ether imide). Polymer Composites 13(3):244, 1992.

55. A. A. Isayev and P. R. Subramanian. Blends of a liquid crystalline polymer with poly ether ketone. Polym. Eng. Sci. 32(2):85–93, 1992.

56. M. Matzner and D. M. Papuga. International Patent WO88/00605, 1980.

57. A. I. Isayev. International Patent WO90/13421, 1990.

58. F. N. Cogswell, B. P. Griffin, and J. B. Rose. European Patent 030417, 1994.

59. B. J. Carvalho, R. E. S. Bretas. Crystallization kinetics of a PEEK/LCP blend. J. Appl. Polym. Sci. 55:233–246, 1995.

60. Y. Zhong, J. Xu, and H. Zeng. Blends of poly(ether ether ketone) with a thermotropic liquid crystalline polyester I. The morphology, crystallisation and melting behaviour. Polym. J. 24(10):999–1007, 1992.

61. Y. Zhong, J. Xu, and H. Zeng. Blends of poly(ether ether ketone) with thermotropic liquid crystalline polyesters: 2. Crystallisation kinetics. Polymer 33(18):3893–3898, 1992.

62. R. E. S. Bretas and D. G. Baird. Miscibility and mechanical properties of poly(ether imide)/poly(ether ether ketone)/liquid crystalline polymer ternary blends. Polymer 33(24):5233–5244, 1992.

63. D. Acierno and C. Naddeo. Blends of PEEK and PET-PHB 60: a preliminary study on thermal and morphological aspects. Polymer 35(9):1994–1996, 1994.

64. H. Niwa and T. Shimokusuzono. Polyimide blend compositions for sliding materials. Japanese Patent 06240273, 1991.

65. T. Noma. Aromatic polyamide resin composition for compression moulding. European Patent 407,821, 1991.

66. J. E. Harris. Nucleating agents for poly(aryl ether ketone) blends and compositions obtained therefrom. US Patent 4910289, 1990.

67. T. Inai and K. Karikaya. Undercoating compositions for metals. Japanese Patent 63,304,068, 1992.

68. T. Inai, L. Karukaya, M. Ikeda, and K. Yamamoto. An undercoat composition and a metal substrate coated with a resin composition. European Patent 343,282, 1989.

69. M. Davis, and P. M. Hatton. Composition of fluoropolymers. International Patent WO 94/05728, 1993.

70. S. Datta and D. J. Lohse. Polymer Compatibilisers. New York: Hanser, 1996.

71. F. Yazaki, K. Suzuki, R. Yosimaya, and Y. Zheng. Effect of polyetheretherketone oligomer on the mechanical properties of polyetheretherketone with aromatic liquid crystalline copolyester blends. Polymers and Composites 1(6):411–420, 1993.

72. G. Qipeng, F. Tianru, C. Tianlu, and F. Zhiliu. Compatibilizing of poly(ethylene ox-
     ide)-b-polystyrene-b-poly(ethylene oxide) in phenolphthalein poly(ether
     ketone)/poly(2,6-dimethyl-1',4-phenylene oxide) blends. Polym. Commun.
     32(1):22–24, 1991.
73. D. G. Baird, S. S. Bafna, J. P. de Souza, and T. Sun. Mechanical properties of in-situ
     composites based on partially miscible blends. Polymer Composites 14:214–223,
     1993.
74. R. E. S. Bretas, D. Collias, and D. G. Baird. Dynamic rheological properties of
     polyetherimide/polyetheretherketone/liquid crystalline polymer ternary blends.
     Polymer Eng. Sci. 34(19):1492–1496, 1994.
75. A. R. Morales and R. E. S. Bretas. Polyetherimide/poly(ether ether ketone)/liquid
     crystalline polymer ternary blends—I. Calorimetric studies and morphology. Eur.
     Polym. J. 32(3):349–363, 1996.
76. A. R. Morales and R. E. S. Bretas. Polyetherimide/poly(ether ether ketone)/liquid
     crystalline polymer ternary blends—II. Approximated interaction parameters and
     phase diagrams. Eur. Polym. J. 32(3):365–373, 1996.
77. J. Hanchi and N. S. Eiss. Synergistic friction and wear of ternary blends of
     polyetheretherketone, polyetherimide and a thermotropic liquid crystalline polymer
     at elevated temperatures. Tribology Trans. 40(1):102–110, 1997.
78. Z. Wu, Y. Zheng, H. Yan, T. Nakamura, T. Nozawa, R. Yosomiya. Molecular ag-
     gregation of PEEK with PES blends and the block copolymers composed of PEEK
     and PES components. Angew. Makromol. Chem. 173:163–181, 1989.
79. Z. Wu, Y. Zheng, X. Yu, T. Nakamura, and R. Yosomiya. Thermal and viscoelastic
     behavior of polyetheretherketone/polyethersulfone blends. Angew. Makromol.
     Chem. 171:119–130, 1989.
80. A. Arzak, J. I. Eguiazábal, and J. Nazábal. Phase behaviour and mechanical proper-
     ties of poly(ether ether ketone)–poly(ether sulphone) blends. J. Mater. Sci.
     26:5939–5944, 1991.
81. T. M. Malik. Thermal and mechanical characterization of partially miscible blends
     of poly(ether ether ketone) and polyethersulfone. J. Appl. Polym. Sci. 46:303–310,
     1992.
82. K. Nakayama and K. A. Malek. Effect of drawing on dynamic viscoelastic proper-
     ties of PEEK and PES blends. Polymat 94—Polymer Blends II Conf. Proc. 543–545,
     1994.
83. Z. Ni. The preparation, compatibility and structure of PEEK–PES blends. Polym.
     Adv. Technol. 5:612–614, 1994.
84. A. Arzak, J. I. Eguiazábal, J. Nazábal. Compatibility in immiscible poly(ether ether
     ketone)/poly(ether sulfone) blends. J. Appl. Polym. Sci. 58:653–661, 1995.
85. I. Fukawa, T. Tanabe, and H. Hachiya. Trans-etherification of aromatic polyether-
     ketone and trans-etherification between aromatic polyetherketone and aromatic
     polyethersulfone. Polym. J. 24:173–186, 1992.
86. N. Fukushima, T. Saitou, and H. Hayashida. Method of orientating thermoplastic
     polyether ether ketone films. Eur. Pat. Appl. EP 059,077, 1982.
87. K. Asai and Y. Suzuki. Thermoplastic resin composition improved in solvent resis-
     tance. Eur. Pat. Appl. EP 247,512, 1987.

88. T. Saito, Y. Suzuki, and K. Asai. Thermoplastic resin composition with improved chemical resistance. Eur. Pat. Appl. EP 224,236, 1987.

89. T. Saito, K. Asai, and Y. Suzuki. Thermoplastic resin composition having an improved chemical resistance. U.S. Patent 4,804,697, 1989.

90. R. D. Birch, and K. J. Artus. Membrane. Eur. Pat. Appl. EP 416,908, 1991.

91. R. D. Birch, and K. J. Artus. Tubular membrane. Eur. Pat. Appl. EP 409,416, 1991.

92. P. A. Damrow, R. D. Mahoney, H. N. Beck, and M. F. Sonnenschein. Process for making a microporous membrane from a blend containing a poly(etheretherketone)-type polymer, an amorphous polymer, and a solvent. U.S. Patent 5,205,968, 1993.

93. L. M. Robeson and J. E. Harris. An article molded from a blend of poly(aryl ether ketone) and a poly(aryl ether sulfone). Eur. Pat. Appl. EP 176,988, 1986.

94. J. E. Harris and L. M. Robeson. Blends of poly(aryl ketone)s. U.S. Patent 4,609,714, 1986.

95. J. E. Harris and L. M. Robeson. Isomorphic behavior of poly(aryl ether ketone) blends. J. Polym. Sci., Polym. Phys. Edn. 25:311–323, 1987.

96. C. K. Sham, G. Guerra, F. E. Karasz, and W. J. MacKnight. Blends of two poly(aryl ether ketones). Polymer 29:1016–1020, 1988.

97. V. L. Rao. Polyether ketones. JMS—Rev. Macromol. Chem. Phys. C35:661–712, 1995.

98. P. C. Dawson and D. J. Blundell. X-ray data for poly(aryl ether ketones). Polymer 21:577–578, 1980.

99. X. L. Ji, W. J. Zhang, and Z. W. Wu. Miscibility of poly(ether ether ketone)/poly(ether diphenyl ether ketone) blends. Polymer 37:4205–4208, 1996.

100. J. E. Harris and P. A. Winslow. Nucleating agents for poly(aryl ether ketones). U.K. Pat. Appl. GB 2,203,744, 1988.

101. J. E. Harris and P. A. Winslow. Nucleating agents for poly(aryl ether ketones). U.S. Patent 4,959,423, 1990.

102. L. M. Robeson. Alloys of a poly(arylene sulfide) and a poly(aryl ketone). Eur. Pat. Appl. EP 062, 830, 1982.

103. T. Mizuno, Y. Teramoto, T. Saito, and J. Wakabayashi. Heat-resistant film and production process thereof. Eur. Pat. Appl. EP 321,215, 1989.

104. G. J. Davies. Plastics alloy compositions. U.K. Pat. Appl. GB 2,108 983, 1983.

105. T. W. Johnson, W. H. Beever, J. E. O'Connor, and J. P. Blackwell. Method of producing poly(arylene sulfide) compositions and articles made therefrom. U.S. Patent 4,690,972, 1987.

106. M. H. Hindi, M. W. Woods, N. Harry, and T. Wakida. Poly(arylene sulfide) composition. U.S. Patent 5,300,552, 1994.

107. R. K. Krishnaswamy, T. Lusk, and D. S. Kalika. Phase behavior, morphology, and crystallization of poly(ether ether ketone)/polyarylate blends. ANTEC Conf. Proc., Society of Plastics Engineers 53:1715–1718, 1995.

108. R. K. Krishnaswamy and D. S. Kalika. Phase behavior, crystallization, and morphology of poly(ether ether ketone)/polyarylate blends. Polym. Eng. Sci. 36:786–796, 1996.

109. L. M. Robeson and J. E. Harris. Blends of a poly(aryl ether ketone) and a polyarylate. Eur. Pat. Appl. EP 170,067, 1986.

110. O. Herrmann-Schönherr, A. Schneller, and U. Falk. High performance polymer blends. New compatible systems. ACS Polym. Prepr. 32:48–49, 1991.

111. P. Maiti, J. Chatterjee, D. Rana, and A. K. Nandi. Melting and crystallization behavior of poly(vinylidene fluoride) samples in its blends with some polyacrylates, poly(vinyl esters) and poly(aryl ether ether ketone). Polymer 34:4273–4279, 1993.

112. S. M. Lefkowitz and D. B. Roitman. The fluorescence of polybenzoxazole dispersed in poly(ether ether ketone). Polymer 35:1576–1579, 1994.

113. E. Alvarez, L. P. DiSano, and B. C. Ward. Molded polybenzimidazole/polyaryleneketone articles and method of manufacture. Eur. Pat. Appl. EP 392,855, 1990.

114. C. A. Gabriel, R. J. Farris, and M. F. Malone. Solution processing of composite fibers containing rodlike and thermoplastic polymers. J. Appl. Polym. Sci. 38:2205–2223, 1989.

115. C. D. Smith, A. Gungor, K. M. Keister, H. A. Marand, and J. E. McGrath. Poly(arylene ether ketone)–poly(arylene ether phosphine oxide) copolymer and blend compositions. ACS Polym. Prepr. 32:93–95, 1991.

116. F. W. Mercer, M. F. Frolx, and T. C. Cheng. Poly(aryl ether ketone) compositions. Eur. Pat. Appl. EP 192,408, 1986.

117. R. D. Mahoney, H. N. Beck, R. A. Lundgard, H. S. Wan, J. Kawamoto, and M. F. Sonnenschein. Microporous membranes from poly(ether-etherketone)-type polymers and low melting point crystallizable polymers, and a process for making the same. Eur. Pat. Appl. EP 492,446, 1992.

118. M. Matzner, B. D. Dean, and D. M. Papuga. Blends comprising poly(aryl ether ketone co-carbonates). U.S. Patent 4,975,470, 1990.

119. G. C. Corfield, G. W. Wheatley, and D. G. Parker. Synthesis and properties of polyetheretherketone–polydimethylsiloxane block copolymers. J. Polym. Sci., Polym. Chem. Edn. 28:2821–2836, 1990.

120. K. Doyama, M. Yamaguchi, M. Ohsuga, K. Yamagata, A. Niki, T. Saito, H. Tsunomachi, and D. Kishimoto. Engineering plastic composition and articles made of the same. Eur. Pat. Appl. EP 420,619, 1991.

121. K. D. Goebel, P. N. Nelson, and C. T. Novak. Polymer blend composition. Eur. Pat. Appl. EP 432,911, 1991.

122. B. Z. Jang, J. Y. Liau, L. R. Hwang, and W. K. Shih. Structure–property relationships in thermoplastic particulate- and ceramic whisker-modified epoxy resins. J. Reinf. Plast. Compos. 8:312–333, 1989.

123. F. Mercer. Aromatic polymer compositions. International Patent WO 86/04073, 1986.

124. S. G. Chu, B. K. Patnaik, and K. S. Shih. Polyindanes as processing aid for engineering thermoplastics. Eur. Pat. Appl. EP 432,696, 1991.

125. H. D. Stenzenberger and P. König. New functionalized poly(arylene-ether ketone)s and their use as modifiers for bismaleimide resin. High Perform. Polym. 5:123–137, 1993.

126. R. J. Kumpf, D. Nerger, C. Lantman, H. Pielartzik, and R. Wehrmann. A novel route to poly(ether ketone)–polycondensate block copolymers. In: P. Stroeve and A. C. Balazs, eds. Macromolecular Assemblies in Polymeric Systems. ACS Sympo Series 493. Washington, D.C.: American Chemical Society, 1992, pp. 300–312.

15

Applications of X-Ray Photoelectron Spectroscopy and Secondary Ion Mass Spectrometry in Characterization of Polymer Blends

Chi-Ming Chan and Jingshen Wu
The Hong Kong University of Science and Technology, Clear Water Bay, Kowloon, Hong Kong

Yiu-Wing Mai
University of Sydney, Sydney, New South Wales, Australia

I. INTRODUCTION

The surface properties of polymers are important for many applications such as adhesion, biomaterials, protective coatings, friction and wear, composites, microelectronics devices, and thin-film technology. In general, special surface properties such as chemical composition, hydrophilicity, roughness, crystallinity, conductivity, lubricity, and cross-linking density are required for the success of these applications. The surface properties of polymers are in general different from their bulk properties. In particular, the surface chemical and physical properties of polymer blends can be significantly different from those predicted by their bulk compositions. To determine the physical and chemical properties of polymer blends, surface analysis techniques such as x-ray photoelectron spectroscopy (XPS), static secondary ion mass spectrometry (SSIMS), atomic force microscopy (AFM), scanning electron microscopy (SEM), attenuated internal reflection infrared spectroscopy (ATIR), and dynamic contact measurements are essential (1–11).

Several important factors must be considered in choosing a surface analysis technique, namely, sampling depth, surface information, analysis environment, and sample suitability. When surface-sensitive techniques are involved, the first important consideration is what surface information is needed. Each technique supplies different and often complementary information. The choice of a suitable technique(s) is affected by many factors, and it is difficult to set up rules to zero in on a choice. When high-resolution, three-dimensional images of surfaces are needed, AFM, and SEM are the appropriate techniques. When chemical analyses demand the most surface-sensitive probe, contact angle measurements and SSIMS are good choices. If quantification and chemical-state information are important, XPS would definitely be a candidate. If one is investigating a small amount (ppm) of impurity in a sample, then SSIMS is a good choice. Before investigators can choose the appropriate technique(s) for the analysis, they need to understand the limitations and capabilities of each technique, and they must be able to identify the information that is relevant to the analysis. In this article, we are going to focus on two of the most powerful surface analysis techniques, XPS and SSIMS, which have been used extensively to characterize surface properties of polymer blends.

II. X-RAY PHOTOELECTRON SPECTROSCOPY (XPS)

X-ray photoelectron spectroscopy, which is also known as electron spectroscopy for chemical analysis (ESCA), is probably the most widely used technique in the characterization of polymer surfaces (1–5). A sampling depth of 3 to 5 nm is typical for XPS. This technique was initially developed by Professor Kai Siegbahn at the University of Uppsala. A sample is irradiated by a beam of x-rays. The interaction between an x-ray photon and the inner-shell electron of an atom causes a complete transfer of the photon energy to the electron. The electron then has enough energy to escape from the surface to the sample. This electron is referred to as the photoelectron. The basic principle of XPS is shown schematically in Fig. 1. The kinetic energy of the photoelectron is measured by an electron energy analyzer. The difference between the kinetic energy of the photoelectron and the x-

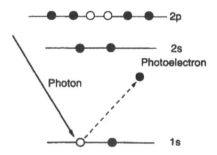

Figure 1 Schematic diagram showing the basic principle of XPS.

ray photon energy yields the binding energy of the inner-shell electron. Knowing the binding energy of the inner-shell electron allows identification of the element. The binding energy of an inner-shell electron is also sensitive to the electronic environment of the atom. When the atom is bonded to atoms of different elements of different electronegativity, the binding energy may increase or decrease. This change in binding energy is called the chemical shift, which can be used to provide structural information for a molecule. For example, when a carbon atom is bonded to different groups of atoms of increasing electronegativity, a systematic shift (chemical shift) in the binding energy of the C1s peak is observed: the more electronegative the group, the higher the binding energy of the C1s peak. Table 1 shows the binding energy for the C1s electrons for several polymers; the binding energy of the C1s electrons gradually increases as the hydrogen atoms are successively replaced by the more electronegative fluorine atoms. From polyethylene to poly(vinyl fluoride), the substitution of a hydrogen atom by a fluorine atom increases the binding energy of the C1s electrons from 285.0 to 288.0 eV for the carbon atoms of the CHF groups. The binding energy of the C1s electrons for the —CH_2— groups also increases to 285.9 eV owing to the inductive effect caused by the CHF groups. When all hydrogen atoms are replaced by fluorine atoms in going from polyethylene to polytetrafluoroethylene, the binding energy of the C1s electrons increases to 292.2 eV. A maximum chemical shift of 8.7 eV is observed in the carbon atom of the CF_3 group, which attaches to the backbone of poly(hexafluoropropylene). The primary shift on a carbon atom induced by a fluorine atom is approximately 2.9 eV per substituent. Since fluorine is the most electronegative element, fluoropolymers give rise to the largest span in the binding energy for the C1s electrons. When other elements are bonded to carbon atoms, the chemical shifts are smaller. For example, in poly(vinyl chloride), the chemical shift of the carbon atom of the CHCl group is significantly smaller than that of the carbon atom of the CHF group in poly(vinyl fluoride) because chlorine is less electronegative than fluorine.

III. STATIC SECONDARY ION MASS SPECTROMETRY (SSIMS)

Static secondary ion mass spectrometry has proved to be a valuable method for determining the composition and structure of polymer surfaces (2,6–11). It complements XPS because SSIMS can differentiate among polymers that have similar XPS spectra. It is also more surface-sensitive than XPS. The typical sampling depth for SSIMS is approximately 0.5 nm. It has sufficient sensitivity to detect amounts less than a monomolecular layer, particularly when a high-resolution time-of-flight mass analyzer is used.

In a SSIMS experiment, the surface is bombarded with a primary ion beam of low current density (typically 10^{-11} to 10^{-8} A/cm^2) to minimize damage. The sputtered material consists largely (>99%) of secondary neutral species, with a

Table 1 Binding Energies of the C1s Core Level in Various Polymers

Polymer	Structure
Polyethylene	$-[CH_2 - CH_2 - CH_2 - CH_2]_n$ 285.0
Poly(vinyl fluoride)	$-[CFH - CFH]_n-$ 288.4
Poly(vinylene fluoride)	$-[CFH - CH_2 - CFH - CH_2]_n-$ 288.0 285.9
Poly(vinylidene fluoride)	$-[CH_2 - CF_2 - CH_2 - CF_2]_n-$ 286.3 290.8
Polytrifluoroethylene	$-[CF_2 - CFH - CF_2 - CFH]_n-$ 289.3 291.6
Polytetrafluoroethylene	$-[CF_2 - CF_2 - CF_2 - CF_2]_n-$ 292.2
Polyhexafluoropropylene	293.7 $-[CF_3 / CF - CF_2]_n-$ 291.8 289.8
Poly(vinyl chloride)	$-[CH_2 - CHCl - CH_2 - CHCl]_n-$ 285.0 286.3

Table 1 Continued

Polymer	Structure
Poly(vinyl alcohol)	285.0 / 286.6; structure: $-[CH_2-CH(OH)]_n-$ with H on carbon
Polysarcosine	287.8 / 286.5; structure: $-[C(=O)-CH_2-N(CH_3)]_n-$
Poly(ethylene terephthalate)	288.9 / 285.0 / 286.6; structure: $-[O-C(=O)-C_6H_4-C(=O)-O-CH_2-CH_2]_n-$
Bisphenol-Polycarbonate	285.0 / 286.6 / 291.2; structure: $-[C_6H_4-C(CH_3)_2-C_6H_4-O-C(=O)-O]_n-$
Polyimide (Pyromellitic dianhydride oxydianiline)	286.0 / 289.0 / 285.0; pyromellitic diimide with oxydianiline

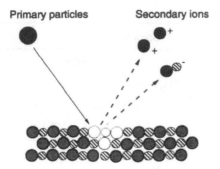

Figure 2   Schematic diagram showing the basic principle of SIMS.

small fraction (<1%) of secondary positive and negative ions. The basic principle of SIMS is shown schematically in Fig. 2. The mass of the ions is measured by a mass analyzer. The quadrupole mass analyzer, the combined electrostatic and magnetic sector analyzer, and the time-of-flight mass spectrometer are analyzers used in SIMS systems. These mass spectrometers have different mass ranges and mass resolutions. The mass resolution ($M/\Delta M$) of a spectrometer is a measure of the ion separating ability of the instrument, where M is the mass of the peak and ΔM is the full width at half-maximum peak height of the peak. The time-of-flight mass spectrometers are very popular in organic material analysis because they offer several advantages: a high transmission over the whole mass range, an unlimited mass number, a high mass resolution, and a simultaneous detection of all masses.

A detailed analysis of the positive and negative ion spectra can provide structural and chemical information about the polymer surface. The use of a highly focused liquid-metal ion beam source allows high-spatial-resolution analyses. A key problem with SSIMS as a surface analysis technique is its limited ability to perform quantitative analyses. In addition, the lack of understanding of the relationship between the structure of a polymer and its ionic fragments obtained in a SSIMS experiment renders the analysis of an unknown sample very difficult. In spite of these deficiencies, in the last several years SSIMS has been applied successfully to characterize surface contaminants, copolymer (12–14) and polymer blend surfaces (15–17), and polymer surfaces modified by various surface treatment processes (18–19).

IV.   SURFACE SEGREGATION IN POLYMER BLENDS

XPS and SSIMS have been applied to study surface properties of polymer blends (15–17,20–26) and copolymers (12–14,27–31) and to characterize specific interactions between the polymer components in blends (32–35). Generally, the bulk and surface compositions of polymer blends are not the same even though they are miscible in the bulk. The significant enrichment of the low-surface-energy component at the surface of a polymer blend is frequently noted. Based on the work of Schmidt and Binder (36), Jones and Kramer (37) derived a simple expression to predict the surface composition of a miscible blend at equilibrium. Their result indicates that a difference of 1 mJm^{-2} in the surface free energy of the blend components can lead to an enrichment of the lower surface energy component at the polymer blend surface.

    X-ray photoelectron spectroscopy has been applied to study the surface composition and structure of polymer blends and copolymers because the binding energy of a core level electron depends on its chemical environment within the molecules. The blend of poly(methyl methacrylate) (PMMA) and polycarbonate (PC) was studied by XPS and SIMS (22). The measured critical surface energies of PMMA and PC are 39 and 45 mJm^{-2}, respectively (38). Hence PMMA is expected to segregate to the surface. Figure 3 shows the XPS spectra of pure PMMA and PC and of blends having 2.5 and 11.8 wt% of PMMA. Curve fitting (1–3) was employed to resolve the overlapping in the spectra, as shown in Fig. 3. The peaks marked in the figure correspond to the carbon atoms in various molecular environments.

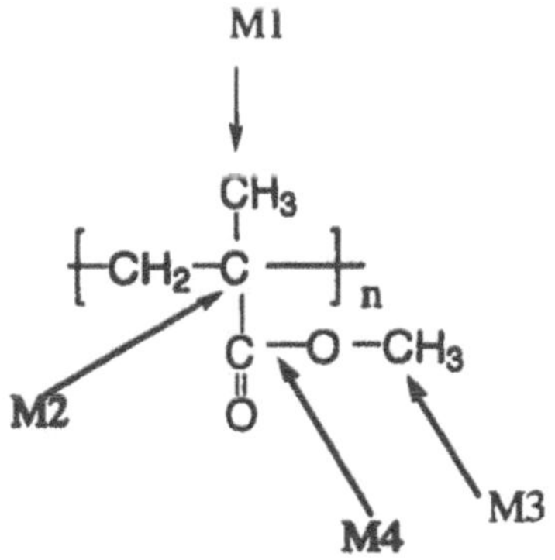

C4 (shake-up structure due to the benzene rings)

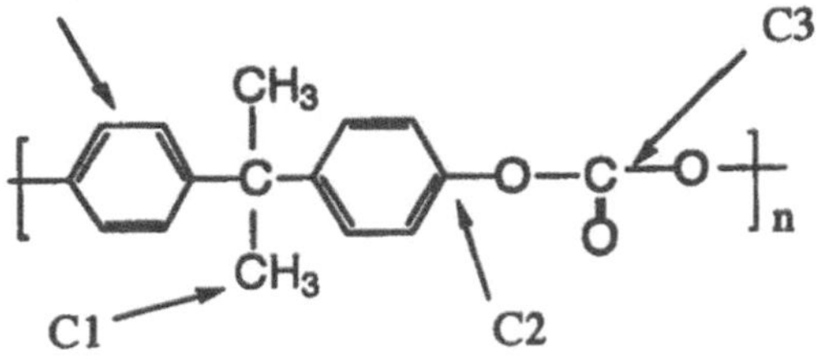

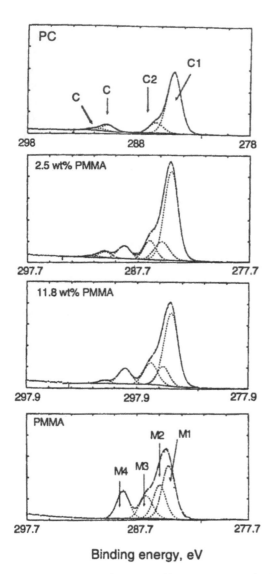

Binding energy, eV

Figure 3   C1s XPS spectra (from to bottom) of pure PC, 2.5 wt% PMMA, 11.8 wt% PMMA, and pure PMMA. (From Ref. 22.)

The PMMA surface concentrations of the blends can be calculated with the carbon/oxygen ratios determined from the XPS spectra by using the equation

$$\left(\frac{C}{O}\right)_{exp} = \frac{XC_{PMMA} + (1 - X)C_{PC}}{XO_{PMMA} + (1 - X)O_{PC}} \tag{1}$$

where X is the molar PMMA surface concentration in the blend and C_{PMMA}, O_{PMMA}, C_{PC}, and O_{PC} are the stoichiometric carbon and oxygen concentrations in pure PMMA and pure PC, respectively. The ester group, which is only found in PMMA, can also be used to calculate the molar PMMA surface concentration. In the blend sample, the fraction of the ester group (E) in the total C1s intensity can be calculated by the equation

$$E = \frac{X}{5X + 16(1 - X)} \tag{2}$$

Equation (2) can be rewritten as

$$X = \frac{16E}{1 + 11E} \tag{3}$$

The PMMA surface concentrations of the blends calculated by using Eqs. (1) and (3) are shown in Fig. 4. The results clearly indicate that the surfaces of the blends are enriched with PMMA.

Another factor that is important in controlling segregation of the low-energy component of the blend to the surface is block length if one of the components of

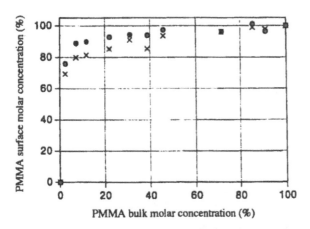

Figure 4 PMMA surface concentration of the blends estimated by means of C/O ratio (circles) and ester peak intensity (crosses) as a function of the PMMA bulk concentration. (From Ref. 22.)

Table 2　Physical Properties of the Copolymers

Polymer	Polysulfone block length M, g/mol	Polysiloxane block length M, g/mol	wt% siloxane
PSFPSX-1	4,900	12,800	72
PSFPSX-3	4,900	4,400	47

the polymer blend is a block copolymer. Patel et al. (24) studied the surface segregation of blends of polysulfone and an alternating block copolymers of polysulfone and polysiloxane. Table 2 shows the physical properties of the two copolymers used in the study. It is recognized that the polysiloxane block that is of lower surface energy will segregate to the surface. The objective of that study is to investigate the effect of block length on the surface composition of the blends. The surface chemical compositions of the blends were measured by XPS and angle-resolved XPS (1–3,5). Figures 5 and 6 show the plots of the surface weight percentage siloxane as a function of its bulk concentration. The XPS data measured at three different exile angles show the changes in the surface chemical composition in the top 6 nm (the data obtained at an exile angle of 15° is the

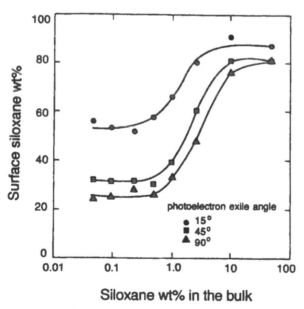

Figure 5　Surface behavior of PSFPSX-3/homopolysulfone blends. (From Ref. 24.)

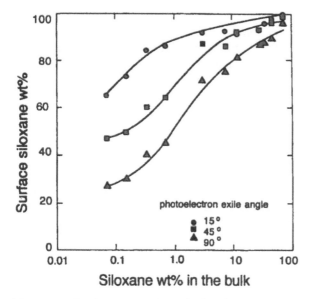

Figure 6 Surface behavior of PSFPSX-1/homopolysulfone blends. (From Ref. 24.)

most surface sensitive). The surface concentrations of siloxane are relatively high (> 20 wt%) even at very low bulk concentrations of siloxane (< 1 wt%). Comparison of Figs. 5 and 6 clearly indicates that the extent of surface segregation of siloxane at a given bulk concentration is much higher for PSPSF-1 because the lengths of the siloxane block in this copolymer are longer. These results clearly reflect that the surface energy and the length of the blocks are the critical parameters controlling the surface composition of polymer blends containing a block copolymer.

V. POLYMER BLEND MISCIBILITY

XPS has recently been applied to characterize specific interactions between the polymer components in blends. In recent years, considerable progress has been made in recognition of the importance of specific interactions in the phase behavior of polymer blends. One of the most important intermolecular interactions is the hydrogen bond between a proton donor and a proton acceptor. The hydrogen bonding has been shown to be responsible for miscibility of many polymer blends. Only a very few studies have been made to investigate the effects of specific interaction such as hydrogen bonding on the surface composition and structure of miscible polymer blends. In a recent study, Goh et al. (33) have applied XPS to study the ionic interaction between sulfonated polystyrene and poly(styrene-co-4-vinylpyridine). Shifts in the binding energy of the N1s and S2p core levels were detected. In addition, their results indicate that poly(styrene-co-4-vinylpyridine), being the lower surface energy component, still segregates to the surface although the ionic interaction between these two polymers is strong.

Poly(vinyl alcohol) (PVAL)/poly(N-vinyl-2-pyrrolidone) (PVP) blends are miscible as indicated by differential scanning calorimetry (DSC) results showing only one glass transition temperature for the blends. It has been suggested that the miscibility of this blend is a result of the formation of hydrogen bonds between PVP and PVAL. XPS was applied to study the formation of hydrogen bonds in these blends (32). Thin films of the homopolymers and the PVAL/PVP blends were prepared by spin-casting a polymer or polymer blend solution on a Si wafer. XPS spectra were obtained at a take-off angle of 45° (the sampling depth was estimated to be about 4.5 nm using an attenuation length of 2.2 nm). Figure 7 shows the C1s spectra for the homopolymers (PVP and PVAL) and their blends. The C1s spectrum of PVP is shown to comprise four peaks by curve fitting. The positions of these four peaks are determined to be at 284.8, 285.2, 286.1, and 287.7 eV, as shown in Table 2. Peaks 1–4 correspond to carbon atoms in PVP with different atomic environments, as shown here.

These results are very similar to the literature values (39). The C1s spectrum of PVAL is shown to comprise two peaks, which are determined to be at 248.8 and 286.3 eV. The high-binding-energy peak corresponds to the carbon atom adjacent to the OH group. The C1s spectra for the blends, as shown in Fig. 7, can be resolved into five peaks, which all correspond to various carbon atoms in the two polymers. However, the C1s spectra for the blends are slightly too complicated for extracting detailed chemical information for the formation of hydrogen bonding between the C=O and OH groups. The O1s spectra for the homopolymers and the blends are shown in Fig. 8. A small peak at approximately 533 eV in the O1s spectrum for PVP is believed to be caused by retained water. The binding energies for the O1s and N1s core levels for the homopolymers were determined and the results are summarized in Table 3. The O1s binding energy difference between the C=O and OH groups (ΔO1s) in these two homopolymers is 1.4 eV, which agrees well with the results of Briggs (39). The O1s spectra for the blends are much broader than those of the homopolymers because of the presence of the two types

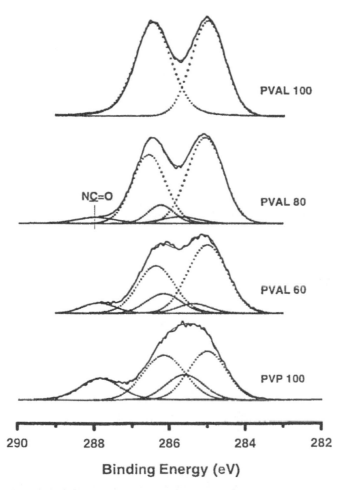

Figure 7 C1s core level spectra for PVAL and PVP homopolymers and their blends: PVAL100 = 100% PVAL; PVAL 80 = 80 wt% PVAL; PVAL 60 = 60 wt% PVAL, and PVP 100 = 100% PVP. (From Ref. 32.)

of oxygen (C=O and OH). When the C=O groups in PVP and the OH groups in PVAL form intermolecular hydrogen bonds, the O1s binding energy of the C=O groups increases because of the electron transfer from the C=O groups to the OH groups, while the O1s binding energy of the OH groups decreases due to the increase in the electron density. As a result, the binding energy differences between the oxygen in the C=O groups and the oxygen in the OH groups of the blends are smaller than that of the homopolymers. The value of the ΔO1s decreases as the PVAL surface volume fraction increases, as shown in Table 4. These XPS results

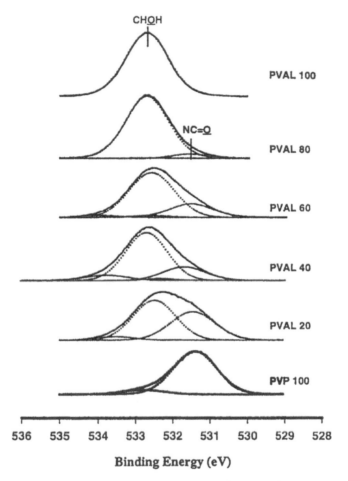

Figure 8 O1s core level spectra for PVAL and PVP homopolymers and their blends: PVAL 100 = 100% PVAL; PVAL 80 = 80 wt% PVAL; PVAL 60 = 60 wt% PVAL, PVAL 40 = 40 wt% PVAL; PVAL 20 = 20 wt% PVAL; and PVP 100 = 100% PVP. (From Ref. 32.)

suggest that the intermolecular hydrogen bonds between C=O and OH groups can cause a change of 0.3–0.4 eV in the O1s binding energy difference between these two functional groups. It is interesting to note that the values for ΔO1s are quite large, even though the surface volume fraction of PVP is very small. The increase in the N1s binding energy due to the formation of the hydrogen bonds is approximately 0.2–0.3 eV. Shifts in the binding energy in polymer blends with strong specific interactions have been observed by Goh et al. (33). Their XPS study on the ionic interactions between sulfonated polystyrene and poly(styrene-

Table 3 C1s, N1s, and O1s Core Levels for the PVAL and PVP Homopolymers

Polymer		C1s (eV)				N1s (eV)	O1s (eV)	$I_{N/C}$	$I_{O/C}$
		1	2	3	4				
PVP	BE[a]	284.8	285.2	286.1	287.7	399.7	531.2	0.16	0.16
	FWHM[b]	1.2	1.1	1.2	1.1	1.2	1.3		
PVAL	BE	284.8	286.3				532.6		0.48
	FWHM	1.1	1.1				1.4		

E = binding energy.
WHM = full width at half-maximum.

co-4-vinylpyridine) indicates that the binding energy of N1s core level of the pyridinium ions (protonated pyridine), formed by the ionic interactions, is about 2.5 eV higher than that of the pyridine.

The surface composition of PVP and PVAL blends was determined by measuring the C/N peak area ratios with the PVP homopolymer as the standard. The results are summarized in Table 5. It is apparent from this table that all PVAL/PVP blends display a surface excess of PVAL. Figure 9 is a plot of the PVAL surface volume fraction versus the PVAL bulk volume fraction. The PVAL surface volume fraction is not linear with respect to the PVAL bulk volume fraction. The surface segregation of PVAL is caused by the difference in the surface energy of the two homopolymers. The solid surface tensions of semicrystalline PVAL and PVP ho-

Table 4 Binding Energies of the O1s Core Levels and the O1s Binding Energy Difference Between the Oxygen Atoms in PVAL and PVP

PVAL, Surface volume fraction	N1s (eV) N	O1s (eV) CH<u>O</u>H	NC=<u>O</u>	Difference of the O1s ΔO1s (eV)
PVAL		532.6		1.4
0.979		532.6	531.2	1.4
0.953		532.6	531.5	1.1
0.886	399.9	532.7	531.5	1.2
0.731	400.0	532.5	531.5	1.0
0.585	399.9	532.6	531.5	1.1
0.422	399.8	532.5	531.5	1.0
0.392	399.8	532.5	531.5	1.0
PVP	399.7		531.2	

Table 5 PVAL Surface Compositions Calculated from C/N Peak Area Ratios

PVAL bulk		PVAL surface
wt %	Volume fraction	Volume fraction
0.0	0.000	0.000
5.0	0.049	0.087
10.0	0.097	0.190
20.0	0.196	0.392
28.4	0.278	0.422
35.0	0.343	0.463
40.0	0.392	0.585
60.0	0.589	0.731
80.0	0.792	0.886
90.0	0.895	0.953
95.0	0.947	0.979
100.0	1.000	1.000

mopolymers were reported to be 37.0 and 53.6 mJm^{-2} (40), respectively. Even though PVAL/PVP blends are completely miscible in bulk because of the formation of hydrogen bonds between PVAL and PVP, the lower surface energy component (PVAL) still segregates to the surface of the blends. These results clearly show that the reduction in the free energy as a result of surface segregation of the lower surface energy component can overcome the increase in enthalpy due to the breaking of some intermolecular hydrogen bonds between PVAL and PVP.

VI.   POLYMER BLEND MORPHOLOGY

The surface morphology of polymer blends is of great academic and industrial interest. Unfortunately, it cannot be studied easily. Time-of-flight secondary ion mass spectrometry (ToF-SIMS) has been widely used in polymer surface characterization owing to the yield of molecular information and its high surface sensitivity. When coupled with a liquid metal ion gun, surface chemical imaging with submicron lateral resolution is possible for polymeric materials. Recent studies have shown that ToF-SIMS can be a very powerful tool to study the surface morphology of polymer blends (15,17). Briggs et al. (15) studied the surface morphology of a poly(vinyl chloride) (PVC)/PMMA blend by ToF-SIMS. The samples for ToF-SIMS studies were prepared by casting from tetrahydrofuran solution. In the negative ion spectra, the well-known PMMA fragments at mass-

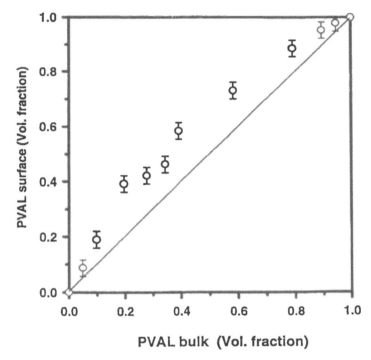

Figure 9   PVAL surface volume fraction vs. PVAL bulk volume fraction. (From Ref. 32.)

to-charge ratio (m/z) 85, 87, 141, and 185 and the PVC fragments at m/z 35 and 37 were identified. From the chemical imaging of the surface of the blend, the outer surface has a thin overlayer of PMMA on the bulk phase-separated morphology.

For the blends of the ethylene-tetrafluoroethylene copolymer (ETFE)/ PMMA, three glass transition temperatures (Tgs) were detected by differential scanning calorimetry (17). To obtain a flat surface for ToF-SIMS analysis, the samples were prepared by a cryomicrotome at $-100°C$. The morphology of the surface of the samples is expected to be identical to that of the bulk because the Tgs of ETFE and PMMA are higher than ambient temperature. If the surface and bulk morphologies are similar, we should be able to identify three phases—a semicrystalline region containing mostly ETFE with a Tg close to that of pure ETFE, an amorphous region containing mostly PMMA with a Tg close that of pure PMMA, and an amorphous region containing significant levels of ETFE and PMMA. The detection of the low (ETFE) and high (PMMA) Tgs is relatively straightforward from the use of differential scanning calorimetry (DSC) and dynamical mechanical analysis (DMA). However, the intermediate Tg appears as a very weak transition in the DSC scans, and its presence is overshadowed by a very broad β transition of PMMA in the DMA curves. It is important to point out that the presence of this intermediate Tg is barely detectable by DSC for the blends obtained by slowly cooling the samples from the melts, and that its detectability is enhanced if the blends are quenched from their melts. The intermediate Tg is associated with an amorphous phase containing significant levels of ETFE and PMMA. Using the Fox equation (41), the weight fraction of ETFE in this phase was estimated to be approximately 0.55.

The most characteristic fragments for ETFE are at $m/z = 12$ (C^+), 31 (CF^+), 51 (CHF_2^+), 59 ($C_2H_4F^+$), 69 (CF_3^+), 77 ($C_3H_3F^+$), 95 ($C_3H_2F_3^+$), and 127 ($C_4H_4H_3^+$, deprotonated monomer). The negative spectrum of ETFE is totally dominated by the peak at $m/z = 19$ (F^-), with some other small characteristic fragments at $m/z = 31$ (CF^-), 38/39 (F_2^-/F_2H^-), and 43 (C_2F^-). In principle, the above fragments can be used to monitor the distribution of PMMA and ETFE. However, as the mass resolution was only moderate in the imaging mode, certain fragments of PMMA cannot be used because they overlap the fragments of ETFE (e.g., at $m/z = 59$, $C_2H_3O_2^+$ of PMMA overlaps with $C_2H_4F^+$ of ETFE). Some other fragments cannot be used owing to their low intensities in the blends. In the following, only the O^-/OH^- images representing PMMA and the F^- images representing ETFE are presented in the negative mode.

Figure 10 displays the $O^- + OH^-$ and F^- images for the blend (the weight ratio of ETFE/PMMA = 20/80). The upper and lower images are for the sample prepared by slow cooling and quenching, respectively. The $O^- + OH^-$ and F^- images are complementary. For the slowly cooled sample, the ETFE particles (phase B) are dispersed in the PMMA matrix (phase A). In most cases, the particles are well defined: the sizes range from about 1 to 20 μm. Spectra are reconstructed using data obtained from phases A and B. The fragments from phases A and B are close to those for PMMA and ETFE, respectively. However, the intensities of the $O^- + OH^-$ peaks are not zero in phase B, and the intensity of the F^- peak is not zero in phase A. These results show that there is a small amount of PMMA and ETFE in phases B and A, respectively. In addition, a small amount of a phase (phase C) that contains high levels of ETFE and PMMA can also be detected. Both PMMA and ETFE fragments are present in the spectrum reconstructed from phase C. These results are in agreement with the DSC measurements (17). For the quenched sample, the surface morphology is quite different, as shown in Fig. 10 (lower images). Although the dispersion of the ETFE particles in the PMMA matrix is still observed, a much larger amount of phase C is detected.

VII. POLYMER BLEND PROCESSING

Segregation of one of the components of a blend or an additive in a homopolymer to the interface between the melt and the die-wall surface has also gained impor-

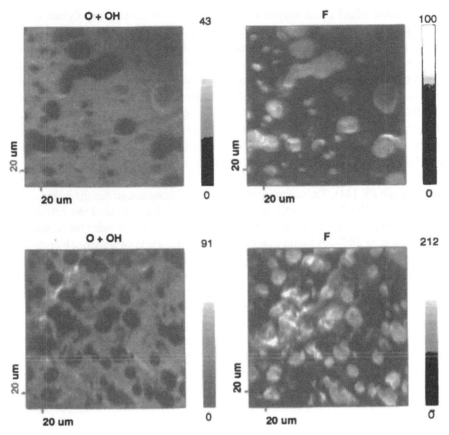

Figure 10 Negative ToF-SIMS images obtained for two samples with a blend composition of ETFE/PMMA = 20/80. The upper and lower images are for the slowly cooled and quenched samples, respectively. Imaged area is 200 μm × 200 μm. (From Ref. 17.)

tance in polymer processing. Reduction in the apparent viscosity of poly(vinyl chloride) with an increase in the concentration of an additive was detected (42). The reduction was attributed to a film of lubricant generated between the surface of a steel die and the polymer melt during extrusion. The lubricant layer was suspected to cause wall slip of the polymer. However, no direct measurement was made to determine the surface concentration of the additive. The migration of one of the components to the interface between the melt and the die wall during extrusion of an incompatible blend presents a very interesting rheological phenomenon (43–52). Shih (43) studied the melt flows and the melt viscosities of an incompatible blend of an ethylene-propylene-1,4,-hexadiene terpolymer and a copolymer of vinylidene fluoride and hexafluoropropylene. A marked reduction in the melt viscosity of either component was observed when a small amount of the other component was present. It was speculated that the die was coated with the minor component of the blend. The reduction of the viscosities was attributed to a slippage between the polymer and the coated capillary. However, it is possible that the die wall was actually coated with the low-energy component and that the reduction of the viscosity was not related to the concentration of the components. In a study of the viscosity of blends of polyethylene (PE) and poly(bis(2,2,2,-trifluoroethoxy) phosphazene) (PPH), it was found that the viscosity of the blend that contained 10% of PPH and 90% PE was much lower than that of pure PE (44). The effect was much more pronounced for the blend containing 30% PPH. Electron microscopic analysis on the surface of the extrudates showed a high concentration of PPH, which acts as a low-viscosity lubricant.

Chan et al. (45–47) found that the viscosity reduction of polymer blends can involve either cohesive failure in the lubricant layer or adhesive failure at the interface between the lubricant layer and the extrudate. The polymer blend of interest typically consists of a low-surface-energy component (a fluoropolymer or fluoroelastomer present in a low concentration) and a thermoplastic. Viscosity reduction is generally observed when the viscosity of the low surface energy component is lower than or similar to that of the thermoplastic. Figures 11 and 12 illustrate the concept of cohesive failure and adhesive failure, respectively, during extrusion of a polymer blend. These models assume that a lubricant layer, which comprises mainly the low-surface-energy component of the blend, is formed on the die wall surface owing to the migration of the low-surface-energy component

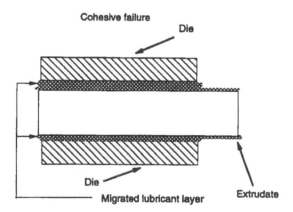

Figure 11 Cohesive failure in the lubricant layer that adheres to the die-wall surface. (From Ref. 46.)

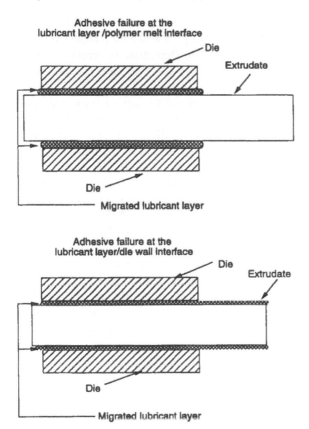

Figure 12 Adhesive failure at the interface between the lubricant layer and the extrudate, and adhesive failure at the interface between the lubricant layer and the die-wall surface. (From Ref. 46.)

to the interface. Figure 11 shows that cohesive failure occurs in the lubricant layer. Figure 12 shows that adhesive failure can occur at the interface between the extrudate and the lubricant layer or at the interface between the lubricant layer and the die wall.

In order to differentiate the different modes of failure as described in Figs. 11 and 12, we have to use an analytical technique that can detect the lubricant layer on the extrudate surface, providing that the lubricant (which is a minor component of a binary polymer blend) has a chemical composition different from the major component of the blend. But if the thickness of the lubricant layer on the extrudate is very small (i.e., 10 nm or less), then the correct mode of failure can be determined only by a surface-sensitive technique such as x-ray photoelectron spectroscopy or secondary ion mass spectrometry, because the sampling depths for SIMS and XPS are typically 0.2 and 3 nm, respectively (1). Therefore these techniques can be applied to differentiate different modes of failure as described in Figs. 11 and 12, even if the lubricant layer is a monolayer thick.

Chan et al. (45) have combined rheological and surface analyses to unravel the mechanism that governs the gradual reduction in the viscosity of incompatible blends of poly(ether ether ketone) and polytetrafluoroethylene (PTFE) during an extrusion. Figure 13 shows the apparent viscosity of the 5%-PTFE sample at various shear rates, normalized with respect to the viscosity of the control at the same rates. It is clear that the viscosity of the 5%-PTFE sample is lower than that of the control at the rates examined and that the rate of reduction of viscosity is higher at the higher rates. The apparent values at the end of these runs (about 30 minutes for the lower rates and 20 minutes for the 500 s^{-1}) are plotted in Fig. 14 along with the flow curves for the control, pure PTFE, and the 5%-PTFE blend. From Fig. 14, it appears that there is no critical rate at which the viscosity changes dramatically from the value of the control. At all rates, the viscosity of the 5%-PTFE sample is lower than that of the control and progressively decreases with an increase in the shear rate. This points to a lubrication effect rather than a wall-slip effect on the viscosity. Similar to the steady-state value curve, the flow curve for the 5%-PTFE sample does not show any critical shear rate.

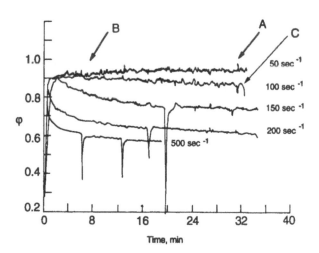

Figure 13 The effect of apparent shear rate on apparent viscosity for the 5%-PTFE sample. Normalized viscosities were used (ϕ = the viscosity of the 5%-PTFE sample over the viscosity of the pure PEEK at the same shear rate). Each discontinuity corresponds to a filling of the reservoir. (From Ref. 45.)

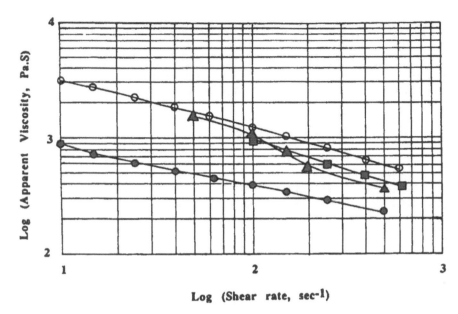

Figure 14 The apparent viscosity of pure PEEK, pure PTFE, and 5%-PTFE sample as a function of apparent shear rate, and the "steady-state" apparent viscosity for the 5%-PTFE sample at different shear rates obtained from Fig. 13. O, pure PEEK; ■, 5%-PTFE sample; ●, pure PTFE; ▲, the "steady-state" apparent viscosity for the 5%-PTFE sample at different shear rates. (From Ref. 45.)

In Table 6, a comparison of the amount of fluorine detected on the extrudate surface appears to confirm this argument. As the apparent shear rate increases from 50 to 500 s^{-1} for the 5%-PTFE sample, the steady-state surface fluorine concentration also increases (from 37.9 atomic % at 50 s^{-1} to 46.3 atomic % at 500 s^{-1}). If there is no accumulation of PTFE at the surface, the amount of fluorine for the 5%-PTFE sample should equal its bulk value, or about 2.0 atomic %. The actual values measured for the extrudates indicate that substantial surface migration of PTFE has occurred for the 5%-PTFE sample.

If reduction in viscosity of samples is related to accumulation of PTFE at the die surface, then the surface of the extrudate should show a gradual increase in the PTFE concentration. The three spectra shown in Fig. 15a–c were taken from extrudates obtained from the first, second, and third runs, respectively, of the 5%-PTFE sample at the apparent shear rate of 500 s^{-1}. The surface chemical compositions were calculated, and the results are summarized in Table 6. If there is no migration of PTFE to the surface, the surface concentration of fluorine of the 1%-PTFE and 5%-PTFE samples should be approximately 0.4 and 2.0 atomic percent,

Table 6 Surface Chemical Composition of Extrudates Determined by XPS

Sample composition	Run	Apparent shear rate s^{-1}	Surface chemical Atomic percent		
			C	O	F
1%-PTFE (bulk composition)			86.1	13.5	0.4
5%-PTFE (bulk composition)			84.9	13.1	2.0
Pure PTFE			33.6		66.7
5%-PTFE (region A in Fig. 13)		50	55.6	6.5	37.9
5%-PTFE (region B in Fig. 13)		50	60.6	7.3	32.1
5%-PTFE (region C in Fig. 13)		100	49.1	15.2	35.7
1%-PTFE	first	500	71.4	10.1	18.5
	second	500	69.4	9.7	20.9
5%-PTFE	first	500	70.1	10.7	19.2
	second	500	46.9	3.7	49.4
	third	500	49.6	4.1	46.3
Control after the 5%-PTFE run	first	500	65.0	9.2	25.8
	second	500	81.1	12.9	6.0
	third	500	82.9	12.8	4.3

respectively. The fluorine concentration of the surface of the 1%-PTFE extrudate from the first run was 18.5 atomic percent, which is substantially higher than that of the bulk. For the 5%-PTFE sample, the fluorine concentration of the extrudate surface increased from 19.2 to 49.4 atomic percent in going from the first to the second run. During the second and the third runs of the 5%-PTFE sample, the surface of extrudates was heavily populated with PTFE, resulting in a dramatic reduction in viscosity (in a pure PTFE surface, the carbon and fluorine concentrations are 33.3 and 66.7 atomic percent, respectively). This definitely supports the postulate that reduction in viscosity of samples is related to the accumulation of PTFE at the die surface.

As shown in Table 6, at the shear rate of 500 s^{-1}, the surface concentration of fluorine of the 5%-PTFE sample increases significantly from the first to the second filling of the barrel of the capillary rheometer and reaches more or less a steady-state value in the third filling. These results mean that an increase in the PTFE concentration at the surface decreases the viscosity of the blend. However, the PTFE surface concentration for the extrudate obtained at 500 s^{-1} (at the end of the first filling) is lower than those obtained at 50 and 100 s^{-1}, and yet the viscosity at 500 s^{-1} is the lowest (cf. Table 6). This apparent discrepancy can be resolved by examining the flow curve of pure PTFE as shown in Fig. 14, which indicates that the apparent viscosity of pure PTFE is a strong function of the

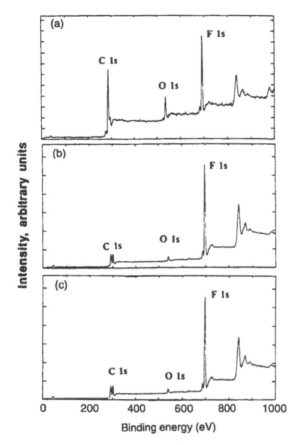

Figure 15 Spectra (a), (b), and (c) were taken from extrudates obtained from the first, second, and third runs of the 5%-PTFE sample at 500 s^{-1}, respectively. (From Ref. 45.)

apparent shear rate. As a result, the apparent viscosity of the 5%-PTFE blend measured at 500 s^{-1} is lower than those measured at 50 and 100 s^{-1} even though the surface PTFE concentration of the extrudate obtained at 500 s^{-1} at the end of the first filling is lower than those of the extrudates obtained at 50 and 100 s^{-1}.

To confirm that the die wall was coated with PTFE, a control sample was run after a run with the 5%-PTFE sample. Figure 16 shows the viscosity of the control as a function of time. The viscosity of pure PEEK at the initial time was much lower than that of the pure sample obtained from a clean die and was very similar to that of the 5%-PTFE sample at the end of the third run. This observation agrees well with the postulate that the die wall is coated with PTFE and that the lubricat-

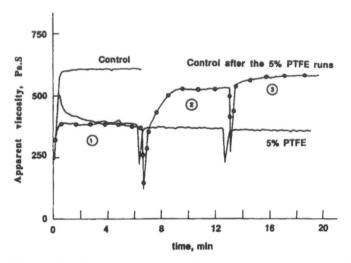

Figure 16 The apparent viscosity of control sample as a function of time. This experiment was performed after a run with a 5%-PTFE sample for 20 minutes at 500 s^{-1}. (From Ref. 45.)

ing effect of PTFE lowers the viscosity of the control sample. As the experiment progressed, the amount of PTFE at the die surface was depleted slowly; hence, the viscosity of the control sample increased gradually. Finally, when all PTFE was depleted from the die wall, the measured viscosity reached the same value as the viscosity of the control sample obtained from a clean die. The surface chemical compositions of extrudates at various regions were calculated and are summarized in Table 6. The surface concentration of fluorine, which is directly related to the PTFE concentration of the extrudate surface, gradually decreased as a function of time. The extrudate from the third run had a very small amount of fluorine, suggesting that the amount of PTFE at the die wall had been almost depleted. The surface chemical composition and viscosity data provide a very convincing argument for the PTFE-coated die-wall theory.

During the extrusion of blends of high-density polyethylene (HHM 5502, Philips Petroleum Ltd.) and Dynamar FX-9613 (a copolymer of vinylidene fluoride and hexafluoropropylene obtained from Dyneon-A 3M-Hoechst company), the viscosity of the blends was found to be significantly lower than that of HHM 5502 (46,47). Figure 17 shows the apparent viscosity of Dynamar FX-9613 and HHM 5502 as well as the steady-state apparent viscosity of three Dynamar/HHM 5502 blends at different shear rates and 200°C. The apparent viscosity of blends is lower than that of HHM 5502 and of Dynamar. Figures 18 and 19 show the normalized apparent viscosity of 0.5% and 2.5% Dynamar blends as a function of time at various shear rates. The normalized apparent viscosity is defined to be the ratio of the apparent viscosity of the blend, which is time dependent, to the apparent viscosity of the neat HDPE at the same shear rate. The normalized steady-state apparent viscosity is defined to be the steady-state apparent viscosity of the blend to the apparent viscosity of the neat HDPE. Figures 18 and 19 clearly indicate that the time to reach the steady state is longer at low shear rates and decreases as the shear rate increases. These results, in general, agree well with those obtained by Chan et al. (45) for the PEEK/PTFE blends.

The results of the XPS analyses of extrudates of the 0.5 and 2.5% blends are shown in Tables 7 and 8, respectively. Positions A and B represent the samples obtained in the unsteady- and steady-state apparent viscosity regions, respectively. The fluorine concentrations on the extrudate surfaces are fairly low. These results also differ from those observed for the PEEK/PTFE blends by Chan et al. (45), who found that a larger viscosity reduction produced a higher fluorine concentration on the extrudate surface. If we assume that the fluoroelastomer, being a low-surface-energy component of the blend, has a stronger tendency to migrate to the die wall surface, then the low fluorine concentrations, especially observed for the extrudates of the 0.5% blends, suggest that adhesive failure did occur at the

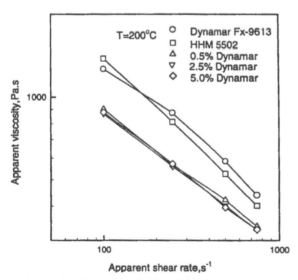

Figure 17 Apparent viscosity vs. apparent shear rate for Dynamar, HDPE HHM5502, and Dynamar/HHM5502 blends for shear rate ranging from 100 to 750 s⁻¹. (From Ref. 46.)

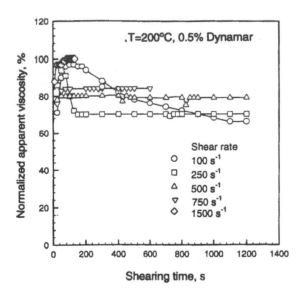

Figure 18 A plot of the normalized apparent viscosity of the 0.5% Dynamar/HHM 5502 blend as a function of time. (From Ref. 46.)

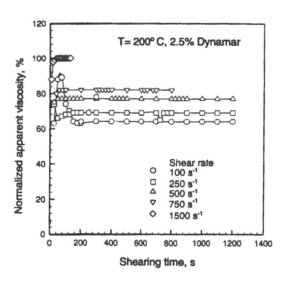

Figure 19 A plot of the normalized apparent viscosity of the 2.5% Dynamar/HHM 5502 blend as a function of time. (From Ref. 46.)

Table 7 Summary of the Surface Chemical Composition of the Extrudates of the 0.5% Dynamar/HHM 5502 Blend and the Pure Dynamar Extrudate

Run	Position[a]	Apparent shear rate (s^{-1})	Surface chemical composition (atomic %)			
			C	N	O	F
1	B	10	98.93	0.00	0.44	0.62
1	A	100	99.09	0.00	0.38	0.53
1	B	100	97.97	0.00	0.69	1.33
1	A	250	99.37	0.00	0.33	0.30
2	B	250	97.48	0.00	0.44	2.06
1	A	500	99.16	0.17	0.43	0.23
3	B	500	96.74	0.00	1.36	1.90
1	A	750	99.51	0.00	0.38	0.11
3	B	750	97.35	0.00	0.84	1.81
1	B	1500	98.70	0.00	0.25	1.06
	Dynamar extrudate		38.8		0.4	60.8

[a] Positions A and B represent the samples obtained in the unsteady- and steady-state apparent viscosity regions, respectively.

Table 8 Summary of the Surface Chemical Composition of the Extrudates of the 2.5% Dynamar/HHM 5502 Blend

Run	Position[a]	Apparent shear rate (s^{-1})	Surface chemical composition (atomic %)			
			C	N	O	F
1	B	10	97.21	0.00	0.32	2.46
1	A	100	95.41	0.69	2.87	1.03
1	B	100	92.44	0.87	4.22	2.48
1	B	250	93.05	0.35	2.73	3.88
1	A	500	94.99	0.11	0.89	4.02
3	B	500	92.67	0.12	0.94	6.27
1	B	750	86.96	0.00	1.31	11.72
1	B	1500	96.51	0.00	0.22	3.27

[a] Positions A and B represent the samples obtained in the unsteady- and steady-state apparent viscosity regions, respectively.

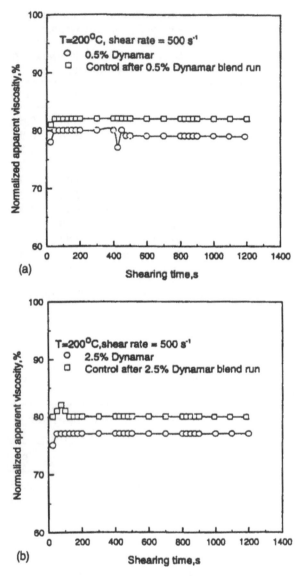

Figure 20 Normalized apparent viscosity of the control sample as a function of time. The experiment was performed without cleaning the die after the third run with the (a) 0.5% and (b) 2.5% Dynamar/HHM 5502 blends at a shear rate of 500 s^{-1}. (From Ref. 46.)

interface between the HDPE extrudate, and that a layer of fluoroelastomer formed on the die-wall surface.

To confirm that the reduction in viscosity is caused by a Dynamar lubricant layer formed on the die-wall surface, neat HHM 5502 was run after the third run for the 0.5 and 2.5% blends without cleaning the die. Figures 20a and b show the normalized apparent viscosity for the blends and the neat HDPE as a function of time. The apparent viscosity of neat HHM 5502, obtained with the "unclean" die, is only slightly higher than that of the blends and much lower than that of neat HHM 5502 obtained with a clean die. These results reveal that during the extrusion of the blend, Dynamar migrated from the melt to the interface and adhered to the die-wall surface. The viscosity reduction was achieved by the HDPE slipping over the fluoroelastomer layer. When neat HHM 5502 was used, the Dynamar layer that had been formed during the extrusion of the blend still functioned as a lubricant layer, which was responsible for observed viscosity reduction of the HDPE. In addition, the viscosity of the HDPE obtained with the "unclean" die is relatively constant even after 1200 seconds of shearing, indicating that the Dynamar layer is strongly adhered to the die-wall surface and has relatively good cohesive strength.

The presence of the Dynamar layer on the die-wall surface is further confirmed by XPS analyses (cf. Table 9) of the extrudates of the HDPE. There is a

Table 9 Summary of the Surface Chemical Composition of the Extrudates of Pure HDPE HHM 5502 After the Extrusion of the Extrusion of the 0.5% and 2.5% Dynamar/HHM 5502 Blends

Run	Position[a]	Apparent shear rate (s^{-1})	Surface chemical composition (atomic %)			
			C	N	O	F
Control runs after the extrusion of the 0.5% blend						
1	A	500	97.59	0.00	0.34	2.08
2	B	500	99.06	0.00	0.18	0.77
Control runs after the extrusion of the 2.5% blend						
1	A	500	88.33	1.21	6.49	3.97
2	B	500	97.68	0.03	1.09	1.20
3	C	500	93.41	0.93	4.25	1.40

[a] Position A represents the samples obtained in the unsteady-state apparent viscosity region. Positions B and C (B and C obtained in the second and third runs, respectively) represent the samples obtained in the steady-state apparent viscosity regions.

small amount of fluorine on the surface of the HDPE extrudate even at the third run, and the fluorine concentration decreases as the run time increases. In fact, the XPS results of the HDPE extrudates obtained with the "fluoroelastomer-coated" die are very similar to the XPS results of the extrudates of the blends. In both cases, the fluorine concentrations are very small, which suggests adhesive failure at the interface between the HDPE melt and the fluoroelastomer layer formed on the die-wall surface.

The above examples show that XPS can be a powerful tool to provide information that will advance our understanding of fundamental mechanisms in polymer processing.

VIII. CONCLUSIONS

We have described some applications of XPS and SSIMS in the surface characterization of polymer blends and copolymers. With further improvement in the spatial resolution of XPS and a better grasp of the fundamentals of the SIMS process, these two important techniques will contribute significantly to the understanding of the surfaces of polymer blends and copolymers in the years to come.

REFERENCES

1. C. M. Chan. Polymer Surface Modification and Characterization. New York: Hanser, 1994, Ch. 3.
2. F. Garbassi, M. Morra, and E. Occhiello. Polymer Surfaces from Physics to Technology. New York: John Wiley, 1998, Ch. 3.
3. J. D. Andrade. X-ray photoelectron spectroscopy. In: J. D. Andrade, ed. Surface and Interfacial Aspects of Biomedical Polymers. New York: Plenum Press, 1985, Ch. 5.
4. G. Beamson and D. Briggs. High Resolution XPS of Organic Polymers. The Sceinta ESCA300 Database. New York: John Wiley, 1992.
5. D. Briggs. In: Practical Surface Analysis, Vol. 1. Auger and X-ray Photoelectron Spectroscopy. D. Briggs and M. P. Seah, eds. New York: John Wiley, 1990.
6. D. Briggs. In: Practical Surface Analysis, Vol. 2. Ion and Neutral Spectroscopy. D. Briggs and M. P. Seah, eds. New York: John Wiley, 1990.
7. C. M. Chan. Polymer Surface Modification and Characterization. New York: Hanser, 1994, Ch. 4.
8. J. C. Vickerman, A. Brown, and N. M. Reed. Secondary Ion Mass Spectrometry, Principle and Applications. Internat. Ser. Monographs on Chemistry, No. 17. Oxford: Oxford Science, 1989.
9. A. W. Czanderna and D. M. Hercules, eds. Ion Spectroscopy for Surface Analysis. New York: Plenum Press, 1991.
10. J. C. Vickerman. Secondary ion mass spectrometry—the surface mass spectrometry.

In: J. C. Vickerman, ed. Surface Analysis, The Principal Techniques. New York, John Wiley, 1997, Ch. 5.

11. A. Benninghoven, F. G. Rüdenauer, and H. W. Werner. Secondary Ion Mass Spectrometry, Basic Concepts, Instrumental Aspects, Applications and Trends, New York, John Wiley, 1987.

12. L. T. Weng, P. Bertrand, W. Lauer, R. Zimmer, and S. Busetti. Quantitative surface analysis of styrene-butadiene copolymers using time-of-flight secondary ion mass spectrometry. Surface Interface Anal. 23:879–886, 1995.

13. B. Briggs and B. D. Ratner. A semi-quantitative SIMS analysis of random ethyl methacrylate: hydroxyethyl methacrylate copolymer films. Polym. Comm. 29:6–8, 1988.

14. L. Lianous, C. Quet, and T. M. Duc. Surface structural studies of polyethylene, polypropylene and their copolymers with ToF SIMS. Surface Interface Anal. 21:14–22, 1994.

15. D. Briggs, I. W. Fletcher, S. Reichlmaier, J. L. Agulo-Sanchez, and R. D. Short. Surface morphology of a PVC/PMMA blend studied by ToFSIMS. Surface and Interface Anal. 24:419–421, 1996.

16. Q. S. Bhatia and M. C. Burrell. Static SIMS study of miscible blends of polystyrene and poly(vinyl methyl ether). Surface Interface Anal. 15:388–391, 1990.

17. L. T. Weng, T. Smith, J. Feng, and C.-M. Chan. Morphology and miscibility of blends of ethylene-tetrafluoroethylene copolymer/poly(methyl methacrylate studied by ToF SIMS imaging. Macromolecules 31:928–932, 1998.

18. H. van der Wel, F. C. B. M. van Vroonhoven, and J. Lub. Surface modification of polycarbonate by U. V. light as studied by ToF-SIMS. Polymer 34:2065–2071, 1993.

19. F. M. Petrat, D. Wolany, B. C. Schwede, L. Wiedman, and A. Benninghoven. In situ ToF-SIMS/XPS investigation of nitrogen plasma-modified polystyrene surfaces. Surface Interface Anal. 21:274–282, 1994.

20. I. W. Fletcher, M. Davies, and D. Briggs. Surface modifications introduced to a polytetrafluoroethylene-filled polycarbonate compound by dry sliding against steel as revealed by imaging XPS. Surface Interface Anal. 18:303–305, 1992.

21. R. D. Boyd and J. P. S. Badyal. Silent discharge treatment of immiscible polystyrene polycarbonate polymer blends surfaces. Macromolecules 30:3658–3663, 1997.

22. J. B. Lhoest, P. Bertrand, L. T. Weng, and J. L. Dewez. Combined time-of-flight secondary ion mass spectrometry and X-ray photoelectron spectroscopy study of the surface segregation of poly(methyl methacrylate) (PMMA) in bisphenol A polycarbonate/PMMA blends. Macromolecules 28:4631–4637, 1995.

23. S. Affrossman, P. Bertrand, M. Hartshorne, T. Kiff, D. Leonard, R. A. Pethrick, and R. W. Richards. Surface segregation in blends of polystyrene and perfluorohexane double end capped polystyrene studied by static SIMS, ISS, and XPS. Macromolecules 29:5432–5437, 1996.

24. N. M. Patel, D. W. Dwight, J. L. Hedrick, D. C. Webster, and J. E. McGrath. Surface and bulk phase separation in block copolymers and their blends Polysulfone/polysiloxane. Macromolecules 21:2689–2696, 1988.

25. H. Inoue, A. Matsumoto, K. Matsukawa, A. Ueda, and S. Nagai. Surface characterization of fluoroakylsilicone-poly(methyl methacrylate) block copolymers and their PMMA blends. J. Appl. Polym. Sci. 40:1917–1938, 1990.

26. H. Inoue, A. Matsumoto, K. Matsukawa, A. Ueda, and S. Nagai. Surface characteri-
     zation of fluoroakylsilicone-poly(methyl methacrylate) block copolymers and their
     PMMA blends. J. Appl. Polym. Sci. 41:1815–1829, 1990.
27. X. Chen, J. A. Gardella, Jr., T. Ho, and K. J. Wynne. Surface composition of a series
     of dimethylsiloxane-urea-urethane segmented copolymers studied by electron spec-
     troscopy for chemical analysis. Macromolecules 28:1635–1642, 1995.
28. L. Li, C.-M. Chan, and L. T. Weng. Effects of the sequence distribution of poly(acry-
     lonitrile-butadiene) copolymers on the surface chemical composition as determined
     by XPS and dynamic contact angle measurements. Macromolecules 30:3698–3700,
     1997.
29. F. P. Green, T. M. Christensen, T. P. Russell, and R. Jerome. Equilibrium surface
     composition of diblock copolymers. J. Chem. Phys. 92:1478–1482, 1990.
30. P. L. Kumler, H. L. Matteson, and J. A. Gardella, Jr. Surface segregation in multi-
     component polymers: poly(acrylonitrile-co-methyl methacrylate) as a model system.
     Langmuir 7:2479–2483, 1991.
31. H. Zhuang and J. A. Gardella, Jr. Solvent effects on the surface composition of
     bisphenol A polycarbonate and polydimethylsiloxane (BPAC-PDMS) random block
     copolymers. Macromolecules 30:3632–3639, 1997.
32. L. Li, C.-M. Chan, and L. T. Weng. The effects of specific interactions on the surface
     structure and composition of miscible blends of poly(vinyl alcohol) and poly(N-
     vinyl-2-pyrrolidone). Polymer 39:2355–2360, 1998.
33. S. H. Goh, S. Y. Lee, J. Dai, and K. L. Tan. X-ray photoelectron spectroscopic stud-
     ies of ionic interactions between sulfonated polystyrene and poly(styrene-co-4-
     vinylpyridine). Polymer 37:5305–5308, 1996.
34. X. Zhou, S. H. Goh, S. Y. Lee, and K. L. Tan. Interpolymer complexation between
     poly(vinylphosphonic acid) and poly(vinylpyridine)s. Polymer 38:5333–5338, 1997.
35. X. Zhou, S. H. Goh, S. Y. Lee, and K. L. Tan. X-ray photoelectron spectroscopic
     studies of interactions between poly(p-vinylphenol) and poly(vinylpyridine)s. Appl.
     Surface. Sci. 119:60–66, 1997.
36. I. Schmidt and K. J. Binder. Model calculations for wetting transition in polymer mix-
     tures. J. Phys. 46:1631–1643, 1985.
37. R. A. L. Jones and E. J. Kramer. The surface composition of miscible polymer blends.
     Polymer 34:115–118, 1993.
38. D. W. Van Krevelen. Properties of Polymers: Their Correlation with Chemical Struc-
     ture; Their Numerical and Prediction from Additive Group Contribution. Amster-
     dam: Elsevier, 1990.
39. G. Beamson and D. Briggs. High Resolution XPS of Organic Polymers: The Scienta
     ESCA 300 Database. John Wiley, Chichester and New York: 1992, p. 96 and p. 192.
40. D. R. Miller and N. A. Peppas. Surface analysis of poly(vinyl alcohol-co-N-vinyl-2-
     pyrrolidone) copolymers by X-ray photoelectron spectroscopy. Macromolecules
     20:1257–1265, 1987.
41. U. Gedde. Polymer Physics. London: Chapman and Hall, 1995, p. 70.
42. J. C. Chauffoureaux, C. Dehennau, and J. van Rijckevorsel. Flow and thermal stabil-
     ity of rigid PVC. J. Rheol. 23:1–24, 1979.
43. C. K. Shih. Rhelogical properties of incompatible blends of two elastomers. Polym.
     Eng. Sci. 16:742–746, 1976.

44. E. M. Antipov, E. K. Borisenkova, V. G. Kulichikhin, N. A. Plate, and A. V. Topchiev. Condis-crystal structure of flexible-chain polymers in polymer blends. Makromol. Chem., Macromol. Symp. 38:275–286, 1990.

45. C.-M. Chan, A. Nixon, and S. Venkatraman. Effects of the low-surface component on the viscosity of poly(ether ether ketone) and polytetrafluoroethylene blends. J. Rheol. 36:807–820, 1992.

46. C.-M. Chan and J. Feng. Mechanism for viscosity reduction of polymer blends: blends of fluoroelastomer and high-density polyethylene. J. Rheol. 41:319–333, 1997.

47. C.-M. Chan and S.-H. Zhu. Effect of interfacial failure on the viscosity of high-density polyethylene. Polymers for Advanced Technologies 8:257–260, 1997.

48. B. Trembly. Sharkskin defects of polymer melts: the role of cohesion and adhesion. J. Rheol. 35:985–998, 1991.

49. C. W. Stewart and J. M. Dealy. Technical note: new information on the mechanism of action of the adhesion promoter reported in "Wall slip of molten high density polyethylene" [J. Rheol. 35:497, 1991]. J. Rheol. 36:967–969, 1992.

50. S. Nam. Mechanism of fluoroelastomer processing aid in extrusion of LLDPE. Intern. Polym. Processing 1:98–101, 1987.

51. S. G. Hatzikiriakos and J. M. Dealy. Wall slip of molten high density polyethylene II. Capillary rheometer studies. J. Rheol. 36:703–741, 1992.

52. S. G. Hatzikiriakos, C. W. Stewart, and J. M. Dealy. Effect of surface coatings on wall slip of LLDPE. Intern. Polym. Processing 8:30–35, 1993.

16

Emulsion Models in Polymer Blend Rheology

René Muller
European Engineering School for Chemistry, Polymers and Materials (ECPM), Strasbourg, France

I. INTRODUCTION

One possibility in the design of new polymeric materials with improved properties is the blending of commercially available polymers. Most polymers pairs are immiscible (1), and for a two-phase system, depending on the blend composition, the minor phase is usually dispersed in the form of inclusions of different sizes in a matrix formed by the major phase. At rest, these inclusions are of spherical shape, which minimizes interfacial energy. In the following we restrict ourselves to this type of dispersion, although more complex morphologies, like cocontinuous structures, can be obtained in a given composition range that depends on the viscosity ratio of the blend components (2–4).

The processing and ultimate properties of multiphase blends depend not only on the corresponding properties of the blend components but also on the size distribution of inclusions and on the interfacial properties. It is therefore of great importance to control the final morphology in an industrial blending process as well as to characterize as precisely as possible the size distribution and interfacial tension of a given polymer blend.

Recent advances in the field of polymer blend rheology, based on the use of emulsion models, allow us to derive quantitative relationships between linear rheological properties on the one hand and morphological and interfacial properties on the other hand. This for instance means that rheological data can in principle

451

be used to evaluate the average particle size and/or the interfacial tension. Such an indirect method may be of particular interest if one considers the experimental problems encountered with the pendant drop or related techniques (5). Alternatively, the deformation of a dispersed drop during flow may be calculated from the same type of model as a function of its size, the rheological properties of both components, and interfacial tension.

The rheology of heterophase media has been extensively studied from a theoretical point of view for several decades. Einstein (6,7) considered the case of suspensions of undeformable spheres in a Newtonian matrix. Taylor (8,9) considered emulsions of two Newtonian liquids, the drops being assumed to remain spherical. The case of Newtonian liquids and deformable drops was later solved by Oldroyd (10,11), whereas Choi and Schowalter considered finite concentration effects (12).

Comparisons of such emulsion models with viscosity or normal stress data obtained from steady shear flow experiments remained difficult since the morphology of the emulsions may change during time under the effect of flow. The idea of using experimental data from linear oscillatory flow for the comparison with emulsion model predictions was first suggested by Scholz et al. (13) and illustrated with data on polypropylene/polyamide (PP/PA6) blends. As a matter of fact, the dynamic moduli G' and G'' can be easily calculated from most emulsion models. Moreover, small amplitude oscillatory strains do not affect the size distribution of dispersed droplets, contrary to steady flows, for which breakup and coalescence phenomena may change the morphology significantly.

II. THE PALIERNE MODEL OF VISCOELASTIC EMULSIONS (14)

A. Theoretical Expression of the Linear Complex Shear Modulus

The most general form of the Palierne model describes the linear viscoelastic response of a dispersion of inclusions that are spherical in the stress-free state. The model may account for viscoelastic behavior of both matrix and inclusions, effects of finite concentration, distribution of size and composition of the inclusions, as well as interfacial tension effects. More precisely, the interfacial tension is considered to be the sum of a static term α (the usually considered equilibrium interfacial tension) and two other frequency dependent contributions $\beta'(\omega)$ and $\beta''(\omega)$ related respectively to the relative variation of local area and to the shear deformation of the interface without change of area. The last two terms then allow us to account for the presence of an interfacial agent like a block copolymer at the interface, whose concentration may locally change with the variations of area induced by the drop deformation, and which may confer to the interface a nonzero shear modulus, due for instance to entanglements between neighboring block copolymer molecules.

In the dilute case, where interactions between the deformations of neighboring particles are neglected, the complex modulus $G^*(\omega)$ of the emulsion can be written as a function of the particle radius R, the volume fraction ϕ of inclusions, the complex moduli $G_m^*(\omega)$ and $G_d^*(\omega)$ of matrix and dispersed inclusions, and the interface properties including the static interfacial tension α and the dynamic contributions $\beta'(\omega)$ and $\beta''(\omega)$:

$$G^*(\omega) = G_m^*(\omega) \left(1 + \frac{5}{2} \phi \, \frac{E(\omega)}{D(\omega)} \right) \tag{1}$$

where

$$E = 2(G_d^* - G_m^*)(19G_d^* + 16G_m^*) + \frac{8\alpha}{R}(5G_d^* + 2G_m^*)$$

$$+ \frac{48\beta'\alpha}{R^2} + \frac{2\beta'}{R}(23G_d^* - 16G_m^*) + \frac{4\beta''}{R}(13G_d^* + 8G_m^*) \tag{2}$$

$$D = (2G_d^* - 3G_m^*)(19G_d^* + 16G_m^*) + \frac{40\alpha}{R}(G_d^* + G_m^*)$$

$$+ \frac{48\beta'\alpha}{R^2} + \frac{32\beta''(\alpha + \beta')}{R^2} + \frac{2\beta'}{R}(23G_d^* + 32G_m^*)$$

$$+ \frac{4\beta''}{R}(13G_d^* + 12G_m^*) \tag{3}$$

In Eqs. (2) and (3), all frequency-dependent quantities like E, D, G_m^*, G_d^*, β', and β'' have to be taken at the same frequency ω.

Palierne (14) was able to take into account finite concentration effects, namely the fact that the strain seen by a given particle is not the macroscopic strain but is modified by the deformation of neighboring particles. The final result of his calculation, based on the Lorentz sphere method in electricity, amounts to

$$G^*(\omega) = G_m^*(\omega) \left(\frac{1 + \dfrac{3}{2}\phi\,\dfrac{E(\omega)}{D(\omega)}}{1 - \phi\,\dfrac{E(\omega)}{D(\omega)}} \right) \tag{4}$$

which indeed reduces to Eq. (1) (dilute case) at small ϕ values.

Polydispersity in size of the dispersed particles can simply be taken into account in the model: for a distribution of particle radius where ϕ_i is the volume fraction of particles of radius R_i, Eqs. (1) and (4) become respectively

$$G^*(\omega) = G_m^*(\omega) \left(1 + \frac{5}{2} \sum_i \phi_i \, \frac{E(R_i, \omega)}{D(R_i, \omega)} \right) \tag{5}$$

and

$$G^*(\omega) = G_m^*(\omega) \left(\frac{1 + \dfrac{3}{2} \sum_i \phi_i \dfrac{E(R_i, \omega)}{D(R_i, \omega)}}{1 - \sum_i \phi_i \dfrac{E(R_i, \omega)}{D(R_i, \omega)}} \right) \tag{6}$$

Additivity of the $\phi E/D$ terms in the above expressions can be extended to a distribution in blend composition, where for instance two or more materials with different viscoelastic and interfacial properties are homogeneously dispersed in the same matrix. Equation (6) could thus be written as

$$G^*(\omega) = G_m^*(\omega) \left(\frac{1 + \dfrac{3}{2} \sum_{i,j} \phi_{ij} \dfrac{E(R_i, G_d^{*j}(\omega), \alpha_j, \beta_j'(\omega), \beta_j''(\omega))}{D(R_i, G_d^{*j}(\omega), \alpha_j, \beta_j'(\omega), \beta_j''(\omega))}}{1 - \sum_{i,j} \phi_{ij} \dfrac{E(R_i, G_d^{*j}(\omega), \alpha_j, \beta_j'(\omega), \beta_j''(\omega))}{D(R_i, G_d^{*j}(\omega), \alpha_j, \beta_j'(\omega), \beta_j''(\omega))}} \right) \tag{7}$$

In Eq. (7), index j corresponds to a summation over the number of dispersed phases, ϕ_{ij} is the volume fraction of particles of phase j with radius R_i, α_j is the static interfacial tension between the matrix and phase j, etc.

The dynamic contributions to the interfacial tension (β' and β'') are not easy to estimate, either theoretically or from experimental data, although Friedrich and coworkers (15) recently identified in compatibilized polystyrene/polymethylmethacrylate (PS/PMMA) blends a long relaxation time phenomenon that may be attributed to interfacial relaxation processes related to the β' and β'' terms. Most attempts to compare dynamic moduli data with the Palierne model have been performed with the assumption of constant interfacial tension ($\beta' = \beta'' = 0$). In this case, the theoretical expression of G^* only contains physical parameters which can in principle be determined by independent measurements: the complex moduli $G_m^*(\omega)$ and $G_d^*(\omega)$ of matrix and dispersed phase (in the case of a binary blend), the interfacial tension α, and the equilibrium blend morphology (ϕ_i, R_i). One obtains

$$G^*(\omega) = G_m^*(\omega) \left(\frac{1 + \dfrac{3}{2} \sum_i \phi_i H_i(R_i, \omega)}{1 - \sum_i \phi_i H_i(R_i, \omega)} \right) \tag{8}$$

with

$$H_i(R_i, \omega)$$

$$= \frac{8\left(\dfrac{\alpha}{R_i}\right)[2G_m^*(\omega) + 5G_d^*(\omega)] + 2[G_d^*(\omega) - G_m^*(\omega)][16G_m^*(\omega) + 19G_d^*(\omega)]}{40\left(\dfrac{\alpha}{R_i}\right)[G_m^*(\omega) + G_d^*(\omega)] + [2G_d^*(\omega) + 3G_m^*(\omega)][16G_m^*(\omega) + 19G_d^*(\omega)]} \tag{9}$$

B. Particular Cases of the Palierne Model

It is interesting to note that Eq. (1) includes several results of the literature as special cases. For rigid spheres dispersed in a viscoelastic matrix ($G_d^*(\omega) \to \infty$), $H_i(R_i,\omega) = 1$ and

$$G^*(\omega) = G_m^*(\omega) \left(\frac{1 + \left(\frac{3}{2}\right)\phi}{1 - \phi} \right) \tag{10}$$

If the matrix is a Newtonian liquid ($G_m^*(\omega) = i\omega\eta_m$), Eq. (10) amounts to Einstein's result for the viscosity of a dilute suspension:

$$\eta = \eta_m \left(1 + \frac{5}{2}\phi \right) \tag{11}$$

Taylor's result (8,9) for the viscosity of an emulsion of undeformable Newtonian droplets in a Newtonian matrix can be obtained by taking $G_m^*(\omega) = i\omega\eta_m$, $G_d^*(\omega) = i\omega\eta_d$, and $\alpha \to \infty$. In the undiluted case, such an emulsion is found to be a Newtonian liquid of viscosity

$$\eta = \eta_m \left(\frac{1 + \frac{3}{2}\phi\dfrac{5p+2}{5p+5}}{1 - \phi\dfrac{5p+2}{5p+5}} \right) \tag{12}$$

where $p = \eta_d/\eta_m$ is the viscosity ratio. Taylor actually obtained the low concentration limit of this expression:

$$\eta = \eta_m \left(1 + \phi\frac{5p+2}{2p+2} \right) \tag{13}$$

In the case of Hookean elastic spheres with shear modulus G, suspended in a Newtonian liquid of viscosity η_m, the complex viscosity η^* of the suspension can be easily expressed if interfacial tension effects are neglected, which is justified whenever $G \gg \alpha/R$ as shown by Eq. (9). In the undiluted case,

$$\eta^*(\omega) = \eta_m \left(\frac{1 + \frac{3}{2}\phi\dfrac{1 - \frac{2}{3}i\omega\tau}{1 + i\omega\tau}}{1 - \phi\dfrac{1 - \frac{2}{3}i\omega\tau}{1 + i\omega\tau}} \right) \qquad \tau = \frac{3\eta_m}{2G} \tag{14}$$

where τ is a characteristic relaxation time. Equation (14) shows that a suspension of elastic spheres in a Newtonian matrix has viscoelastic properties. In the low concentration limit, the above expression reduces to the result by Fröhlich and Sack (16):

$$\eta^*(\omega) = \eta_m \left(1 + \frac{5}{2}\phi\right)\left(\frac{1 + i\omega\frac{3\eta_m}{2G}\left(1 - \frac{5}{2}\phi\right)}{1 + i\omega\frac{3\eta_m}{2G}\left(1 + \frac{5}{3}\phi\right)}\right) \tag{15}$$

whose low frequency limit is Einstein's equation for undeformable spheres. In the more general case of a dispersion of viscoelastic spheres in a viscoelastic matrix where interfacial tension effects can be neglected, the complex modulus takes the form

$$G^*(\omega) = G_m^*(\omega)\left(\frac{1 + \frac{3}{2}\phi\frac{[G_d^*(\omega) - G_m^*(\omega)]}{2G_d^*(\omega) + 3G_m^*(\omega)}}{1 - \phi\frac{[G_d^*(\omega) - G_m^*(\omega)]}{2G_d^*(\omega) + 3G_m^*(\omega)}}\right) \tag{16}$$

which legitimates Dickie's result (17). If the phases are elastic solids ($G_d^*(\omega) = G_d$, $G_m^*(\omega) = G_m$), Kerner's expression (18) for incompressible materials ($\nu = 0.5$) is obtained. It can be noticed that as soon as interfacial tension effects are neglected, the linear rheological properties of the dispersion only depends on the volume fraction ϕ of dispersed particles and not on their radius or radius distribution; see Eqs. (12), (14), and (16).

We consider now emulsions of deformable Newtonian droplets in a Newtonian matrix, where the deformation of the drops is controlled by interfacial tension. If all droplets have the same radius, the complex viscosity of the emulsion can be expressed from Eq. 8 as

$$\eta^*(\omega) = \eta_m \left(\frac{1 + \frac{3}{2}\phi\frac{5p + 2}{5p + 5}}{1 - \phi\frac{5p + 2}{5p + 5}}\right)\left(\frac{1 + i\omega\tau_2}{1 + i\omega\tau_1}\right) \tag{17}$$

where

$$\tau_1 = \frac{R\eta_m}{40\alpha}\frac{(19p + 16)(2p + 3)}{(p + 1)}\left(\frac{1 - 2\phi\frac{p - 1}{2p + 3}}{1 - \phi\frac{5p + 2}{5p + 5}}\right) \tag{18}$$

and

$$\tau_2 = \frac{R\eta_m}{40\alpha}\frac{(19p + 16)(2p + 3)}{(p + 1)}\left(\frac{1 + 3\phi\frac{p - 1}{2p + 3}}{1 + \frac{3}{2}\phi\frac{5p + 2}{5p + 5}}\right) \tag{19}$$

are characteristic relaxation and retardation times. Equation (17) shows that an emulsion of two Newtonian liquids with interfacial tension possesses bulk viscoelastic behavior. In the low-frequency limit, the dispersed drops are less and less deformed by the flow of the surrounding matrix liquid, and the zero-shear viscosity, $\omega = 0$ in Eq. (17), reduces as it should to Taylor's expression for undeformable drops. At small concentrations, Eqs. (17) to (19) simplify to

$$\eta^*(\omega) = \eta_m\left(1 + \frac{5}{2}\,\phi\,\frac{5p + 2}{5p + 5}\right)$$

$$\cdot \left(\frac{1 + i\omega\,\dfrac{R\eta_m}{40\alpha}\,\dfrac{(19p + 16)(2p + 3)}{(p + 1)}\left(1 - \dfrac{3\phi}{10}\,\dfrac{(19p + 16)}{(2p + 3)(p + 1)}\right)}{1 + i\omega\,\dfrac{R\eta_m}{40\alpha}\,\dfrac{(19p + 16)(2p + 3)}{(p + 1)}\left(1 + \dfrac{\phi}{5}\,\dfrac{(19p + 16)}{(2p + 3)(p + 1)}\right)}\right) \qquad (20)$$

which is equivalent to Oldroyd's (10,11) result taken at the first order in concentration. Comparing Eq. (15) and (20) shows that both a suspension of elastic spheres in a viscous matrix and an emulsion of viscous spheres in a viscous matrix show bulk viscoelastic properties. In the former case, the bulk viscoelasticity comes from the elastic nature of the dispersed spheres, whereas in the latter it comes from the interfacial tension, which provides an elastic force tending to restore the deformed drop to its equilibrium spherical shape. In a polymer blend both phenomena are present (viscoelasticity of the phases and interfacial tension) and contribute in a coupled way to the bulk viscoelastic properties. This coupling is rigorously accounted for by the Palierne model.

III. INFLUENCE OF EMULSION PARAMETERS ON LINEAR VISCOELASTIC FUNCTIONS

A. Emulsions of Newtonian Liquids

Figures 1 to 3 illustrate the influence of the viscosity ratio p, volume fraction ϕ, and interfacial tension α on the dynamic functions $G'(\omega)$ and $\eta'(\omega)$. The particle radius would influence the shape of the curves in the same way as $1/\alpha$, since the viscoelastic properties only depend on the ratio α/R. In a general way, the storage modulus increases first with frequency as ω^2 (slope of 2 in the logarithmic plots) before it reaches a plateau value. At the same time, the dynamic viscosity η' decreases from a low-frequency Newtonian limit η_0 (the Taylor expression) to a high-frequency Newtonian limit given by

$$\eta_\infty = \eta_m\left(1 + 5\phi\,\frac{p - 1}{2p + 3}\right) \qquad (21)$$

in the low-concentration limit. It is found that $\eta_\infty > \eta_m$ when $p > 1$ and $\eta_\infty < \eta_m$ when $p < 1$, whereas $\eta_0 > \eta_m$ whatever the value of the viscosity ratio.

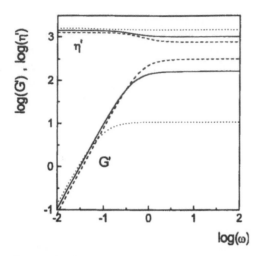

Figure 1 Real part of complex modulus (G' in Pa) and complex viscosity (η' in Pa·s) for an emulsion of two Newtonian liquids according to Eq. (20). $\eta_m = 10^3$ Pa·s, $R = 10$ μm, $\phi = 0.2$, $\alpha = 10^{-2}$ N/m. $p = 0.1$ (dashed lines), $p = 1$ (solid lines), $p = 10$ (dotted lines).

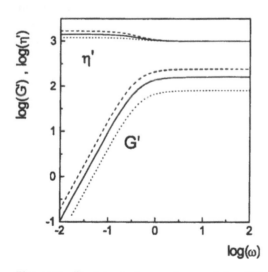

Figure 2 Real part of complex modulus (G' in Pa) and complex viscosity (η' in Pa·s) for an emulsion of two Newtonian liquids according to Eq. (20). $\eta_m = 10^3$ Pa·s, $R = 10$ μm, $\alpha = 10^{-2}$ N/m, $p = 1$. $\phi = 0.3$ (dashed lines), $\phi = 0.2$ (solid lines), $\phi = 0.1$ (dotted lines).

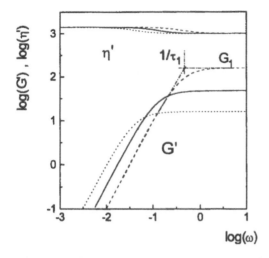

Figure 3 Real part of complex modulus (G' in Pa) and complex viscosity (η' in Pa·s) for an emulsion of two Newtonian liquids according to Eq. (20). η_m = 10^3 Pa·s, $R = 10$ μm, $\phi = 0.2$, $p = 1$. $\alpha = 10^{-2}$ N/mm (dashed lines), $\alpha = 3 \times 10^{-2}$ N/m (solid lines), $\alpha = 10^{-3}$ N/m (dotted lines).

As can be seen in the figures, the elastic part of the complex modulus is much more affected by the emulsion parameters like p, ϕ, or α than the viscous part. This already indicates that viscoelastic measurements may be much more sensitive than viscosity data for the characterization of molten polymer blends.

The plateau modulus for G', G_1, as well as the characteristic relaxation time corresponding to the transition between the two regimes (see Fig. 3) can be calculated from Eqs. (17) to (19). The latter is equal to τ_1 given by Eq. (18) and corresponds to the time required by a deformed droplet to recover its spherical equilibrium shape. This time is typically of the order of seconds for polymer blends, which falls in the frequency range accessible with commercial rheometers. On the other hand, G_1 takes the following simple form

$$G_1 = \frac{20\alpha\phi}{R(2p + 3)^2}\left(1 - 2\phi\,\frac{p - 1}{2p + 3}\right)^{-2} \tag{22}$$

It is found that G_1 increases with volume fraction of dispersed phase and interfacial tension but decreases with increasing viscosity of dispersed drops. On the other hand, the relaxation time τ_1 increases with p, decreases with α, and depends only slightly on ϕ.

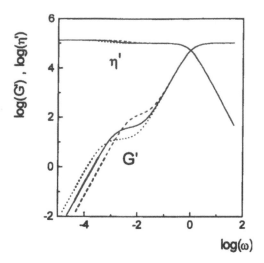

Figure 4   Real part of complex modulus (G' in Pa) and complex viscosity (η' in Pa·s) for an emulsion of two viscoelastic liquids according to Eqs. (8) and (9). Both phases are single relaxation time Maxwell liquids. $\tau_m = \tau_d = 1$ s, $G_m = G_d = 10^5$ Pa, $R = 10$ μm, $\phi = 0.2$, $\alpha = 10^{-2}$ N/m (dashed lines), $\alpha = 3 \times 10^{-2}$ N/m (solid lines), $\alpha = 10^{-3}$ N/m (dotted lines).

B. Emulsions of Viscoelastic Liquids

In Figs. 4 and 5, we have shown the real part of the complex modulus and complex viscosity for a blend of two viscoelastic liquids, whose linear behavior is represented by a Maxwell model. Depending on the values taken by the relaxation times and moduli of both phases, as well as by the emulsion parameters (R, α, and ϕ) a secondary plateau corresponding to G_1 in the Newtonian case may be observed at low frequencies for G'. Figure 5 shows that for some values of the parameters, this plateau can hardly be observed, and for experimental data may be completely hidden in the experimental noise. A detailed discussion on the secondary plateau modulus and associated relaxation times for an emulsion of viscoelastic liquids has been previously published (19). In particular, the conditions for which this plateau can actually be determined experimentally have been examined.

C. Distribution of Particle Radius

Real blends are most often characterized by a nonuniform distribution of particle size. If the distribution is known, it may be taken into account for the calculation of the complex modulus of the emulsion, as shown by Eq. (8). However, it would

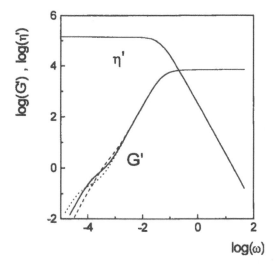

Figure 5 Real part of complex modulus (G' in Pa) and complex viscosity (η' in Pa·s) for an emulsion of two viscoelastic liquids according to Eqs. (8) and (9). Both phases are single relaxation time Maxwell liquids. $\tau_m = \tau_d = 20$ s, $G_m = 5000$ Pa, $G_d = 10^5$ Pa, $R = 10$ μm, $\phi = 0.2$, $\alpha = 10^{-2}$ N/m (dashed lines), $\alpha = 3 \times 10^{-2}$ N/m (solid lines), $\alpha = 10^{-3}$ N/m (dotted lines).

make the model even more straightforward to use if the summation of the H_i terms over the distribution could be replaced by a single term involving some average radius. To determine what average should be taken, we consider the zero-frequency limit of $H_i(\omega)$ for the case of Newtonian phases. The first-order expansion in ω is

$$H_i(\omega) = \frac{5p + 2}{10(p + 1)} - i\omega \frac{R_i \eta_m}{4\alpha} \left[\frac{19p + 16}{10(p + 1)}\right]^2 \qquad (23)$$

It appears that in the terminal zone of an emulsion of Newtonian liquids,

$$\sum \phi_i H_i (\omega, R_i) = \phi H(\omega, \overline{R}_v) \qquad (24)$$

where $\phi = \sum \phi_i$ is the total volume fraction of dispersed phase and $\overline{R}_v = (\sum \phi_i R_i)/\phi$ is the volume-average particle radius. To estimate the error made by using Eq. (24) over the whole frequency range, the influence of polydispersity at constant $\overline{R}_v$ has been tested for various distributions (19,20). It could be shown that Eq. (24) was a good approximation even for viscoelastic phases, providing the polydispersity is not too high (typically lower than 2).

IV.   COMPARISON WITH LINEAR VISCOELASTICITY DATA ON POLYMER BLENDS

A.   Comparison Between Data and Emulsion Model

If all parameters of a given blend (dynamic moduli of phases, morphology, blend composition, and interfacial tension) are known, direct comparison between experimental data and Eqs. (8) and (9) is possible. As described by Graebling and Muller (19,21), the system formed by polydimethylsiloxane/polyoxyethylene (PDMS/POE) blends allows us to separate the fabrication of the POE inclusions from the blending itself. A precise and reliable characterization of the particle size distribution is therefore possible by direct observation of the inclusions by optical microscopy prior to their incorporation into the PDMS matrix. Figures 6 and 7 show that for such a blend, quantitative agreement is obtained between the G' and η' data and the Palierne emulsion model. In Fig. 7, the data are shown in the form of a Cole–Cole plot with linear scales, where the droplet shape relaxation mecha-

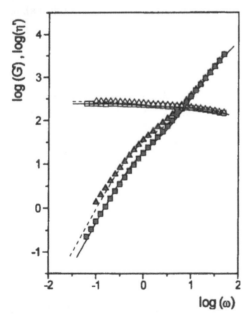

Figure 6   Real part of complex modulus (G' in Pa: ▲, ■) and complex viscosity (η' in Pa·s: △, □) for two PDMS/POE blends with composition in weight 85/15 (▲, △) and 92.5/7.5 (■, □). Details on the polymers are given in Ref. 19. Solid and dashed lines: emulsion model. (From Ref. 19.)

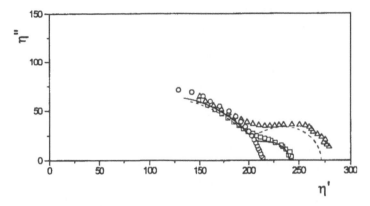

Figure 7 Cole–Cole plot of the data of Fig. 6. △: 85/15 blend; □: 92.5/7.5 blend; ○: pure matrix. Solid and dashed lines: emulsion model for 92.5/7.5 and 85/15 blends, respectively. (From Ref. 19.)

nism observed at low frequencies is seen to be precisely accounted for by the emulsion model. Further comparisons were carried out in the literature on various systems such as PS/PMMA (19,20,22), PS/poly(styrene-co-acrylonitrile) (23), PP/PA6 (13), PS/polyethylene, PP/ethyl vinyl acetate (EVA), and polyethylene terephtalate glycol/EVA (24). Whereas for uncompatibilized blends, the agreement is usually quite satisfactory, discrepancies have been observed for compatibilized blends (25,26) and submicronic cross-linked inclusions with grafted interfaces (22) at high concentrations of dispersed phase. In this case, the approximate Eqs. (8) and (9), where constant interfacial tension is assumed, almost systematically underestimate the G' values at low frequencies. Several reasons may explain these differences. First, if copolymer is present at the interface, the dynamic interfacial effects represented by the β' and β'' terms in the Palierne model can no longer be neglected, and recent data on compatibilized blends (15,27) have shown the presence of a distinct relaxation mechanism (characterized by a relaxation time λ_β) occurring at still lower frequencies than that attributed to the shape relaxation of deformed drops (characterized by τ_1) and that was attributed to dynamic interfacial effects. Another explanation could be the smaller size of inclusions in compatibilized blends or in polymers containing rubber latex particles. The distance between particles may then become small enough to allow aggregation or entanglements between chains grafted at the interface. In this case of course one of the basic assumptions of the Palierne model falls down, namely that interactions between dispersed particles are only of hydrodynamic nature, which means that they occur through the deformation of the matrix phase assumed to be present between all particles and to keep its bulk rheological properties.

B. Use of Viscoelasticity Measurements to Characterize Morphology or Interfacial Properties

It is clear from Eqs. (8) and (9) that if the viscoelastic properties of the phases are known over an appropriate frequency range, information on the morphology or the interfacial tension can in principle be obtained from viscoelastic data of a blend.

For uncompatibilized blends, knowing the particle size distribution allows us to determine the interfacial tension as the value that leads to the best agreement between experimental G' data (the most sensitive parameter) and values calculated from Eqs. (8) and (9). This method has been successfully used in the literature (28–31). For a given blend, it basically requires (a) that the particle size distribution be precisely known and (b) that the low-frequency plateau in G' associated with the shape relaxation of dispersed inclusions appear in the range of frequency and modulus accessible to experimental measurements. A more detailed discussion on the required conditions for the blend parameters can be found in the literature (19). Alternatively, if the interfacial tension is known, a similar approach leads to information on the particle size distribution (20). This method has been used to characterize by small oscillation measurements the morphology obtained after steady shear flow (32,33).

For compatibilized blends, only a static interfacial tension can be simply obtained with the above described method. In general, relaxation processes due to dynamic interfacial effects are difficult to measure since they appear at still lower frequencies. Their precise interpretation appears today to be speculative since two frequency-dependent functions $\beta'(\omega)$ and $\beta'(\omega)$ are then involved in the Palierne model, instead of a single constant (α).

V. ANALYSIS OF DROP DEFORMATION IN STEADY FLOWS

A. Literature Survey and Prediction of Palierne Model

The problem of the deformation of a fluid drop suspended in another fluid undergoing flow is of particular relevance in the case of immiscible polymer blends, for which final properties depend on the blend morphology induced by processing. Since the pioneering work by Taylor (8,9), this problem has been considered by many authors from both theoretical and experimental points of view. For Newtonian systems, the deformation process of a drop of radius R is controlled by the viscosity ratio $p = \eta_d/\eta_m$ and the capillary number $Ca = R\eta_m \dot{\gamma}/\alpha$, where $\dot{\gamma}$ is the shear rate. The capillary number can be understood as the ratio between the hydrodynamic stress $\sigma = \eta_m \times \dot{\gamma}$ acting to deform the drop and the interfacial stress α/R tending to minimize the surface energy and to keep the drop in its equilibrium spherical shape. Above a critical value Ca_c of the capillary number, the droplet is elongated and breaks up into smaller droplets due to growing disturbances at the interface. For capillary numbers below Ca_c, the drop reaches a steady, only slightly deformed (by an amount of the order of Ca) equilibrium shape. The value of Ca_c has been shown (34) to depend strongly on the viscosity ratio p as well as on the type of flow (simple shear or planar extension).

Taylor's approach is basically a first-order theory and is therefore restricted to small drop deformations. It was extended by Chaffey and Brenner (35), who obtained second-order solutions in the drop deformation. The case of general time-dependent flows was considered by Cox (36) and Barthès-Biesel and Acrivos (37). Though restricted to small drop deformations, the Palierne model can be used, as we will see, for arbitrary time-dependent flows and should be well adapted to the case of polymer blends, where the phases are generally viscoelastic fluids.

An original experimental method for measuring the time-dependent droplet deformation in constant strain-rate uniaxial extensional flow has been recently described in the literature (38). The method makes it possible to work directly with polymer melts: it involves observation of the elongated droplets at room temperature after elongational flow at high temperature in the molten state and quenching.

During uniaxial elongational flow, a specimen of initial length L_0 is elongated and the deformation λ is defined as the ratio of the specimen length at a given time t to the initial length. For dilute blends, λ is also the bulk deformation of the surrounding matrix far away from a given drop. The experimental quantity that will be compared to model predictions is the drop deformation λ_d defined as the ratio of the major drop axis (in the flow direction) to the initial drop diameter. Since the theories by Taylor and Palierne are linear, they will only predict the drop shape in the small deformation range. In this case, the Cauchy strains $\lambda - 1$ and $\lambda_d - 1$ are equivalent to the Hencky strains $\varepsilon = \mathrm{Ln}(\lambda)$ and $\varepsilon_d = \mathrm{Ln}(\lambda_d)$.

The Palierne theory, written out for uniaxial harmonic bulk strain, is

$$\varepsilon^* = \lambda^* - 1 = \varepsilon_0 e^{i\omega t} \tag{25}$$

which allows us to calculate the harmonic drop deformation λ_d^* (39):

$$\frac{\varepsilon_d^*}{\varepsilon^*} = \frac{\lambda_d^* - 1}{\lambda^* - 1} = \frac{5G_m^*(19G_d^* + 16G_m^*)}{(2G_d^* + 3G_m^*)(19G_d^* + 16G_m^*) + \dfrac{40\alpha}{R}(G_d^* + G_m^*)} = A^*(\omega) \tag{26}$$

where $G_m^*(\omega)$ and $G_d^*(\omega)$ are taken at the frequency of the harmonic strain. Equation 26 shows that in general, there is a phase difference between the bulk oscillatory strain and that of the drop. To compare the theory with experimental results obtained for a start-up elongational flow at constant strain rate $\dot{\varepsilon}$, we need the relation between λ and λ_d when

$$\varepsilon(t) = \lambda(t) - 1 = \dot{\varepsilon}t \tag{27}$$

This requires us to calculate the Fourier transform of $A^*(\omega)$ and to integrate twice with respect to time (39). Analytical expressions can be obtained for some particular cases.

1. Newtonian Phases Without Interfacial Tension

In this case $G_m^*(\omega)=i\omega\eta_m$ and $G_d^*(\omega)=i\omega\eta_d$ where η_m and η_d are the (constant) viscosities of matrix and dispersed phase. If, on the other hand, $\alpha=0$, Eq. (26) reduces to

$$\frac{\lambda_d^* - 1}{\lambda^* - 1} = \frac{5\eta_m}{2\eta_d + 3\eta_m} = \frac{5}{2p + 3} \tag{28}$$

$A^*(\omega)$ is found to be real and constant, which means that the deformation of drop and matrix are affine for any type of flow. For constant strain rate flow and high viscosity ratios, Taylor's result (9) is recovered.

Equation (28) predicts that highly viscous drops ($p>1$) deform less than the surrounding matrix ($\lambda<\lambda_d$). For isoviscous drop and matrix ($p=1$), the system cannot be distinguished from a single continuous phase when interfacial tension is neglected: as expected, the deformation is then the same in the two phases ($\lambda=\lambda_d$). The prediction of Eq. (28) for low viscosity drops ($p<1$) may appear more surprising: in this case, the drop should deform more than the surrounding matrix with a limiting ratio of drop versus matrix deformation of 5/3 for vanishing drop viscosity. As will be seen below, these predictions are well confirmed by experimental data.

2. Newtonian Phases with Interfacial Tension

With $G_m^*(\omega)=i\omega\eta_m$, $G_d^*(\omega)=i\omega\eta_d$, and $\alpha\neq0$, the response to a start-up flow is easily shown to be a single exponential (39):

$$\lambda_d(t) - 1 = \frac{19p + 16}{16p + 16}\frac{2\dot{\varepsilon}\eta_m R}{\alpha}\left[1 - \exp\left\{-\frac{t}{\tau}\right\}\right] \tag{29}$$

where τ is a characteristic relaxation time for the drop deformation process:

$$\tau = \frac{R\eta_m}{40\alpha}\frac{(19p + 16)(2p + 3)}{(p + 1)} \tag{30}$$

which is also the low concentration limit of the mechanical relaxation and retardation times in Eqs. (18) and (19). The equilibrium droplet elongation at long times is easily shown to be identical to the classical result by Taylor:

$$D = \frac{\lambda_d - \lambda_d^{-1/2}}{\lambda_d + \lambda_d^{-1/2}} = \frac{3}{2}\frac{\eta_m R\dot{\varepsilon}}{\alpha}\frac{19p + 16}{16p + 16} \tag{31}$$

by considering only the first-order terms in ε_d. At short times ($t<<\tau$), Eq. (29) reduces to Eq. (28), which means that in the initial stage of the start-up flow, when the drop deformation is still very small, interfacial stress is negligible compared to hydrodynamic stress.

3. Viscoelastic Phases Without Interfacial Tension

If we assume that both phases are Maxwell fluids with single relaxation times, then

$$G_m^*(\omega) = \frac{i\omega\eta_m}{1 + i\omega\tau_m} \quad \text{and} \quad G_d^*(\omega) = \frac{i\omega\eta_d}{1 + i\omega\tau_d} \tag{32}$$

where η_m, η_d and τ_m, τ_d are respectively the zero-shear viscosities and relaxation times of matrix and dispersed phase. The response to start-up flow can still be calculated analytically if $\alpha = 0$ (39) by

$$\lambda_d(t) - 1 = \dot{\varepsilon}\,\frac{5}{2p + 3}\left[t + \frac{2p}{2p + 3}\,(\tau_d - \tau_m)\left(1 - \exp\left\{-\frac{t}{\tau_v}\right\}\right)\right] \tag{33}$$

where $p = \eta_d/\eta_m$ is the ratio of zero-shear viscosities. The relaxation time τ_v of the drop deformation process for viscoelastic phases,

$$\tau_v = \frac{2p\tau_m + 3\tau_d}{2p + 3} \tag{34}$$

is now an average of the relaxation times of the phases, which is close to that of the matrix ($\tau_v \cong \tau_m$) for highly viscous drops and to that of the inclusion ($\tau_v \cong \tau_d$) for low-viscosity drops.

The model basically predicts that it is easier, in the initial stage of the start-up flow, to deform a viscoelastic drop than a purely viscous drop with the same zero-shear viscosity. On the other hand, a drop is less deformed by a viscoelastic matrix than by a viscous matrix with the same zero-shear viscosity. At long times, the phases eventually behave as Newtonian liquids having the zero-shear viscosities of the Maxwell fluids. In this range, the increase of λ_d-1 as a function of λ-1 is linear with a slope of $5/(2p+3)$, p being the ratio of zero-shear viscosities, but shifted by a constant amount with respect to the Newtonian affine law: deformation of drops having a higher relaxation time than the matrix is underestimated by the Newtonian approximation, whereas that of drops having a lower relaxation time is overestimated. A more detailed discussion can be found in Ref. 39.

When viscoelasticity of the phases and interfacial tension are considered together, or when more than one relaxation time is taken for either phase, the time-dependent drop deformation $\lambda_d(t)$ is a sum of exponentials whose coefficients must be calculated numerically from the interfacial tension, drop radius, and viscosities and relaxation times of the phases (39).

B. Comparison with Experimental Results

We took PS as the continuous transparent phase and high-density polyethylene, HDPE, or PMMA as the dispersed phase. In order to vary the viscosity ratio p over

Table 1   Composition of Blends and Corresponding Viscosity Ratios

Blend	Continuous phase	Dispersed phase	Viscosity ratio at constant strain rate		Viscosity ratio at constant stress
			$0.01\ s^{-1}$	$6 \times 10^{-4}s^{-1}$	
B1	PS1	PE2	0.005		0.003
B2	PS2	PE2	0.03	0.03	0.0025
B3	PS1	PE1	0.1		0.033
B4	PS2	PE1	0.63	0.74	0.5
B5	PS1	PMMA	1.38		1.8
B6	PS2	PMMA	13.0		18.7

For the ratio at constant stress, we took the stress in the matrix at a strain rate of $0.01\ s^{-1}$.
Source: Ref. 40.

a wide range (see Table 1) we selected two PS samples (PS 1 and PS 2), two HDPE samples (PE1 and PE2), and one PMMA sample. The master curves at 150°C for the dynamic viscosity of all samples are given in Fig. 8. The interfacial tension α at 150°C for the pairs PS/PE and PS/PMMA are 5 mN/m and 1 mN/m respectively. All specimens were stretched at 150°C with strain rates of $\dot{\varepsilon} = 10^{-2}$ s^{-1} or $6 \times 10^{-4}\ s^{-1}$. Experimental details can be found in Ref. 38.

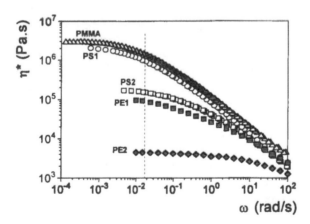

Figure 8   Dynamic viscosity at 150°C as a function of frequency for PS1 (O), PS2 (□), PE1 (■), PE2 (◆), and PMMA (△). The dotted and dashed lines correspond to the experimental values of strain rate: (···) $\dot{\varepsilon} = 6 \times 10^{-4}s^{-1}$, (- - -) $\dot{\varepsilon} = 0.01s^{-1}$. (From Ref. 40.)

1. Results for High Capillary Numbers—Newtonian Analysis

In this first series of experiments, specimens of the different blends were stretched at $\dot{\varepsilon} = 10^{-2}\,\text{s}^{-1}$ up to values of λ between 1.1 and 5. The data in Fig. 9 show that the droplet deformation λ_d increases almost linearly with the macroscopic deformation at least up to $\lambda = 3$. Our experimental results agree with Eq. (28): for $p > 1$ the ratio of droplet over matrix deformation decreases and eventually reaches zero at large p values, whereas for $p < 1$ the ratio of droplet over matrix deformation is higher than 1 and tends to 5/3 at vanishing p. The overall agreement between the small deformation Newtonian model and the experimental data is quite satisfactory, especially at $p \gg 1$ and $p \ll 1$, though the theory is only valid to the first order in

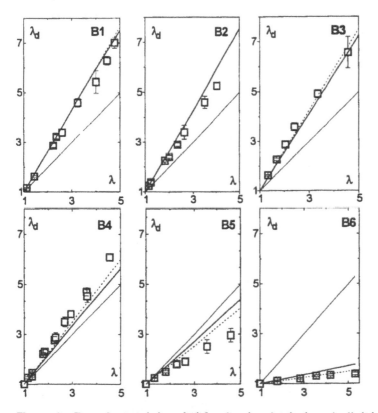

Figure 9 Experimental data ($\square$) for the droplet (λ_d) vs. bulk (λ) deformation in all blends. Thick solid lines: theoretical curves according to Eq. (28) with viscosity ratio taken at constant strain rate. Dashed lines: theoretical curves according to Eq. (28) with viscosity ratio taken at constant stress (see Table 1). Fine solid lines: affine deformation ($\lambda_d = \lambda$). (From Ref. 40.)

deformation and our experiments are in a range of relatively large deformations. If the ratio p of viscosities is taken at constant stress (actually the stress in the matrix far away from the drop) rather than at constant strain rate, the agreement between the model and the data is slightly improved as shown by the dotted lines in Fig. 9.

2. Results for Low Capillary Numbers—Newtonian Analysis

Figure 10 shows the drop deformation in a specimen of blend B2 stretched at $\dot{\varepsilon} = 6 \times 10^{-4} \, \text{s}^{-1}$ up to bulk deformations of $\lambda = 1.82$ and $\lambda = 2.4$, respectively. Since drops of different sizes are present in the sample, a single experiment gives access to different values of the capillary number $Ca = 2R\eta_m\dot{\varepsilon}/\alpha$. The data, plotted as a function of Ca, show a linear increase of λ_d with radius for small droplets ($Ca < 0.5$), whereas the deformation of larger drops ($Ca > 1.5$) becomes independent of their size. Figure 10 also shows the theoretical curves according to the linear Newtonian model [Eq. (29)], which for a given elongation λ predicts

$$\lambda_d - 1 = \frac{19p + 16}{16p + 16} \, Ca \left[1 - \exp\left\{ -\frac{\lambda - 1}{Ca} \frac{80(p + 1)}{(19p + 16)(2p + 3)} \right\} \right] \quad (35)$$

Close agreement is observed between the data and the Newtonian model for the lowest value of λ, which is related to the fact that both components behave nearly

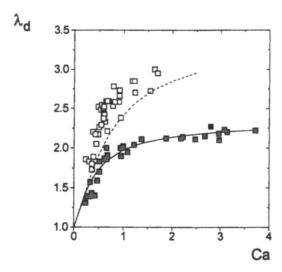

Figure 10 Drop deformation versus capillary number for blend B2 stretched at $\dot{\varepsilon} = 6 \times 10^{-4} \, \text{s}^{-1}$ up to bulk deformations $\lambda = 1.82$ (■) and $\lambda = 2.41$ (□) at 150°C. Solid and dashed lines: Eq. (35) for $\lambda = 1.82$ and $\lambda = 2.41$, respectively. (From Ref. 40.)

as Newtonian liquids for the deformation rate of the test (see Fig. 8). At higher deformations, the model underestimates the drop elongation. This seems to indicate that the range of validity of the linear approximation is more restricted for low capillary numbers than for high capillary numbers.

3. Viscoelastic Analysis

Though a good overall agreement was found between the data and the Newtonian model, some discrepancies remain, in particular for viscosity ratios of the order of unity at high capillary numbers (40). The description of the data can usually be improved by using the linear viscoelastic model instead of the Newtonian approximation. For this purpose, the dynamic mechanical response of each phase has to be fitted with a Maxwell type model (one or several relaxation times for each phase).

However, it should be pointed out that the predictions of the viscoelastic model are quite sensitive to the choice of the relaxation times and moduli. Therefore it is often not obvious, when the linear Newtonian approach fails to describe the data, how to discriminate between the influence of viscoelastic and that of non-linear effects. Finite element numerical simulations should allow us to improve our understanding of this type of problem.

VI. CONCLUSION

Introducing the Palierne model into the rheological studies on polymer blends has represented a major breakthrough in our understanding of the structure–properties relationships of these systems. As a matter of fact, this model allows us to account rigorously for the viscoelastic nature of the phase polymers and for finite concentration effects, particle size distribution, and interfacial tension. For uncompatibilized blends, the agreement with linear viscoelastic data has been found satisfactory and allows us to use dynamic mechanical measurements in the melt for the determination of interfacial tension or morphological parameters. The experimental method is easy, and there is no requirement for special conditioning of the blend samples. For compatibilized blends, the recent data by Friedrich (15,27) demonstrate that viscoelastic data can be sensitive to interfacial relaxation phenomena, like variations of local concentration of compatibilizer due to drop deformation, re-equilibration of these variations by diffusion, or the existence of a surface shear modulus at the interface. However, there is still a need for thorough experiments on well-characterized systems with interfacial agent, which would allow us to improve the interpretation of bulk viscoelastic data.

The case of aggregating particles, or systems with very small interparticle distances, where direct interactions between dispersed particles may occur, does not fall within the assumptions of the Palierne model, which usually underestimates the values of the storage modulus for these systems. However, the difference between the model predictions and the experimental data can be usefully interpreted in terms of interparticular interactions (22). Also, the model is basically a linear model and is therefore restricted to small deformations. Recent attempts have been made to derive more general models accounting for large drop deformations and nonlinear effects like breakup and coalescence (41). The approach by Lee and Park (41) is essentially based on the model by Doi and Ohta (42) and on the Palierne model, but it remains empirical to some extent since, unlike the Palierne model, it contains adjustable parameters.

Finally, emulsion models also give predictions for the deformation of phases during steady or time-dependent flows. The most striking result is that data in uniaxial elongational flow at relatively high particle deformation are in very good agreement with linear models like those of Taylor or Palierne. Numerical simulations are a promising alternative to account both for nonlinear viscoelastic behavior of the phases and nonlinear effects of the drop deformation processes.

REFERENCES

1. LA Utracki. Polymer Alloys and Blends. Munich: Hanser, 1989, pp. 29–124.
2. DR Paul, JW Barlow. Polymer blends (or alloys). J Macromol Sci-Rev Macromol Chem C18:109, 1980.
3. LH Sperling. Interpenetrated Polymer Networks and Related Materials. New York: Plenum, 1981, Ch. 2.
4. A Bouilloux, B Ernst, A Lobbrecht, R Muller. Rheological and morphological study of the phase inversion in reactive polymer blends. Polymer 38:4775–4783, 1997.
5. S Wu. Polymer Interface and Adhesion. New York: Marcel Dekker, 1982, pp. 266–274.
6. A Einstein. Ann. Phys. 19:289, 1906.
7. A Einstein. Ann. Phys. 34:591, 1911.
8. GI Taylor. The viscosity of a fluid containing drops of another fluid. Proc R Soc London A132:41–48, 1932.
9. GI Taylor. The formation of emulsions in definable fields of flow. Proc R Soc London A146:501–523, 1934.
10. JG Oldroyd. The elastic and viscous properties of emulsions and suspensions. Proc R Soc London A218:122–132, 1953.
11. JG Oldroyd. The effect of interfacial stabilizing films on the elastic and viscous properties of emulsions. Proc R Soc London A232:567–577, 1955.
12. SJ Choi, WR Schowalter. Rheological properties of nondilute suspensions of deformable particles. Phys Fluids 18:420–427, 1975.
13. P Scholz, D Froelich, R Muller. Viscoelastic properties and morphology of two-phase polypropylene/polyamide 6 blends in the melt. Interpretation of results with an emulsion model. J Rheol 33:481–499, 1989.

14. JF Palierne. Linear rheology of viscoelastic emulsions with interfacial tension. Rheol Acta 29:204–214, 1990.
15. RE Rieman, HJ Cantow, C Friedrich. Interpretation of a new interface governed relaxation process in compatibilized blends. Macromolecules 30:5476–5484, 1997.
16. H Fröhlich, R Sack. Proc R Soc London A185:415, 1946.
17. RA Dickie. Heterogeneous polymer–polymer composites I. theory of viscoelastic properties and equivalent mechanical model. J Appl Polym Sci 17:45–63, 1973.
18. EH Kerner. Proc Phys Soc 69:808, 1956.
19. D Graebling, R Muller, JF Palierne. Linear viscoelastic behavior of some incompatible blends in the melt. Interpretation of data with a model of emulsion of viscoelastic liquids. Macromolecules 26:320–329, 1993.
20. C Friedrich, W Gleinser, E Korat, D Maier, J. Weese. Comparison of sphere-size distributions obtained from rheology and transmission electron microscopy in PMMA/PS blends. J Rheol 39:1411–1425, 1995.
21. D Graebling, R Muller. Rheological behavior of polydimethylsiloxane/polyoxyethylene blend in the melt. Emulsion model of two viscoelastic liquids. J Rheol 34:193–205, 1990.
22. M Bousmina, R Muller. Linear viscoelasticity in the melt of impact PMMA. Influence of concentration and aggregation of dispersed rubber particles. J Rheol 37:663–679, 1993.
23. W Gleinser, H Braun, C Friedrich, HJ Cantow. Correlation between rheology and morphology of compatibilized immiscible blends. Polymer 35:128–135, 1994.
24. C Lacroix, M Aressy, PJ Carreau. Linear viscoelastic behavior of molten polymer blends: a comparative study of the Palierne and Lee and Park models. Rheol Acta 36:416–428, 1997.
25. B Brahimi, A Ait-Kadi, A Ajji, R Jerôme, R Fayt. Rheological properties of copolymer modified polyethylene/polystyrene blends. J Rheol 35:1069–1091, 1991.
26. Y Germain, B Ernst, O Genelot, L Dhamani. Rheological and morphological analysis of compatibilized PP/PA blends. J Rheol 38:681–697, 1994.
27. RE Riemann, HJ Cantow, C Friedrich. Rheological investigation of form relaxation processes in polymer blends. Polymer Bulletin 36:637–643, 1996.
28. D Graebling, R Muller. Determination of interfacial tension of polymer melts by dynamic shear measurements. Colloids and Surfaces 55:89–103, 1991.
29. D Graebling, A Benkira, Y Gallot, R Muller. Dynamic viscoelastic behavior of polymer blends in the melt. Experimental results for PDMS/POE, PS/PMMA and PS/PEMA blends. Eur Polym J 30:301–308, 1994.
30. H Gramespacher, J Meissner. Interfacial tension between polymer melts measured by shear oscillations of their blends. J Rheol 36:1127–1141, 1992.
31. C Lacroix, M Bousmina, PJ Carreau, BD Favis, A Michel. Properties of PETG/EVA blends: viscoelastic, morphological and interfacial properties, part I. Polymer 37:2939–2947, 1996.
32. I Vinckier, P Moldenaers, J Mewis. Relationship between rheology and morphology of model blends in steady shear flow. J Rheol 40:613–631, 1996.
33. M Minale, P Moldenaers, J Mewis. Effect of shear history on the morphology of immiscible polymer blends. Macromolecules 30:5470–5475, 1997.
34. HP Grace. Dispersion phenomena in high viscosity immiscible fluid systems and ap-

plication of static mixers as dispersion devices in such systems. Chem Eng Commun 14:225–277, 1982.

35. CE Chaffey, H Brenner. A second-order theory for shear deformation of drops. J Colloid Interf Sci 24:258–269, 1967.

36. RG Cox. The deformation of a drop in a general time-dependent fluid flow. J Fluid Mech 37:601–623, 1969.

37. D Barthès-Biesel, A Acrivos. Deformation and burst of liquid droplet freely suspended in a linear shear field. J Fluid Mech 61:1, 1973.

38. I Delaby, B Ernst, Y Germain, R Muller. Droplet deformation in polymer blends during uniaxial elongational flow. Influence of viscosity ratio for large capillary numbers. J Rheol 38:1705–1720, 1994.

39. I Delaby, B Ernst, R Muller. Drop deformation during elongational flow in blends of viscoelastic fluids. Small deformation theory and comparison with experimental results. Rheol Acta 34:525–533, 1995.

40. I Delaby, B Ernst, D Froelich, R Muller. Droplet deformation in immiscible polymer blends during transient uniaxial elongational flow. Polym Eng Sci 36:1627–1635, 1996.

41. HM Lee, OO Park. Rheology and dynamics of immiscible blends. J Rheol 38:1405–1425, 1994.

42. M Doi, T Ohta. Dynamics and rheology of complex interfaces I. J Chem Phys 95:1242–1248, 1991.

17

Microstructure of Multiphase Blends of Thermoplastics

Barbara A. Wood
DuPont Co., Wilmington, Delaware

I. INTRODUCTION

In the human world, expectant parents often wonder about the appearance of their future child. Will he or she have the mother's eyes or hair? The father's nose? When two or more polymers are blended, there may be similar uncertainty: how will they combine? Can the best features of each component be preserved? At what cost? Of all the different techniques used in polymer blend characterization, microstructure visualization using microscopy is seemingly the most literal and direct. The simplicity of assigning chemical descriptions to areas in micrographs of a blend according to image contrast and rationalizing its physical properties from how it "looks" is both appealing and potentially deceptive. How does the casual observer separate issues of processing history from the underlying miscibility of the component polymers? How well does the microscopical sampling technique represent the anisotropic microstructures created by common plastics processing techniques such as extrusion and injection molding? Surface versus bulk differences in three-dimensional parts? Optimizing the end-use properties of a blend and its cost/benefit ratio through control of microstructure requires familiarity with microscopy techniques and their limitations. Together with other techniques that yield microstructure information, such as thermal analysis and scattering techniques, microscopy can help exploit the potential of blends as tailored materials with desirable combinations of appearance, mechanical or other properties, and price.

The types of microscopy currently applied to synthetic polymer blends can be categorized as light, electron, and scanning probe microscopies. Another way to categorize the approaches, which may make more sense in light of recent developments in microscopy, is whether the technique is optical versus nonoptical regarding the role of lenses in image formation. Both the compound light microscope and the transmission electron microscope (TEM) are "optical" in the sense that they use lenses to focus and magnify the images: glass lenses in the case of the light microscope, magnetic lenses in the case of the TEM. Both instruments can function as an optical bench in the sense that diffraction can also be performed. The scanning electron microscope (SEM), a more recent innovation, is considered nonoptical in its image formation. A secondary electron detector used in combination with a rastered incident beam is used to generate images sequentially, even though a column of magnetic lenses is needed to produce the focused incident beam. Newer microscopies such as atomic force microscopy (AFM) based on deflections of fine tips use no lenses in their sequential image formation. Applications of AFM and other new scanning microscopy techniques to polymer blends are still under development (1); a tapping mode of tip operation has been found useful in imaging soft latex blend films without scraping the surface (2). Laser scanning confocal microscopy (LSCM), which produces images of small focal depth to allow three-dimensional reconstruction (3), has been applied to an SBR/PB blend system (4). Finally, scanning thermal microscopy, which detects differences in thermal conductivity or thermal diffusivity in a heated sample, has been applied to polymer blends (5).

The bulk of the microscopy results described in this chapter are from fairly recent TEM images of thermoplastics. TEM imaging of thin sections prepared using ultramicrotomy and selective staining techniques offers an unmatched combination of contrast and spatial resolution for blend characterization. The application of the techniques to industrial blends has been reviewed for TEM (6) and SEM (7). Microscopical observations of blends have been extremely useful in developing our understanding of toughening mechanisms and morphology development during polymer processing, which will be explored in separate sections in this chapter. Thermodynamic issues in blends such as competition between phase separation and crystallization and surface/interface/adhesion are hard to isolate, but some morphological aspects are discussed in Sect. III on miscibility, interfaces, and adhesion. Finally, challenges for the future such as recycling and environmental protection issues which place special demands on polymer morphology are described.

Together with consideration of microscopy techniques, it should also be noted that desktop computers and scanners have become so inexpensive and easy to use that quantitative image analysis, once the domain of experts with dedicated systems, is now (1997) within the reach of many more microscopy users. Calculations that were very recently too computationally painful for the average personal computer are now accomplished quickly with a variety of free and commercially available software packages.

While it has long been recognized that polymer blends are metastable systems, and while the issue of morphology versus phase-separation mechanism has been discussed extensively for systems with miscibility, it is only relatively recently that transient as well as final morphologies in incompatible blends have been thoroughly documented, as in the work of Sundararaj, Macosko, and Shih (8). The field of blend morphology includes in-situ as well as static observations, though the experimental difficulties associated with observation during a dynamic process have been overcome to a greater extent for scattering techniques than for microscopy. Small-angle light scattering (SALS) (9) and synchrotron small-angle x-ray scattering (SR-SAXS) (10) have been used to gain insight into toughening mechanisms. X-ray microscopy using synchrotron radiation to image ultrathin sections of phase-separated polymer blends with 35 nm resolution and elemental information from energy absorption edges has been reported (11). TEM tensile specimen holders have been available for many years, but only recently have notched thin sections of PC/PET blends subjected to tensile pull been successfully imaged by Rightor and coworkers (12). Cracks progressing through versus around PET domains could be distinguished, but the point was made that the state of stress in thin films is different from that of three-dimensional bulk specimens (13).

Morphology can and should address the temperature dependence of failure mechanisms in blends. Many examples of morphology observations are now in the literature for specimens generated by testing at subambient temperatures [see, for example, the work of Sue and Yee (14) on nylon 66/PPO] and high temperature impact [see Wu and Mai (15) on PBT/PC/impact modifier tested at 70°C]. Since toughening mechanisms in polymer blends involve changes in morphology, and recycling polymer blends will inevitably alter their morphology, it is appropriate to broaden the arena of blend microstructure to include not only the present state for a particular material but a "cradle to grave" analysis of all the future possible structures it may assume in its life cycle.

II. BLEND MORPHOLOGY AND TOUGHNESS

The relationship between polymer blend morphology and impact resistance of blends as described in terms of an Izod or other impact test value or ductile/brittle transition temperature has been discussed in the literature since the 1950s (16), yet it is impossible to distill it down to a single generic set of morphology requirements for toughening. Part of the problem in generalizing structure/property relationships can be blamed on the imperfections of the property tests: for example, the Izod test, which fails to separate initiation and propagation steps in crack growth. Will even the most perfect, complete morphology information correlate with a bad test? Possibly not. On the other hand, if a test is suitable for a particular end use, and a window of suitable blend microstructure is found to satisfy that test, it is well worth keeping track of morphology.

Over time, an evolution in thought on polymer toughening can be seen, beginning with the belief that dispersed rubber particles, particularly in an epoxy matrix, act as crack stoppers by absorbing the energy of a growing crack. For ductile polymers, recognition gradually grew that the rubber particles do not arrest crack growth so much as allow the material to fail in a more controlled (and possibly noncatastrophic) manner through multiple deformation mechanisms. Yield occurring in the matrix, and hence the yield stress of the matrix, as well as properties of the rubber that promote rubber particle cavitation and/or debonding, are now seen as important factors in toughening semicrystalline resins.

The development of toughening theories for nylon/rubber blends has been reviewed in detail (17), but toughening is worth summarizing and updating because aspects of morphology are frequently invoked. The work of Souheng Wu addressed the distribution of the rubber phase in toughened blends looking first at particle size (18) and subsequently at interparticle distance and finally a percolation threshold for the onset of toughness. A critical particle size d_c and surface-to-surface interparticle distance T_c (or critical center-to-center interparticle separation L_c) corresponding to the brittle–tough transition in rubber-toughened nylon 66 were related to rubber volume fraction and subjected to a minimum adhesion requirement (19). Interparticle distance T was subsequently renamed matrix ligament thickness τ, with $\tau_c = 0.30$ μm reported for the nylon rubber system (20). Ligaments thicker than τ_c were considered to be unable to yield and absorb energy during crack growth. TEM images were used to draw distinctions between well-dispersed and flocculated rubber particles corresponding to tough versus brittle blends and to show holes from cavitation after impact fracture. The ligament thickness approach led to the conclusion that uniform rubber particle sizes are more effective than heterogeneous distributions in toughening. Margolina and Wu (21) then proposed a percolation model in which the brittle–tough transition corresponded to the onset of a yielding process requiring multiple connectivity of thin ligaments. Their model, which used spherical stress volumes centered around rubber particles in a scaling law with a critical exponent, was received with hostility (22), which was returned (23). However, Wessling (24) suggested that a dispersion/flocculation phase transition model for conductivity in carbon black filled polymers also applies to the brittle–tough transition in nylon/EPDM.

Internal cavitation of rubber particles to relieve triaxial stress has become accepted in the toughening field, yet the idea that rubber particles "blow up" and develop internal cavities is so strange that many investigators have resorted to microscopy as well as dilatometry to prove it. The first TEM images of cavitation in nylon blends were actually of surface replicas prepared from microtomed areas taken near fracture surfaces (25). TEM of actual sections showing internal rubber particle cavitation produced by tensile deformation was published in 1985 by Ramsteiner and Heckmann (26). TEM images of rubber particle cavitation produced by impact fracture were published by Wu (21) in 1988. Voiding in only the larger particles (>0.2 μm) of a notched tensile-impact deformed nylon-6/butadiene rubber blend was shown using TEM of a selectively stained section. Organization of cavitated rubber particles into rows described as dilatation bands by Lazzeri and Bucknall (27) was shown using TEM images of fractured toughened nylon 6. The voided particles promote matrix yield at the crack tip, but some degree of initial yielding without cavitation is also possible.

The usual morphology of neat nylon 6 or 66 crystallized during a molding process is spherulitic, but for toughened compositions, the matrix morphology is generally ignored in favor of rubber particle size distribution. A fresh approach was taken by Muratoglu et al. (28), who used TEM images of nylon 66 crystalline lamellae taken at nylon/grafted elastomer interfaces to show a dependence of the lamellar organization on the elastomer particle spacing. Oriented parallel bundles of lamellae with hydrogen-bonded planes parallel to the rubber–matrix interface, which connect closely spaced particles, were described as having lower slip resistance and lower yield stress than regions of randomly oriented nylon crystals. The toughening mechanism is enabled by easy slip of nylon crystals in the interparticle region followed by rubber particle cavitation. Although the overlap of the oriented lamellae in tough material with ligament thicknesses below 0.3 μm is described as percolation and compared to Wu's critical interparticle distance, the authors stop short of invoking a phase transition.

The effect of elastomers on nylon matrix morphology can actually be seen on a much lower level of resolution than the lamellae, because spherulites may or may not be observed in the matrix. A TEM series (Fig. 1) shows spherulites in neat nylon 66 (a); smaller spherulites in lightly toughened Zytel® 408 based on nylon 66 (b); no spherulites but some retention of order in an experimental nylon 66/Surlyn® 9020 ionomer blend (c); and a very disordered matrix texture in an experimental blend of nylon 66 and EPDR-g-MA (d). This series is showing the increasingly effective suppression of spherulitic order by the various elastomers. The suppression is also reflected by the onset of crystallization observed by DSC in cooling traces. The $T_{c,cool}$ for the EPDR-g-MA toughened blend in (d) is 236°C versus only 230°C for the nylon 66 control (a). The disappearance of spherulites together with the higher onset temperature for nylon crystallization shows the profuse nucleation of the matrix by EPDR-g-MA. Of the series, the only material that would be considered tough by the Izod impact test is (d) at 19.7 ft-lb/in, but since (d) has the smallest average rubber particle size, direct comparisons are inappropriate.

Concentration of a reactive rubber such as SEBS-g-MA can also determine whether a nylon 66 matrix will be spherulitic. Majumdar et al. (29) showed that SEBS-g-MA particle size is essentially constant in the range 1 to 20% due to steric

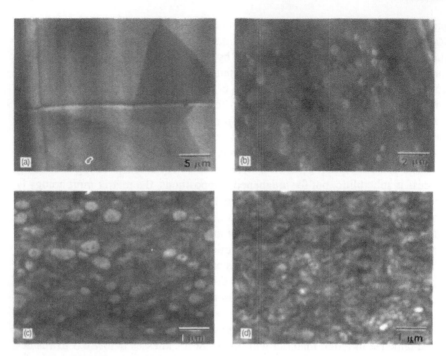

Figure 1 (a) Nylon 66. (b) Zytel® 408 toughened nylon resin. (c) 80% nylon 66/20% Surlyn® 9020 ionomer resin. (d) 80% nylon 66/20% EDPR-g-MA elastomer.

stabilization generated by graft copolymer formation at the interface. Their TEM images clearly show 2–6 μm spherulites in the case of 1% SEBS-g-MA that are absent in the 20% blend for both nylon 6 and nylon 66. Wu and coworkers (30) used WAXD of the matrix of nylon 6 blends to show that a fraction of α-form crystalline structure coexists with γ-form in blends with SEBS-g-MA, and that the ratio of γ- to α-forms decreased with increasing concentrations of maleated SEBS.

Even very similar ionomers can have different effects on the nylon 66 matrix morphology. The effect on fracture mechanics of the blends can be examined in detail using a J-integral fracture toughness test (31) in which stable crack growth in notched specimens is monitored versus the energy required to grow the crack. Blends containing 20% of Surlyn 9320 versus 8320 had fitted parameters C_1 and C_2 for their J-R curves (Table 1). The yield stress values bracket the value of 50.0 MPa reported for an 80/20 blend of nylon 66 with Fusabond® N polymer modifier (32).

Table 1　Fracture Toughness of Nylon/Ionomer Blends

Elastomer type	C_1	C_2	Yield stress (MPa)
Surlyn ® 9320	70	0.71	47
Surlyn ® 8320	56	0.63	53

Courtesy of D. D. Huang, DuPont Co., Wilmington, Delaware.

TEM images (Fig. 2) clearly showed that only the Surlyn® 9320 blend, which had a slightly higher mean particle diameter (0.3 versus 0.2 μm), was spherulitic, with fluffy, indistinct spherulite boundaries compared to the sharp, straight boundaries seen for Zytel® 408 (Fig. 1b.) The lower yield stress value for the spherulitic blend brings to mind the work of Starkweather and Brooks (33), who showed yield stress decreasing with increased spherulite size for dry nylon 66 samples of approximately equal crystallinity.

DSC traces of the two ionomer blends to monitor the crystallinity of the nylon matrix revealed initial melting temperatures and onset of crystallization temperatures on cooling within a degree of each other. Higher magnification TEM images (Fig. 3) of phosphotungstic acid stained sections reveal lamellar organization, showing bundles of fairly straight lamellae for (a) versus a jumbled, disorganized structure for (b), reflecting the difference in spherulitic organization. Due to negative staining in which the heavy metal is selectively absorbed by the amorphous phase of the nylon, the nylon crystals appear as white lines.

Distinct differences were seen in the deformation morphology. TEM of sections microtomed from the whitened region generated in fracture toughness tests show cavitation and fibrils in Fig. 4a, the nylon 66/Surlyn® 8320 blend. The fibrils originate from the nylon ligaments separating adjacent rubber particles. The alignment of the cavitated particles and the noncircular hole shapes suggest matrix shear yielding subsequent to cavitation. In contrast, the blend with Surlyn® 9320 (Fig. 4b) shows hole formation at the particle/matrix interface of the largest holes, as well as internal holes in some of the other particles, but not the ligaments.

On the lamellar level, interaction between crystals and toughener of the type cited by Muratoglu can be seen in many types of nylon blends. The nylon/EPDR-g-MA blend shown in Fig. 1d is shown at higher resolution in Fig. 5. Lamellae viewed edge-on appear to emanate from rubber domains like porcupine needles.

The nylon lamellae are also seen in a blend with functionalized Kraton G1901X triblock polymer. Double staining of the section (phosphotungstic acid followed by RuO vapor) reveals both the nylon crystals and the block copolymer microdomain morphology (Fig. 6.)

Nylon 6 blends with functionalized styrenic block copolymers (34) have been an interesting model system for demonstrating the loss of effectiveness of the copolymer particles as impact modifiers at particle diameters below 0.1 μm. For

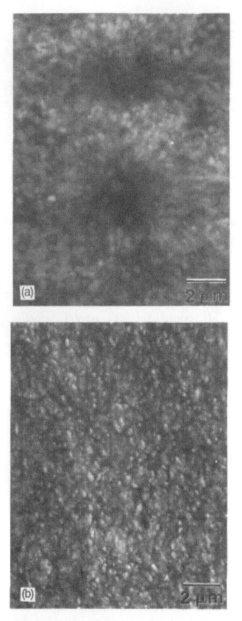

Figure 2   (a) 80% nylon 66/20% Surlyn® 9320 ionomer resin. (b) 80% nylon 66/20% Surlyn® 8320 ionomer resin.

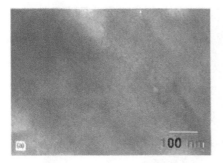 

Figure 3 (a) 80% nylon 66/20% Surlyn® 9320 ionomer resin, phospho-tungstic acid stain. (b) 80% nylon 66/20% Surlyn® 8320 ionomer resin, phos-photungstic acid stain.

retention of tensile properties, however, these finely dispersed particles were preferred. As well as maximum and minimum particle sizes for SEBS-g-MA in nylon 6, Oshinski and coworkers (35) recognized that the nylon matrix crystallinity is affected by the MA content of the elastomer mixture chosen. Heat of fusion ratios relative to neat nylon 6 decreased with increasing SEBS-g-MA content for 20% rubber blends, implying a depressed crystallization rate of nylon-6 due to the grafting reaction. Similar investigations of SEBS-g-MA in nylon 66 (36) resulted in complex elastomer particle shapes as revealed by TEM, with no obvious mini-

 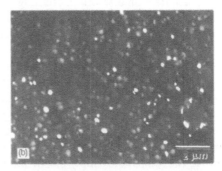

Figure 4 (a) 80% nylon 66/20% Surlyn® 8320 ionomer resin, whitened region after impact fracture test, phosphotungstic acid stain. (b) 80% nylon 66/20% Surlyn® 9320 ionomer resin, whitened region after impact fracture test, phosphotungstic acid stain.

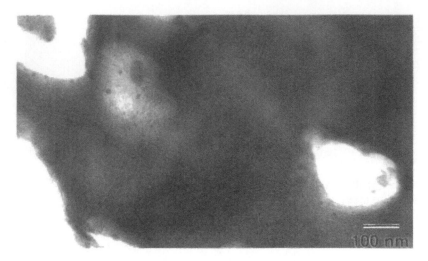

Figure 5   80% nylon 66/20% EDPR-g-MA elastomer, phosphotungstic acid stain.

mum size for toughening. No crystallinity dependence on the reactive elastomer level was seen for nylon 66.

Toughener interaction with matrix morphology can be much milder than the disappearance of spherulites seen in nylon. In polypropylene, spherulites were reduced in diameter but not eliminated by the addition of 20% EPDM (37). The fractional crystallinity of the matrix was unchanged. Optical microscopy of highly crystalline acetal polymer (POM) in blends with ionomers has shown a decrease in spherulite diameter with increasing modifier level, but DSC cooling traces do not suggest any nucleating effect (38). Improved impact properties obtained from adding a 25% neutralized terpolymer were credited to enhanced interfacial adhesion from ion–dipole interactions.

The effect of irregularly shaped elastomer domains on cavitation appears to be minimal. Functionalized Kraton G1901X in nylon 66 is shown in the TEM images (Fig. 7); Fig. 7a shows an undamaged area stained with RuO_4 to darken styrene-rich regions, and 7b shows a region damaged in an Izod impact test and stained with phosphotungstic acid, which affects the nylon matrix. Voids of circular cross section are seen inside a few of the larger elastomer particles contained in 7b. The round holes imply that no massive matrix deformation occurs after cavitation.

TEM imaging of deformed specimens has been applied for blends other than nylon with rather different conclusions. In PC/PE blends, Sue and coworkers (39) sampled a subfracture surface zone by TEM and optical microscopy and con-

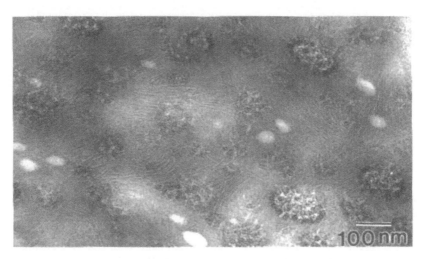

Figure 6 80% nylon 66/20% Kraton G1901X triblock elastomer, phospho-tungstic acid and ruthenium tetroxide stain.

cluded that particle debonding occurs, followed by shear banding in the PC matrix. Synergistic toughening, with favorable interactions between shear yielding, microcavitation, and debonding of star SBS and methacrylate-styrene copolymer, was reported by Yamaoka (40).

More different types of toughening and deformation mechanisms can be seen by widening the arena to thermoset epoxy matrix materials. Fracture surface im-

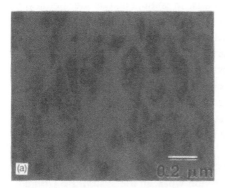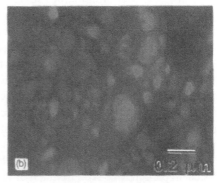

Figure 7 (a) 80% nylon 66/20% Kraton G1901X triblock elastomer, ruthenium tetroxide stain. (b) 80% nylon 66/20% Kraton G1901X triblock elastomer, whitened region after impact fracture test, unstained.

ages of epoxy/CTBN led to a rubber crack-bridging model in the work of Kunz-Douglass et al. (41). A critical comparison of dispersed and highly aggregated core/shell rubber particle morphology in a DGEBA epoxy matrix showed that a connected rubber microstructure obtained by limiting the mixing resulted in a higher fracture toughness and easier propagation of shear bands versus a blend with more nearly uniform distribution (42).

TEM observations of the interaction between crazing and rubber particles of various sizes in HIPS have been useful in explaining the toughening behavior. Okamato and coworkers (43) combined TEM observations with finite element modeling of stress around rubber particles to conclude that Izod impact strength could be improved by preparing HIPS with a bimodal distribution of rubber particle size. The large particles are effective craze initiators; the small particles separated by thin ligaments, while capable of initiating minute crazes, also participate in a craze extension mechanism in which the yield zone around large particles extends to the surrounding small particles and bridges them with crazes. Questions of particle size and second phase volume fraction in HIPS can be challenging in HIPS because of the salami particles: composites with inclusions of matrix polymer. Physical separation via dissolution and centrifugation was compared to image analysis of TEM photos obtained by cutting HIPS sections at four different thicknesses by Maestrini et al. (44).

A polybutadiene reinforced PMMA system that could be prepared in a variety of morphologies including the HIPS type morphology was described by Hill (45). Toughening by lamellar rubber—especially with short, disordered lamellae—resulted in a very rough fracture surface, presumably because the lamellae can redirect the propagating crack, and higher toughness than a comparable sample with multicellular rubber particles.

Relationships between particle size, interparticle distance, and cavitation have been the foci of numerous investigations. In some, such as the model proposed by Dompas and Groeninckx (46), a strict adherence to a minimum particle size for relief of triaxial stress and a prediction that large particles cavitate early in the deformation process while smaller particles cavitate later led to light scattering experiments on PVC/methyl methacrylate-butadiene-styrene graft copolymer blends under strain (47). The stress whitening intensity was interpreted to show that cavitation of the largest particles occurs early and that resistance to cavitation increases for smaller particles. When the same blend was prepared with monodisperse particle size (48), a critical particle size of 150 nm required for cavitation was found.

Not all investigators concerned with toughening have had the single-minded focus on particle size seen in much of the nylon work. In the system PBT/butadiene-co-acrylonitrile rubber, Hourston and coworkers (49) noticed irregularly shaped rubber domains in TEM images of the blends and suggested that the shape of the domains would give rise to much more complex stress fields than spherical particles and would promote shear yielding.

III. MISCIBILITY, INTERFACES, AND ADHESION IN BLENDS

Blends that exhibit partial miscibility offer varied matrix morphologies according to the details of the molecular weight distribution and thermal history. In PC/PBT/core/shell impact modifier blends, the effect of introducing partially miscible PC was described as a decrease in the PBT spherulite size and the modification of amorphous interlamellar regions of the PBT, enhancing ductility and improving low-temperature toughness (50). Details of the PC/PBT phase diagram were proposed by Delimoy et al. (51) based on TEM images, SEM of surfaces selectively etched to remove PC, optical microscopy, and thermal analysis. For some PC-rich compositions, the dispersed phase domain size and the matrix crystallization were linked because of competition between spinodal decomposition and crystallization via homogeneous nucleation. The PBT lamellae that appear as white lines in the matrix of a TEM image (Fig. 8) are visible owing to the contrast obtained by staining the mixed amorphous phase of PC/amorphous PBT with RuO_4 vapor. Staining PBT alone by the same method does not produce the same type of contrast, and the lamellae are not revealed.

Partial miscibility may also play a role in the scratch-resistant PP-EP rubber blends developed by Toyota researchers (52). Ethylene——olefin block copoly-

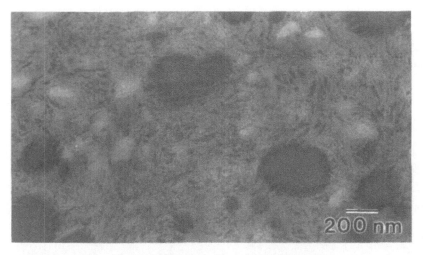

Figure 8   Toughened PC/PBT, ruthenium tetroxide stain.

mers added to reinforce the EPR phase showed the least melting point depression for a composition that crystallized in long lamellar crystals within the EPR. In contrast, the best balance of scratch resistance and tensile properties was obtained when a lower density ethylene copolymer was selected. TEM images did not resolve the crystals clearly, but apparently they were detected by SAXS.

In systems with some miscibility, microscopy of artificially created interfaces between slabs of different polymers has proven useful. Convincing TEM images of diffusion bonding at a straight, sharp PBT/PC interface were generated by spraying a PC solution onto a molded PBT surface (53). The growth of fibrillar PBT crystals across the interface and into the PC phase as a function of annealing time was shown, but debonding of the layers always occurred at the original interface. In experiments on the system PMMA/PVF published by Wu et al. (54), TEM images were used to monitor the movement of sputtered gold particles toward the faster diffusing phase in order to deduce diffusion coefficients. Recently, the interface between a thermoplastic and heavily brominated thermoset was monitored with electron microprobe line scans and images using a wavelength dispersive x-ray spectrometer to follow the bromine (55).

Interestingly, when the existence of an interphase or mixed interface layer between blend components is postulated, the evidence usually does not come from direct observation by microscopy techniques. Wenig and Wasiak (56) calculated an interface thickness in i-PP/EPDM from interfacial tensions after concluding from thermal analysis that movement of the EPDM T_g with concentration was due to i-PP molecules penetrating the EPDM. Wenig and Asresahegn subsequently

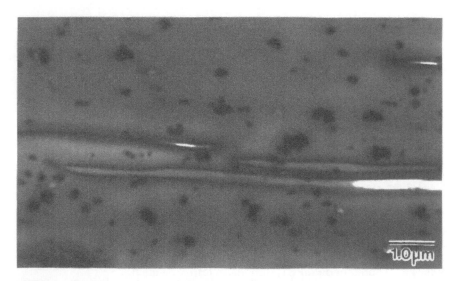

Figure 9 Opaque polyester label stock (3M).

used hot stage optical microscopy to measure spherulite growth rates in the same blend and concluded that the incorporation of EPDM into growing i-PP spherulites affects nucleation more than the growth rate at any particular temperature, and that the enhanced secondary nucleation rate is a consequence of a reduction in the surface free energy of i-PP crystals (57).

In the territory between idealized "blend" interfaces from prepared slabs and compounded molding resins for three-dimensional parts, there are some polyester blends used in film applications. Lamellar morphologies and interfaces parallel to the plane of the film are typically seen. TEM images compare an opaque polyester film contained in a 3M label stock (Fig. 9) with a Toyobo white opaque film (Fig. 10). Both contain TiO_2 particles and a second polymer in addition to the polyester matrix. Mediocre adhesion between what appears to be polyolefin in Fig. 9 and polystyrene in Fig. 10 promote voiding, which contributes to the opacity of the film.

Multicomponent fibers as well as films can have complex multiphase structures and associated internal interfaces. A flash-spun plexifilament fiber is shown with two different staining treatments in Fig. 11. A liquid epoxy mixture of sufficiently low viscosity to enter the pore spaces was used to embed the material, providing support for sectioning after thermal cure.

Morphology can provide indirect but valuable information about the nature of internal interfaces in blends. Sharp versus diffuse interfaces can be distinguished in extreme cases but are difficult to quantify. Sectioning blends for TEM can induce microdebonding when interfaces are weak. The partial debonding is a sample preparation artifact in the sense that it is not a true representation of the three-

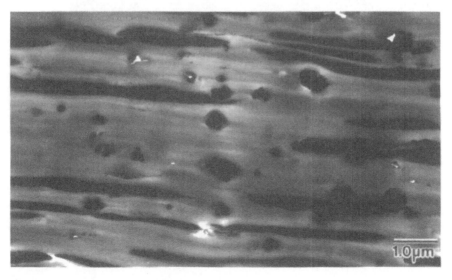

Figure 10 Opaque polyester film (Toyobo)

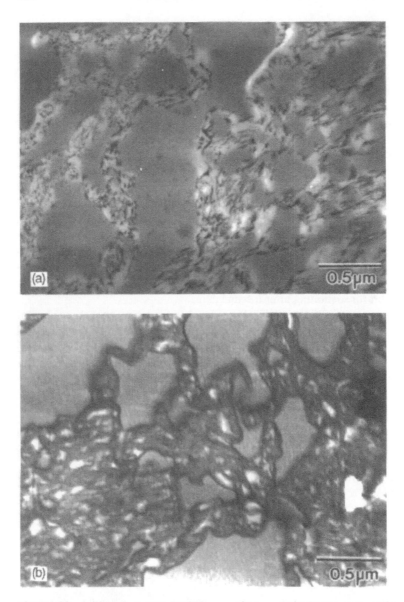

Figure 11 (a) Flash-spun plexifilament fiber, phosphotungstic acid stain. (b) Flash-spun plexifilament fiber, ruthenium tetroxide stain.

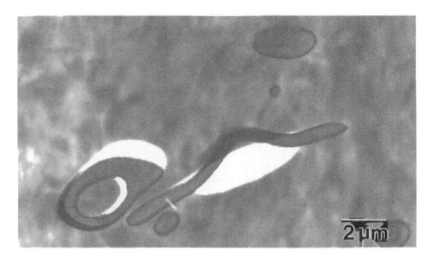

Figure 12   PP/EVOH extruded film sectioned perpendicular to flow direction, ruthenium tetroxide stain.

dimensional solid, but it is highly reproducible. In a PP/EVOH copolymer blend the debonding clearly distinguishes the uncompatibilized blend (Fig. 12) from the compatibilized blend (Fig. 13).

The strong staining of the interfaces between immiscible polymers by RuO_4 is shown in the example above. In a recycled blend of SAN with an ethylene

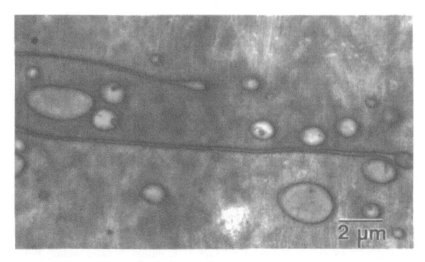

Figure 13   Compatibilized PP/EVOH extruded film sectioned perpendicular to flow direction, ruthenium tetroxide stain.

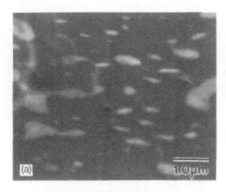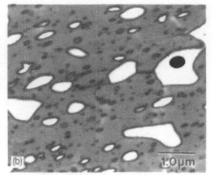

Figure 14 (a) SAN/ethylene copolymer blend, unstained. (b) SAN/ethylene copolymer blend, stained.

copolymer, the contrast difference between an unstained section (Fig. 14a) and a RuO$_4$ stained section (Fig. 14b) is dramatic. Deposition of ruthenium oxide from the RuO$_4$ vapor at interfaces does not necessarily correspond to the segregation of a third stainable species at the interface but rather appears to be a general phenomenon due to the dirt, chain ends, and different chemical potential found at interfaces.

The influence of interfacial contact on properties in immiscible blends has been examined by Leclair and Favis (58). Contraction on cooling of a crystalline matrix on a solidified amorphous dispersed phase results in pseudoadhesion behavior. Tensile modulus matched the theoretical values for perfect adhesion in blends containing up to 30% amorphous polymer. The role of polymeric compatibilizers as microbridges or nanobridges for load transfer is discussed in the work of Armat and Moe (59) for a polyethylene/nylon 6 blend compatibilized using maleic anhydride functionalized styrene-(ethylene-co-butylene)-styrene block copolymer.

Examples of exploiting the interfacial structure of immiscible polymer blends may be found in the blends with designed electrical properties generated by Gubbels and coworkers (60). Using image analysis of optical micrographs of PS/PE blends containing carbon black, specific interfacial area was determined as a function of molding conditions. Selective extraction of PS was performed to reveal the three-dimensional morphology of the PE phase. The best morphology for a conductive material at low carbon content (0.4 wt%) was cocontinuous PS and PE with the carbon black selectively localized at the interface.

IV. PROCESSING EFFECTS ON BLEND MORPHOLOGY

Processing effects on polymer microstructure as revealed by microscopy have been recently reviewed (61). Some similarities can be seen between the morphology work associated with toughened blends and their deformation mechanisms and morphology studies associated with investigations of processing effects and blend rheology. In both cases, there is little doubt about the initial and final morphology of the system, but the sequence of events undergone by the material during deformation or melt mixing can be controversial—and difficult to interrupt for direct observation. For many blends, an important question is whether morphology and its response to processing steps is primarily controlled (a) by the relative viscosities of the melt components or (b) by interfacial tension and interfacial compatibilization by reaction or additives.

Jordhamo and coworkers (62) proposed a simple criterion for predicting the composition at which phase inversion occurs for a binary blend with volume fraction ϕ and viscosity η:

$$\frac{\eta_1}{\eta_2} \cdot \frac{\phi_1}{\phi_2} \sim 1$$

defining criteria for a bicontinuous morphology. Varying both viscosity and volume fraction for a single binary blend could result in a considerable number of microscopy samples, but the Jordhamo relationship does not describe any time-dependent behavior, so monitoring morphology versus time as well as composition in the search for a phase inversion could require a staggering number of observations.

TEM images can provide clues to previous morphologies in the melt state as shown in Fig. 15, a nylon (stained dark)/modifier "bullseye" or "onion" that suggests a lamellar precursor to the dispersed droplet morphology. Nylon inclusions within modifier particles are often interpreted as evidence for previous phase inversion.

Other phase inversion investigations have examined transient as well as final morphologies for one or more compositions. Compounding studies by Shih (63,64) and by Macosko (65,66) and coworkers have elucidated the mechanisms of morphology development during processing. The order of melting of the components is important, because a minor phase that melts first can coat solid pellets of the major components and form a matrix phase initially but subsequently become dispersed after melting of the other component and phase inversion. Selective dissolution of one phase by Soxhlet extraction followed by SEM imaging of the remaining structure has been used extensively in these morphology development studies.

A classic stereology approach to determining surface-to-volume ratios from image analysis, the Chalkey parameter determination, has been useful in characterizing incompatible blend morphologies. Quintens et al. (67) used the Chalkey parameter with SEM images of PC/SAN to quantify dramatic changes in microstructure with annealing time.

Like droplet morphologies, immiscible blends in bicontinuous morphologies can be stabilized by the addition of interfacially modifying block copolymers.

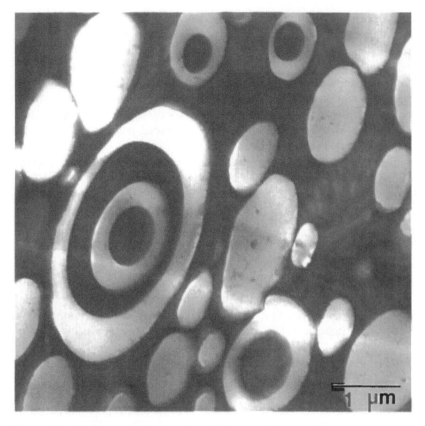

Figure 15 Nylon/experimental modifier blend, phosphotungstic acid stain.

Mekhilef (68) and coworkers added SEBS copolymer to PS/PE, which lowered interfacial tension fivefold and prevented the breakup of the PE phase on annealing. Nylon 6/EPR blends with maleic anhydride functionalized EPR or SEBS even appear to obey the Jordhamo equation, because the highest molecular weight PP generated a blend with a nylon 6 continuous phase, the lowest molecular weight PP yielded a PP continuous phase, and intermediate molecular weight led to a co-continuous structure (69).

Envelope formation in ternary blends in which a dispersed phase is encapsulated by another polymer was explained in terms of interfacial effects by Hobbs et al. (70). Surface and interfacial tensions of polymer melts were used in Harkin's equation for the spreading coefficient λ_{ij} applied to complex polymer blends. TEM images of various blends of PMMA, PC, PBT, PS, and SAN verified predictions that encapsulation would occur for negative values of λ_{ij}. The analysis was subsequently extended to blend systems containing preformed impact modifier particles (71). Interfacial segregation of preformed core/shell particles added to PC/SAN is treated similarly in the work of Debier et al. (72) and correlated with improved impact resistance.

A characteristic surface defect in injection-molded PC/ABS blends described by Hobbs (73) was attributed to oscillation of the melt front from the top and bottom surfaces of the mold, producing out-of-phase streaks on opposite sides of the part. Once understood, the visual defect could be addressed by modifying the molding conditions.

When incompatible blends are molded, morphology can vary drastically throughout the part according to distance from the gate. An end-gated tensile bar of a polyolefin/SEBS/polyester blend showed larger dispersed phase particles near the gate (Fig. 16a) versus the end of the bar opposite the gate (Fig. 16b). The sectioning direction was perpendicular to the flow. The polypropylene matrix freezes and arrests the coalescence of SEBS copolymer-wrapped polyester opposite the gate to produce submicron domains. Particle sizes in the precompounded pellets used to make the bar were up to an order of magnitude smaller. In an LCP/polyester elastomer blend, Jang and Kim (74) used SEM to show differences in the morphology of injection-molded test specimens according to gate position in the mold. Direct gated bars had higher tensile strength compared to bars molded through the side gate at the same temperature across the entire composition range of the blend.

Melt spinning of blends into fibers can result in droplet or fibril morphologies depending on the processing conditions and the deformability of the dispersed phase. Padsalgikar and Ellison (75) used a reduced capillary number κ^* to describe four regions of droplet behavior: droplets not deforming, droplets deforming but not breaking, droplets deforming and possibly breaking up, and droplets deforming affinely into stable filaments. Morphologies were verified experimentally using SEM of fracture surfaces of PS/PP fibers.

Extrusion, molding, and spinning are not the only processing techniques for blends. Photopolymers can be prepared with specific morphologies that impart useful optical properties or permit novel means of fabricating finished products for printing and patterning. Many photopolymer systems might be rigorously classified as interpenetrating networks (IPNs) or semi-IPNs (76) rather than blends since one of the polymers is photopolymerized in the presence of the other to produce a fine phase-segregated morphology. Photocure of a photopolymerizable monomer dispersed in a polymeric binder is a type of processing, and a phase segregated polymer blend morphology is frequently the result. Growth in spherical monomer-rich domains with exposure to a mercury metal halide source was reported in a poly(styrene/sec-butyl maleate/trimethylolpropane) triacrylate binder/monomer system by Maerov (77). Polymer dispersed liquid crystals intended for holographic gratings (78) and polymer-stabilized liquid crystals for

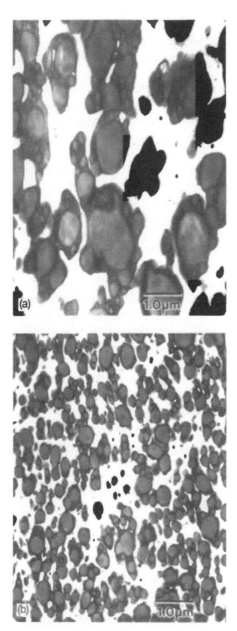

Figure 16 (a) Polyolefin/SEBS/polyester blend, near gate end of injection molded bar, sectioned perpendicular to flow direction, ruthenium tetroxide stain. (b) Polyolefin/SEBS/polyester blend, opposite gate end of injection molded bar, sectioned perpendicular to flow direction, ruthenium tetroxide stain.

flat-panel displays (79) are generally polymers mixed with low-molecular-weight liquid crystals, but a polymer/polymer blend system designed for diffraction gratings produced in an electric field has also been reported (80). In the case of the holographic transmission gratings recorded in polymer dispersed liquid crystals (PDLC), phase segregation of photomonomer and a liquid crystal and the details of the domain morphology controlled the switching behavior of the hologram. The grating pattern written into the material produces an anisotropically cured system in which LC diffusion is forced by the spatially varying molecular weight in the acrylate undergoing photocure.

V. FUTURE CHALLENGES

Aging effects on polymers have been studied extensively since the 1960s. A major change underway in the use of polymers, though, adds a completely new dimension to polymer aging: reincarnation, or the expectation that many or most polymer products will be recycled. Postconsumer recycling (PCR) offers cheap, plentiful feedstocks for making blends. Multilayer packaging films and bottles produced by coextrusion and coinjection-molding techniques will generate blends automatically on recycling unless extraordinary steps such as pyrolysis or other conversion back to monomers is attempted. Blends themselves will also have to be recycled or reclaimed to some extent. The anticipated effect of recycling on the morphology and properties of Noryl GTX nylon 66/PPO alloys was described from observations of changes induced by repeated extrusions and abusive molding (81) (allowing up to 20 minutes of extra holdup time in an injection-molding machine). PPO domain size histograms derived from image analysis of transmission optical micrographs were generated for unaged versus aged, abusively molded Noryl. Expectations that blends must retain some of their properties in recycled form introduces a whole new layer of competitive pressure on blends producers.

The recycling of multilayered packaging film produced by coextrusion will presumably necessitate giving up the layered structure, at least if melting and mixing steps are involved. Figure 17 shows a multilayer HIPS-based rigid food container that has been melted with polymeric compatibilizers. Little or none of the original layer structure is retained.

Recycling is not the only area in which environmental progress makes new demands on blends and blend morphology. The ban on potentially ozone-depleting chlorofluorocarbon gases used as refrigerants and blowing agents has prompted the reformulation of blends used in refrigerators. Blisters and cracks in HIPS used in refrigerator interiors induced by the replacement blowing agent are shown in SEM images by Maestrini et al. (82). In addition to mechanical properties and barrier properties with sufficient chemical resistance, some polymer

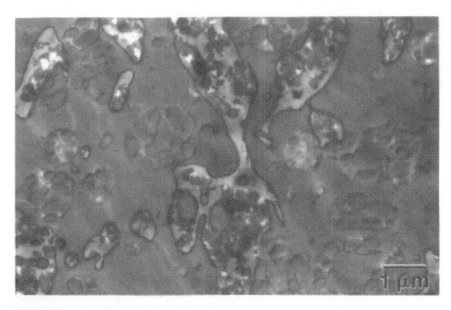

Figure 17 Compatibilized HIPS-based multilayer recycled packaging, ruthenium tetroxide stain.

blends need to have morphologies that are optimized for their interaction with light. For many outdoor applications, blends are expected to resist photodegradation. Assuming the part is opaque, this means that sufficient levels of UV stabilizers added must end up at or near the surface of a finished part. However, an additive intended for one component of a blend may migrate into the "wrong" phase and fail to confer protection.

In 1984, Ulrich Seiffert and Petr Walzer of the Research Division of Volkswagenwerk in Wolfsburg published *The Future for Automotive Technology* (83), a prediction of the automobiles that will be built in the year 2000. Included in a section on alternative materials are ambitious metals replacements such as axles and connecting rods from fiber-reinforced plastics, but there is no mention of a plastic automotive fuel tank. Subramanian (84,85) has reported a platelet or lamellar morphology of nylon in polyethylene in barrier blends intended for prevention of evaporation of hydrocarbons though fuel tanks. The laminar morphology of barrier blends of HDPE with as little as 4% polyamide (to reduce permeability) has made blow-molded fuel tanks practical (86). Ultrasound analysis serves as a nondestructive probe that can be correlated with microscopy of the desired morphology. Similar morphologies were induced in PET/ethylene-methacrylic acid copolymer blends to raise barrier properties (87).

Figure 18   Nylon/EVOH barrier film, phosphotungstic acid stain.

Nylon/EVOH barrier blend films (Fig. 18) have an intrinsic layered structure that could be degenerated by melting with loss of orientation. Presumably, loss of all or most of the barrier efficiency would accompany the loss of the layers. The EVOH lamellae grow perpendicular to the elongated domains, which have molecular orientation imposed by the processing direction.

Blends may be designed thoughtfully or totally by accident, but morphology analysis is laborious and hence never accidental. The study of blend morphology can include any system that will become a blend at some point in its life cycle, regardless of whether it originates in the form of a blend. Waste may be comingled at various stages, and multilayer rigid and flexible packaging that is melted down is destined for "blendhood" in the absence of practical separation techniques. Coextrusion and coinjection molding will only increase the number of potential blends generated by recycling efforts.

In addition, any microstructural changes exhibited by a blend in its use or testing are fair game for the morphologist. A toughened plastic damaged by an impact test or tensile pull has a different morphology from the undisturbed material, with crazes, cavitation holes, or other features that can be imaged, measured, described, and if need be quantified. If a component selectively degrades away, such as starch in garbage bags, the morphology as well as the chemical composition has changed.

The examples discussed have shown how sampling and sample preparation steps such as staining for TEM can affect the outcome of morphology studies. It is desirable to find a balance between examining a single polymer blend composition to death with multiple samplings of skin, core, gate, as a function of injection rate, pressure, hold time, and mold temperature versus blindly assuming that every morphology can be understood from a single sampling and one staining treatment, if any. Advances in processing techniques and equipment will surely affect the morphology of the blends of the future and introduce many questions. Changes and upgrades in polymer processing, such as high-throughput compounding extruders, virtually guarantee the future employment of polymer morphologists, as does the use of morphological criteria in patents on new or improved blends.

REFERENCES

1. OL Shaffer, R Bagheri, JY Qian, V Dimonie, RA Pearson, MS El-Aasser. Characterization of the particle–matrix interface in rubber-modified epoxy by atomic force microscopy. J Appl Polym Sci 58:465–484, 1995.
2. AA Patel, J Feng, MA Winnik, G Vancso, C McBain. Characterization of latex blend films by atomic force microscopy. Polymer 37:5577–5582, 1996.
3. AE Ribbe, T Hashimoto, H Jinnai. Complex image generation in the laser scanning confocal microscope of a polymer blend system. J Mat Sci 31:5837–5847, 1996.
4. V Verhoogt, J van Dam, P deBoer, A Draaijer, PM Houpt. Confocal laser scanning microscopy: a new method for determining the morphology of polymer blends. Polymer 34:1325–1329, 1993.
5. A Hammiche, M Song, HM Pollock, DJ Hourston, M Reading. Phase separation of miscible blends: a study by SThM and M-T DSC. Polym Mater Sci Eng 75:275–276, 1996.
6. BA Wood. Transmission electron microscopy of polymer blends. In: K. Finlayson, ed. Advances in Polymer Blends and Alloys Technology 3. Lancaster: Technomic, 1992, pp 24–36.
7. VK Berry. Characterization of polymer blends by low voltage scanning electron microscopy. Scanning 10:19–27, 1988.
8. U Sundararaj, CW Macosko, C-K Shih. Evidence for inversion of phase continuity during morphology development during polymer blending. Polym Eng Sci 36:1769–1781, 1996.
9. M Okamoto, Y Shinoda, T Kojima, T Inoue. Toughening mechanism in a ternary polymer alloy: PBT/PC/rubber system. Polymer, 34:4868–4873.
10. Y Okamoto, H Miyagi, T Uno, Y Amemiya. Impact toughening mechanisms in rubber-dispersed polymer alloy. Polym Eng Sci 33:1606–1610, 1993.
11. AP Smith, JH Laurer, HA Ade, SD Smith, A Ashraf, RP Spontak. X-ray microscopy and NEXAFS spectroscopy of macrophase-separated random block copolymer/homopolymer blends. Macromolecules 30:663–666, 1997.
12. EG Rightor, GP Young, K Sehanobish, JC Conboy, CP Bosnyak. Real time study of failure events in polymers. Part I. Advanced methods for crack propagation studies of bulk polymers at the submicrometre level. J Mat Sci 30:2077–2082, 1995.

13. EG Rightor, K Sehanobish, GP Young, JC Conboy, JW Wilchester, CP Bosnyak. Real time study of failure events in polymers. Part II. Influence of polycarbonate-polyester morphology on the failure micromechanisms. J Mat Sci 30:2083–2090, 1995.
14. H-J Sue, AF Yee. Failure mechanisms in alloy of polyamide 6,6/polyphenylene oxide under severe conditions. J Mat Sci 26:3449–3456, 1991.
15. J Wu, Y-W Mai. Fracture toughness and fracture mechanisms of PBT/PC/IM blend. Part II. Toughening mechanism. J Mat Sci 28:6167–6177, 1993.
16. EH Merz, GC Claver, M Baer. Studies on heterogeneous polymeric systems. J Polym Sci 22:325–344, 1956.
17. H Keskkula, DR Paul. Toughened nylons. In: M. Kohan, ed. Nylon Plastics Handbook. Munich: Munser, 1995, pp 414–431.
18. S Wu. Impact fracture mechanisms in polymer blends: rubber-toughened nylon. J Polym Sci Polym Phys Ed 21:699–716, 1983.
19. S Wu. Phase structure and adhesion in polymer blends: a criterion for rubber toughening. Polymer 26:1855–1863, 1985.
20. S Wu. A generalized criterion for rubber toughening: the critical matrix ligament thickness. J Appl Polym Sci 35:549–561, 1988.
21. A Margolina, S Wu. Percolation model for brittle-tough transition in nylon/rubber blends. Polymer 29:2170–2173, 1988.
22. RJ Gaymans, K Dijkstra. Comments on "Percolation model for brittle-tough transition in nylon/rubber blends". Polymer 31:971, 1990.
23. S Wu, A Margolina. Reply to comments. Polymer 31:972–973, 1990.
24. B Wessling. Electrical conductivity in heterogeneous polymer systems. V. Further experimental evidence for a phase transition at the critical volume concentration. Polym Eng Sci 31:12001–1206, 1991.
25. SY Hobbs, RC Bopp, VH Watkins. Toughened nylon resins. Polym Eng Sci 23:380–389, 1983.
26. F Ramsteiner, W Heckmann. Mode of deformation in rubber-modified polyamide. Polym Commun 26:199–200, 1985.
27. A Lazzeri, CB Bucknall. Applications of a dilatational yielding model to rubber-toughened polymers. Polymer 36:2895–2902, 1995.
28. OK Muratoglu, AS Argon, RE Cohen, M Weinberg. Toughening mechanism of rubber-modified polyamides. Polymer 36:921–930, 1995.
29. B Majumdar, H Keskkula, D Paul. Morphology development in toughened aliphatic polyamides. Polymer 35:1386–1398, 1994.
30. C-J Wu, J-F Kuo, C-Y Chen. Rubber toughened polyamide 6: the influences of compatibilizer on morphology and impact properties. Polym Eng Sci 33:1329–1335, 1993.
31. DD Huang. The application of fracture mechanics to material selection. Polym Eng Sci 36:2270–2274, 1996.
32. V Flaris, E McBride, BL Glazar. A review on polyamide toughening. Proc 2d Int'l Conf Compounding '96, Philadelphia, PA, 1996.
33. HW Starkweather, RE Brooks. Effect of spherulites on the mechanical properties of nylon 66. J Appl Polym Sci 1:236–239, 1959.
34. MJ Modic, LA Pottick. Modification and compatibilization of nylon 6 with functionalized styrenic block copolymers. Polym Eng Sci 33:819–826, 1993.

35. AJ Oshinski, H Keskkula, DR Paul. Rubber toughening of polyamides with functionalized block copolymers: 1. Nylon-6. Polymer 33:268–283, 1992.

36. AJ Oshinski, H Keskkula, DR Paul. Rubber toughening of polyamides with functionalized block copolymers: 2. Nylon-6,6. Polymer 33:284–293, 1992.

37. BZ Jang, DR Uhlmann, JB Vander Sande. Rubber-toughening in polypropylene. J Appl Polym Sci 30:2485–2504, 1985.

38. J Horrion, S Cartasegna, PK Aggarwal. Morphology, thermal, and mechanical properties of polyacetal/ionomer blends. Polym Eng Sci 36:2061–2068, 1996.

39. HJ Sue, J Huang, AF Yee. Interfacial adhesion and toughening mechanisms in an alloy of polycarbonate/polyethylene. Polymer 33:4868–4871, 1992.

40. I Yamaoka, Toughened polymer blends composed of a ductile styrene-butadiene-styrene matrix with brittle methyl methacrylate-styrene particles. Polymer 36:3359–3368, 1995.

41. S Kunz-Douglass, PWR Beaumont, MF Ashby. A model for the toughness of epoxy-rubber particulate composities. J Mat Sci 15:1109–1123, 1980.

42. RA Pearson, AF Yee. Influence of particle size and particle size distribution on toughening mechanisms in rubber-modified epoxies. J Mat Sci 26:3828–3844, 1991.

43. Y Okamoto, H Miyagi, M Kakugo, K Takahashi. Impact improvement mechanism of HIPS with bimodal distribution of rubber particle size. Macromolecules 24:5639–5644, 1991.

44. C Maestrini, M Merlotti, M Vighi, E Malaguti. Second phase volume fraction and rubber particle size determinations in rubber-toughened polymers: a simple stereological approach and its application to the case of high impact polystyrene. J Mat Sci 27:5994–6016, 1992.

45. RG Hill. The role of microstructure on the fracture toughness and fracture behaviour of rubber-reinforced acrylics. J Mat Sci 29:3062–3070, 1994.

46. D Dompas, G Groeninckx. Toughening behaviour of rubber-modified thermoplastic polymers involving very small rubber particles: 1. A criterion for internal rubber cavitation. Polymer 35:4743–4749, 1994.

47. D Dompas, G Groeninckx, M Isogawa, T Hasegawa, M Kadokura. Toughening behaviour of rubber-modified thermoplastic polymers involving very small rubber particles: 2. Rubber cavitation behaviour in poly(vinyl chloride)/methyl methacrylate-butadiene-styrene graft copolymer blends. Polymer 35:4750–4759, 1994.

48. D Dompas, G Groeninckx, M Isogawa, T Hasegawa, M Kadokura. Toughening behaviour of rubber-modified thermoplastic polymers involving very small rubber particles: 3. Impact mechanical behaviour of poly(vinyl chloride)/methyl methacrylate-butadiene-styrene graft copolymer blends. Polymer 35:4760–4765, 1994.

49. DJ Hourston, S Lane, HX Zhang. Toughened thermoplastics: 3. Blends of poly(butylene terphthalate) with (butadiene-co-acrylonitrile) rubbers. Polymer 36:3051–3054, 1995.

50. MEJ Dekkers, SY Hobbs, VH Watkins. Toughened blends of poly(butylene terephthalate) and BPA polycarbonate. Part 2. Toughening mechanisms. J Mat Sci 23:1225–1230, 1988.

51. D. Delimoy, B Goffaux, J Devaux, R Legras. Thermal and morphological behaviours of bisphenol A polycarbonate/poly (butylene terephthalate) blends. Polymer 36:3255–3266, 1995.

52. T Nomura, T Nishio, K Iwanami, K Yokomizo, K Kitano, S Toki. Characterization of microstructure and fracture behaviour of polypropylene/elastomer blends containing small crystal in elastomeric phase. J Appl Polym Sci 55:1307–1315, 1995.
53. SY Hobbs, VH Watkins, JT Bendler. Diffusion bonding between BPA polycarbonate and poly(butylene terephthalate). Polymer 31:1663–1668, 1990.
54. S Wu, H-K Chuang, CD Han. Diffuse interface between polymers: structure and kinetics. J Poly Sci Polym Phys Ed 24:143–159, 1986.
55. HT Oyama, JJ Lesko, JP Wightman. Interdiffusion at the interface between poly(vinylpyrrolidone) and epoxy. J Polym Sci Polym Phys Ed 35:331–346, 1997.
56. W Wenig, A Wasiak. Interactions between the components in isotactic proplyene blended with EPDM. Colloid Polym Sci 271:824–833, 1993.
57. W Wenig, M Asresahegn. The influence of rubber-matrix interfaces on the crystallization kinetics of isotactic polypropylene blended with ethylene-propylene-diene terpolymer (EPDM). Polym Eng Sci 33:877–888, 1993.
58. A Leclair, BD Favis. The role of interfacial contact in immsicible binary polymer blends and its influence on mechanical properties. Polymer 21:4723–4728, 1996.
59. R Armat, A Moet. Morphological origin of toughness in polyethylene-nylon-6 blends. Polymer 34:977–985, 1993.
60. F Gubbels, S Blacher, E Vanathem, R Jerome, R Deltour, F Brouers, Ph Teyssie. Design of electrical conductive composites: key role of the morphology on the electrical properties of carbon black filled polymer blends. Macromolecules 28:1559–1566, 1995.
61. LC Sawyer, DT Grubb. Polymer applications 5.3. Engineering resins and plastics. In: Polymer Microscopy. 2d ed. London: Chapman and Hall, 1996, pp 219–229.
62. GM Jordhamo, JA Manson, LH Sperling. Phase continuity and inversion in polymer blends and simultaneous interpenetrating networks. Polym Eng Sci 26:517–524, 1986.
63. C-K Shih. Mixing and morphological transformations in the compounding process for polymer blends: the phase inversion mechanism. Polym Eng Sci 35:1688–1694.
64. C-K Shih. Fundamentals of polymer compounding: the phase inversion mechanism during mixing of polymer blends. Adv Polym Technol 11:223–226, 1992.
65. U Sundararaj, CW Macosko, C-K Shih. Evidence for inversion of phase continuity during morphology development during polymer blending. Polym Eng Sci 36:1769–1781, 1996.
66. CE Scott, CW Macosko. Morphology development during the initial stages of polymer–polymer blending. Polymer 36:461–470, 1995.
67. D Quintens, G Groeninckx, M Guest, L Aerts. Phase morphology coarsening and quantitative morphological characterization of a 60/40 blend of polycarbonate of bisphenol A (PC) and poly(styrene-co-acrylonitrile) (SAN). Polym Eng Sci 30:1484–1490, 1990.
68. N Mekhilef, BD Favis, PJ Carreau. Morphological stability, interfacial tension, and dual-phase continuity in polystyrene–polyethylene blends. J Polym Sci Polym Phys Ed 35:293–308, 1997.
69. A Gonzalez-Montiel, H Keskkula, DR Paul. Impact-modified nylon 6/polypropylene blends: 1. Morphology–property relationships. Polymer 36:4587–4603, 1995.
70. SY Hobbs, MEJ Dekkers, VH Watkins. Effect of interfacial forces on polymer blend morphologies. Polymer 29:1598–1603, 1988.

71. MEJ Dekkers, SY Hobbs, I Bruker, VH Watkins. Migration of polymer blend components during melt compounding. Polym Eng Sci 30:1628–1632, 1990.
72. D Debier, J Devaux, R Legras, D Leblanc. Influence of a core/shell rubber phase on the morphology and the impact resistance of a PC/SAN blend (75/25). Polym Eng Sci 34:613–624, 1994.
73. SY Hobbs. The development of flow instabilities during the injection molding of multicomponent resins. Polym Eng Sci 36:1489–1494, 1996.
74. SH Jang, BS Kim. Morphology and mechanical properties of liquid crystalline copolyester and polyester elastomer blends. Polym Eng Sci 35:528–537, 1995.
75. AD Padsalgikar, MS Ellison. Modeling droplet deformation in melt spinning of polymer blends. Polym Eng Sci 37:994–1002, 1997.
76. DA Thomas, LH Sperling. Interpenetrating polymer networks. In: DR Paul, S Newmans, eds. Polymer Blends. Vol. 2. New York: Academic Press, 1978, pp 1–33.
77. SB Maerov. Photopolymer morphology. J Imaging Sci 30:235–241, 1986.
78. TJ Bunning, LV Natarajan, V Tondiglia, RL Sutherland, DL Vezie, WW Adams. The morphology and performance of holographic transmission gratings recorded in polymer dispersed liquid crystals. Polymer 36:2699–2708, 1995.
79. CJ Rajaram, SD Hudson, LC Chien. Effect of polymerization temperature on the morphology and electrooptic properties of polymer-stabilized liquid crystals. Chem Mater 8:2451–2460, 1996.
80. D Winoto, SH Carr. Grating diffraction of blends involving NLO random copolymers. Macromolecules 29:5149–5156, 1996.
81. JJ Laverty, T Ellis, J O'Gara, S Kim. Effect of processing on the recyclability of nylon 66/poly(2,6 dimethyl-1,4 phenylene oxide) (PPO) alloys. Polym Eng Sci 36:347–357, 1996.
82. C Maestrini, A Callaioli, M Rossi, R Bertani. Styrenics materials and cyclopentane: problems and perspectives. J Mat Sci 31:3747–3761, 1996.
83. U Seiffert, P Walzer. The Future of Automotive Technology. London: Frances Pinter, 1984, pp 112–117.
84. PM Subramanian. Polymer blends: morphology and solvent permeability. Polym Prepr (Am Chem Soc Div Polym Chem) 30:28–29, 1988.
85. PM Subramanian. Polymer blends: morphology and permeability barrier properties. Polym Sci (Symp Polym '91) 2:847–854, 1991.
86. PM Subramanian. Polymer blends. Morphology and solvent barriers. ACS Symp Ser (Barrier Polymer Structure) 423:252–265, 1990.
87. PM Subramanian. Polyethylene terephthalate blends. Effect of selective crosslinking on permeability and morphology. Int Polym Process 3:33–37, 1988.

18

Study on Morphology and Toughening Mechanisms in Polymer Blends by Microscopic Techniques

Jingshen Wu and Chi-Ming Chan
Hong Kong University of Science and Technology, Clear Water Bay, Kowloon, Hong Kong

Yiu-Wing Mai
University of Sydney, Sydney, New South Wales, Australia

Morphology of polymer blends can affect substantially their physical, chemical, and mechanical properties. A polymer blend with a well-tailored morphology may possess a combination of properties that cannot be achieved by a homogeneous polymer alone. For this reason, the understanding of the morphology–property relationships for polymer blends is of particular importance in the development of new blends, and considerable research efforts have been devoted to their morphology characterization and property evaluation. Possible close relationships between the morphology and properties of polymer blends have been pursued extensively. This chapter contains a brief overview of the basic principles of optical and electronic microscopes and their applications in morphology characterization; the various methods developed for microscopical specimen preparation; and the technical details of some contrast enhancement techniques. It also reviews the unique contributions of polymer microscopy for morphology characterization and the investigation of toughening mechanisms.

I. INTRODUCTION

Polymer blends are multicomponent in nature, so depending on the physical and chemical properties of the components, they can be arranged into a variety of morphologies using different processing conditions. Research shows that many important properties of a polymer blend, such as thermal stability, chemical resistance, yield strength, elastic modulus, and fracture toughness, depend critically on the nature of this microscopical arrangement (1–3). A well-designed and controlled morphology can result in polymer blends with property combinations that cannot be easily achieved by a homogeneous polymer. Driven by this economic and technical importance, many research efforts have been devoted in the past few decades to establishing the morphology–property relationships for polymer blends. Obviously, without a deep understanding of the relationships, it is impossible to develop new polymer blends with the desired combination of properties through morphology control. The required combination can only be achieved through serendipity.

Morphology characterization of polymer blends plays a critical role in the establishment of the morphology–property relationships. Various analytical techniques have been employed, modified, and developed for this purpose. Among these are thermal analysis, microscopy, and spectroscopy. The information sought from the techniques includes size and shape of both discrete and continuous phases, internal structure of each phase, interfacial boundary condition, miscibility between the phases, and so on. Microscopy has a unique position in the morphology characterization of polymer blends because of its power to reveal the microstructures directly with visible images. Both optical and electronic microscopes are nowadays considered essential equipment in a materials characterization laboratory. Many successful examples, as will be reviewed in the following text, prove that microscopy is probably the most important technique in morphology characterization. On the other hand, however, due to the compositional nature of polymer materials, lack of contrast is a frequently encountered problem in polymer blend microscopy. Due to weak contrast, details of microstructure for a polymer blend may not be easily discerned without well-prepared specimens. In fact, polymer blend microscopy is very demanding on specimen preparation, and one of the major challenges is how to enhance phase contrast effectively without introducing artifacts. In this chapter, microscopical techniques developed for morphology characterization of polymer blends will be reviewed and the contrast enhancement techniques will be highlighted.

Polymer blends are normally designed and manufactured to modify certain properties in order to meet the requirements of specific end applications. One of the properties most often to be improved is the fracture toughness. This is simply because many homogeneous polymers are prone to brittle fracture. Thus toughening is necessary before they can be used reliably. The term "fracture toughness" refers to a measure of fracture resistance of a material, which is represented often by the energy required to extend a crack by a unit area. Therefore, toughness is directly related to the energy absorption processes such as shear yielding, crazing, cavitation, and crack bridging. Crazing and shear yielding are two major events frequently observed during the deformation of polymeric materials. Both crazing and shear yielding absorb large amounts of work. In brittle fracture the energy dissipation processes occur only in a localized area around the crack tip, and the total amount of plastic energy absorbed is therefore low. To increase the fracture toughness it is necessary to enhance the volume of material in which plastic deformation processes occur. In conventional polymer blends this is achieved by adding soft rubber particles to a rigid matrix. It is well known that the fracture toughness of a brittle polymer can be increased substantially by adding a small proportion of rubber. The nature of rubber toughening has been extensively studied in the past, and several toughening mechanisms related to various energy absorption processes have been suggested.

The study of toughening mechanisms for polymer blends relies very much on good experimental techniques. An appropriate analytical technique should be able to provide information on the deformation history of the polymer blends before and during fracture, as well as information on the size and nature of the damage zone in front of a crack tip. In the past, both optical and electron-microscopical techniques were employed to acquire this information. Many special sample preparation methods were also developed to make appropriate specimens that can preserve and deliver rich information about the deformation history of a tested sample. In Sec. V of this chapter a brief review of the application of microscopical technique in toughening mechanism study of polymer blends will be given.

II. OPTICAL MICROSCOPY AND ITS APPLICATION IN MORPHOLOGY CHARACTERIZATION

Optical microscopy is the most valuable technique for the study of polymer blends because it is relatively simple and cheap. Optical microscopes are readily available in most research laboratories. In many cases, a general picture of a polymer blend can be achieved through an overall visual inspection with optical microscopes at relatively low magnifications. Much basic information on the morphology and deformation behavior of the polymer blends can be obtained at this stage. Optical microscopical observations also provide useful information to help identify the most important locations for further study under high-magnification electron microscopes.

A typical professional optical microscope consists of an objective, a condenser, and an eyepiece (ocular) mounted in a tube supported by a stand. Light is supplied by an illuminator and condensed by a condenser onto the specimen. Operation in transmitted or incident light is possible, and some microscopes are capable of both modes simultaneously. In transmitted optical microscopy (TOM), light is supplied by an illuminator and focused on the specimen by a condenser lens under the specimen stage. A collimated light beam passes through the thin sample and into the objective lens of the microscope, forming an enlarged image. The thickness requirement of the sample for TOM falls into the range of 1 to 40 μm for most polymer materials. For reflected optical microscopy (ROM), the light source is placed above the objective and the objective lens serves as the magnifier as well as the condenser. An illuminating light beam is first provided through the objective lens and then is reflected from the specimen surface back into the objective to reveal surface topography. ROM can be used to examine variations in surface structure when the specimen is opaque or a thin specimen is not obtainable.

The major disadvantages of the optical microscope are limited resolution and small depth of field. Resolution can be defined as the minimum distance between two points such that the two points are perceived as separate in the image. The wavelength of visible light and the quality of the objective lens determine the resolution of an optical microscope. In general, point-to-point resolution for optical microscopy is at best approximately 0.2 μm, but practically, an order of 0.5 μm is expected. Depth of field is the thickness of the object that is simultaneously in focus, which is inversely proportional to the magnification and thus to the resolution. Practically speaking, the depth of field required in TOM is the thickness of the specimen. Therefore, for high-magnification TOM study, thin specimens with uniform thickness are required. In ROM, the surface roughness of the specimen must fit into the objective depth of field. Since the best depth of focus is obtained at the lowest magnifications, it is common practice that OM micrographs are taken at low magnifications with good depth of focus and then enlarged to increase the image magnification while preserving the depth of field.

Insufficient contrast is often found in optical microscopy. Because of weak contrast, two points on an object are still not visible even when the distance between the two points is much larger than the resolution of the equipment. Under this circumstance, contrast enhancement is needed to overcome the problem. Optical contrast arises from a number of sources. Color, opacity, reflective index, orientation, absorption or dichromatic differences have been exploited to improve the contrast. Besides, there are also a number of techniques used in polymer blends characterization to improve the contrast. Some of them are briefly introduced below.

Dark field image can be used to increase the phase contrast in both transmitted and reflected optical microscopy. To generate a dark field image, the central beam of the directly transmitted or reflected light, which is the main light source in bright field illumination, is prevented from entering the objective by an opaque central plate, forming a dark background. When a sample is viewed in the dark field, things that reflect or scatter light, such as surface roughness, phase boundary, or fillers, are highlighted against the dark background.

Phase contrast microscopy has been increasingly employed in the study of morphology of polymer blends. In many cases, thin sections of polymer blends give bright field images with little or no contrast between the components. Identification of blend morphology can be very difficult or impossible. This is because when the illuminating light passes through a thin section of a two-component polymer blend, phase shift caused by the differences in the refractive indices of the two components cannot be detected by either the human eye or the photographic emulsion. Using a special condenser with a phase retarding absorption ring, the phase contrast microscope converts the phase changes to light and dark image regions by interference between the two lights with different phases. It produces image intensity differences, which are visible to the human eye. Phase contrast microscopy is used for both reflected and transmitted light, though it has been most effectively applied in transmitted optical microscopy in polymer blend studies. In reflected light, phase contrast can provide useful information on polished surfaces when the components have the same reflecting power and thus are indistinguishable in a bright field. In general, fine surface structure, fillers, boundaries, phase separations, surface imperfections, and etching results can be revealed using a phase contrast microscope with either transmitted or reflected light.

Polarized optical microscopy is most useful for studying anisotropic materials. A polarizing microscope has two polarizers crossed to each other. The illuminating light is plane polarized by one of the polarizers before it impinges on the samples. When the polarized light passes through anisotropic regions of a specimen, it will be rotated to a certain extent, depending on the orientation and thickness of the region. With another polarizer placed above the objective lens, this rotation of polarized light can be used for contrast enhancement between phases in a polymer blend. Since optical anisotropic elements, such as localized shear yielding zones and crazes, have higher molecular orientation, polarized optical microscopy is especially powerful in investigating the deformation mechanisms of polymer blends.

In *differential interference microscopy,* the illuminating light is first polarized and then split into two beams by a prism. Two beams are displaced at the specimen and recombined by another prism after passing through the specimen. Interference of the two beams is used to enhance image contrast on the boundaries of regions having different optical properties. The interference pattern can also be used to measure the specimen thickness in TOM or the specimen roughness in ROM.

For *fluorescence microscopy,* a mercury or xenon discharge lamp is used as the light source. The incident light is filtered to obtain a narrow band of excitation wavelength. When some molecules in a specimen are excited with the energy from the illumination light, the molecules retain some of the energy for a period of time and release the energy via fluorescent radiation. The wavelength of the emitted radiation depends on the material investigated and can be analyzed and discerned with various filters.

Specimen preparation is critical in optical microscopical studies. The major task of all sample preparation techniques for morphology and toughening mechanism studies is to provide artifact-free and high-phase-contrast specimens. In fact, image contrast can be greatly enhanced through proper specimen preparation techniques.

A. ROM Sample Preparation and Contrast Enhancement

The basic ROM sample preparation technique in morphology studies is relatively simple. It includes sectioning, embedding in a resin, grinding, polishing, and sometimes, etching. The major concern of etching has been to remove selectively one of the blend components so as to increase the roughness of sample surfaces and lead to higher image contrast. Liquid etching is the most widely used method for surface modification for polymer blends. Organic solvents, oxidizing acids, bases, or mixtures of different solvents have long been used as liquid etchants. Table 1 (4) contains some examples of the liquid etchants that are frequently used in polymer sample preparation. Plasma can also be utilized to remove selectively one of the components in a polymer blend because the amorphous regions on polymer surfaces are more susceptible to plasma than crystalline regions. Due to the difference in plasma etching rate of the two phases, exposing a polished surface directly to plasma for a period of time may result in sufficient high roughness on the surface, leading to good image contrast.

B. TOM Sample Preparation and Contrast Enhancement

Microtomy gives easy access for TOM thin-sample preparation. Samples with thicknesses of 1–40 μm can be cut with steel or glass knives. The thin specimens may be stained with chemicals, like osmium tetroxide, lead citrate, or uranyl acetate, to improve its contrast. Figure 1 is an optical micrograph taken from a thin section of PBT/PC blend cut on a microtome and stained with osmium tetroxide. The black PC domains and white PBT phase can be clearly identified without difficulty.

Though the specimens obtained via microtomy may be adequate for microstructure study, they are not favorable for toughening mechanism study. This is because the damaged zone under investigation may be further deformed by the large cutting force introduced in microtomy. It may destroy the deformation patterns preserved in the damage zone by introducing additional plastic deformation, further crack propagation, and surface scratching and/or other unexpected artifacts.

Table 1 Chemical Agents for Etching of Polymers and Polymer Blends

Polymer	Etchant	Temperature (°C)	Time (min.)
Polypropylene	Refluxing vapors of benzene	80.1	
	Nitric acid	120	0–360
	6 M Chromic acid	70	0–90,000
	Chromic-sulfuric	82 ± 2	1–7
	Chromic-sulfuric-phosphoric	60–80	0–60
	Chromic acid or >95% sulfuric acid	100	≥30
	Chromic-sulfuric	50–80	
	Chromic-sulfuric	100	32
	Chromic-sulfuric-phosphoric	70–75	
	Chromic acid	82–88	15
	Benzene-chloroform-xylene	80–85	2–3
	Chromic-sulfuric	70 ± 1	0–360
	Chromic acid	78	0–16
	Chromic-sulfuric	70	30
Polyethylene	Refluxing vapors of benzene	≅80.1	
	Carbon tetrachloride	≅76.8	1–2 sec
	Fuming nitric acid	60–90	15–960
	Toluene or xylene or decalin	various	
	Chromic-sulfuric	20–25	10–15
	Chromic-sulfuric	70 ± 1	0–360
	Chromic sulfuric	80	1
	90% HNO_3 and red fuming HNO_3	60	≥180
Other polyolefins	6 M Chromic acid	70	720
Polyethylene terephthalate (PET)	n-Propylamine	Room	
	50% aqueous sodium hydroxide	120	5
	Aluminum trichloride or Aluminum tribromide or Titanium tetrachloride	25–125	0.15–5
	Acidic potassium	70	2
	permanganate, water	180 and 130 (9.9 and 2.7 atm water vapor pressure respectively)	0–500
	n-Propylamine	Room	0–2160
	100 g Sodium hydroxide/1 water	80	90
	8.2–53% aq. KOH	27–120	≥6
	42% aq. n-Propylamine	30	
	n-Propylamine	Room	60–1540
Polyamides	Dilute acetic acid	Room	
	Chromic-sulfuric	75–80	3–5
	0.25–0.50 M Iodine–potassium iodide aqueous solution	20–45	0.5–5

Table 1 Continued

Polymer	Etchant	Temperature (°C)	Time (min.)
	Liquid oxygen	-183	Until evaporation
	Xylene	65–75	2–4
	Dilute formic acid	Room	0–20
Fluorocarbon polymers and chlorine-containing polymers	Carbon tetrachloride sodium in liquid ammonia or naphthalene-tetrahydrofuran	25–35 —	2–5 0–1
Polycarbonates	Dilute sulfuric acid	80	
	5% Aqueous sodium hydroxide	—	300
Acrylonitrile-butadiene-styrene (ABS)	Chromic sulfuric	60	5–15
	Chromic sulfuric	55–60	15–20
	Chromic sulfuric	65	15
	Chromic sulfuric	60	15
	Chromic sulfuric-phosphoric	12–110	0.25–30
	Chromic sulfuric-sulfuric	65	10
	Aqueous solutions of potassium hydroxide and potassium manganate	40	10
	Chromic acid	82–88	15
	Chromic-sulfuric-phosphoric	50–70	1–15
	Chromic-sulfuric	50–70	0.5–10
Polysulfones	Chromic acid	82–88	15
Polystyrene (PS)	Chromic-sulfuric	65	0–15
	Amylacetate	Room	10–16 sec
	Chromic-sulfuric and chromic-sulfuric-phosphoric	80	5–180
Polyvinyl chloride (PVC)	5% Potassium hydroxide in isopropanol	70	180
Epoxy resins	5% Sodium hydroxide	20–30	60
	Chromic-sulfuric	65	15
	Molten eutectic solution of 60% potassium hydroxide and 40% sodium hydroxide	200–230	1–2
	Chromic-sulfuric-phosphoric	25	10
	Chromic-sulfuric	40	3
	Ethylene glycol-DMF	20–40	3–20
	Cone sulfuric acid	50	3–13, 5
	Boiling acetone	≅56.2	15–240
	1 molal $CrO_3/H2O$	70	480
Phenolics	Fuming sulfuric acid containing 0.5g $I_2O_6/1$	30	2–4
	Boiling methanol	≅64.7	0–120
Polyether–polyurethane foams	Concentrated sulfuric acid	Room	1

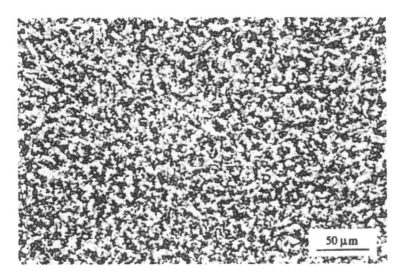

Figure 1 TOM photograph taken from a PBT/PC thin-section specimen cut by a microtome and stained with OsO_4. Cocontinuous structure is clearly seen.

The petrographical thin-sectioning method (5) provides a better technique for preparation of optical transparent thin sections for morphology as well as toughening mechanisms study. As demonstrated in Fig. 2, the bulk sample containing a damaged zone is first cut and embedded in a fast-cure and optically clean epoxy resin. After being fully cured at room temperature, one face of the sample is ground, wet polished, and glued using a fast-curing epoxy on a glass microscope slide with the polished face down. The opposite face of the sample is then cut, ground, and subsequently polished until it is thin enough to transmit light. Figure 3 is a TOM photograph taken from a thin specimen of a rubber/epoxy blend prepared by the petrographical thin-sectioning method. The morphology of the blend was disclosed completely. Another example is shown in Fig. 4. The sample is a deformed PBT/PC/rubber blend prepared by this technique. TOM observation using polarized light reveals the shear bands clearly. The rubber particle cavitation of a rubber-toughened epoxy was also detected using bright field optical microscopy with the thin sections so prepared, as shown in Fig. 5.

It should be mentioned here that although epoxies are widely used in sample preparation as embedding materials, epoxide resins may attack certain polymers. As a result, the morphology of the polymers may be changed during the embedding process and lead to unexpected conclusions. In this case, different types of embedding materials should be employed.

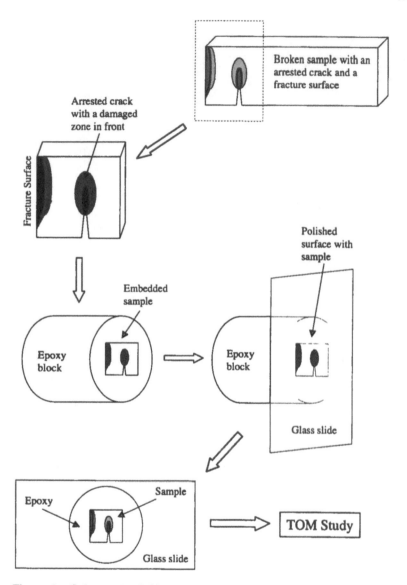

Figure 2 Schematic of the petrographical thin-sectioning technique for TOM specimen preparation.

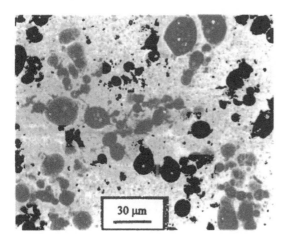

Figure 3 TOM photograph taken from a thin section of rubber-toughened epoxy blend. The TOM specimen was prepared using the petrographical thin-sectioning technique. Morphology of the blend can be seen.

Etching techniques can also be applied to the thin-section samples to discover the morphology of multicomponent polymer blends; phase contrast can be improved by selectively dissolving one phase using a volatile solvent on a microtomed sample (6,7). Use of a hot stage with thin-section samples can effectively improve image contrast by raising the temperature of the samples to near the melting point of one of the components.

III. SCANNING ELECTRON MICROSCOPY

Because of the resolution limitation of optical microscopy, the scanning electron microscope (SEM) has been widely employed in the microscopical study of polymer blends. The essential feature of a scanning electron microscope is that the specimen is examined by scanning a finely focused electron beam across the surface. The electrons emitted from the specimen due to the interaction between the primary electron beam and the specimen are detected and amplified. The signal is then used in the image-forming process. The final image is formed point by point and displayed as a TV type picture.

SEM sample preparation for surface observation is usually simple. The basic requirement is that the sample must be conductive, otherwise surface charge may cause distorted images with bright spots. For polymer specimens, an electrically conductive layer is provided by coating a thin layer of metal on the top of the specimen surface, which may be connected to the specimen stub by silver paint or a

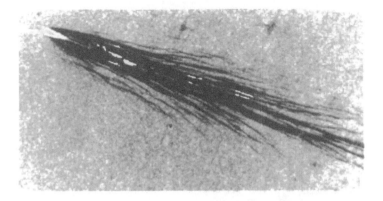

a

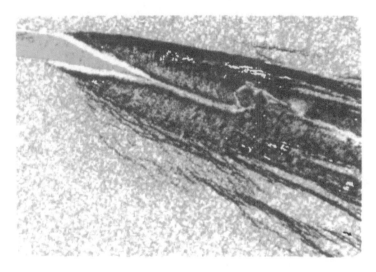

b

Figure 4 TOM photographs taken from a thin section of a rubber-toughened PBT/PC blend. The specimen contains an arrested crack tip with a damaged zone. The TOM sample was prepared using the petrographical thin-sectioning technique. Structure of the damaged zone was revealed.

carbon paste. Gold and platinum are two of the frequently used materials for metal coating. The thickness of the conductive layer depends on the metal used and the resolution required. For high-resolution dedicated SEM imaging, a thickness of about 5 nm of sputtered platinum is recommended. Thick gold coating tends to be granular, cracked and nonuniform (8–10).

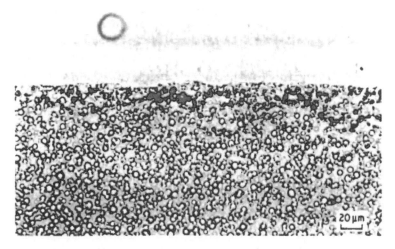

Figure 5 TOM photograph taken from a thin section of a rubber-toughened epoxy blend, showing the deformation zone underneath a fracture surface with extensive rubber particle cavitation. (From Ref. 44.)

Two types of contrast are available in the SEM image, i.e., compositional contrast and topographic contrast. Compositional contrast originates from the intensity of backscattered electrons. Backscattered electrons are primary beam electrons that have been elastically scattered by nuclei in the sample and escape from the surface. Thus the intensity of backscattered electrons is decided by the sample composition, i.e., the size of compositional elements of each component in the sample. Variation in the intensity of backscattered electrons reflects composition change in the samples and consequently provides the image with compositional contrast.

In the polymer blend study, because most polymers are composed of low-atomic-number elements and exhibit little variation in electron density from one polymer to another, compositional contrast of polymer blends is generally poor, and contrast enhancement is necessary in morphology characterization. Recently, the selective staining technique has been adopted in SEM sample preparation. In recent work reported by Graciela et al. (11–12) and Brown et al. (13), morphology of several polyolefin blends such as PP/EPDM, PP/PB, HIPS, PP/EPR, and LDPE/PP were successfully revealed using SEM with stained specimens. The specimen was first treated with heavy metal oxides, i.e., OsO_4 or RuO_4, and then trimmed with a microtome to produce a flat, smooth, and defect-free surface. During the staining process, one of the phases in the blend was selectively stained with the heavy metal oxide due to its higher reactivity with the chemical. As a consequence, the composition of this phase was changed to one containing much more of large-size heavy metal elements. The scattering capacity of the phase is therefore increased, resulting in higher electron intensity. The compositional contrast between the phases was enhanced. Figure 6 shows the morphology of a HIPS specimen stained with RuO_4. The white regions represent the rubbery domains, which contain double bonds and readily react with RuO_4 to increase electron scattering capacity, leading to higher electron density. The black one is the PS phase, which has a relatively lower reactivity with RuO_4. The contrast of the image is high.

Topographic contrast of a specimen is provided mainly by secondary electrons. Secondary electrons are those electrons emitted from the sample with energy less than 50 eV. Because of this low energy, the secondary electron signal only comes from a region very close to the surface. It is therefore highly sensitive to surface morphology and provides an SEM image with topographic contrast. The average lateral spatial resolution of a secondary electron image is high and normally less than 10 nm. It is used exclusively for examination of sample surfaces and widely employed in morphology and toughening mechanism study of polymer blends.

Since topographic contrast depends greatly on surface texture, techniques such as chemical or plasma etching or cryogenic fracture are commonly used in order to generate different surface textures between phases to enhance the topographic contrast. As with ROM sample preparation, both chemical and plasma

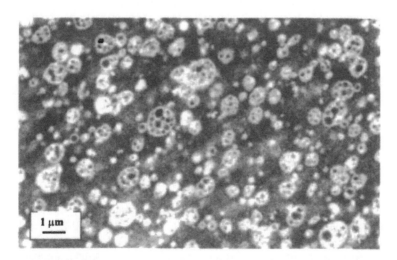

Figure 6 SEM micrograph taken from a RuO_4 stained surface of HIPS. Opposite the TEM micrograph, the white regions represent rubbery domains because of their high electron scattering power after reaction with RuO_4. (From Ref. 4.)

Figure 7 SEM micrograph taken from a plasma-etched PBT/PC specimen showing a cocontinuous structure. The SEM observation is consistent with the TOM photograph in Fig. 1.

etching techniques have been utilized in SEM sample preparation to remove selectively one component and increase surface roughness. Figure 7 is a SEM micrograph taken from a plasma-etched surface of a PBT/PC blend. The amorphous PC phase was removed during exposure to the plasma. The micrograph clearly shows that the PBT/PC blend has a cocontinuous structure. This result is consistent with the TOM observation on the same PBT/PC blend shown in Fig. 1.

Cryogenic fracture of a multiphase polymer blend can create fracture surfaces with high roughness and can be used to obtain information about the interior structure of bulk polymer blends. In the preparation of the cryogenic fracture surfaces for a polymer blend with two components, the operation temperature is selected to be lower than the glass transition temperature T_g of one component but higher than the T_g of the other. After the sample is treated at this temperature for a sufficiently long time, it is fast fractured immediately. Since the polymer component with a higher T_g is in its glass state when fracture occurs, the fracture of the component must be in a brittle mode and the fracture surface of the component will be quite smooth and featureless. On the other hand, the component with lower T_g is most probably in its plastic state when fracture takes place. Fracture of this component will be in either a semibrittle or a semiductile mode depending on the fracture toughness of the component at the operation temperature. The fracture surface of the component will be quite different when compared with the fracture surface of the component with higher T_g. SEM observation on the fracture sur-

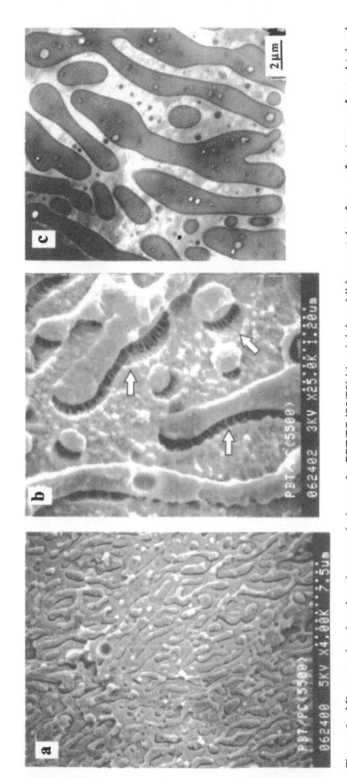

Figure 8 Micrographs showing the morphology of a PBT/PC (50/50) blend. (a) and (b) were taken from a fracture surface obtained using the cryogenic fracture method. Boundary condition between PBT and PC was revealed in (b), as indicated by arrows. A TEM micrograph (c) shows the same morphology but interface information is not available.

faces generated by this technique can disclose the phase structure of polymer blends clearly. In some cases, it can reveal interfacial boundary conditions between two phases, which is not readily available with other techniques. Examples of SEM study using this technique are shown in Fig. 8. The SEM micrographs were taken from the same PBT/PC blend used in OM study. The bulk PBT/PC sample was first treated in a low-temperature bath with liquid nitrogen/ethanol mixture at a temperature of $-30°C$ for 20 minutes. The frozen sample was then fast fractured by a pendulum impact tester at a speed of 3 m/s. Fracture surfaces so generated were coated with a thin layer of gold before SEM examination. The morphology of the PBT/PC blend can be clearly seen in Fig. 8a, which is consistent with the OM micrograph as shown in Fig. 1 and the TEM micrograph as shown in Fig. 8c. Moreover, the interfacial boundary condition between PC and PBT can be found in the SEM micrograph, Fig. 8b. In this picture, debonding between PBT and PC is observed, and there are fibrils at the interface. This detailed morphological information is valuable to the understanding of the deformation behavior of the polymer blend, but it is difficult to obtain with either optical microscopy or transmission electron microscopy, as will be discussed below.

IV. TRANSMISSION ELECTRON MICROSCOPY

Transmission electron microscopy (TEM) has been used widely in the investigation of morphology and toughening mechanisms of polymer blends. As with transmitted optical microscopy (TOM), the TEM sample is supported by a small copper grid and illuminated by an electron beam, which is formed using very high accelerating voltages (120–400 kV) in a high vacuum chamber and focused by a series of apertures and magnetic lenses. The image of the sample is viewed on a fluorescent screen. The biggest advantage of TEM is its extraordinarily high resolution, which enables small particle characterization and microscopic identification of phases. These features greatly enhance the general understanding of the relationship between microstructure and properties. Figure 9 is a TEM micrograph of a 50/50 nylon/PPO blend. The detailed morphology of the blend is shown clearly. The fine structure of the inclusions can also be read without any problem.

On the other hand, TEM requires complex specimen preparation and specialized equipment. This is because the penetrating power of an electron beam is low and therefore a polymer sample of less than 100 nm in thickness is usually required by TEM. The ultrathin sections are produced by ultramicrotomy using glass or diamond knives. Some high-toughness polymer blends, that is, blends with soft unsaturated rubbery particles, may be too soft to be sectioned at room temperature. In this case, chemical staining and hardening should be carried out prior to microtoming: the rubbery particles can be hardened by exposure to certain chemical agents to make sectioning easier. In the case of polymer blends with

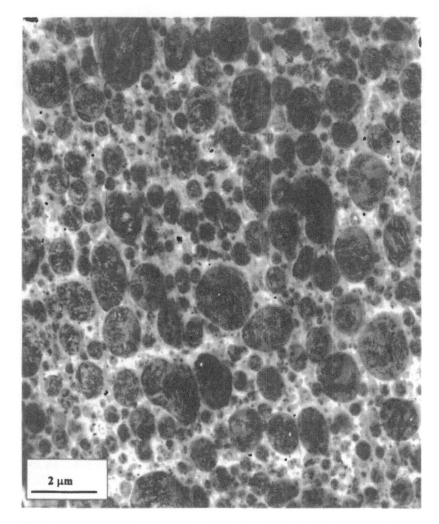

Figure 9   TEM micrograph of a rubber-toughened nylon/PPO blend. Detailed structure of the PPO inclusions can be clearly seen. (Courtesy of A. F. Yee.)

saturated rubbery particles, chemical hardening may be unavailable. Low-temperature microtomy, that is, cryomicrotomy, can be used to obtain satisfactory thin sections (14).

With TEM, electron scattering differences are the primary source of contrast. In the case of ordered or crystalline materials it gives diffraction contrast, which is strongly dependent on crystal orientation. In amorphous materials mass thickness contrast results. Similar to compositional contrast in SEM, mass thickness contrast is based on the difference in electron scattering power of individual phases of the specimens. For samples of even thickness (i.e., with the same number of atoms per electron path), the contrast is formed by differences in atomic size. Dark regions in the bright field image are regions having larger atoms with higher electron-scattering capacity. Obviously, mass thickness contrast of the TEM image can also be improved through selectively staining one of the components with electron-dense chemicals such as osmium tetroxide (OsO_4) or ruthenium tetroxide (RuO_4). (15–37).

OsO_4 has been employed in the microscopy of polymer materials for many years. It is commercially available in small ampoules as crystals that are readily dissolved in water. Premixed OsO_4 solutions are also available from special suppliers. In polymer blend staining, 1–2% OsO_4 aqueous solution is preferred because it is easy to handle and relatively stable. Staining polymer blends with an OsO_4 aqueous solution may take days to weeks because the penetration is rather slow. To speed up the staining process, OsO_4 vapor may be utilized at elevated temperatures, and the staining time can be reduced to a few hours. Great care must be taken during OsO_4 vapor staining because OsO_4 vapor is highly toxic and vapor pressure is high. The complete staining procedure must be carried out in a hood with good ventilation.

Ruthenium tetroxide, RuO_4, is a staining reagent with stronger oxidizing ability than that of OsO_4. It can oxidize not only unsaturated double bonds but also aromatic rings that OsO_4 cannot stain. RuO_4 is quite unstable, hence freshly prepared RuO_4 solution ($\sim$1%) is required in sample staining. One method that ensures fresh RuO_4 staining is to carry out RuO_4 preparation and sample staining simultaneously. In this method, RuO_4 is prepared through the following reaction:

$$2RuCl_3 \cdot 3H_2O + 8NaClO \rightarrow 2RuO_4 + 8NaCl + 3Cl_2 + 3H_2O$$

The advantage of this method is that it uses cheap, stable, and nontoxic $RuCl_3 \cdot 3H_2O$ rather than expensive, unstable, and highly toxic RuO_4. Moreover, this staining process is fast. Previous work showed that it took only 5–15 min to get a satisfactory staining effect for PBT/PC polymer blends' ultrathin sections (38). Since RuO_4 is also volatile and highly toxic, appropriate care must be taken during the reaction and staining. Apart from OsO_4 and RuO_4, other staining agents are also available for increasing TEM image contrast. Which agent should be used in sample preparation depends on the functional group(s) of the polymer to be stained. In Table 2 some frequently used staining agents are listed together with polymers and functional group(s) that the agents can stain (15–37).

An example of morphology study of the PET/PC/elastomer blend using TEM and a staining technique is given in Fig. 10. The ultrathin specimen of the blend was sliced from the top of a trimmed block using a cryogenic Leica Ultracut R ul-

Table 2 Chemical Agents for Staining of Polymers and Polymer Blends

Polymer	Group to stain	Staining agent	Referen
Unsaturated hydrocarbons: ABS; ASA; SBS; HIPS; impact PVC; polybutadiene; rubber; isoprene	—C=C—	Osmium tetroxide ruthenium tetroxide or ebonite	(15–26)
Acids	—COOH	Hydrazine, then osmium tetroxide	(28)
Alcohols Aldehydes	-OH -COH	Osmium tetroxide or ruthenium tetroxide or silver sulfide	(15, 16) (22–25) (29)
Amides (nylon)	-CONH$_2$ -CONH-	Phosphotungstic acid or tin chloride	(30) (31)
Amines	-NH$_2$	Osmium tetroxide or ruthenium tetroxide	(15, 16) (22–25)
Aromatics: Aromatic PA PPO Polystyrene (PS)	Aromatics	Ruthenium tetroxide or silver sulfide or mercuric trifluoro- acetate	(22–25) (29) (32) (33)
Bisphenol-A based epoxy resin	Bisphenol-A	Ruthenium tetroxide	(31)
Esters (butyl acrylate) (polyesters) (ethylene-vinyl acetate)	-COOR	Hydrazine, then osmium tetroxide or phosphotungstic acid or silver sulfide or methanolic NaOH	(34) (29) (28)
Ethers	-O-	Osmium tetroxide or ruthenium tetroxide	(15, 16) (22–25)
Saturated hydrocarbons (PE; PP)	-CH$_2$-CH$_2$-	Chlorosulfonic acid or phosphotungstic acid or ruthenium tetroxide	(27) (35) (22–25)

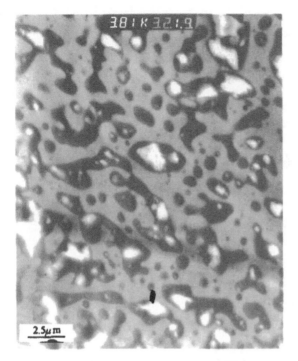

Figure 10 TEM micrograph of a PET/PC/elastomer blend stained with OsO₄. Three phases are discerned without difficulty.

tramicrotome. The block was first stained with a 2% OsO_4 water solution for a few days before cutting. The thickness of the specimen was approximately $60 \sim 100$ nm. From the micrograph, it is seen that the PET phase (gray) is the continuous phase. The black dispersing partiles with irregular shapes are PC domains. The length of the PC domain is less than 10 μm and the width is about 1 μm. The elastomer particles are white and exclusively located inside the PC domains.

Other than staining with heavy metal oxidizing agents, adequate differences in scattering can also be achieved by electron etching of one phase to produce a sufficiently high mass thickness difference. In recent work done by Zhu and Chan (39), a novel technique making use of electron irritation was developed to improve phase contrast for PVC/SBR and PVC/NBR/SBR blends. In the research, thin sections of the blends were bombarded with an electron beam at an accelerating voltage of 100 kV for various times. Figure 11 shows the TEM micrographs of the PVC/SBR blend taken at two different electron irritation times. The dispersed

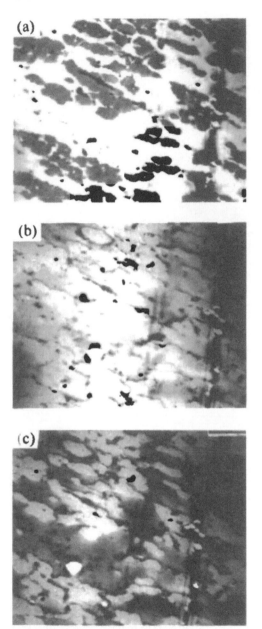

Figure 11 TEM micrographs of a PVC/SBR blend taken from a microtomed thin section after different times of electron irradiation. Phase inversion due to the electron irradiation was found.

dark regions in Fig. 11a represent the PVC domains because the chloride (Cl) scatters electrons more effectively than other light elements such as H, C, and O. SBR, which has only C and H, appears transparent. On the other hand, since PVC is very sensitive to electron irradiation due to the cleavage of the C—Cl bonds, as the beam irritation continues, the loss of Cl in PVC reduces its scattering power. The contrast between the rubber and the PVC decreases, Fig. 11b. If irradiation time is sufficiently long, most of the Cl in PVC will disappear. The PVC regions will be thinner than the SBR regions because Cl in PVC accounts for 63 wt%. Consequently, a contrast inversion, as shown in Fig. 11c, is observed due to the contrast caused by the difference in mass.

No contrast inversion is observed in a series of TEM micrographs obtained as a function of irradiation time for the PVC/NBR/SBR blend, as shown in Fig. 12. But dark rings at the interface between the PVC/SBR phases are observed. The rings become increasingly visible as irradiation continues. Since the rings are darker than both the PVC and the rubber regions, it is believed that the PVC at the interface is more resistant to the electron beam exposure. This indicates that rubber concentration at the interface between the PVC and SBR phases is high and provides a more effective stabilization against electron beam irradiation. With this technique, interface geometry can be identified and its thickness can be measured. Interface characterization has long been considered a challenge to morphology study of polymer blends.

Some caution must be taken in the electron microscopical study of polymer blends consisting of components sensitive to electron radiation and/or temperature. For example, the morphology of polyolefin-based blends may be altered after long-term exposure to an electron beam. In TEM study electron radiation can seriously damage the ultrathin specimens and result in unexpected artefacts.

V. TOUGHENING MECHANISMS STUDY BY MICROSCOPICAL TECHNIQUES

Over the past few decades, there have been extensive efforts devoted to the understanding of the various toughening mechanisms associated with rubber-toughened multiphase polymer blends. With the help of microscopical techniques, as well as other analytical equipment, a number of breakthroughs were achieved. The most dominant and important plastic deformation events involved in the fracture of a rubber-toughened polymer blend have been identified by microscopy as crazing, cavitation, shear banding, and diffuse shear yielding. Extensive research on toughening mechanisms has answered, to some extent, several fundamental questions such as (1) what the cause of stress whitening is, this being a frequently observed phenomenon in the fracture of toughened blends; (2) whether deformation of rubber particles or of the matrix is the major energy absorption event; (3) how

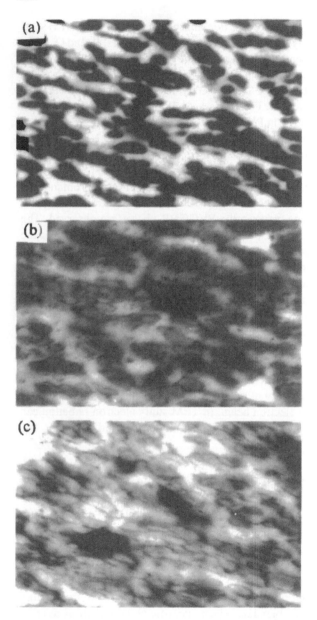

Figure 12 TEM micrographs of a PVC/SBR/NBR blend. Interface geometry and thickness become visible after certain time of electron irradiation (black rings).

multiple crazing is initiated and terminated in a polymer blend; and (4) how shear bands form and benefit polymer blends, and so on. Many toughening models based on these mechanisms have been proposed, and they have greatly enhanced our understanding of polymer blends. This section will review the contributions of microscopical techniques to establish the various toughening models that have been proposed in the past.

A. Rubber Particle Stretching and Tearing Model

In 1956, optical microscopy was first employed by Merz et al. (40) in the study of toughening mechanisms for a rubber-modified high-impact polystyrene (HIPS). Optical microscope observation of the fractured samples revealed that there was a whitened area underneath the fracture surface. Since a volume increase of the specimen was also observed during the fracture of the specimen, the authors therefore suggested that stress whitening found was a consequence of massive tiny cracks induced by stress concentration around the rubber particles in the HIPS. They further suggested that the function of the rubber particles was to bridge the opposite surfaces of a propagating crack. In so doing, the toughness was increased because of extra energy absorption by the rubber particle stretching and tearing as the crack faces were separated. This mechanism emphasizes the importance of the rubber particle bridging effect and neglects the contribution of matrix deformation.

Further evidence of rubber particle stretching and tearing was reported by Kunz-Douglass et al. in 1980 (41). In fracture tests with a rubber-toughened epoxy, an optical microscope was adopted to monitor crack growth in situ. The optical microscopical observation confirmed that the rubber particles between the crack surfaces were stretched as the crack opened up and finally fractured. According to this finding, a quantitative model was later proposed in 1981 (42). The model suggests that rubber particle stretching, tearing, and bridging are the major sources of fracture toughness. In 1985, the rubber particle stretching and bridging behavior was further confirmed by Bandyopadhyay et al., by an in situ fracture test in a scanning electron microscope (43), and by Pearson and Yee in 1991 (44), with an optical microscope, Fig. 13.

B. Massive Crazing Model

In 1965, Bucknall and Smith (45) studied the toughening mechanisms of a HIPS using an optical microscope and disclosed that the stress whitening, which had been attributed to the large number of microcracks in Merz's toughening model, was actually the result of massive crazes induced by the rubber particles. The numerous rubber particles inside the toughened HIPS were found to have two major functions: (1) they initiate a large number of crazes due to the high stress concen-

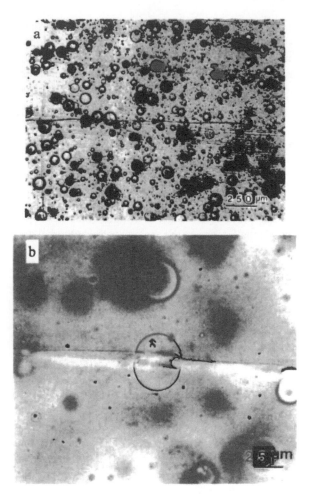

Figure 13 TOM photographs of a rubber-toughened epoxy blend showing crack bridging effect from rubber particles (From Ref. 44.)

tration in the matrix around the particles, and (2) the large-size rubber particles simultaneously stabilize and prevent crazes from growing into harmful cracks. Clearly, the proposed toughening mechanism stresses the key function of matrix deformation–massive crazing.

In 1971, with the help of TEM, Kambour and Russell (46) further confirmed that the massive crazing mechanism was indeed the predominant toughening mechanism in HIPS. As clearly illustrated in Fig. 14, crazes in a HIPS specimen

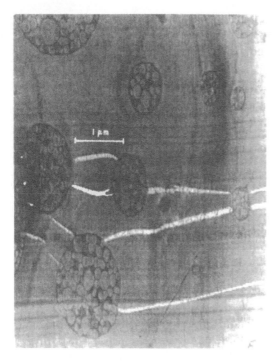

Figure 14 TEM micrograph of a deformed HIPS. Crazes initiated at equators of rubber particles and terminated by neighboring ones. (From Ref. 46.)

started at the equators of the rubber particles and later ended in neighborhood rubber particles. Their excellent TEM work provided firm evidence to support the toughening model proposed by Bucknall.

C. Shear Yielding Model

At the same time when Bucknall proposed the massive crazing model, Newman and Strella (47) discovered that shear yielding of the matrix material in a rubber-toughened SAN blend played a dominant role and absorbed a major proportion of the total fracture energy. The Newman–Strella proposal is now commonly known as the rubber particle induced matrix shear yielding model. In the development of this model, optical microscopy of a compression-molded, hand-stretched ABS sheet was performed. The results revealed that a large number of rubber particles in the specimen were permanently deformed with an aspect ratio of 3. This phenomenon was considered to imply the occurrence of shear yielding of the matrix.

The importance of matrix shear yielding in toughening of polymer blends was later discussed by Wu (48,49). Using SEM, Wu studied the fracture surfaces of several PA-6/rubber blends containing different volume fractions of rubber and found that shear yielding was the prevailing energy dissipation process for those blends having a rubber content higher than a critical value. Based on this observation, Wu suggested that there exists a critical interparticle distance for a rubber-particle–toughened polymer blend. When the interparticle distance of a polymer blend is smaller than the critical distance, the matrix material between two rubber particles will undergo shear deformation and dissipate a large amount of the fracture energy. Wu's model is now known as the critical interparticle distance model and is widely accepted by the industry.

D. Crazing–Shear Yielding Model

It is well accepted that a polymer blend can be toughened by more than one mechanism. An extensive microscopic study on a HIPS/PPO blend carried out by Bucknall and coworkers (50–53) showed that a combination of massive crazing and localized shear yielding, i.e., shear bands formation, was responsible for the enhanced toughness found in HIPS/PPO blends. In the development of the crazing-shear yielding model, SEM and TEM were exploited to obtain experimental evidence. Figure 15 shows a SEM photograph of a tensile-deformed, chromic-acid-etched sample, and Fig. 16 is a TEM micrograph taken from a replica of the

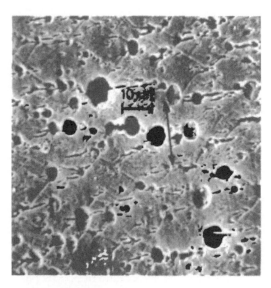

Figure 15 SEM micrograph of a tensile-deformed, chromic-acid-etched HIPS/PPO (50/50) blend. (From Ref. 1.)

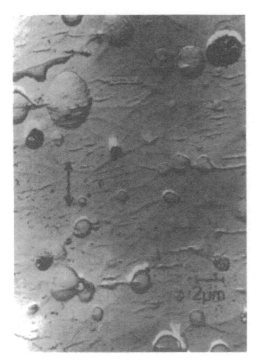

Figure 16   TEM micrograph of a replica of a HIPS/PPO (50/50) blend. (From Ref. 50.)

same specimen. These two photographs clearly display that both shear bands and massive crazes were initiated during the deformation of the HIPS/PPO blend. The crazes grew in a direction perpendicular to the maximum principal tensile stress, whereas the shear bands developed along a path 45° to the applied stress. The interaction between the crazes and the shear bands stabilized the growing crazes and prevented them from developing into harmful cracks. Evidently, the shear bands are not only a source of energy absorption but also act as terminators of the growing crazes. The establishment of this model partially answers why a relatively large particle size is needed for HIPS than for ABS. This is because the former mainly undergoes simple crazing and it is necessary to have large particles to stabilize and stop craze growing. The latter, however, possesses both massive crazes and shear bands, which can stabilize the crazes during fracture propagation.

E.   Cavitation–Shear Yielding Model

As an alternative to the crazing-shear yielding model, the cavitation enhanced matrix shear yielding model was suggested by Yee. Cavitation is an important plas-

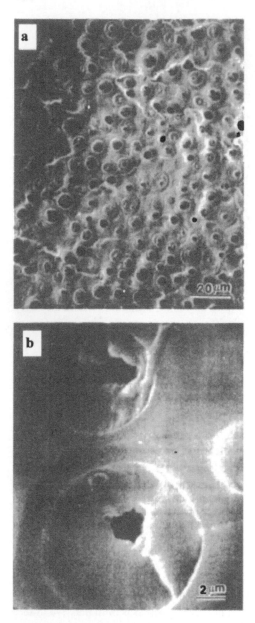

Figure 17   SEM micrographs taken from a fracture surface of rubber-toughened epoxy. Rubber particle cavitation is evident. (From Ref. 54.)

tic deformation frequently observed during the fracture of many ductile materials. Formation of cavities absorbs a large amount of energy and is a major source of toughness. Strong evidence of rubber particle cavitation and matrix shear yielding was obtained using both optical and electron microscopes by Pearson and Yee (44,54,55). Figure 17 are SEM micrographs taken from a fracture surface of a rubber/epoxy blend. Cavitation of rubber particles and matrix shear deformation are unambiguously seen. The key points of the cavitation shear yielding model, as concluded by Yee (54,55), are (1) the materials in front of a crack tip are subjected to a triaxial tensile stress, which promotes brittle failure, and (2) the cavitation process must take place prior to massive shear yielding. Cavitation is probably an essential condition for shear yielding to occur, because the matrix material around the crack tip is prone to brittle fracture without the cavitation process to relieve the triaxial tensile stress field.

To verify the necessity of cavitation prior to matrix shear yielding, the deformation behavior of a rubber/epoxy blend was investigated by Yee and coworkers thoroughly (56). A double-edge–double-notch four-point bend (DEDN-4PB) specimen was employed to probe the deformation history of the materials in the vicinity of a crack tip under different stress states. As illustrated in Fig. 18, the DEDN-4PB specimen has four identical blunt V-notches, two on each side. Under four-point bending, two notches on one side are subject to tension and another

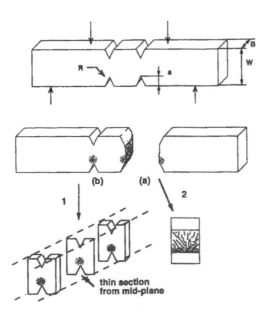

Figure 18 Schematic of double-edge–double-notched four-point bend (DEDN-4PB) specimen preparation. (From Ref. 55.)

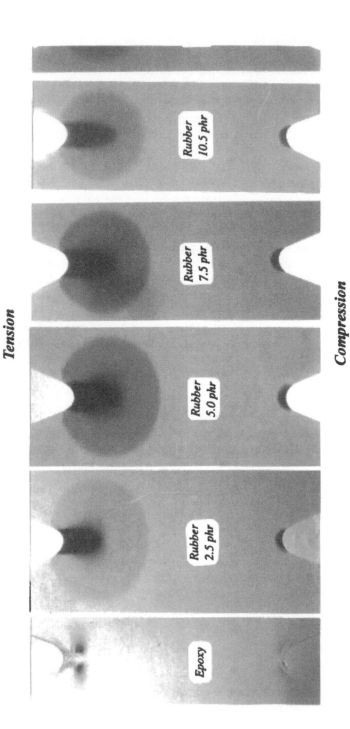

Figure 19   Optical micrographs taken from DEDN samples of rubber-toughened epoxy blends. (From Ref. 55.)

two are under compression. After one pair of notches fails the other pair is in a subcritical condition with damaged zones in front. Optical microcopies of the damage zones are then carried out with a thin section of the broken sample containing the damaged zones. The thin sections may be prepared using the petrographic thin-sectioning technique introduced in Sec. 2. Figure 19 shows the TOM photographs obtained by Yee and coworkers using this technique. Interpretation of the photographs leads to the following points: (1) The plastic zone size in the compressive region is much smaller than that in tensile region. (2) The plastic zone size in the compressive region is invariant with rubber volume fraction but that in the tensile region increases systematically when more rubber is added. (3) A cavitation zone, which is spherical and outside the plastic zone, is clearly seen in the tensile region, but no cavitation zone is visible in the compressive region. These experimental findings imply that the rubber particles in the tensile region were under a triaxial tensile stress and that particle cavitation occurred first. The triaxial stress at the crack tip was released through the cavitation process and enabled massive matrix shear yielding. In the compressive region the rubber particles were subjected to a triaxial compressive stress, hence rubber particle cavitation was suppressed. Without cavitation the triaxial stress field maintained itself and the plastic deformation in that zone was therefore limited.

Direct evidence of such a functional relationship between cavitation and shear yielding has also been reported by Wu and Mai (57). Using a TEM with a single-edge–double-notch four-point-bend (SEDN-4PB) specimen, as shown in Fig. 20, the toughening mechanisms of a PBT/PC/rubber blend were revealed.

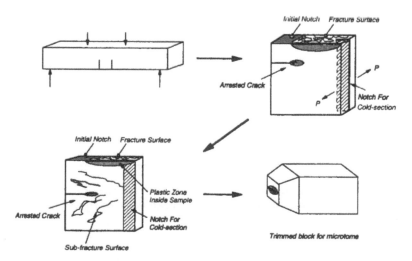

Figure 20 Schematic of single-edge–double-notched four-point bend (SEDN-4PB) specimen preparation.

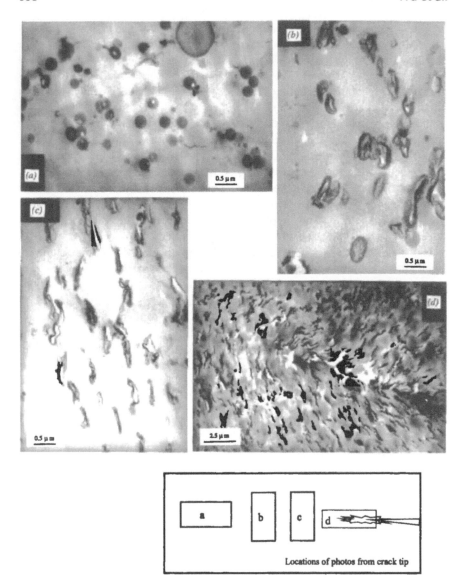

Figure 21   TEM micrographs taken from an ultra-thin section of PBT/PC/rubber blends. The sequence of toughening events can be seen from the set of micrographs.

Similar to the DEDN-4PB technique, the SEDN-4PB specimen has two identical notches on one side. Under four-point bending, one of the two notches propagates through the entire ligament and the other is stopped at a subcritical condition. Thus the unbroken notch preserves information about the damage zone generated just prior to unstable crack propagation. In Wu and Mai's study (57), the material around the surviving notch was thin-sectioned using the ultramicrotome technique and stained with OsO_4 to enhance contrast between phases. The photographs shown in Fig. 21 were taken from different locations on an ultrathin section cut from a SEDN-4PB specimen containing an arrested crack tip. From the micrographs, the sequence of deformation events can be identified. Figure 21a was taken from a location relatively far away from the crack tip. It represents the first step of the sequence. In this step, rubber particle cavitation occurred without visible matrix shear deformation because the cavitated particles are all spherical, which implies that the matrix around the particles was stretched uniformly in all directions. Figure 21b was taken from an area closer to the crack tip. It gives the second step of the sequence. In this step, cavitated particles were obviously distorted in the loading direction, indicating that some shear deformation occurred in the matrix between the cavitated particles. Extensive shear deformation of the matrix between the cavities was found in the third step of the sequence. As demonstrated by the TEM micrograph taken from a location immediately in front of the crack tip, Fig. 21c, the rubber particles were elongated in the principal stress direction several hundred percent, and the internal fracture surfaces of the particles were closed laterally and formed the black rubbery strips. Even more severe deformation of the rubber particles was observed around the blunted crack tip, Fig. 21d, which suggests the occurrence of heavy plastic flow of matrix in the fourth step of the sequence.

The above-described cavitation-shear yielding mechanism has been proven to be dominant for many rubber-toughened polymer blends (58–62).

F. Debonding–Shear Yielding–Crack Bridging Model

Rigid–rigid polymer blends generally consist of two or more rigid polymer components with very little or no rubbery phase. Successful commercial examples have proven that a well-designed rigid–rigid polymer blend combines high toughness, high modulus, and good chemical resistance, leading to an integrated property combination that cannot be achieved by traditional rubber-toughened polymers. Despite their importance, toughening mechanisms for rigid–rigid polymer blends have not received as much attention as those for rubber-toughened polymer blends. The mechanisms suggested for the toughening of this unique class of materials, essentially, stem from the application of existing theories for conventional rubber-toughened polymer blends (64–73).

In recent work by Wu and coworkers (38), a new toughening mechanism for rigid–rigid polymer blends was discovered using SEM and TEM. The polymer

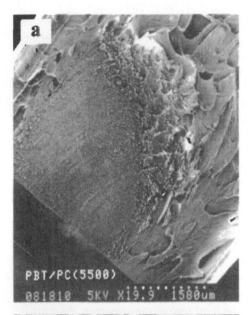

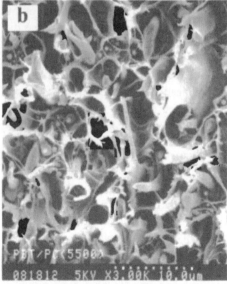

Figure 22 SEM photographs taken from the fracture surface of a PBT/PC (50/50) blend. (a) The entire fracture surface shows two distinct zones. Extensive cavitation and plastic deformation are observed. (b) Bridging domains were stretched into fibrils before breaking.

blends studied were rigid PBT/PC blends having no rubbery phase. Fracture toughness of a PBT/PC blend containing 40% PBT and 60% PC was found to be five to seven times higher than that of homogeneous PC. The toughening mechanisms found with this blend are illustrated in Figs. 22 and 23. The entire fracture surface shown in Fig. 22a consists of two distinct zones, viz., a cavitation zone in the middle core area and a shear zone in the outer skin area. The formation of the cavitation zone implies that there had been a triaxial stress field before the initial crack started to propagate under loading. A close view of the cavitation zone, as shown in Fig. 22b, reveals that widespread cavitation occurred at the boundary between the PBT and PC domains. Some of the very large voids were formed via the coalescence of neighboring small holes. Evidently, the triaxial stress state in the center area was relieved through this cavitation and coalescence mechanism, which followed by general shear deformation. When the micrographs are studied more closely, it is noted that there are many fiberlike broken pieces standing in the holes and surrounded by highly deformed material, as indicated by the arrows in Fig. 22b. It seems that either PBT or PC had been stretched into fibrils during the crack opening process and broken after fracture of the surrounding material. In other words, crack bridging might have happened during crack propagation.

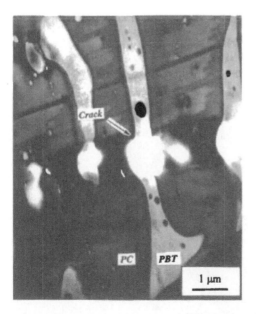

Figure 23 TEM photograph of an arrested crack in a PBT/PC blend. Crazes were found at crack tip and stabilized by PC domains (black phase). Crack bridging of PC domains are also observed.

To verify the hypothesis of crack bridging, TEM with a single-edge–double-notched four-point-bend (SEDN-4PB) specimen was employed for further analysis. Figures 23 and 24 are the TEM photographs taken from a RuO$_4$ stained ultrathin section containing an arrested crack tip. From these pictures the following conclusions can be drawn: (1) Crack bridging did occur during the fracture of the

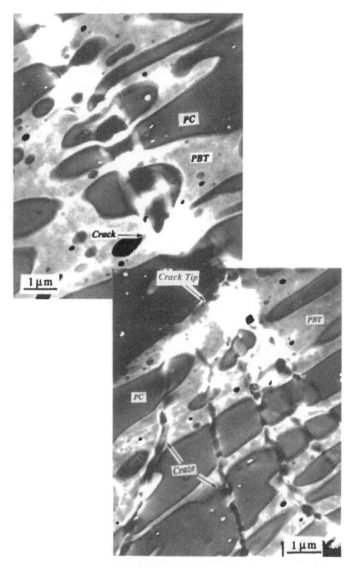

Figure 24 TEM micrograph showing that the bridging domains (PC) were stretched and deformed as the crack opened up.

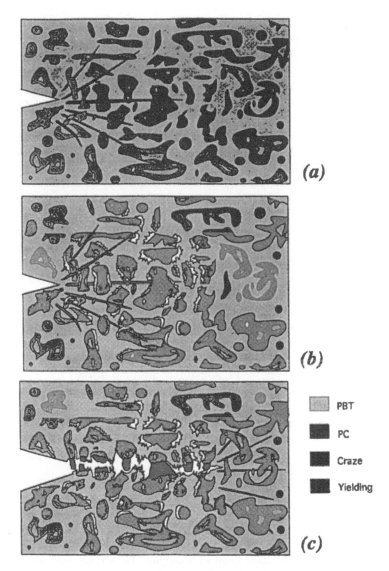

Figure 25 Schematic of toughening mechanisms found in PBT/PC blends.

blends. (2) It was the PC domains that were stretched into fibers and that bridged the crack surfaces. As shown in Figs. 23 and 24, the black PC domains link up the crack surfaces, and shear yielding is found in these bridging particles. It is also clear that the PBT domains were already broken at this stage. (3) The crazes formed in front of the propagating crack tip were stabilized by the bridging PC domains prior to PBT/PC interfacial debonding. In Fig. 24 several crazes are found to initiate and radiate from the crack tip, but no debonding is visible.

Based on this microscopical observation, the authors proposed the debonding–shear yielding–crack bridging model for these rigid–rigid polymer blends. The model attributes the substantial increase in toughness of the PBT/PC blend to the coordination of several toughening mechanisms, namely, debonding between PBT and PC, shear deformation of both PBT and PC, and crack bridging of PC. The authors described the toughening processes as follows. When a cracked sample is subjected to loading, a triaxial stress field will build up in front of the crack tip, which may cause brittle failure of the sample if it is not released. In the traditional rubber-toughened polymer blends, the triaxial stress may be released via rubber particle cavitation. But for the rigid–rigid polymer blends, cavitation stresses of the rigid components are too high for cavitation to occur. In the PBT/PC blend, the triaxial stress first initiates crazes in the PBT phase, but the crazes are stabilized by the PC domains. The stabilization of the crazes enables the triaxial stress to rise and reach the interfacial strength of the PBT/PC phase. As a consequence, interfacial debonding occurs, and voids are formed at the boundary, as shown in Fig. 25a. The localized tiny voids will continue to expand under the increasing triaxial stress, until the polymer ligaments between the neighboring voids become thin enough to undergo shear deformation, Fig. 25b. At this stage massive shear deformation takes place in both PBT and PC and absorbs a tremendous amount of energy. Further increase in the applied load finally causes the failure of the PBT domains. However, the PC domains at this time are only stretched into fibers and continue to bridge the crack surfaces, Fig. 25c. Apparently, the bridging domains not only prevent crazes from growing into harmful cracks but also bridge the crack faces and continuously transfer stress to the material in the vicinity of the crack faces after the main crack has passed by. All these toughening mechanisms definitely increase the damaged plastic zone size and enhance the total fracture toughness.

VI. SUMMARY

Several microscopical techniques, which are widely used in morphology characterization and toughening mechanism studies for polymer blends, are introduced in this chapter. The operation principles of various microscopes are discussed briefly. Particular attention has been given to specimen preparation methods and contrast enhancement techniques. Some recently developed experimental procedures were also reviewed in detail. Examples of both morphology characterization and toughening mechanism studies were presented with a large number of supporting micrographs. A summary of the unique contributions of polymer microscopy in establishing various toughening models for polymer blends was given.

REFERENCES

1. C. B. Bucknall. Toughened Plastics. London: Applied Science Publishers, 1977.
2. D. R. Paul and S. Newman. Polymer Blends. Vol. 1. London, New York, and San Francisco: Academic Press, 1978.
3. D. R. Paul and S. Newman. Polymer Blends. Vol. 2. London, New York, and San Francisco: Academic Press, 1978.
4. J. S. Mijovic and J. A. Koutsky. Etching of polymeric surfaces: a review. Polym. Plast. Technol. Eng. 9:139–179, 1977.
5. M. T. Takemori and A. F. Yee. Fractography and damage. In: K. Takahashi and A. F. Yee, eds. Impact Fracture of Polymers—Materials Science and Testing Techniques. Fukuoka-shi, Japan: Kyushu University, 1992, pp. 331–392.
6. B. D. Favis and J. P. Chalifoux. Effect of viscosity ratio on the morphology of polypropylene/polycarbonate blends during processing. Polym. Eng. Sci. 27:1591–1600, 1987.
7. E. Martuscelli, C. Silvestre, R. Greco, and C. Ragosta. Properties of polystyrene-polyolefin alloys. II. Processing morphology relationship. In: E. Martuscelli, R. Palumbo, and M. Kryszewski, eds. Polymer Blends. New York: Plenum Press, 1980, pp. 295–318.
8. T. Braten. High resolution scanning electron microscopy in biology: artifacts caused by the nature and mode of application of coating material. J. Microsc. (Oxford) 113:53–59, 1978.
9. I. M. Watt, A comparison of gold and platinum sputtered coating for scanning electron microscopy. In: J. M. Sturgess, ed. Proceedings of the Ninth International Electron Microscopy Congress of Electron Microscopy. Toronto: 1978, pp. 94–95.
10. H. S. Slayter. High resolution metal coating of biopolymers. Scanning Electron Microsc. 13:171–182, 1980.
11. G. Goizueta, T. Chiba, and T. Inoue. Phase morphology of polymer blends: scanning electron microscope observation by backscattering from a microtomed and stained surface. Polymer 33:886–888, 1992.
12. G. Goizueta, T. Chiba, and T. Inoue. Phase morphology of polymer blends: 2. SEM observation by secondary and backscattered electrons from microtomed and stained surface. Polymer 34:253–256, 1993.
13. G. M. Brown and J. H. Butler. New method for the characterization of domain morphology of polymer blends using ruthenium tetroxide staining and low voltage scanning electron microscopy (LVSEM). Polymer 38:3937–3945, 1997.
14. J. Dlugosz, M. Folkes, and A. Keller. Macrolattice based on lamellar morphology in SBS copolymer. J. Polym. Sci.: Part B: Polym. Phys. 11:929–938, 1973.
15. K. Kato. Osmium tetroxide fixation of rubber latices. J. Polym. Sci., Part B, 4:35–38, 1966.
16. K. Kato. The osmium tetroxide procedure for light and electron microscopy of ABS plastics. Polym. Eng. Sci. 7:38–39, 1967.
17. H. Yoshimoto, S. Sagae, M. Matsuo, S. Uemura, and Y. Ishida. Crystalline texture of polychloroprene as revealed in ultrathin sections. Kolloid-Z. Z. Polym. 236:116–118, 1970.
18. J. A. Odell, J. Dlugosz, and A. Keller. Ultralthin sectioning of rubbery block copoly-

mers: thickness test and observations of separate microphase units. J. Polym. Sci., Part A-2, Polym. Phys. Edn. 14:861–867, 1976.

19. I. D. Fridman and E. L. Thomas. Morphology of crystalline polyurethane hard segment domains and spherulites. Polymer 21:388–392, 1980.

20. G. Gillberg, L. C. Sawyer, and A. L. Promislow. Tire chord adhesion—a TEM study. J. Appl. Polym. Sci. 28:3723–3743, 1983.

21. R. W. Smith and J. C. Abdries. New methods for electron microscopy of polymer blends. Rubber Chem. Technol. 47:64–78, 1974.

22. J. S. Trent, J. I. Scheinbeim, and P. R. Couchman. Transmission electron microscope studies of polymers stained with ruthenium and osmium tetroxide. Polym. Sci. Technol. 22:205–213, 1983.

23. J. S. Trent, J. I. Scheinbeim, and P. R. Couchman. Electron microscopy of PS/PMMA and rubber-modified polymer blends: use of ruthenium tetroxide as a new staining agent. J. Polym. Sci., Polym. Lett. Edn. 19:315–319, 1981.

24. J. S. Trent, J. I. Scheinbeim, and P. R. Couchman. Ruthenium tetroxide staining of polymers for electron microscopy. Macromolecules 16:589–598, 1983.

25. J. S. Trent, J. I. Scheinbeim, and P. R. Couchman. Ruthenium tetroxide staining of polymers: new preparation methods for electron microscopy. Macromolecules, 17:2930–2931, 1984.

26. R. E. Cohen and A. R. Ramos. Viscoelastic properties of homopolymers and diblock copolymers of butadiene and polyisoprene. In: Multiphase Polymers, Symposium Sponsored by the Division of Polymer Chemistry at the American Chemical Society 175th Meeting, American Chemical Society, Washington D.C., 1979, pp. 237–255.

27. G. Kanig. Contrast method for electron microscopic studies of polyethylene Kolloid-Z. Z., Polym. 251:782–783, 1973.

28. G. Kanig, New electron microscopic study on the morphology of polyethylene. Prog. Colloid Polym. Sci. 57:176–191, 1975.

29. M. C. R. Sotton, Technique for revealing microporous structures: application to fibrous high polymer. Acad. Sci. Ser. B. 270:1261–1264, 1970.

30. K. Hess, E. Gutter, and H. Mahl. Preparation of Perlon for electron microscopic observation of longitudinal period. Naturwissenschaften 46:70–71, 1959.

31. L. C. Sawyer and D. T. Grubb. Polymer Microscopy. London and New York: Chapman and Hall, 1987.

32. M. G. Dobb, D. J. Johnson, A. Majeld, and B. P. Saville. Microvoids in aramid-type fibrous polymers. Polymer 20:1284–1288, 1979.

33. S. Y. Hobbs. Polymer microscopy. J. Macromol. Sci.-Rev. Macromol. Chem. C19(2):221–265, 1980.

34. E. M. Belavtseva and K. Z. Gumargalieva. Investigation of synthetic polymers and molecular crystals by the negative contrast method. Zavodsk. Lab. 29:966–968, 1963.

35. C. W. Hock. Transmission electron microscopy of polypropylene lamellae delineated by staining. J. Polym. Sci. A2 5:471–478, 1967.

36. H. J. Sue, E. I. Garcia-Meitin, B. L. Burton, and C. C. Garrison. A Novel staining technique for studying toughening mechanisms in saturated acrylic rubber-modified polymers. J. Polym. Sci.: Part B: Polym. Phys. 29:1623–1631, 1991.

37. H. Janik, E. Walch, and R. J. Gaymans. Ruthenium tetroxide staining of polybutylene terephthalate (PBT) and polyisobutylene-b-PBT segmented block copolymers. Polymer 33:3522–3524, 1992.

38. J. S. Wu, A. F. Yee, and Y.-W. Mai. Fracture toughness and fracture mechanics of

polybutylene-terephthalate/polycarbonate/impact modifier blends. Part III: Fracture toughness and mechanics of PBT/PC blends without impact modifiers. J. Mat. Sci. 29:4510–4522, 1994.

39. X. H. Zhu and C.-M. Chan. A novel method to detect the increase in the rubber concentration at the interface between the poly(vinyl chloride) (PVC) and styrene-butadiene copolymer (SBR) phase in compatibilized PVC/SBR blends. Macromolecules 31:1690–1693, 1998.

40. E. H. Merz, G. C. Claver, and M. J. Baer. Heterogeneous polymeric systems. J. Polym. Sci. 22:325–341, 1956.

41. S. Kunz-Douglass, P. W. R. Beaumont, and M. F. Ashby. A model for the toughness of epoxy-rubber particulate composites. J. Mater. Sci. 15:1109–1123, 1980.

42. S. C. Kunz and P. W. R. Beaumont. Low temperature behaviour of epoxy-rubber particulate composites. J. Mater. Sci. 16:3141–3152, 1981.

43. S. Bandyopadhyay. Crack propagation studies of bulk polymeric materials in the scanning electron microscope. J. Mater. Sci. Lett. 3:39–43, 1984.

44. R. A. Pearson and A. F. Yee. Influence of particle size and particle size distribution on toughening mechanisms in rubber-modified epoxies. J. Mater. Sci. 26:3828–3844, 1991.

45. C. B. Bucknall and R. R. Smith. Stress-whitening in high impact polystyrene. Polymer, 6:437–446, 1965.

46. R. P. Kambour and R. R. Russell. Electron microscopy of crazes in polystyrene and rubber modified polystyrene: use of iodine-sulphur eutectic as a craze reinforcing impregnant. Polymer 12:237–246, 1971.

47. S. Newman and S. Strella. Stress–strain behaviour of rubber-reinforced glassy polymers. J. Appl. Polym. Sci. 9:2297–2310, 1965.

48. S. Wu. Impact fracture mechanisms in polymer blends: rubber-toughened nylon. J. Polym. Sci. (Phys.) 21:699–716, 1983.

49. S. Wu. Phase structure and adhesion in polymer blends: a criterion for rubber toughening. Polymer 26:1855–1863, 1985.

50. C. B. Bucknall and D. Clayton. Dilatometric studies of crazing in rubber-toughened plastics. Nature (London) Phys. Sci. 231:107–108, 1971.

51. C. B. Bucknall and D. Clayton. Rubber-toughening of plastics. Part 1: Creep mechanisms in HIPS. J. Mater. Sci. 7:202–210, 1972.

52. C. B. Bucknall, D. Clayton, and W. E. Keast. Rubber-toughening of plastics. Part 2: Creep mechanisms in HIPS/PPO blends. J. Mater. Sci. 7:1443–1453, 1972.

53. C. B. Bucknall, D. Clayton, and W. E. Keast. Rubber-toughening of plastics. Part 3: Strain damage in HIPS and HIPS/PPO blends. J. Mater. Sci. 8:514–524, 1973.

54. R. A. Pearson and A. F. Yee. Toughening mechanisms of elastomer-modified epoxies. Part 2: Microscopy studies. J. Mater. Sci. 21:2475–2488, 1986.

55. R. A. Pearson and A. F. Yee. Toughening mechanisms in elastomer-modified epoxies. Part 3: The effect of cross-link density. J. Mater. Sci. 24:2571–2580, 1989.

56. A. F. Yee, D. Li, and X. J. Li. The importance of constraint relief caused by rubber cavitation in the toughening of epoxy. J. Mater. Sci. 28:6392–6398, 1993.

57. J. S. Wu and Y.-W. Mai. Fracture toughness and fracture mechanisms of PBT/PC/IM blends. Part II: Toughening mechanisms. J. Mat. Sci. 28:6167–6177, 1993.

58. R. J. M. Borggreve, R. J. Gaymans, J. Schuijer, and J. F. Ingen Housz. Brittle-tough transition in nylon-rubber blends: effect of rubber concentration and particle size. Polymer 28:1489–1496, 1987.

59. R. J. M. Borggreve and R. J. Gaymans. Impact behaviour of nylon-rubber blends 4. Effect of the coupling agent, maleic anhydride. Polymer 30:63–70, 1989.

60. R. J. M. Borggreve, R. J. Gaymans, and J. Schuijer. Impact behaviour of nylon-rubber blends: 5. Influence of mechanical properties of the elastomer. Polymer 30:71–77, 1989.

61. R. J. M. Borggreve, R. J. Gaymans, and H. M. Eichenwald. Impact behaviour of nylon-rubber blends: 6. Influence of structure on voiding process; toughening mechanism. Polymer 30:78–83, 1989.

62. Y. Seo, S. S. Hwang, K. U. Kim, J. Lee, and S. I. Hong. Influence of the mechanical properties of the dispersed phase upon the behaviour of nylon/rubber blends: crosslinking effect. Polymer 34:1667–1676, 1993.

63. I. Narisawa and M. Ishikawa. Crazing in semicrystalline thermoplastics. Adv. Polym. Sci. 91/92:353–391, 1990.

64. T. Kurauchi and T. Ohta. Energy absorption in blends of polycarbonate with ABS and SAN. J. Mater. Sci. 19:1699–1709, 1984.

65. K. Koo, T. Inoue, and K. Miyasaka. Toughened plastics consisting of brittle plastics and ductile matrix. Polym. Eng. Sci. 25:741–746, 1985.

66. Y. Fujita, K. Koo, J. C. Angola, T. Inoue, and T. Sakai. Toughened plastics: toughening phenomenon and its mechanism in polymer blends consisting of ductile polymer and brittle polymer. Kobunshi Ronbunshu 43:119–131, 1986.

67. T. Fukui, Y. Kikuchi, and T. Inoue. Elastic-plastic analysis of the toughening mechanism in rubber-modified nylon: matrix yielding and cavitation. Polymer 32:2367–2371, 1991.

68. J. C. Angola, Y. Fujita, T. Sakai, and T. J. Inoue. Compatibilizer-aided toughening in polymer blends consisting of brittle polymer particles dispersed in a ductile polymer matrix. Polym. Sci., Part B: Polym. Phys. 26:807–816, 1988.

69. M. E. J. Dekkers, S. Y. Hobbs, and V. H. Watkins. Morphology and deformation behaviour of toughened blends of poly(butylene terephthalate), polycarbonate and poly(phenylene ether). Polymer 32:2150–2154, 1991.

70. M. E. J. Dekkers, S. Y. Hobbs, and V. H. Watkins. Toughened blends of poly(butylene terephthalate) and BPA polycarbonate. Part 2: Toughening mechanisms. J. Mater. Sci. 23:1225–1230, 1988.

71. S. Y. Hobbs, M. E. J. Dekkers, and V. H. Watkins. Toughened blends of poly(butylene terephthalate) and BPA polycarbonate. Part 1: J. Mater. Sci. 23:1219–1224, 1988.

72. S. Y. Hobbs and C. F. Pratt. The effect of skin-core morphology on the impact fracture of poly(butylene terephthalate). J. Appl. Polym. Sci. 19:1701–1722, 1975.

73. S. Y. Hobbs, V. H. Watkins, and J. T. Bendler. Diffusion bonding between BPA polycarbonate and poly(butylene terephthalate). Polymer 31:1663–1668, 1990.

19

Deformation Mechanisms in Toughened PMMA

Philippe Béguelin, Christopher J. G. Plummer, and H. H. Kausch
Ecole Polytechnique Fédérale de Lausanne, Lausanne, Switzerland

I. INTRODUCTION

Poly(methylmethacrylate) (PMMA), which has a glass transition temperature T_g of about 100°C, is one of the more brittle amorphous thermoplastic materials. The molecular mobility in the glassy state is low, and the polymer chains are unable to undergo large-scale molecular motions in response to rapidly applied external stresses or impacts. Thus although it has relatively good global creep properties at ambient temperature, PMMA is brittle and notch sensitive. Fracture proceeds by the formation of isolated crazes, associated with stress concentrations, a phenomenon that has been extensively described in the literature (1–6).

Owing to its transparency, PMMA is often used in applications that require good optical properties. However, its brittleness is a limiting factor in many cases (such as in the transport business). Indeed, even for high molecular weights, the energy dissipated during the fracture of commercial grades of unmodified PMMA is significantly lower than for polycarbonate (PC), which is one of its main competitors for optical applications. Nevertheless, as with many other polymers, a significant improvement in the toughness of PMMA can be obtained by combining it with a discrete secondary phase that has a subambient T_g. Since this modifier is rubbery at room temperature it introduces soft, incompressible domains into the matrix. The size of these domains should be chosen to maximize their influence on the intrinsic deformation mechanisms of the matrix so as to optimize the rate of energy dissipation during crack development. This approach is currently used to improve the toughness of a wide range of thermoplastics, ranging from amorphous and semicrystalline thermoplastics to highly cross-linked thermosets (notably the epoxy-based thermosets).

A. Toughening Strategies

Several different routes are used for the preparation of rubber-toughened polymers. The formation of cross-linked rubbery regions by phase separation during bulk polymerization has been commonly employed in the manufacture of high-impact polystyrene (HIPS) (7), for example, as well as certain acrylonitrile-butadiene-styrene copolymers (ABS) and carboxy-terminated nitrile rubber (CTBN) toughened epoxy resins (8–11). The structure of the materials obtained by this route depends on the chemical and physical changes that occur during the polymerization process. Since the final morphology of the modifier phase is of primary importance, preformed particles of different composition and size provide an attractive alternative to the in-situ polymerization of the modifier phase. The advantage of this route is that it allows independent control of the size, the morphology, and the content of the modifier particles.

1. Preformed Spherical Particles

Preformed monodisperse modifier particles prepared by emulsion polymerization (12–15) are widely used to toughen a variety of matrices. Modern rubber-toughened acrylics are manufactured using this technique, spherical particles being added either directly to the acrylic monomer before polymerization or to molten low-molecular-weight PMMA using a screw extruder. The particle dispersion in the matrix depends on the mixing conditions and on the chemical nature of the particle surface.

Efficient stress transfer between the modifier phases and the matrix is in most cases essential for effective toughening. The rubbery phase (typically a cross-linked copolymer based on either butadiene or n-butyl acrylate) therefore needs to adhere to the matrix (16,17). This can be achieved using a core–shell particle structure, illustrated in Fig. 1, which not only promotes good adhesion to the matrix but also reduces particle agglomeration. The external shell is made from macromolecules that are thermodynamically compatible with the matrix in the melt (MMA-based copolymers are often used), grafted on to the rubbery domains of the particles. The strong polymer–polymer interaction resulting from the interpenetration and entanglement of the external shell with the matrix ensures adequate adhesion.

Different morphologies and sizes of the preformed particles have been studied and adapted for practical applications. The simplest is the two-layer morphology described above (referred to henceforth as 2L). However, commercial grades of rubber-modified PMMA are often based on multilayer particle structures such as the three layer (3L) particles also shown in Fig. 1. The 3L particles consist of a

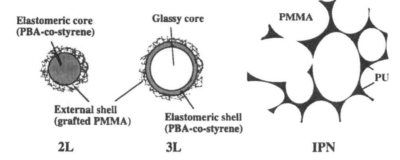

Elastomeric core
(PBA-co-styrene)

Glassy core

PMMA

External shell
(grafted PMMA)

Elastomeric shell
(PBA-co-styrene)

PU

2L　　　　　　**3L**　　　　　　**IPN**

Figure 1 Schematic representation of the modifier morphologies referred to in the text: 2L, soft core particles; 3L, hard core–soft shell particles; IPN, an interpenetrating PU/PMMA network.

glassy PMMA core surrounded by a rubbery n-butyl acrylate inner shell and a grafted PMMA outer shell. The interest in such multilayer structures stems from the systematic study by Lovell and coworkers of the influence of the particle morphology on the mechanical properties of modified PMMA (18,19). They showed that important quantities such as the Young modulus E and the yield stress σ_y decrease when the rubber content is increased. For a given particle volume fraction and external diameter, the overall rubber content is lower for the 3L particles than for the 2L particles, limiting the degradation in static mechanical properties. Moreover, as will be discussed later, the 3L morphology results in improved toughness at impact loading rates (20–22).

2. Interpenetrating Polymer Networks

The impact modification of PMMA by the mechanical blending of a low-molecular-weight matrix with preformed particles can only be achieved by a process involving shear deformation, such as extrusion or injection molding. Since shear deformation is not associated with the casting process, which is another common means of manufacturing acrylic-based products, the in-situ formation of interpenetrating polymer networks (IPNs) during casting represents an alternative means of producing rubber-toughened materials.

The IPNs were developed in the 1960s by Sperling and coworkers (23–25). The term IPN refers to the permanent interlocking of two or more different polymers, formed by interstitial polymerization. The synthesis involves mixing all the monomers or prepolymers, together with their corresponding cross-linking agents, either simultaneously [to give "simultaneous interpenetrating networks" (SINs)] or sequentially. The final morphology of an IPN generally consists of a fine dispersion of distinct phases, as opposed to an interpenetrating network at the molecular level. The length scales characteristic of this morphology depend on the kinetics of phase separation (25,26) and may range from a few nanometers up to several microns.

Several groups have synthesized and characterized acrylic IPNs based on polyurethane (PU)/PMMA networks (27–35). In some cases, a wide range of PU contents has been considered (33,34), while in others, the PU content did not exceed 50% (26–31). The studies of Allen and coworkers (27–32) focused on network formation and on thermal and mechanical characterization. They found that the impact properties depend on the fraction of PU and its cross-link density. The T_g of the PU phase is shifted towards the ambient temperature when its cross-link density is increased, but for significant enhancement of the impact properties, it should remain below $-30°C$. Poor impact performance is also observed in cases where the PU phase is grafted to the PMMA phase (30).

As with particle-modified PMMA, the static properties of the IPNs degrade when the rubber content is increased, so that the optimization of such systems is difficult and may depend on the nature of the application. Although in most prior studies, the PU content was varied by steps of 5 to 10%, Heim et al. (35) have synthesized a range of SINs in which the PU content increases gradually up to 10%. They report a significant improvement in impact properties with only 5–6% PU as the secondary phase. They also considered the influence of the kinetics of the network formation on the final morphology and that of the cross-link density of the rubbery phase on the mechanical properties. Their results show that the morphology of the cocontinuous networks (and therefore the optical and mechanical properties of the final material) are determined by the cross-link density of the PU network at the onset of the MMA polymerization.

B. Acrylic Matrices

Although unmodified ultra-high-molecular-weight PMMA (several million g/mol) has better fracture properties than that of lower M_W (2), the impact-modified PMMAs are based on low M_W matrices. Indeed, when preformed particles are used, a low matrix melt viscosity is essential, and this can only be achieved if the M_W does not exceed 100,000–200,000 g/mol. The PMMA phase of the PU/PMMA IPNs is cross-linked during the formation of the network, and the M_W is consequently expected to be effectively infinite (and the PMMA hence insoluble).

C. Optical Properties

To maintain good optical properties, the rubbery domains in modified PMMA should be smaller than the wavelength of visible light, and/or the refractive index of the secondary phase should match that of the matrix. Thus the size and the nature of the rubbery domains are of primary importance, not only for the mechanical properties but also for the optical properties, which places relatively severe constraints on the choice of the modifier. Transparent PU/PMMA interpenetrating polymer networks can also be synthesized, given appropriate morphologies and compositions (32,35). Nevertheless, as will be seen below, both preformed particle modified PMMAs and IPNs are often observed to undergo severe stress whitening during plastic deformation, even when they are highly transparent in the undeformed state.

D. Mechanical Properties

As we have already seen, the mechanical properties of rubber-modified PMMA are highly dependent on microstructural factors and on the mechanical interactions between the modifier phase and the glassy matrix. Several authors have studied the influence of the modifier phase content (16,36–40). In these studies, the highest values of the toughness were reported for modifier phase volume fractions of between 25 and 40%. The optimum size of the particles has also been investigated and found to be of the order 250 nm (16–19). Nevertheless, owing to the different morphologies, testing methods, and strain rates used in the different studies, it is difficult to compare the results in detail.

Since the main purpose of modification is to improve the impact properties, while maintaining as far as possible the static properties and creep resistance of unmodified PMMA, mechanical characterization should cover a wide range of test speeds. Furthermore, such an approach is helpful in understanding the basic mechanisms responsible for the energy dissipation at impact loading rates, the occurrence of transitions in the deformation mechanisms as a function of test speed or temperature, as well as the role of the morphology of the modifier phase.

II. THE INFLUENCE OF MORPHOLOGY ON MECHANICAL PROPERTIES

In what follows, we describe the mechanical properties and micromechanisms of tensile deformation and fracture in bulk PMMA toughened with different modifier morphologies and contents, including a PU/PMMA-based IPN. Rather than undertake a systematic study of the morphological parameters relevant to rubber toughening (such as the size, shape, and volume fraction of the rubbery phase), our aim here is to provide a better understanding of the micromechanisms of deformation, both of the modifier and of the surrounding matrix. In order to investigate the influence of the strain rate on these micromechanisms, the test speed has been varied over several decades, with the highest test speeds encompassing impact conditions.

A. Morphological Characterization

Thin sections (about 200 nm thick) were prepared from RuO$_4$-stained samples of each type of material, using a Diatome 45° diamond knife and the Reichert-Jung Ultracut E ultramicrotome, and observed by transmission electron microscopy (TEM) (Philips EM 300 at 100 kV). The morphologies of the undeformed samples are shown in Fig. 2: (a) 15% 2L particles (2L15), with a cross-linked butyl acrylate-co-styrene interior and a PMMA shell, and an overall diameter of about 160 nm; (b) 20%, and (c) 30%, of 3L particles with a PMMA core, surrounded by an inner shell of cross-linked n-butyl acrylate-co-styrene and an outer shell of grafted PMMA, and an overall diameter of about 250 nm (3L20 and 3L30, respectively). Fig. 2d shows the structure of an IPN containing 6 volume % PU chosen for the present investigation, and which is similar to that described by Heim and coworkers (35). The matrix molecular weight average of the particle-modified materials was determined by gel permeation chromatography to be about 130,000 g/mol, whereas the IPN was insoluble, confirming the matrix to be highly cross-linked.

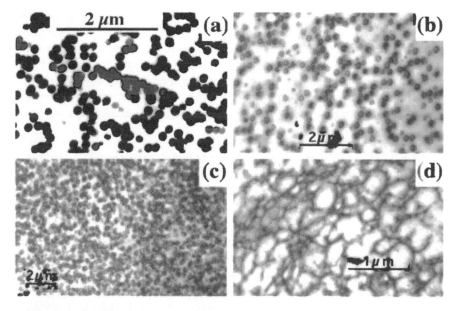

Figure 2 TEM micrographs of thin sections stained with RuO$_4$: (a) 15% 2L particles (2L15); (b) 20% and (c) 30% 3L particles ((2L30) and (3L30)); (d) 6% PU/PMMA IPN.

B. Tensile Behavior

To compare their tensile properties, dog bone specimens of 3L30 and the IPN were deformed in tension over an extended range of crosshead speeds (10^{-4} ms^{-1} $\leq v \leq 10$ ms^{-1}), using a Schenck servohydraulic high-speed tensile test apparatus and a remote optical system to measure the displacements. The experimental approach developed in our laboratory is based on the reduction of the acceleration at high rates (> 0.1 ms^{-1}) to avoid initial transient dynamic effects in the specimen; thus it allows reliable force–displacement data to be obtained for test speeds of up to 10 ms^{-1} (41,44–46).

Typical stress–strain behavior at different strain rates is shown in Fig. 3. In both materials, E and σ_y increased drastically with test speed, whereas the ultimate strain decreased. Thus the overall viscoelastic properties in tension were clearly highly strain-rate-dependent in each case.

C. Stress Whitening

As pointed out in Sec. I.C, tensile deformation in PMMA modified by spherical particles is often characterized by intense stress whitening. Here, in-situ optical measurements of (white) light transmitted by the specimen have been performed using an optoelectronic setup (44). In Fig. 4 the stress σ and the transmitted light T are shown as functions of the longitudinal strain ε in a typical measurement of this type carried out on a 3 mm thick 3L30 specimen deformed in tension at a strain rate of 4×10^{-4} s^{-1}. Stress whitening, as reflected by a decrease in T, could be detected at very low strains (about 1%), but T only began to decrease substantially near the yield point of the stress–strain curve. The overall behavior can be summarized in terms of the three distinct domains indicated in Fig. 4:

Domain I: In the linear and reversible nonlinear region of the stress–strain curve, there is little loss in transmittance, and a cross-shaped diffraction pattern is observed. In this domain, only slight stress whitening can be discerned with the naked eye.

Domain II: The transmittance decreases drastically at the yield point of the stress–strain curve, and the diffraction pattern disappears.

Domain III: In the plateau region of the stress–strain curve, the specimen appears totally white and only minimal transmittance can be detected by the optoelectronic device.

Surprisingly, the PU/PMMA IPN shows very similar behavior to 3L30 over a wide range of test conditions. The yield strain ε_y and the threshold defined by the intercepts of domains I and II, ε_s (as shown in Fig. 4) have been plotted for both materials as functions of strain rate $\dot{\varepsilon}$ in Fig. 5. From these plots, it is clear that

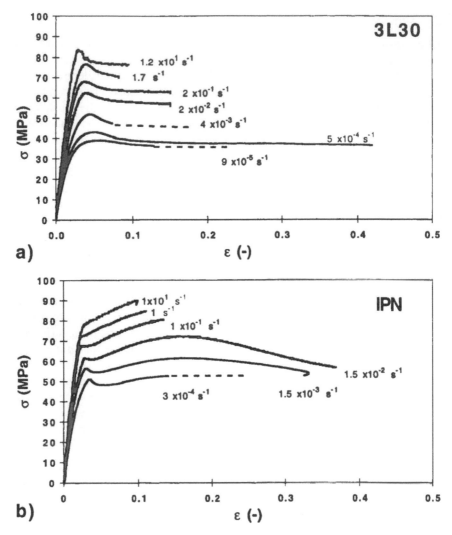

Figure 3 Typical stress σ versus strain ε curves over six decades of strain rate in (a) 3L30 and (b) the PU/PMMA IPN.

stress whitening is closely related to the yielding process in both types of materials at all strain rates. Thus, in general, stress whitened regions observed in toughened PMMAs can be identified with regions that have undergone yield. The damage in the materials can nevertheless be rapidly healed by heating the matrix above T_g, under which conditions the stress whitening disappears.

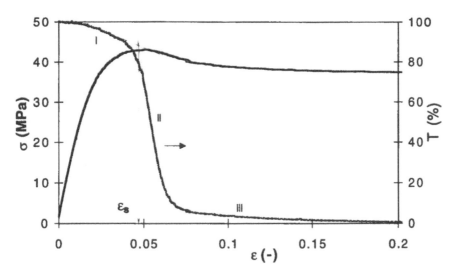

Figure 4 Plot of stress σ and optical transmission T against strain ε for 3L30 tested at a strain rate of $4 \times 10^{-4}\ \text{s}^{-1}$.

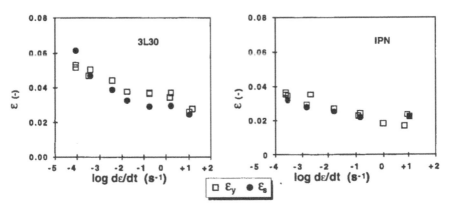

Figure 5 Comparison of the yield strain ε_y and the threshold strain ε_s at which T begins to decrease rapidly, plotted versus the log of the strain rate $\dot{\varepsilon}$.

D. Fracture Behavior

Although impact tests are critical for the assessment of the potential of rubber-toughened PMMA for many applications, and are widely represented in the literature, data from standard methods (the Izod and Charpy tests, for example) are often difficult to interpret in terms of the intrinsic response of a given material to impact loading, owing to dynamic effects (42,43). Here, fracture parameters were obtained from precracked notched compact tension (CT) specimens deformed over an extended range of crosshead speeds (10^{-4} ms^{-1} $\leq v \leq$ 10 ms^{-1}) using the same procedures as referred to in Sect. II.B (44–46). The specimens were machined from 4.5 to 10 mm thick extruded or compression-molded sheets containing different volume fractions of 2L or 3L modifier particles. The notching and precracking were carried out as described elsewhere (22), care being taken to avoid introducing residual stress to the crack tip in the more ductile materials. Sample thicknesses were chosen to ensure that there was no significant variation in the measured fracture mechanics parameters for further increases in thickness. The critical stress intensity factor for crack initiation, K_{IC}, was derived from the maximum force, and the critical strain energy release rate, G_{IC}, from the area under the force displacement curve at crack initiation (22).

Results from fracture mechanics analysis of the CT tests are shown in Figs. 6 and 7. Figure 6 compares the performance of 3L30 with that of samples containing 15, 30, and 45% of 2L particles (2L15, 2L30, and 2L45, respectively) as well as that of unmodified PMMA, illustrating the improved performance of the 3L particles mentioned in Sect. I.A. All the 2L-particle-filled materials underwent a

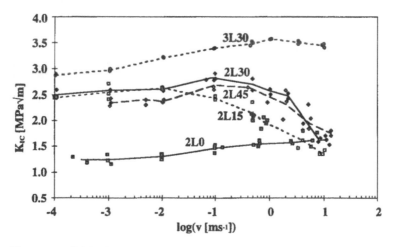

Figure 6 Critical stress intensity factor K_{IC} plotted versus the log of the test speed v for neat PMMA (2L0), for PMMA modified with 15, 20 and 30% 2L particles (2L15, 2L30, 2L45), and for 3L30.

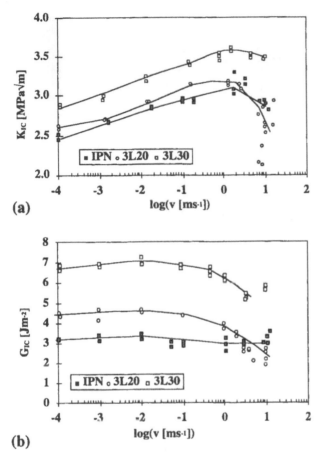

Figure 7 (a) Critical stress intensity factor K_{IC} plotted against the log of the test speed, v, for 3L20, for 3L30, and for the IPN. (b) Critical strain energy release rate G_{IC} plotted against log (v).

tough–brittle transition as impact speeds were approached, reflected by a sharp decrease in K_{IC} and fully unstable crack propagation (although no substantial decrease in K_{IC} occured in the unmodified PMMA (2L0) at the highest test speeds). In Fig. 7, results for 3L20 and 3L30 are compared with results for the IPN. No tough–brittle transitions were seen in either the IPN or the 3L30 at high loading rates, but a tough–brittle transition did occur in 3L20 at a crosshead speed of 10 ms^{-1}. This transition was again reflected by a sharp drop in K_{IC} as shown in Fig. 7a. G_{IC}, plotted in Fig. 7b, decreased at crosshead speeds greater than 1 ms^{-1} in both 3L-particle-modified PMMAs, but there was little evolution in that of the

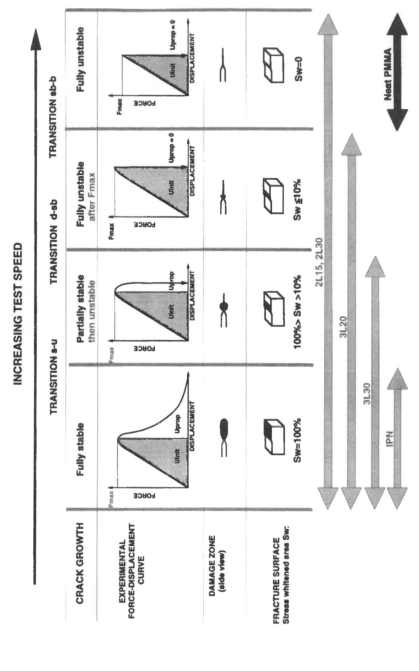

Figure 8 Schematic of the stages of evolution of the fracture behavior with test speed.

IPN over the whole range of strain rates considered here. Moreover, the IPN showed stable crack growth even at the highest test speeds, whereas the decrease in fracture resistance at high speeds in the 3L30 was accompanied by a transition to semistable crack propagation. In other words, the crack became unstable, but only after initial development of a damage zone, accompanied by crack blunting. Similar transitions also appeared as precursors to the fully unstable behaviour observed in the more fragile materials at high strain rates.

The damage zone in the CT tests was again typically characterized by intense stress whitening around the crack tip. In each case, the size of the stress whitened zone decreased progressively as the test speed was increased, and the zone disappeared altogether when the crack entered unstable propagation. This was reflected by progressive embrittlement as the strain rate was increased, presumably because the time available for the damage mechanisms associated with stress whitening became too limited to permit their full development (leading to a decrease in energy dissipation at the crack tip).

A schematic representation of the different stages of the evolution of the fracture behavior with test speed (such as is observed in the more brittle samples) is given in Figure 8. At low test speeds, where the propagation was fully stable, the whitened zone covered the entire fracture surface. As the test speed increased, and the crack became semistable, the whitened area of the fractured ligament (which corresponded to the stable crack growth area) and the energy of propagation decreased continuously. On further increases in test speed, crack propagation became fully unstable beyond the maximum in the force–displacement curve, and the additional energy required to fracture the specimen dropped sharply. At these speeds, the stress whitened zone was restricted to a process zone that developed ahead of the precrack during the initiation process. Finally, at the highest speeds, unstable fracture initiated in the linear portion of the force–displacement curve, coinciding with the total disappearance of toughening effects and stress whitening at the crack tip. Under these conditions, the toughness of the modified systems that showed a full ductile–brittle transition (2L15, 2L30, and 3L20) fell below that of unmodified PMMA in some cases.

III. CONTRIBUTIONS OF ELASTIC, CAVITATIONAL, AND SHEAR DEFORMATION

To distinguish the different processes contributing to the overall tensile deformation, low-speed tests (2 to 200 mm/min, corresponding to strain rates of 5×10^{-4} and 4×10^{-2} s^{-1}, respectively) have been performed at room temperature on 3L20, 3L30, and the IPN. The volume change during the deformation process was measured using simultaneous longitudinal and lateral extensometry, following the approach proposed by Frank and Lehmann in 1986 (47). The elastic strain ε_{elast}

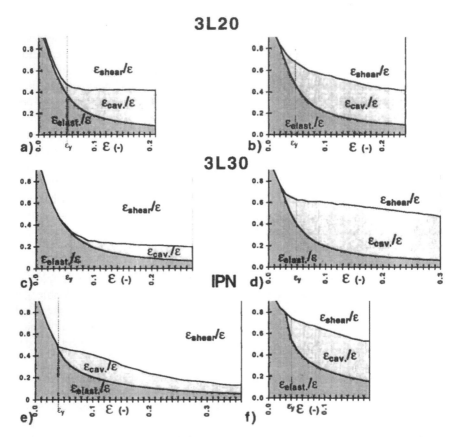

Figure 9 Separation of the different contributions to the total strain: ratio of the elastic strain $\varepsilon_{elast}/\varepsilon$, of the cavitational strain $\varepsilon_{cav}/\varepsilon$, and of the shear strain $\varepsilon_{shear}/\varepsilon$ to the total strain, plotted against the total strain ε: (a) 3L20 at $\dot{\varepsilon} = 5 \times 10^{-4}$ s^{-1} (2 mm/min); (b) at $\dot{\varepsilon} = 3.5 \times 10^{-2}$ s^{-1} (200 mm/min); (c) 3L30 at $\dot{\varepsilon} = 5 \times 10^{-4}$ s^{-1} (2 mm/min); (d) at $\dot{\varepsilon} = 3.5 \times 10^{-2}$ s^{-1} (200 mm/min); (e) PU/PMMA IPN at $\dot{\varepsilon} = 5 \times 10^{-4}$ s^{-1} (2 mm/min), and (f) at $\dot{\varepsilon} = 4.3 \times 10^{-2}$ s^{-1} (200 mm/min).

was calculated assuming that E did not change significantly during the deformation process. Based on the same assumption, the permanent cavitational strain ε_{cav} was calculated by

$$\varepsilon_{cav} = (1 + \varepsilon)(1 + \varepsilon_{lat})^2 - 1 - \varepsilon_{elast}(1 - 2\nu_{elast}) \tag{1}$$

where ε is the longitudinal strain, ε_{lat} the lateral strain, and ν_{elast} the Poisson ratio measured in the elastic region. The shear strain ε_{shear} was taken to be the nonelastic and noncavitational strain, that is,

$$\varepsilon_{shear} = \varepsilon - \varepsilon_{elast} - \varepsilon_{cav} \tag{2}$$

As shown in Figure 9, the permanent cavitational strain was clearly detectable using this approach. Dilatational deformation mechanisms occurred with all types of morphology under the above test conditions, so that one may conclude that a spherical modifier morphology is not necessary for substantial cavitational strain to develop in these materials. Although some cavitation was detected prior to yielding at ε_y in 3L20 and 3L30 at the higher test speed (Figs. 9b and d), cavitation was in all cases closely related to the overall yielding of the material. In the regime of strain rates under consideration, the amount of permanent cavitational strain increased with the strain rate in 3L30 and IPN. These measurements therefore confirm the tendency, reported elsewhere, for increased volume deformation to occur as the strain rate is increased (20,47–49). Thus in 3L30 strained to 30% at a strain rate of $4 \times 10^{-2} s^{-1}$, the volume strain was about 12%, which compares with only 3% at a strain rate of $5 \times 10^{-4} s^{-1}$. Since the elastic contribution to the total strain at this level of deformation is relatively minor, simple (isovolumetric) shear deformation of the matrix must also play an important role. This is equally true of the IPN, which showed similar behavior to that of 3L30 in simple tension at room temperature. On the other hand, 3L20 showed relatively little change within the range of strain rates investigated. For this particular material, cavitational deformation was already significant at the lower strain rate, the volume strain being of the order of that measured in 3L30 and the IPN at the higher strain rate.

A decrease in stress whitening has also been reported for particle-modified PMMAs strained in tension either at low strain rates or under creep conditions, as well as in samples tested at higher temperatures (48,49). Low levels of stress whitening are linked not only to a decrease in cavitation of the modifier particles but also to the suppression of crazing, as suggested by small-angle x-ray scattering (49,50) and to some extent by TEM of thin sections (18).

IV. TEM OBSERVATION OF THE DEFORMATION MECHANISMS

The micromechanisms of deformation were studied by TEM [Philips EM300 (100 kV), Philips CM 20 (200 kV)]. Thin sections were prepared from bulk specimens deformed in tension and stained in RuO_4 as described in Sect. II.A.

A. PMMA Modified with Spherical Particles

As suggested in the previous section, in tensile tests at low speeds (and/or high temperatures), both crazing and cavitation are apparently suppressed in simple tension. This is difficult to demonstrate unambiguously by TEM because of relaxation of unloaded samples, which may result in craze closure either prior to or during sectioning, microtome damage (which may result in crazelike features), and, not least, beam damage. Nevertheless, limited modifier cavitation at high temperatures appears to be confirmed by systematic changes in the appearance of thin sections of unstained samples containing 40 volume % of 3L particles (3L40), as shown in Fig. 10a and b. The sample deformed at 60°C showed little stress whitening optically. In fact some crazing was visible in samples strained at room temperature, as shown in Fig. 10c. This was more apparent in samples strained at higher speeds and/or containing lower particle volume fractions and stained in RuO_4 immediately after deformation, i.e., prior to extensive relaxation and craze closure (Fig. 10d). Although RuO_4 staining may obscure cavitation of the elastomer owing to precipitation of RuO_2 within the voids, it has been found to be of great help in revealing the existence of matrix crazing.

To investigate crack tip microdeformation in the CT specimens, a limited amount of stable crack tip propagation was in some cases assured by testing two nominally identical specimens mounted in series so that failure of one specimen resulted in instantaneous unloading of the second (45). In the specimens of interest, the notch was slightly wedged open during subsequent staining in aqueous

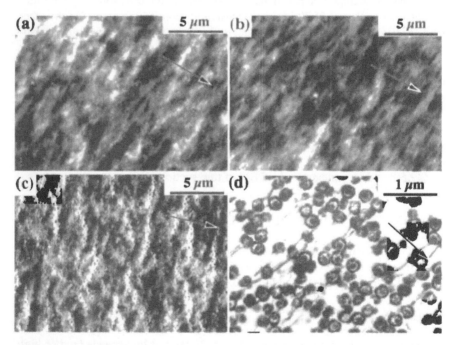

Figure 10 TEM of thin sections from samples deformed in simple tension: (a) 40 volume % 3L (3L40) deformed to 27% at 40°C and at 10^{-3} s^{-1}; (b) as in (a), but deformed at 60°C; (c) as in (a), but deformed at room temperature; (d) thin section from 3L30 strained at 10^{-2} s^{-1} at room temperature and stained with RuO_4.

RuO$_4$, and thin sections (about 100 nm in thickness) were taken from around the crack tip for TEM investigations. The stain was able to penetrate over relatively large distances into the voided/crazed crack tip damage zone. Given the large deformation gradients in the vicinity of the crack tip, this method gave access to the entire sequence of microdeformation events up to final fracture.

An overview of the different stages of crack tip deformation in 3L20 obtained in this way is illustrated in Fig. 11. It is apparent from these TEM observations that cavitation in the 3L-particle-modified materials results from the voiding of the elastomeric shell, and that this is intimately linked to the onset of crazing in the intervening regions of the PMMA matrix. This initial damage is driven by the negative hydrostatic pressure applied to the particles by the remote stress, consistent with the appearance of small cavities at the poles of the elastomeric shells of the

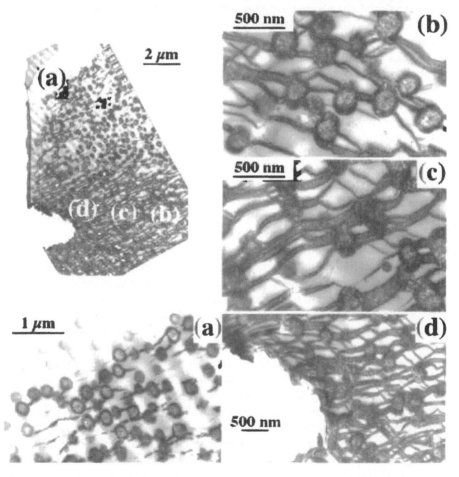

Figure 11 Crack tip deformation in a notched CT sample of 3L20 deformed in mode I at 1 mms^{-1} and stained in RuO$_4$.

particles, visible in Fig. 11a and b. The abrupt increase of stress in the matrix surrounding the particles subsequent to cavitation of the shells may then favor initiation of crazes at the particle equators (it is difficult to establish a causal relationship here, but it is clear that the appearance of a large number of crazes coincides
with the onset of cavitation).

The crazes then propagate in the matrix perpendicular to the principal stress
axis (see Fig. 11b) by the usual mechanism of growth at constant stress in the craze
fibrils, as in unmodified PMMA. However, once the crazes penetrate regions
whose stress field is strongly influenced by the presence of neighboring particles,
they deviate from their initial plane of propagation. In the highly strained regions
close to the crack tip (Fig. 11c), many of the crazes are observed to bifurcate
("craze splitting"), with the trajectories of the craze tips rotating towards the (cavitated) poles of neighboring particles (where the local principal stress axis is also
rotated owing to the presence of the free surface). The net result of this is that isolated cavities become linked by the matrix crazes, creating a continuous network
of voided material, which will in turn isolate intervening regions of undeformed
matrix from each other (see Fig. 11d). The accompanying triaxial constraint release may also favor plastic flow in the matrix, leading to high levels of global
plastic deformation and blunting at the crack tip prior to crack tip advance, which
presumably involves chain scission or chain slip in the craze fibrils (where the true
stresses are relatively high).

It is not clear yet whether the mechanism of craze splitting is related specifically to the presence of 3L particles, as opposed to 2L particles (Fig. 12c), which
appear to remain active in terms of stress concentration well after their cavitation.
Nevertheless, craze splitting should be efficient in terms of energy dissipation, because it significantly increases the area of fracture as well as contributing to the
establishment of a continuous network of crazes and cavities.

For the relatively low 3L particle contents, the mechanical interactions between the particles are initially weak. Damage is clearly initiated at the particle
poles (cavitation of the rubbery shell) and equators (crazing of the matrix) (Fig.
12a). For particle contents of 30% or more, however, as shown in Fig. 12b, the initial damage mechanisms are more uniformly distributed around the particles,
since the local stress state is more strongly influenced by neighboring particles.
Thus several crazes emanating from the same particle are often observed, and further craze splitting is less frequent, presumably because the craze trajectories are
dictated by particle–particle interactions from the outset.

Widespread crazing has been also been reported elsewhere to occur at crack
tips in notched specimens of particle-modified PMMA, where the local triaxiality
and strain rates are generally much higher than in simple tension (19,51,52).
Based on our own results, we can summarize the three main stages of crack tip
damage development as follows (in order of increasing total deformation):

1. Cavitation of the rubbery shell or core (for 3L and 2L particles, respectively) and craze initiation in the matrix

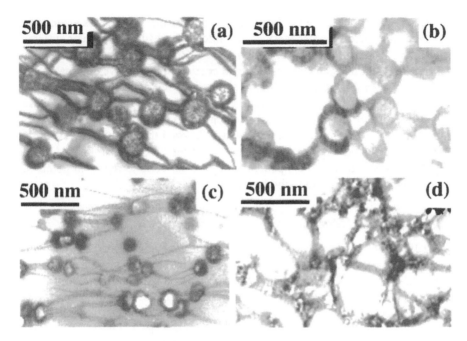

Figure 12 Cavitated particles and interparticle crazing in (a) 3L20, (b) 3L30, (c) 2L15; (d) initiation of the cavitation in the rubbery phase of the IPN (all samples stained in RuO_4 prior to sectioning).

2. Craze thickening and growth at constant stress
3. Global matrix flow (followed by crack propagation)

Clearly some or all of these mechanisms may be suppressed, particularly at the highest test speeds, for which tensile tests on thin films have suggested that unstable crack propagation intervenes before widespread particle cavitation and crazing can occur (53).

B. Discussion

The most obvious distinctions between the 3L-particle-modified PMMAs and the 2L-particle-modified PMMAs lie in the nature of the particle cavitation. In the case of the 2L particles, cavitation occurs in the particle center, generally giving rise to the formation of a single void. In the 3L particles, on the other hand, and at relatively low particle contents, cavitation initiates in the polar regions of the elastomeric shell as observed in the bulk samples (giving at least two voids per particle). However, particularly in particle clusters, 3L particles may show voiding within the whole of the elastomeric shell. The individual voids are separated by ligaments of rubber linking the glassy outer shell/matrix to the glass inner core. Moreover, as noted by other workers, for microtomed films of similar systems (54) there is evidence for fine fibrillation of the matrix at the 3L particle poles. As deformation continues, there is widespread crazing in the intervening matrix regions, with several narrow crazes being initiated at each rubber particle. The particles are then incorporated within the crazed regions of matrix, and as these widen, the particle shell become highly elongated, which suggests that they continue to be load bearing at deformations well beyond those at which cavitation takes place. This behavior contrasts with that of the 2L particles, for which the large voids associated with the particle cores are expected to represent points of substantial weakness within individual crazes.

C.   Deformation Mechanisms in IPNs

In the IPNs, no craze formation is observed. High-magnification TEM observations show that cavitation exists (Fig. 11d) but is initiated in star-shaped regions of the PU rubbery phase (corresponding to the network nodes). The volume of the matrix elements isolated by the rubbery ligaments is in the same order of magnitude as that observed in the crazed particle-modified PMMA. Thus one anticipates that close to the crack tip, the behavior is analogous to stage 3 of the crack tip damage development mechanism described in the previous section, namely global plastic deformation of the matrix, facilitated by release of the local triaxial constraints. The initial cavitation of the PU is predominantly influenced by the glass transition, since cavitation and constraint release require substantial mismatch in the elastic response of the two phases as well as high compliance of the elastomer (the PMMA can be assumed always to be in the glassy state for the present test conditions). It is therefore not surprising that the ductile–brittle transition appears to have been shifted towards strain rates beyond the highest strain rates attained in this study, since the nominal T_g of PU of about $-40°C$ is lower than that of the butyl-acrylate-co-styrene-based elastomers in the particle-filled materials (for which T_g is about $-10°C$). Moreover, to achieve connectivity of the crazes and cavitated particles, and hence to reach an equivalent stress state to that of the cavitated IPN, the matrix of the particle-filled materials must undergo considerable crazing, whose kinetics will in turn be influenced by the temperature and strain response of the matrix.

V.   CONCLUSIONS

There has been considerable effort to develop models for toughening in glassy polymers toughened with elastomeric particles, much of which has centered on criteria for cavitation in isolated particles in an infinite matrix. Even given that cavitation of the second-phase particles is the overriding consideration, there remain questions relating to particle–particle interactions at high filler contents for which simple analytical approaches are inadequate, and for which numerical approaches are also limited (applicable in the elastic limit only and restricted to relatively small numbers of particles). Our aim here has not been to seek to validate such approaches, but to report on what occurs in the damaged material well beyond the damage threshold and the onset of linearity. Surprisingly, different morphologies and/or modifier contents result in similar low-rate toughness and a similar macroscopic appearance of the damaged (stress whitened) zone. This has been observed both in dog bone tensile specimens and in the damaged zone surrounding a stable crack generated during fracture tests. As demonstrated by measuring the light transmittance during the tensile tests, the abrupt increase in irreversible stress whitening corresponds to the overall yielding of the material. In-situ measurements of the volume change in specimens submitted to uniaxial tension have shown cavitational strain to be an important deformation mechanism, tending generally to be favored at high strain rates. Furthermore, at room temperature, the micromechanisms of deformation in the matrix appear similar for the materials modified by spherical particles (either 2L or 3L particles). Widespread crazing of the matrix, triggered by the cavitation of the rubbery phase of the particles, is believed to be the predominant damage mechanism. Results from tensile tests on thin films (55), and to a certain extent from the bulk samples, indicate that damage initiation in the form of crazes is sensitive to clustering of the modifier particles. Further, interparticle interactions clearly have a strong incidence on the way the damage zone and in particular crazes develop. Finally, based on the results for the IPNs, it appears that crazing itself is not a prerequisite for toughening, but that toughening does require the existence of an interconnected network of voided material and/or material with a low shear modulus (either crazes or elastomer) close to the crack tip in both. This may in turn be an indication of the important role of matrix shear in energy dissipation as the crack tip advances. Nevertheless, at low-strain-rate and/or high-temperature tensile tests, and in particular at high rubber contents, shearing of the matrix may occur at stress levels too low to initiate cavitation in the rubbery phase and crazing in the glassy matrix.

The progressive embrittlement observed in notched spherical-particle-modified materials when the test speed is increased is believed to be linked to the kinetics of the damage development. Transitions from stable to unstable (or even partially unstable) crack propagation may be seen as the response of the material at time scales over which shearing and flow of the damaged matrix (interconnected by crazes) are kinetically limited. Similarly, fully brittle behavior at high rates, when it occurs, is linked to the inability of the material to undergo cavitation and widespread crazing of the matrix. These phenomena are closely related to the interparticle distance (and hence the particle size, for a given particle content) either through changes in the hydrostatic stress in the rubbery phase or through particle–particle interactions, which control the local stress in the glassy matrix. Moreover, the mobility of the matrix (which has not been studied here) should be considered an important factor in the control of the tough-to-brittle transition and the potential of a given material to dissipate energy during fracture. This is a major difficulty when toughening strategies for one type of matrix material are extended to another.

ACKNOWLEDGMENTS

Thanks are due to Ph. Heim and Ph. Tordjeman of the Groupe de Recherche de Lacq, Elf Atochem, France, for providing some materials used in this study, and to K. Takahashi and M. Todo of the Research Institute for Applied Mechanics, Kyushu University, Japan for their collaboration.

REFERENCES

1. Williams J. G., Marshall G. P. Environmental crack and craze growth phenomena in polymers. Proc. R. Soc. A342:55, 1975.
2. Kausch H. H. Polymer Fracture. 2d ed. Berlin: Springer-Verlag, 1987.
3. Döll W., Könczöl L. Micromechanics of fatigue under static and fatigue loading: optical interferometry of crack tip craze zones. In: Kautsch H. H., ed. Crazing in Polymers, 2. Berlin: Springer-Verlag, 1990, p. 137.
4. Schirrer R. Optical interferometry: running crack-tip morphologies and craze material properties. In: Kautsch H. H., ed. Crazing in Polymers, 2. Berlin: Springer-Verlag, 1990, p. 215.
5. Miller P., Buckley D. J., Kramer E. J. Microstructure and origin of cross-tie fibrils in crazes. J. Mat. Sci 26:4445, 1991.
6. Brown H. R. A molecular interpretation of the toughness of glassy polymers. Macromolecules 24:2752, 1991.
7. Echte A. Rubber toughened styrene polymers—a review. In: Riew C. K., ed. Rubber-Toughened Plastics. Advances in Chemistry Series 222. Washington, D.C.: American Chemical Society, 1989, pp. 3–64.
8. Pearson R. A. Sources of toughness in modified epoxies. Ph.D. thesis, Univ. of Michigan, 1990.
9. Montarnal S., Pascault J.-P., Sautereau H. Controlling factors in the rubber-toughening of unfilled epoxy networks. In: Riew C. K., ed. Rubber-Toughened Plastics. Advances in Chemistry Series 222. Washington, D.C.: American Chemical Society, 1989, pp. 194–223.
10. Borrajo J., Riccardi C. C., Moschiar S. M., Williams R. J. J. Effect of polydispersity on the miscibility of epoxy monomers with rubbers. In: Riew C. K., ed. Rubber-Toughened Plastics. Advances in Chemistry Series 222. Washington, D.C.: American Chemical Society, 1989, pp. 320–328.
11. Huang Y., Hunston D. L., Kinloch A. J., Riew K. C. Mechanisms of toughening ther-

moset resins. In: Riew C. K., Kinloch A. J., eds. Rubber-Toughened Plastics. Advances in Chemistry Series 233. Washington, D.C.: American Chemical Society, 1993, pp. 1–35.

12. Imperial Chemical Industries Ltd., Patent Br. 965,786, 1964.
13. Imperial Chemical Industries Ltd., Patent Br. 1,093, 1967.
14. Röhm and Haas, Patent Br. 1,340,025, 1973.
15. Röhm and Haas, Patent Br. 1,414,187, 1975.
16. Wrottecki C., Heim P., Gaillard P. Rubber toughening of poly(methyl methacrylate). Part I: Effect of size and hard layer composition of the rubber particles. Polymer Engineering and Science 31(4):213, 1991.
17. Yang J., Soo Lee M., Cho K. The effect of particle size and interfacial adhesion on fracture behaviour of toughened poly(methyl metacrylate). Proceedings of the American Chemical Society, Division of Polymeric Materials: Science and Engineering, Spring Meeting, San Diego, CA, 1994, p. 145.
18. Lovell P. A., Sheratt M. N., Young R. J. Mechanical properties and deformation micromechanics of rubber-toughened acrylic polymers. In: Riew C. K., Kinloch A. J., eds. Rubber-Toughened Plastics II. Advances in Chemistry Series 252. Washington, D.C.: American Chemical Society, 1996, pp. 211–232.
19. Lovell P. A. Rubber-toughened acrilic materials. TRIP 4(8):264–272, 1996.
20. Hooley C. J., Moore D. R., Whale M., Williams M. J. Fracture toughness of rubber modified PMMA. Plastics and Rubber Processing and Applications 1:345–349, 1981.
21. Béguelin Ph, Julien O., Monnerie L., Kausch H. H. The loading rate dependence of the fracture toughness of rubber modified poly(methyl methacrylate). Proceedings of the American Chemical Society Division of Polymeric Materials: Science and Engineering, Spring Meeting, San Diego, CA, 1994, p. 147.
22. Julien, O., Béguelin, Ph., Monnerie, L., Kausch, H.-H. Loading rate dependence of the fracture toughness of rubber modified poly(methyl methacrylate). In: Riew C. K., Kinloch A. J., eds. Rubber-Toughened Plastics II. Advances in Chemistry Series 252. Washington, D.C.: American Chemical Society, 1996, pp. 233–248.
23. Sperling L. H., Friedman D. W. Journal of Polymer Science Part A2 7:425, 1969.
24. Sperling L. H. Interpenetrating Polymer Networks and Related Materials, New York: Plenum Press, 1981.
25. Sperling L. H. Interpenetrating polymer networks: an overview. In: Klempner D., Sperling L. H., Ultracki L. A., eds. Interpenetrating Polymer Networks. Advances in Chemistry Series 239. Washington, D.C.: American Chemical Society, 1994, pp. 3–38.
26. Mishra V., Sperling L. H. Metastable phase diagrams for simultaneous interpenetrating networks of a PU and PMMA. Polymer 36(18):3593–3595, 1995.
27. Allen G., Bowden M. J., Blundell D. J., Jeffs G. M., Vyvoda J., Hutchinson F. G. Composites formed by interstitial polymerisation of vinyl monomers in PU elastomers: 1. Preparation and mechanical properties of PMMA based composites. Polymer 14:597–603, 1973.
28. Allen G., Bowden M. J., Blundell D. J., Jeffs G. M., Vyvoda J., White T. Composites formed by interstitial polymerisation of vinyl monomers in polyurethane elastomers: 2. Morphology and relaxation processes in methyl methacrylate based composites. Polymer 14:604–616, 1973.

29. Allen G., Bowden M. J., Lewis G., Blundell D. J., Jeffs G. M. Composites formed by interstitial polymerisation of vinyl monomers in PU elastomers: 3. The role of graft copolymerization. Polymer 15:13–18, 1974.

30. Allen G., Bowden M. J., Lewis G., Blundell D. J., Jeffs G. M., Vyvoda J. Composites formed by interstitial polymerisation of vinyl monomers in PU elastomers: 4. Preparation, properties and structure of acrylonitrile and styrene based composites. Polymer 15:19–27, 1974.

31. Allen G., Bowden M. J., Todd M., Blundell D. J., Jeffs G. M., Davies W. E. Composites formed by interstitial polymerisation of vinyl monomers in polyurethane elastomers: 5. Variation of modulus with composition. Polymer 15:28–32, 1974.

32. Blundell D. J., Longman G. W., Wignall G. D., Bowden M. J. Composites formed by interstitial polymerisation of vinyl monomers in PU elastomers: 6. Low angle X-ray scattering and turbidity. Polymer 15:33–36, 1974.

33. Hur T., Manson J. A., Hertzberg R. W., Sperling L. H. Fatigue behavior of acrylic IPN. Journal of Applied Polymer Science 39:1933–1947, 1990.

34. Akay M., Rollins S. N., Riordan E. Mechanical behaviour of sequential polyurethane-poly(methyl methacrylate) interpenetrating polymer networks. Polymer 29:37–42, 1988.

35. Heim Ph., Wrotecki C., Avenel M., Gaillard P. High impact cast sheets of poly(methylmethacrylate) with low levels of polyurethane (IPN). Polymer 34(8):1653, 1993.

36. Mauzac O., Schirrer R. Effect of particle volume fraction on crack-tip crazes in high impact poly(methyl methacrylate). Journal of Applied Polymer Science 38:2289–2302, 1989.

37. Gloagen J. M., Steer P., Gaillard P., Wrotecki C., Lefebvre J. M. Plasticity and fracture initiation in rubber-toughened poly(methyl methacrylate). Polymer Engineering and Science 33(12):48–753, 1993.

38. Gloagen J. M., Lefebvre J. M., Wrotecki C. Critical energy for crack initiation in rubber-toughened poly(methyl methacrylate). Polymer 34(2):443–445, 1993.

39. Mauzac O., Schirrer R. Crack-tip damaged zones in rubber-toughened amorphous polymers: a micromechanical model. J. Mat. Sci. 25:5125–5133, 1990.

40. O. Julien. Etude de la relation entre la morphologie et les propriétés mécaniques des polyméthacrylate de méthyl renforcés au choc. Ph.D. thesis, Université Pierre et Marie Curie (Paris VI), 1995.

41. Béguelin Ph., Barbezat M., Kausch H. H. Mechanical characterization of polymers and composites with a servohydraulic high-speed tensile tester. J. Phys. III France 1:1867, 1991.

42. Kessler S. L., Adams, G. C., Driscoll S. B., Ireland D. R. Instrumented impact testing of plastics and composite materials. ASTM STP 936, Philadelphia, 1987.

43. Zanichelli C., Rink M., Pavan A., Ricco T. Experimental analysis of inertial effects in impact testing of polymers. Polymer Engineering and Science 30:18, 1990.

44. Béguelin Ph. Approche expérimentale du comportement mécanique des polyméres en sollicitation rapide. Ph.D. thesis No. 1572. EPFL, Lausanne, 1996.

45. Béguelin Ph. A technique for studying the fracture of polymers from low to high loading rates. In: Williams J. G., Pavan A., eds. Impact and Dynamic Fracture of Polymers and Composites. ESIS 19. London: Mechanical Engineering Publications, 1995, pp. 3–19.

46. Béguelin Ph., Fond C., Kausch H. H. The influence of inertial effects on the fracture of rapidly loaded compact specimen. Part A: Loading and fracture initiation. To be published in Int. J. Fracture, 1998.

47. Frank O., Lehmann J. Determination of various processes in impact modified PMMA at strain rates up to 10^5%/min, Colloid and Polymer Science 264, 1986.

48. Bucknall C. B., Partridge I. K., Ward M. V. Rubber toughening of plastics, Part 7. Kinetics and mechanisms of deformation in rubber-toughened PMMA. J. Mat. Sci. 19:2064, 1984.

49. Schirrer R., Fond C., Lobbrecht A. Volume change and light scattering during mechanical damage in PMMA toughened with core shell particles. J. Mat. Sci. 28:6799, 1996.

50. He C., Butler M. F., Donald A. M. Deformation of rubber toughened PMMA: an in situ synchrotron study. Proc. European Conference on Macromolecular Physics, Surfaces and Interfaces in Polymers, June 1–6, Lausanne, 1997.

51. Mauzac O., Schirrer R. Crack-tip damaged zones in rubber-toughened amorphous polymers: a micromechanical model. J. Mat. Sci. 25:5125–5133, 1990.

52. Mauzac O., Schirrer R. Effect of particle volume fraction on crack-tip crazes in high impact poly(methyl methacrylate). Journal of Applied Polymer Science 38:2289–2302, 1989.

53. Plummer C. J. G., Béguelin Ph., Kausch H. H. Microdeformation in core-shell particle modified polymethymethacrylate. To be published in Colloids and Surfaces, 1998.

54. Starke J. U., Godehardt R., Mischler G. H., Bucknall C. B. Mechanisms of cavitation in rubber-toughened PSAN modified with three-stage core-shell particles. J. Mat. Sci. 32:1855, 1997.

55. Plummer C. J. G., Béguelin Ph., Kausch H. H. On the influence of particle morphology on microdeformation in rubber modified poly(methyl methacrylate). Polymer 37(1):7–10, 1996.

20

Impact Toughening Mechanisms in Glassy Polymers

Y. Okamoto, H. Miyagi, H. Kihara, and S. Mitsui
Sumitomo Chemical Company, Sodegaura-shi, Chiba, Japan

I. INTRODUCTION

It is well known that the impact toughness of glassy polymers can be improved by the dispersion of rubber particles (1,2), and the toughening mechanism of these systems has been studied extensively (3–7). In the case of high-impact (polystyrene HIPS), the absorption of impact energy is thought to increase mainly by the generation and extension of crazes in the PS matrix (8). The effect of the rubber content and rubber particle size on the toughening of HIPS has been studied from both the scientific and the technical perspective. In monomodal rubber particle HIPS (monomodal HIPS), the optimum rubber particle diameter is known to be 1–2 μm (9,10).

From a practical point of view, HIPS with a dual population of rubber particle sizes (bimodal HIPS) has been studied for its unique impact properties (11,12). It has been shown that better impact toughness can be achieved in an optimum bimodal HIPS than in a monomodal one with the same rubber content (13). However, few studies have been concerned with the cause of the increase in impact loading of bimodal HIPS. A model was recently presented to explain the craze extension mechanism in bimodal HIPS by using the finite element method (FEM) (14). This model assumed that the small particles initiated crazes that propagated in the direction of large particles, where they were stopped.

In 1990, a new toughening model was proposed by Gebizlioglu et al. (15). They examined the mechanism in rubber-dispersed polystyrene samples that included low Mw polybutadiene (PB) rubber droplets. They concluded that the liquid PB in these pools acts as a plasticizing agent under the prevailing negative pressure of the craze tip and craze borders, resulting in a greatly increased propensity for crazing at low stresses and thus avoiding early craze fracture from extrinsic flaws. Furthermore, Argon et al. (16) also provided a theoretical model of the above phenomenon, and craze flow stress based on the model has shown excellent agreement with experimental measurement.

In our study (17), the crazing process in bimodal HIPS was studied by observation with a transmission electron microscope (TEM). Nonlinear stress analysis around rubber particles was also performed by using FEM.

We recently discovered that cavitation can occur in high-impact polystyrene (HIPS) with rubber particles of rubbery core/rigid polymer shell and studied the cavitation mechanism in this system (18). Here we provided direct verification of the rubber sorption model by microscopic work by observing the effect of annealing on crazing, and we showed by FEM analysis that sorption of rubber components to craze fibrils would plasticize the craze. In addition, we find impact improvement in a system of polystyrene with the addition of rigid particles several microns in diameter for concentrations of 0.01 ~ 1 wt%.

II. RUBBER-TOUGHENED POLYSTYRENE

A. Experimental

Bimodal HIPS specimens were obtained by injection molding, following the blending of two monomodal HIPS that had rubber particles of different average diameters, or different structure. Characteristic data of monomodal HIPS used in this work are summarized in Table 1. Rubber particle sizes were estimated through TEM observation, and gel content was determined from extracts taken after centrifuging the dissolved samples with mixture solvents (methyl ethyl ketone/methanol = 10/1 in volume ratio).

Table 1 Summary of Material Characteristics of Monomodal HIPS as Base Polymer for Bimodal HIPS

Sample	Structure of rubber particles	Gel[a] (wt%)	Dp[b] (μm)
RC-HIPS	Rubbery core/solid shell	20.0	0.15
SC-HIPS	Solid core/rubbery shell	23.0	0.20
L-HIPS	Salami structure	23.2	4.9
M-HIPS	Salami structure	18.3	1.0

[a] Gel: rubber particle phase content in HIPS (wt%).
[b] Dp: rubber particle size (μm).

Impact strength was measured by the V-notched Izod impact test (based on ASTM D256). The fracture events (craze or cavitation) in a stress whitened area beneath an impact fracture surface were observed by TEM. The specimens were microtomed into thin sections in a direction perpendicular to the fracture surface after staining with osmium tetraoxide (OsO4) (19).

Nonlinear stress analysis around rubber particles in HIPS was carried out for uniaxial (y-axis) tensile loading using the finite element method (FEM). A two-dimensional model was employed under the following conditions: (1) the plane strain condition, (2) von Mises' criterion for yielding (20), and (3) the presence of rubber particles with a uniform rubbery substance, i.e., neither a "salami" nor a "core-shell" structure for simplicity, where Mises' equivalent stress could be expressed as

$$\sigma = \tfrac{1}{2}[(\sigma_x - \sigma_y)^2 + (\sigma_y - \sigma_z)^2 + (\sigma_z - \sigma_x)^2 + 6\,\tau_{xy}{}^2]^{1/2} \tag{1}$$

On the equatorial plane in our model, tensile directional normal stress was much higher than other stress components. That is, equivalent stress on the equatorial plane is roughly expressed by σ_y.

On the other hand, it is considered that dilatational stress acts as the main factor in crazing. Therefore we may qualitatively regard the yielding zone on the equatorial plane as the crazing zone.

For the calculation, the FEM program ABAQUS version 4.8(HK & S Inc.) together with a Cray supercomputer (Model X-MP EA/116SE) owned by Sumitomo Chemical Co. was employed.

B. Results and Discussion

Figure 1 shows the relationship between the Izod impact strength and the blend ratios of two different monomodal HIPS for two bimodal systems (SC-HIPS/L-HIPS systems and L-HIPS/M-HIPS systems.) Three points are evident: First, the impact toughness of monomodal HIPS was improved by the blending of two different monomodal HIPS. Second, the extent of the improvement effect depends on the particle size ratio in the bimodal system. The greater the difference between the particle size of the constituent bimodal HIPS, the higher the Izod impact loading. Third, as also found by Hobbs (13), the maximum value of the impact loading exists on the SC-HIPS-rich side in each bimodal system.

It has been recognized that the impact toughness of HIPS is attributable to the absorption of the applied energy by the generation of crazes (1–10), and we carefully observed craze generation in HIPS specimens that had been subjected to different impact energy loadings. TEM observations were made of whitened regions in front of the initial notch tip in bimodal HIPS samples consisting of SC-HIPS and L-HIPS, where the blend ratio was 9/1 in weight. Figure 2 shows typical examples of these observations. Impact energies applied to each specimen were (a)

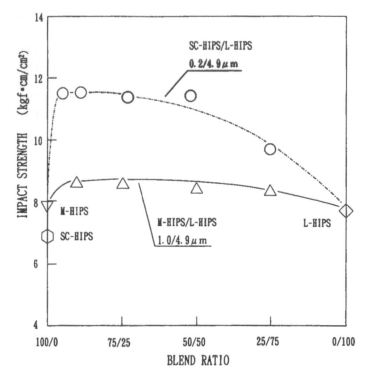

Figure 1 Impact strength vs. blend ratio of two monomodal HIPS in SC bimodal system.

0.123, (b) 0.346, (c) 0.526, and (d) 0.981 J (corresponding to fracture) in bimodal HIPS and (e) 0.981 J in monomodal HIPS of L-HIPS.

As indicated in Figs. 2a and b by arrows, the minute crazes are generated from large particle surfaces at the initial stage of craze growth. It should be noted that few crazes were observed initiating from small rubber particles. With greater impact loading, the number of crazes generated from large particles was not increased, but rather the crazes became longer (Fig. 2c). Crazes initiated from the large particles extended bridgelike to neighboring small rubber particles parallel to the equatorial plane. It was not clear, however, whether there were crazes initiated from small particles just ahead of the growing craze tip. Finally, each of the crazes generated from a large particle grew extensively until the length reached two or three times the equatorial axis of the particle (Fig. 2d). Also in this stage, few crazes were observed to initiate from small rubber particles.

A comparison of crazing behavior between bimodal HIPS and monomodal HIPS of L-HIPS after fracture (Fig. 2d and e) revealed that whilst the number of

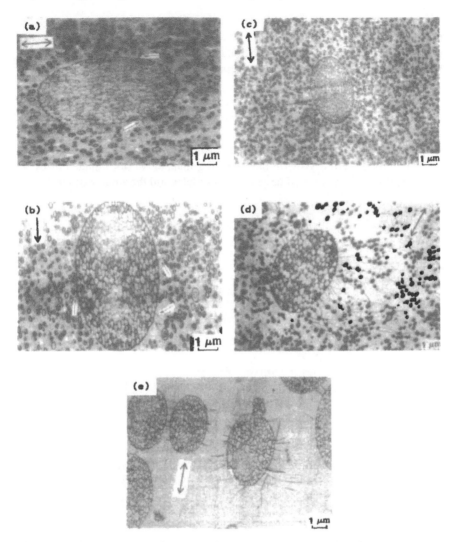

Figure 2 Transmission electron micrographs of crazes. Applied impact energies are (a) 0.123, (b) 0.346, (c) 0.526, and (d) 0.981 J (fracture) in SC bimodal HIPS, and (e) 0.981 J (fracture) in L-HIPS. (Small arrows point to minute crazes; large arrows indicate tensile direction.)

crazes was almost the same, the length was much longer (more extensive) in bimodal HIPS than in monomodal HIPS. It is therefore considered that the improved impact of bimodal HIPS is due to the unique growth of long crazes generated at the periphery of the large rubber particles and not to the increase in the number of crazes generated.

To investigate this mechanism further, we carried out a stress analysis, based on nonlinear FEM analysis, around rubber particles in bimodal HIPS. Figure 3 shows the yielding process changing with an increase of applied strain around rubber particles. Yielding is indicated by crosshatched areas. Figure 3 suggests that crazes in bimodal HIPS grow in the following manner: in (a), minute crazes are generated from the surfaces of large rubber particles and the small particle A clos-

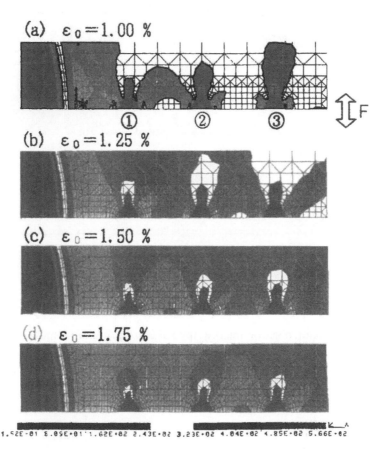

Figure 3 Steps in the expansion of the yielding zone with an increase in applied strain. Dark zones correspond to particles.

est to the large particle; in (b), the crazes connect with each other, and other crazes generated from particle B develop. With an increase in the applied strain in (c) and (d), they grow in a manner similar to that above. Combining TEM observations (Fig. 2) with nonlinear FEM analysis (Fig. 3), the manner of the craze initiation and extension is illustrated schematically in Fig. 4, a diagrammatic representation of the mechanism described above.

The stress convergence length, which corresponds to the effective stress field for craze generation of rubber particles, is proportional to the particle size. The effective stress field of a large particle is more extensive than that of a small one. On the other hand, the ligament thickness, which is the distance between two rubber particle surfaces in a monomodal HIPS, is proportional to the rubber particle size, as described by Wu (21). In bimodal HIPS, the long stress convergence length caused by large rubber particles and the short ligament thickness caused by a large number of small particles will give longer crazes and consequently a higher impact strength.

We have investigated another bimodal system (RC-bimodal) of RC-HIPS and ordinary HIPS. The same type of impact improvement for the RC-bimodal system was observed for the SC-bimodal system of SC-HIPS. The same TEM technique was also applied to investigate the deformation sequence in RC-bimodal HIPS.

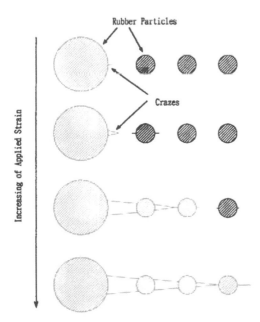

Figure 4 Schematic model of the craze extension with an increase in applied strain.

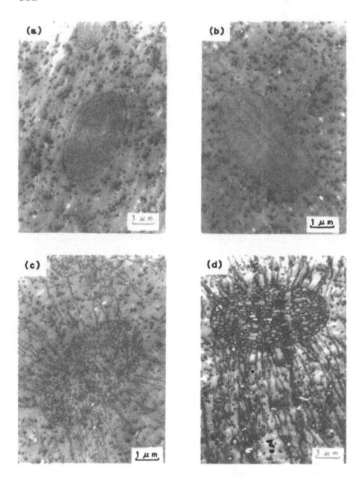

Figure 5 Sequence of plastic deformation in RC bimodal HIPS. Impact energies applied to each specimen were (a) 0.123, (b) 0.346, (c) 0.526, and (d) 0.981 J in RC bimodal HIPS.

Figure 5 shows TEM observations of whitened regions in front of the initial notch tip in RC-bimodal-HIPS samples consisting of RC-HIPS and L-HIPS, where the blend ratio was 7 to 3 by weight. In Fig. 5a, we can see small crazes generated from a salami-structured large rubber particle, but no cavitation in the RC rubber particles is yet visible. In Fig. 5b, the number of crazes has increased, and they have grown long, and a few cavitations are visible at this stage, the size of the cavitated rubber particles being almost the same as those of other rubber particles. In Fig. 5c, the crazes have further increased in number and length. Many cavitated RC rubber particles can also be seen, but only on the lines of matured crazes. Finally, in Fig. 5d, the number and lengths of crazes are developed so that we can see a great many cavitated particles; the size and shape of these do not differ much from noncavitated ones.

Parker et al. reported a cavitation mechanism in the polycarbonate matrix system (22) in which cavitation occurred first and induced the shear yielding. Here, by contrast, the cavitation mechanism was induced by crazing, which occurred first.

As compared with SC-bimodal HIPS, the following differences can be noted for the RC-bimodal HIPS material: (1) Cavitation is apparent in RC-bimodal HIPS. (2) The cavitations occurred inside only those rubber particles on the craze lines. (3) The crazes are both long and numerous.—From these results, it can be concluded that the craze extension mechanism of RC-bimodal HIPS and that of SC-bimodal HIPS differ somewhat.

Deformed samples of RC-bimodal HIPS were subjected to different levels of annealing at 100°C, stained by OsO$_4$, and observed by TEM (Fig. 6). After annealing, the crazes tended to disappear, but there are many stained submicron spots on the lines of eliminated crazes, especially for the RC-bimodal HIPS. Considering the results, these spots may be rubber components that have been sorbed into crazes. Thus the following cavitation model was suggested (to explain Fig. 7): (1) Crazes are initially generated from large rubber particles. This generation is the same as that in SC-bimodal HIPS. (2) PB rubber sorption into crazes occurrs by negative pressure of microvoids in the crazes and results in cavitation of RC rubber particles. On the basis of this model, rubber components sorbed into crazes

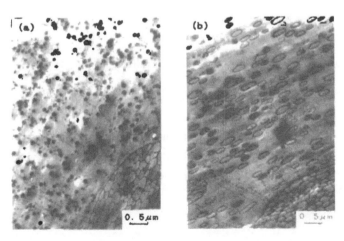

Figure 6 Transmission electron micrographs of fractured samples annealed at 100°C for 36 h after fracture for (a) RC bimodal HIPS and (b) SC bimodal HIPS.

① craze generation

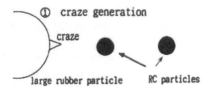

large rubber particle   RC particles

② extension of craze

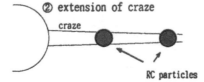

(note)
negative pressure generated
in the craze

③ PB rubber of RC particles sorption into craze

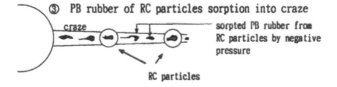

sorpted PB rubber from
RC particles by negative
pressure

④ generation of cavitation in the RC particles

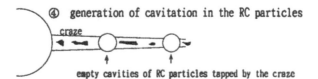

empty cavities of RC particles tapped by the craze

⑤ disappearance of craze and cavitation by annealing

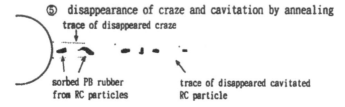

sorbed PB rubber   trace of disappeared cavitated
from RC particles   RC particle

Figure 7 Schematic model of new cavitation mechanism for bimodal system of RC-HIPS and ordinary HIPS.

would plasticize the crazes. Using the stress–strain curves that express the different levels of plasticization of the matrix, we calculated the area of the yielding zone by FEM analysis. The area of the yielding zone increases with decreases in yield stress of the matrix. We therefore conclude that sorption of rubber components into crazes results in enlargement of the yielding zone of the matrix. In other words, the sorption of rubber components enhances the tendency for crazes to extend and be more numerous. Sorption of rubber components to craze fibrils would thus plasticize the craze and result in a long and thick craze.

III. DISPERSION OF RIGID PARTICULATES (27)

A. Experimental

PS pellets including particulates (Table 2) were regranulated using an extruder after blending of PS pellets [Sumibright E183; Sumitomo Chemical Co., Mw =

Table 2 Particulates Included in PS and Their Average Diameters

	Particulate	Average diameter (μm)
Inorganic	Tricalcium phosphate	3.9[a]
particulate	Barium sulfate	2.5[b]
	Calcium carbonate	1.15[c]
	Silicon dioxide	0.8[d]
		1.4[d]
		2.0[d]
		3.4[d]
		5.0[d]
Organic	Cross-linked	2.0[e]
particulate	PMMA beads	30.0[f]
	Cross-linked	1.7
	PS beads	6.0[f]
		11.0[f]
	Cross-linked	26.2[g]
	polydivinylbenzene beads	44.0[g]

Tohoku Chemical Industries, Ltd.
[b] Nippon Chemical Industrial Co., Ltd. "AD Barium sulfate."
Nitto Chemical Industry Co., Ltd. "Calcium carbonate NS#200."
[d] Mizusawa Chemical Industry Co., Ltd. "Shilton AMT-08, -15, -20S, -30, -50."
Nippon Shokubai Kagaku Kogyo Co., Ltd. "MA-1002."
Sumitomo Chemical Co., Ltd. "Fine Pearl PM-3030E, PB3006E, PB3011E."
[g] Japan Synthetic Rubber Co., Ltd. "S2467(p)-03, S2467(p)-08."

300,000, MFR (melt flow rate) = 2.4 g/10 min]. The particulate was used at levels of 0.1 wt%. To investigate the dependence of the amount of particulate, barium sulfate (average diameter 2.5 μm) and cross-linked PS beads (average diameter 1.7 μm) were added to PS at percentages from 0.01 to 20 wt%.

PS (E183N) and AS (acrylonitrile-styrene copolymer) resin (BS303: Sumitomo Dow Co., acrylonitrile content = 30 wt%, MFR = 3.5 g/10 min) without the particulate were used as reference materials.

Plates of 90X150X2 mm were injection molded using the above samples molded at 230°C, an injection speed = 80 cm/s, an injection pressure = 88.2 MPa) with a nonvented injection molding machine (J150E; Japan Steel Works). The injection-molded plates were cut into 50X50X2 mm sample specimens.

The falling ball impact test was carried out according to JIS K7211, with a steel ball of 28.8 g; the 50% failure height was measured. The impact mark on the sample specimen after the falling ball impact test was observed using a laser microscope (1LM11; Lasertec Co.). With the real-time deformation technique (23) developed by Sumitomo Chemical Co., the PS including the particulate was observed by means of a transmission electron microscope (H-8000; Hitachi Co.).

B. Results and Discussion

1. The Effect of Dispersion of Rigid Particulates

Table 3 shows the results of the falling ball impact strength (50% failure height) of the injection-molded plates of PS with the various dispersed particulates; see Table 2. Almost all the falling ball impact loadings of PS with particulates (whether organic or inorganic) were higher than those without particulates.

The degree of increase of falling ball impact strength on PS with each dispersed particulate correlated with the average diameter of the dispersed particulate. Figure 8 shows the dependence of the falling ball impact strength of PS on average particulate diameter. The improvement in falling ball impact strength reaches a maximum at an average particulate diameter of 2 μm. At the same time, Fig. 8 shows a falling ball impact strength on AS resin indicated by a dotted line. The maximum value of the falling ball impact loading of PS dispersed with particulates of average diameter of about 2 μm was higher than that of AS resin.

Figure 9 shows the falling ball impact loading of PS dispersed with barium sulfate (average particle diameter = 2.5 μm) and cross-linked PS beads (average particle diameter = 1.7 μm), of which the amount added was 0.01 to 20 wt% (based on total weight), respectively. The falling ball impact loading increases with the amount of particulate under 1 wt% and decreases gradually above 1 wt% of particulate.

2. Mechanism of Impact Improvement

Many flaws occurred around the point of impact. Figure 10 is a photograph of the flaws on PS with dispersed barium sulfate particulate of 0.1 wt% (magnified 200

Table 3   Falling Ball Impact Strength on the Injection-Molded Plates of Various PS-Dispersed Particulates (Added Particulate Amount = 0.1 wt%)

	Particulate	Average diameter (μm)	Falling ball impact strength [50% failure height (cm)]
No addition	—	—	28.5
Inorganic particulate	Tricalcium phosphate	3.9	68.0
	Barium sulfate	2.5	58.8
	Calcium carbonate	1.15	69.6
	White carbon	0.8	67.0
		1.4	75.2
		2.0	87.3
		3.4	80.5
		5.0	75.0
Organic particulate	Cross-linked PMMA beads	2.0	77.7
		30.0	30.7
	Cross-linked PS beads	1.7	74.0
		6.0	61.6
		11.0	45.6
	Cross-linked polydivinylbenzene beads	0.3	26.2
		0.8	44.0

times using a laser microscope). The flaws disappear almost completely by annealing at 110°C for 5 min. Therefore it is possible to deduce that the flaws are crazes, not cracks. Consequently, we believe that the impact improvement is caused by crazes generated from the particulates. The particulate is the origin of craze initiation, and the crazes absorb part of the energy of impact. It is known that for a dispersion of soft particulate (rubber) in a hard matrix (PS), such as HIPS, the applied stress is concentrated around the rubber particles, and crazes initiate from these stress concentration points.

Yamaoka and Kimura (24) have reported impact loading improvement by dispersing methacrylate styrene copolymer (MS) particles in a styrene butadiene copolymer (SBS) matrix. They concluded that the impact improvement mechanism was related to the development of microcracks from peeling layers between MS and SBS.

Fu and Wang (25) have investigated the system of dispersion of $CaCO_3$ particles in a PE matrix. They concluded that voids generated near the $CaCO_3$ particles were related to the absorption of impact energy. Wang et al. (26) investigated the behavior of stress concentration and craze initiation of PS with large steel or

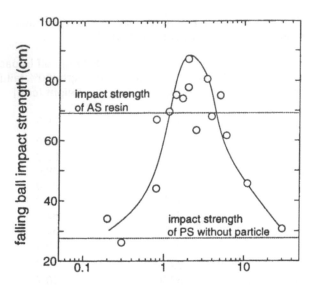

Figure 8 Falling ball impact strength (50% failure height) vs. average particle diameter for PS including particulate from Table 3. Upper dotted line indicates AS resin and lower dotted line indicate PS without particles.

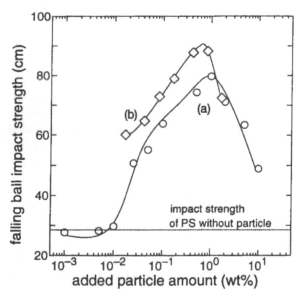

Figure 9 Falling ball impact strength (50% failure height) vs. added particle
amount for (a) barium sulfate (average particle diameter = 2.5 μm) and (b)
cross-linked PS beads (average particle diameter = 1.7 μm). Dotted line indicate PS without particles.

Figure 10 Laser microscope photograph of the flaws on PS-dispersed barium sulfate particulate, 0.1 wt%, magnified 200 times. The impact point is out of this view area.

rubber balls. They reported that for rubber balls, the crazes initiated perpendicular to the stretch direction, and on the other hand, for large steel balls, crazes initiated at a direction ~40° from the stretch direction.

For the dispersion of more rigid particles in the rigid matrix, two craze initiation mechanisms work to improve the impact strength of PS. In the first, the crazes initiate from the particulates themselves by stress concentration (26). In the other, the crazes initiate from voids that are formed between the particulate and the matrix (24,25). To examine more details of craze initiation from the particulates, we have observed the real-time deformation of PS with dispersed barium sulfate particulate using a TEM.

Figure 11 shows the typical craze initiation. It reveals that the void is formed around the surface of the particulate. Figure 11a shows a void formed in the direction of applied strain, whilst Fig. 11b shows a void formed at a direction of ~45° toward the strain. In both cases, the crazes extended perpendicular to the applied strain. It is considered that inorganic particulates, such as barium sulfate with PS, are dispersed without adhesion at the boundary. Therefore, voids are formed easily by the applied stress.

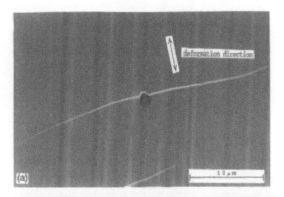

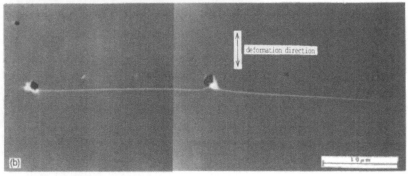

Figure 11 TEM photograph of real-time deformation of PS-dispersed barium sulfate particulate (1 wt% concentration).

3. Analytical Results of Finite Element Method

To investigate the craze initiation mechanism from the rigid particle, we carried out a stress analysis around the particulate in PS using the finite element method (FEM). Figure 12 shows the models used for calculation. Figure 12a is the system with dispersed softer particles in a PS matrix (soft particle model). Figure 12b is the system with dispersed harder particles in a PS matrix (hard particle model). Figure 12c is the system with dispersed harder particles in a PS matrix with apeeling layer around the particles. The systems in Figs. 12a and 12b were given the boundary conditions that the particle adhere completely to the matrix. A stretching strain was adopted on the FEM analysis, as shown in Fig. 12.

Figure 13 shows the stress concentration factor of the maximum principal stress $C(\theta)$, expressed as

$$C(\theta) = S_m(\theta) \frac{\text{maximum principal stress}}{\text{applied stress}}$$

$$= \frac{S_m(\theta)}{2.73 \text{ GPa} \times 2\%} = \frac{S_m(\theta)}{54.5 \text{ Pa}}$$

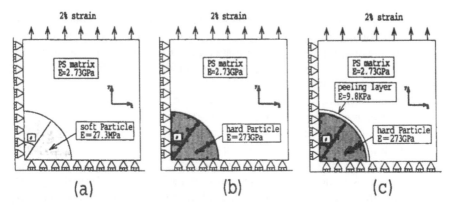

Figure 12 FEM analysis model of stress analysis around the particulate in PS (*E*-modulus of elasticity). (a) The system with dispersed softer particles in a PS matrix (soft particle model). (b) The system with dispersed harder particles in a PS matrix (hard particle model). (c) The system with dispersed harder particles in a PS matrix with a peeling layer around the particles.

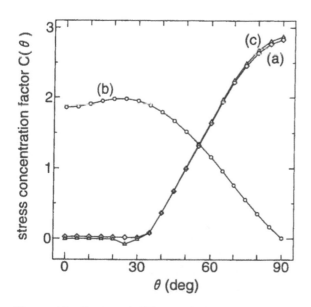

Figure 13 Result of FEM analysis: concentration coefficient of maximum principal stress (θ is used in Fig. 6). (a) The system with dispersed softer particles in a PS matrix (soft particle model). (b) The system with dispersed harder particles in a PS matrix (hard particle model). (c) The system with dispersed harder particles in a PS matrix with a peeling layer around the particles.

In Fig. 12a, the highest stress concentration point was at $\theta = 90°$, and in the case of Fig. 12b, the highest stress concentration point was at $\theta = 40°$. These results agree with the observation of craze initiation points from the rubber or steel balls reported by Wang et al. (26). On the other hand, in Fig. 12c, the highest stress concentration point was at $\theta = 90°$.

From the results of the FEM and the observation of real-time deformation, the craze initiation mechanism is proposed as follows.

Initially, the maximum stress concentration arises around a particle at $40°$ from the strain direction (from the result of FEM model, Fig. 12b). The void is formed at that maximum stress concentration point. Successively, the stress concentration occurs around the void and craze extended perpendicular to the applied strain (from the result of FEM model, Fig. 12c). The adhesion at the boundary of the PS and the particle is assumed to be almost zero. The peeling layer forms at the strain direction such as in Fig. 11, and successively crazes propagate from the peeling layer perpendicular to the applied strain.

IV. CONCLUSIONS

In bimodal HIPS, the long stress convergence length caused by large rubber particles and the short ligament thickness caused by a large number of small particles will give longer crazes and consequently a higher impact loading.

A new cavitation model is suggested: First, crazes are initially generated from large rubber particles. This generation is the same as that in bimodal HIPS. Second, rubber sorption into crazes occurs by negative pressure of microvoids in the crazes and results in cavitation of rubber particles; sorption of rubber components to craze fibrils would thus plasticize the crazes.

Further, we have investigated the system of PS with rigid particles and found that the impact strength of PS increased. Falling ball impact loading increases with the amount of particulate under 1 wt% and decreases gradually with particulate content greater than 1%. The maximum value of falling ball impact loading of PS-dispersed particulate occurred at an average particulate diameter of ~ 2 μm. In the best system of PS and rigid particles, the impact loading of the PS is higher than that of AS resin. From real-time deformation and FEM analysis of PS including the rigid particles, it is clear that the impact improvement mechanism is related to the generation of voids or a peeling layer around the rigid particles and successive extension of the craze from that void or peeling layer.

REFERENCES

1. C. B. Bucknal. Toughened Plastics. London: Applied Science, 1977.
2. H. Keskkula. Rubber-Toughened Plastics. New York: A.C.S. 1989, pp. 289–299.

3. R. F. Boyer and H. Keskkula. Encycl. Polym. Sci. Technol. 13:392, 1982.
4. D. J. Angier and E. M. Fettes. Rubber Chem. Technol. 36:1164, 1965.
5. J. D. Moore. An electron microscope study of the microstructure of some rubber-re-inforced polystyrenes. Polymer 12:478, 1971.
6. A. M. Donald and E. J. Kramer. Craze initiation and growth in high-impact polystyrene. J. Appl. Polym. Sci. 27:3729, 1982.
7. Y. Okamoto, H. Miyagi, T. Uno, and Y. Amemiya. Impact toughening mechanisms in rubber-dispersed polymer alloy. Polym. Eng. Sci. 33(24):1606, 1993.
8. C. B. Bucknall and R. R. Smith. Stress-whitening in high-impact polystyrene. Polymer 6,437, 1965.
9. C. B. Bucknall. In: D. R. Paul and S. Newman, eds. Polymer Blends. New York: Academic Press, 1978, Chap. 2, p. 99.
10. A. M. Donald and E. J. Kramer. Internal structure of rubber particles and craze breakdown in high-impact polystyrene. J. Mater. Sci. 17:2351, 1982.
11. BASF. U.S. Patent 4,493,922. Thermoplastic impact-resistant modified polystyrene moulding composition containing soft particles of differing size and shape.
12. Sumitomo Chemical Co. U.S. Patent Appl. 07/223,599. Rubber-modified polystyrene resin composition.
13. S. Y. Hobbs. The effect of rubber particle size on the impact properties of high impact polystyrene blends. Polym. Eng. Sci. 26:74, 1986.
14. C. Wrotecki and F. X. de Charentenay. Deform. Yield Frac. Polym. 7:51, 1988.
15. O. S. Gebizlioglu, H. W. Beekham, A. S. Argon, R. E. Cohen, and H. R. Brown. New mechanism of toughning glassy polymers. 1. Macromolecules 23:3968, 1990.
16. A. S. Argon, R. E. Cohen, O. S. Gebizlıouglu, H. R. Brown, and K. J. Kramer. A new mechanism of toughening glassy polymers. 2. Macromolecules 23:3975, 1990.
17. Y. Okamoto, H. Miyagi, M. Kakugo, and T. Takahashi. Impact improvement mechanisms of high-impact polystyrene with bimodal distribution of rubber particle. Macromolecules 24:5639, 1991.
18. Y. Okamoto, H. Miyagi, and S. Mitsui. New cavitation mechanism of rubber dispersed polystyrene. Macromolecules 26:6547, 1993.
19. K. Kato. Polym. Lett. 4:35, 1966.
20. Von Mises. Z. Angew. Math. Mech. 8:161, 1928.
21. S. Wu. Phase structure and adhesion in polymer blends. Polymer 26:1855, 1985.
22. D. S. Parker, H. J. Sue, J. Huang, and A. F. Yee. Toughening mechanism in core-shell rubber-modified polycarbonate. Polymer 31:2267, 1990.
23. T. Tamori, S. Okamura, T. Kamino, Y. Okamoto, and H. Miyagi. Japa. Soc. Electron Microsc. Prepr. 49:84, 1993 (in Japanese).
24. I. Yamaoka and M. Kimura. KinoZairyo 11(7):26, 1991 (in Japanese).
25. Q. Fu and G. Wang. Polyethylene toughened by rigid inorganic particles. Polym. Eng. Sci. 32:94, 1992.
26. T. T. Wang, M. Matsuo, and T. K. Kwei. Criteria of craze initiation in glassy polymers. J. Appl. Phys. 42:4188, 1971.
27. S. Mitsui, H. Kihara, S. Yoshimi, and Y. Okamoto. Impact strength improvement of PS by dispersion of rigid particulate. Polym. Engin. Sci. 36:2241, 1996.

21

New Strategies for the Tailoring of High-Performance Multiphase Polymer-Based Materials

Ph. Teyssié
University of Liège, Sart-Tilman, Liège, Belgium

INTRODUCTION

After an explosive growth during and after the second World War, the development of new homo- and copolymers per se became less and less favored in the industrial world, although a few successful examples are still emerging today. This trend is due to two main reasons. First, long and risky development processes represent a heavy investment burden that is often economically prohibitive. Second, the duration of such materials (often 4–5 years) prevents any market "agility," a key asset in our very competitive world.

As a result, applications people, faced with urgent requests for a broad set of properties in a single material, started to consider the possibility of blending already known polymers to meet such challenges. They were confirmed in that approach in that polymers are probably the most individualistic species in chemistry; they are usually immiscible and thus provide the multiphase morphology indispensable for ensuring additivity of properties. A wave of research resulted from such premises in the 1950s and 1960s, but researchers were soon confronted with disappointing results because of two main drawbacks. First, it was difficult to stabilize the submicron morphology often required for a homogeneous bulk mechanical behavior of the resulting material (even when that morphology was attainable owing to the use of appropriate processing tools). Second and worse, the interfacial adhesion in those materials was in most cases too weak to ensure long-range stability and resistance to large mechanical deformations demanded in modern technological applications.

After that period of "brute force" blending, the following decades (the 1970s and 1980s) accordingly saw the implementation of more sophisticated strategies, leading to much better products, for which was coined the general name of "polymer alloys." Based on the same concepts as in simple blends, these new approaches involved additional techniques. Among them, one can cite the use of "compatibilizers" (usually segmented copolymers that ensure small phase size and good interfacial adhesion), the control of reactive processing (yielding similar results, through covalent bond formation between phases), and generation of (semi-)interpenetrated networks. Quickly and with a good dose of naïveté, the method was seen as "a gold rush to quick bucks," the main fallacy being that it would represent a low-cost alternative for developing new materials. At that time, alloys were even predicted to be the dominant force in the 1990, for such developments, and the term "plasturgy" was adopted by analogy with metallurgy.

However, it soon appeared that optimization was a key but narrow bottleneck costing time and money, and that there were no such things as "universal" compatibilizers or strategies. Each blend was a case in itself, requiring a careful adaptation to the technical application needs. It is clear nowadays that the success of an alloy will strictly depend on its cost/performance profile in *actual* applications. And this is true for still realistic goals, such as recycling materials, upgrading commodity resins, enhancing properties of engineering plastics, and making higher performance polymers more cost-competitive (1).

The trend that logically emerges in the 1990s is thus a case-by-case approach based on a broad and deep understanding of the structure/morphology/bulk property relationships (and particularly of the interface/interphase situation), thus implying a sometimes painful optimization. And last but not least, it is an approach that is constantly driven by application requirements and customer needs.

In line with this trend, much more basic research has been carried out, leading to new concepts and techniques to enable the generation of better-performing, necessarily multiphase, polymer-based materials. It is the purpose of this contribution to illustrate some of these very promising achievements with a series (of course a nonexhaustive one) of examples chosen out of our own experience. They are representative of general strategies, and (in line with what has been said above) technologically significant, i.e., oriented towards specific applications, requirements, and needs. In this respect, one can now speak of really "functional" materials, specifically tailored for well-defined uses.

I.   DIRECT TAILORING OF MECHANICAL AND RHEOLOGICAL PROPERTIES OF MULTIPHASE MATERIALS

Interfacial compatibility and/or compatibilization is obviously the key issue in the field, and it is still a difficult challenge in many cases. The first part of this analysis will thus be devoted to a number of specific approaches to this general problem.

A. Use of Dispersed Segmented Polymer Additives

This time-honored method, a key one in the 1970s and 1980s, still represents an efficient tool for the tailoring of multiphase materials. The segmented polymers are graft, comb, or preferably diblock structures, synthesized either independently, often by "living" polymerization methods (ionic or radical), or in situ by reactive processing, a preferred approach at the industrial level. One should note that their action, described in a general way in Fig. 1, does not imply that the copolymer segment structures should be similar to the ones of the phases to be compatibilized. The only necessary requirement is that the interactions between A and PX, and B and PY, as provided by entanglements or/and dipolar forces, be strong enough to provide for both decrease of interfacial tension (small phase dimension) and increase of interfacial adhesion. Also, rather small amounts of those additives are usually required (ca. 0.5 to 3%). The whole field has recently been thoroughly reviewed (2–3).

As was stated in the introduction, this kind of technology has been increasingly applied to more sophisticated problems, requiring precise tailoring of the final functional (smart) material to yield a precisely optimized additive. A recent typical example is the improved dispersion of liquid crystals in transparent polymer-based panels, i.e., "switchable windows," which opacify under the influence of an electric field modifying the distribution morphology of the crystals (4). Another important one is a new and general approach to the stabilization of small-particle dispersions (see Sect. II). Interestingly, more exotic engineering plastic alloys can also be tailored using block copolymers, when combined with other additives: a good example is a blend of polyarylethersulfone (PAES) and

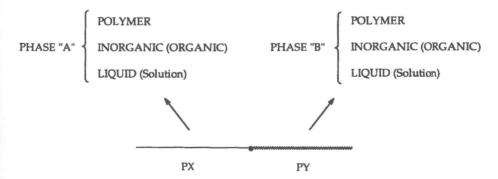

Figure 1 Phase bridging by segmented copolymers.

polyamide 4.6, added with a PAES-b-PA6 block copolymer plus a C_2—C_3 rubber grafted with maleic anhydride (5).

B. Tailoring of Surfaces Using Block Copolymers

It is often desirable that the surface of a given material chosen for its specific and satisfying physicomechanical properties will display another set of properties very different from those of the bulk sample. Typical requirements include adhesion, paintability, hydrophobicity or hydrophilicity, friction, electrostatic properties, and biocompatibility. Traditionally, surface modification has involved physical posttreatment, which is usually cumbersome and expensive (e.g., plasma photochemistry). More recently, that difficult but very general problem has been approached by blending the base material with a small amount of a diblock copolymer, consisting of one segment (A) similar to the bulk material and another immiscible one (B) displaying a very different interaction parameter and surface tension. Obviously, such a copolymer tends to form micelles dispersed in the bulk of the material. However, that equilibrium situation can be displaced and the copolymer attracted to the surface, when that surface is in contact with a medium having a favorable interaction parameter with segment B. Under those conditions, that surface will finally display the characteristics of polymer B, e.g., surface tension, dipolar (or ionic) interactions, adhesivity. This occurs provided there is sufficient surface coverage, which also requires a sufficient rate of migration to the surface (a function of temperature and copolymer molecular weight).

A well-studied example is the coverage on the atmosphere side of a commodity plastic, i.e., polystyrene (or polymethylmethacrylate, PMMA), with a layer of polydimethylsiloxane (PDMS) provided by a PS-PDMS diblock copolymer. Complementary characterization techniques, involving wettability, contact angles, XPS, and SIMS (including depth profiles and surface imaging) all support a complete in-depth surface coverage by the PDMS block, even with copolymer concentrations as low as 0.2% (6).

The reverse problem, i.e., the accumulation of a copolymer-contained polar segment at the interface between a low-surface-energy polymer matrix and a high-surface-energy substrate, is a much tougher one, although it is even more interesting in technological terms. Its feasibility has been demonstrated by thorough basic investigations in Japan (7), and an interesting practical version has been implemented at CERM, in which a PS-b-PMMA copolymer in a PS matrix migrates towards Al plaques under 200°C (8). Despite the interest of these achievements, they still suffer in that when separated from the polar substrate (which can also be water, an attractive variation), thermodynamics take over and the polar segment "reconforms" towards the inside of the sample (even if not migrating back at a temperature $< T_g$). Trying to prevent that phenomenon by physical cross-linking (i.e., by irradiation) met with only limited success, and in any event it relies on methods that one tried to avoid by using copolymers in the first place. Nevertheless, this approach remains of interest when the substrate is to be maintained at the polymer surface, e.g., providing an adhesive interphase between polymer and metal.

C. Strategies Other than the Use of Segmented Polymer Additives for Creating Efficient Interfaces

The generation and optimization of a segmented copolymer additive is always a demanding task, often hindered by technical or managerial barriers. Accordingly, in all cases where it appeared feasible, people have tried other approaches. Although not numerous, successful examples exist based on the *use and organization of dipolar or/and ionic forces*. This kind of strategy, which should be better implemented in the future, is illustrated here by three typical examples.

1. Alloys of Polyvinylidenefluoride (PVDF) and Polyamides (PA)

As observed by a set of techniques, PVDF and polyamide 6 (PA6) are immiscible under normal conditions, a situation that can be explained by much stronger intermolecular interactions within the PA6 component (H-bonding). Nevertheless, an unexpectedly fine phase morphology is currently observed in their blends (ca. 0.5 μm), which might be accounted for by possible cross-interactions between the two phases. Such interactions have been suggested by the mutual solubility of PVDF and polycaprolactam above PVDF melting point and the existence of a negative interaction energy density (B) that is temperature and concentration-dependent. This is most probably the basic reason for the totally unusual mechanical properties of such blends, which exhibit a *true synergism,* i.e., blend properties much better than those of the individual components, within a significant composition range. It is particularly true for elongation at break, with PA6 being the minor phase. Interestingly, the phase size is minimized in those areas of synergism. Still more striking, the lap shear strength measured between two plaques of those two polymers (compression molded at high temperature) is extremely high for immiscible partners and very similar to that of a homopolymer (9).

2. Alloys of PVDF and Polycarbonate (PC)

The same concerns about the cost of preformed block copolymers and limitations of reactive processing in their in-situ formation were incentives to extend this new compatibilization strategy. Obviously, the problem was particularly acute in the case of PVDF/PC blends. To solve it, interfacial interactions were promoted by the use of a third homopolymer, in this case PMMA, which is totally miscible with PVDF in the melt. When plates of PC and PVDF/PMMA blends were compression molded at high temperature, the interfacial toughness G (in N/m, as measured with dual cantilever geometry) was found to increase from a very small value (below 2, nonreproducible) up to ca. 80 when the PMMA content in PVDF went up from 0 to 40%, and then to reach a plateau value (which is nothing but the interfacial adhesion measured between PC and pure PMMA). It clearly shows that PMMA migrates and accumulates at the interface, substituting a PC/PMMA interface for the original PC/PVDF one. The thermodynamic driving force for this phenomenon has been found in the smaller final interfacial tension (10).

3. In Conclusion

We clearly have here a number of results that open new prospects for the tailoring of polymer alloys, avoiding the use of any additive. One should also remind oneself here of the possibility of creating strong interfaces by using ionic interactions between positively and negatively charged polymers. Terminal functionalization, easily implemented by living anionic or radical polymerization, is often sufficient to reach that goal (11). Although the cases described above remain specific, the underlying general strategy is broad enough to extend these successful examples to many more challenges of interest.

D. Tailoring of Polymer Melt Viscosities by Blending

In current processing technologies, melt viscosity (η_m) is another key parameter, from both a technical feasibility and an economic (speed and temperature) viewpoint. Optimizing viscosity is however a very challenging task, owing to three main difficulties. First, the modification approach adopted most often, the inclusion of an additive, should not be detrimental to the final bulk properties of the material. Second, that additive must be able to promote either a decrease or an increase (negative or positive deviation) of η_m. Third, its addition should also promote much larger viscosity variations than those obtained with classical processing aids (i.e., ideally at least one order of magnitude). Fortunately, it has been shown that blending with moderate amounts of carefully optimized polymers (ca. 5%) can solve most of the encountered problems, as illustrated in the two following case studies.

1. The Design of CPVC (Chlorinated PVC) Melt Viscosity

CPVC is an interesting material known for a number of desirable properties: high T_g (proportional to Cl content), high solvent resistance, nonflammability, and its economical nature. However, its high η_m requires a high processing temperature, sometimes close to degradation conditions.

Interestingly enough, it has been found that the addition of some polyalkylacrylates or methacrylates (PRAs or PRMAs) in reasonable amounts, i.e., ca. 5%, was able to decrease the M.V. by a factor ranging from 5 to 20, depending essentially on the chain flexibility of the additive, but only if such combinations are immiscible. In these biphasic blends, the additive is finely and homogeneously dispersed throughout the major matrix (ca. 1 μm), which means that important bulk properties of the latter, such as the heat-distortion temperature, are preserved. Moreover, the rheological behavior of the system versus shear rate is not modified, i.e., the $\bar{n}$ exponent remains the same in the observed power law $\eta_m = \gamma^{\bar{n}}$.

Finally, it appears that this desirable control can be obtained using most PRAs and PRMAs immiscible with, and of a much lower dynamic viscosity than, CPVC. The additive supposedly acts in a kind of roll-bearing mechanism and must also be able to promote a well-balanced interfacial adhesion (low-viscosity PE has no effect). A precise tailoring of that parameter (by varying alkyl chain length, or by copolymerization with MMA) can promote additional impact strength. This successful approach has also been applied to PVC (12).

2. The Design of Aliphatic-Aromatic Polyamide Melt Viscosity

A low Newtonian melt viscosity is thus usually a prerequisite for fast and easy processing. However, for some techniques, such as molding or extrusion blowing, it happens that the η_m may be too low. This is the case with polyamides (PA) from adipic acid and m-xylene diamine ($\eta_{m260°C} = 170$ Pa·s), for which an $\eta_{m260°C} >$ 850 Pa·s is required at low shear rates ($\gamma < 100$ s^{-1}), particularly for avoiding blur formation as well as a time- and polymer-consuming finishing step.

In agreement with current hypotheses, it was shown that PA rheofluidity can be dramatically decreased by mixing with low amounts (typically 5%) of copolymers of MMA and methacrylic acid salts. The copolymer composition must however be optimized (around 12% MAA), since the η_m increase is negligible in cases of both complete immiscibility and miscibility. A remarkable maximum is observed when the compatibility is improved just to a point where a finely dispersed biphasic blend is obtained. H-bonding, chain-branching of PA terminal NH$_2$ groups onto pendant copolymer COOH groups, and ion-dipole interactions are at the origin of this behavior. The nature of the counter-cation also has a striking influence: alkaline cations, particularly Cs, are the most efficient, leading to an η_m increase over 30 times at low γ. These blends exhibit a welcome rheothinning behavior, which can be explained by the slip of the different phases at increasing shear rates (13).

It is relevant to note that conversely, ion-dipole interactions may be at the origin of a very interesting *increase* of bulk or solution viscosity upon increasing shear rate, i.e., a rheothickening behavior, in different types of ion-containing (co)polymers (14).

In conclusion, blending with small amounts of additives, which have no detrimental influence on the material bulk mechanical properties, can be of prime interest for the highly desirable control of processing rheology. As a useful empirical rule, η_m usually shows a negative deviation from the logarithmic additivity rule when there is no strong specific interaction between phases, while positive deviation is witnessed in the cases of strong interactions and for emulsion type blends with low interfacial tension and long relaxation times relating to the structured morphology.

II. PARTICLE STABILIZATION IN LIQUIDS: A NOVEL GENERAL STRATEGY FOR THE TAILORING OF POLYMER ALLOYS

In the design of multiphase polymer-based materials, one is often confronted with the problem of having to disperse small particles homogeneously, i.e., at or below 100 nm, in a polymer matrix of a given structure, while ensuring interfacial adhesion and noncoalescence of these particles. Among all the research efforts of the last decade, a new strategy has emerged that is able to meet the challenge in a general way, i.e., under very different conditions and for the production of vastly different types of materials (15). It requires a well-controlled two-step approach. In the first step, a fine dispersion of the submicron particles is obtained with efficient mixing. This dispersion is stabilized by a well-tailored block copolymer (linear diblock, comb) in a liquid medium (see Fig. 2), thereby generating the morphology suitable for the future material requirements. It has to be stressed here that the particles may be of very different nature (polymer latex, organic pigment, or inorganic filler) and that the liquid medium is sometimes organic and sometimes aque-

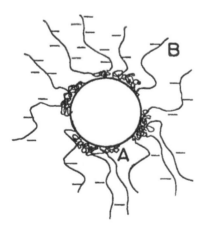

Figure 2 (Electro)steric stabilization of particle dispersion by segmented copolymers: (A) hydrophobic block adsorbed on particle; (B) "hairy" block (charged or not) interacting with the medium.

ous depending on the material and its application characteristics. In the second step, this liquid medium, i.e., monomer, prepolymer, or polymer solution, is then converted into a "solid" polymer phase by an appropriate chemical or physical treatment such as evaporation, polymerization, or cross-linking.

This very broad and versatile approach allows a "nanometric" tailoring of the morphology and bulk final properties of successful multiphase materials. Its basis, which is really the establishment of an adherent "steric crown" around the particle, has already been successfully tested in the stabilization of carbon black particles dispersed in liquid paraffin, i.e., a toner (16). This will be illustrated here by three exemplary cases, all implemented at the technological level (in either a small pilot or a production unit).

A. Production of Flexible Polyurethane Multiphase Foams with Improved Load-Bearing (and Flammability)

Flexible polyurethane foams of very low density enjoy an important and diversified application market, a good deal of which is in upholstery. Despite a broad set of desirable properties, their load-bearing capabilities and flammability are often unsatisfactory.

The load-bearing drawback can be alleviated by the presence of a high cohesive-energy density polymer, i.e., SAN. However, the dispersion of such a polymer additive was not easy to achieve in a perfectly homogeneous manner and with good adhesion at the submicron scale, and clear-cut structure/morphology/properties relationships had not been previously determined. These problems were solved by synthesizing a double-comb copolymer, the backbone of which was composed of PS containing an optimized amount of isoprene units as well as of hydroxyethylmethacrylate ones (a few % of each). Polyether-polyol blocks of controlled length were then grafted onto these OH groups via classical ethylene and propylene oxide plus glycerol anionic polymerization using KOH. This "prestabilizer"-containing polyol was then used as a dispersion medium for SAN polymerization, during which the unsaturated isoprene units of the backbone favor the "grafting onto" of SAN, yielding a double-comb copolymer that is the actual steric stabilizer. The whole process thus leads to reinforcing SAN particles covered with this stabilizer, which ensures an excellent dispersion and interfacial adhesion between the SAN particles and the polyol, now going into the foaming process. The resulting well-controlled foam morphology with high interphase adhesion, as visualized by SEM (Fig. 3), provides for an excellent set of mechanical properties that are often even better than those of classical PU foams and include a 50 to 100% improvement of the load bearing (17). These results have been successfully extrapolated into an industrial process that produces several thousand tons a year. It must be

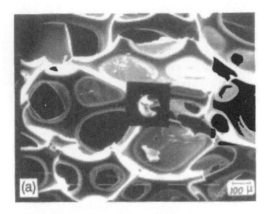

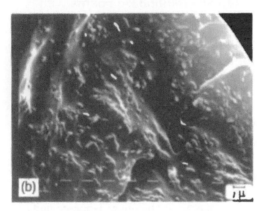

Figure 3 Scanning electron microscopy of a flexible polyurethane foam reinforced with randomly dispersed adhering SAN particles.

stressed however that such a behavior was reached only after a careful in-depth
optimization, involving the $C\!\!=\!\!C$ content, the number of polyether branches, the
particle size, and last but not least the medium viscosity. The general necessity of
such feedback multistage optimization processes will be discussed in the conclusion.

It is also timely to remember here that fillers such as aluminum hydrate could
also be successfully integrated in PU foams, thanks to a filler pretreatment based
on a strategy similar to the one described above. Filler contents as high as 35 wt%,
the grains being homogeneously distributed and enveloped at the cell corners, provide of course for a significantly decreased flammability (18).

B. Stabilization of a Latex Dispersion in Blends and/or in Emulsion Polymerization

The same general concept as above was also applied to the stabilization of latex particle dispersions, but this time in an aqueous medium instead of an organic one. For this purpose a poly(alkylmethacrylate-b-methacryloylhydroxyethylsulfonic acid) was synthesized that could provide for an "electrosteric stabilization crown" around the particles and a protection against pH and ionic strength variations due to the acid strength.

This diblock copolymer was used as an "emulsifier" for the radical poly-me-rization of (meth)acrylates in aqueous medium, leading to a very interesting type of latex (19) in which the thickness of the electrosteric crown, as measured by photon correlation spectroscopy, corresponds to the extended chain length of the polysulfonic acid block. Moreover, the use of allyl methacrylate in the synthesis of the first block allows a cross-linking of that "emulsifier" within the core of the latex, preventing any further leaching. This latex has been used in combination with gelatin in film-casting processes, efficiently playing the role of a "plastifier" thanks to the good interactions between the polysulfonic acid block and the protein functional groups (20). The overall approach ensures the obtaining of the desired mechanical properties, together with better, nonfoaming processing of some photographic layers.

Besides having exceptional stability (a critical coagulation concentration about 30 times that of the corresponding low-molecular-weight surfactant), such particles may act as a *molecular core-shell*. In particular, they offer a functional surface (sulfonic groups) onto which different types of desirable compounds can be coupled, such as dyes and U.V. stabilizers.

C. Stabilization of Double Dispersions: Inorganic Particles in Hydrophilic Polymers

The stabilization of a fine dispersion of an inorganic compound such as TiO_2 in an aqueous medium is a particularly demanding challenge, and it is still more difficult when another compound such as carbon black (CB) has to be simultaneously dispersed under the same conditions. Moreover, ideally at least, coalescence has to be avoided, and the dispersion has to maintain both primary particle sizes.

Although the above-described amphiphilic sulfonic block copolymer already gave interesting results related to pure water, the presence of polyvinylalcohol (PVOH), a matrix of general interest, led to disappointing performances. However, such alloys could be optimized by using another strong polyelectrolytic block, i.e., a poly(MMA-b-quaternized dimethyaminoethylmethacrylate). This ensured a very good stabilization of CB dispersions in aqueous PVOH. In confirmation of the necessity of finely tuned optimization, it was shown that replacement of PMMA by PS in the hydrophobic block still improved the situation versus TiO$_2$. Both copolymers were thus used together and found to be efficient in surprisingly low amounts, ca. 0.5% (despite the unavoidable equilibrium between free micelles and adsorbed species). Those remarkably stable and fine double dispersions in PVOH were useful in half-tone reprographic techniques (21). A similar strategy can also be used for paints (22).

D. Synthesis of Core-Shell Latexes from Amphiphilic Block Copolymers

The interest of functional latexes, covered with an electrosteric crown of a strong polyelectrolyte as described above, prompted a more in-depth study of their production. Amazingly, it was found that the rate of polymerization was maximum, and sometimes much higher than with a corresponding low-MW surfactant, when a radical was used having the same electrical charge as the polyelectrolyte block (23). This interesting effect, witnessed for both anionic and cationic types of polyelectrolyte block, has been ascribed to the much higher mobility of the charged radical in the hydrophilic crown (despite electrostatic repulsion), preventing radical deactivation.

Conversely, opposite charges slow down the radical, which has more time to deactivate before entry, as confirmed by the fact that in this case, a higher MW block (ca. 20,000) may even inhibit the polymerization. Quite expectedly, the nature of the polyelectrolyte block has no significant influence on the rate of polymerization by a neutral radical.

In conclusion the general strategy outlined in this section is extremely powerful and can certainly meet many challenges in the field of multiphase materials. Although very precise in its multistep conception, it is quite straightforward and practical, and even more so recently as a number of the necessary diblock copolymers have become available from "living" radical polymerization techniques.

III. THE IMPORTANCE AND THE CONTROL OF (CO)CONTINUOUS MORPHOLOGY

Most bulk properties of a blend are usually determined by those of a major continuous phase. There are however two types of interesting situations where desirable properties are generated by a different morphology: the cocontinuous interweaving of two phases and the continuous spider-web-like organization of a very minor phase. Moreover, both types of morphologies can promote true synergism, i.e., generating particular properties much better than those expected from the additivity laws, and sometimes even better than those of any of the components. A theoretical approach to such a behavior has been reviewed (24) based on general scaling relationships from percolation theory. Here we present four typical cases showing the breadth and the potential of this area in basic experimental research on morphology and in its promising applications.

A. Stabilization, Quantitative Characterization, and Synergistic Behavior of Cocontinuous Morphology in Polystyrene (PS)–Polyethylene (PE) Blends

In a successful attempt to produce very-high-impact PS materials with improved aging, PS/PE alloys in an 80/20 ratio had a few wt% of poly(styrene-b-hydrogenated butadiene) equimolar diblocks added. Interestingly, it was found that the tensile properties of the resulting alloys were clearly superior when a tapered diblock (fuzzy interface) was used instead of a pure one with a sharp interface (a structural variation easily controlled by anionic polymerization). Within a range of compositions (and particularly at 80 PS/20 PE), true synergism was even observed, giving σ_B (stress at break) values about 20% higher than those measured for pure PS!

An explanation for such exceptional behavior is probably to be found in the cocontinuous morphology observed by TEM at that blend composition. It was thus of prime importance to be able to characterize quantitatively such cocontinuous structures as well as the diblock efficiency in stabilizing them well above the T_g and T_m of the components (ca. 200°C, i.e., processing conditions). These two questions had received only limited attention until then, mainly because none of the established or new observation methods were able to give that kind of answer ((S)TEM, AFM, solid state NMR, imaging IR, and so on). Thus multiscaling (multifractal) image analysis (MSA) of digitized representations of TEM micrographs of thinly sectioned blends were used. From such analysis, multidimensional $f(\alpha)$ spectra of local "singularities" could be obtained by classical mathematical methods (see Ref. 25), giving a quantitative evaluation of the cocontinuous structure homogeneity (see Fig. 4). Rewardingly, it turned out that the samples with less multifractal area had the best mechanical properties, i.e., the ones added with the tapered diblock (25a).

Moreover, this valuable method proved equally powerful in the analysis of the blend stability against shear and temperature (25b). As confirmed (when feasible) by other classical image treatments (standard granulometry, opening size granulometry distribution), the MSA method clearly showed striking differences (impossible to evaluate when simply looking at the TEM micrographs) between the two types of copolymers, the tapered one again being much more efficient and

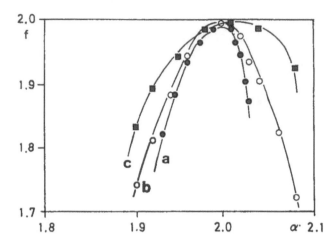

Figure 4 Singularities spectra (from multifractal image analysis) of 20 PE/80 PS blends, containing 10% of tapered or pure P(HBD-b-S) copolymer: (a) tapered—initial blend, and annealed 60 min, 180°C; (b) pure—initial; (c) pure—annealed 5 min, 180°C.

able to protect a homogeneous cocontinuous structure for hours at 180°C. This differential behavior was explained by the capability of the tapered diblock to produce some diffuse interface, whilst having less of a tendency to segregate and form its own micelles.

B. Phase Cocontinuity in Hybrid (Ceramer) Materials

During the last decade, special attention has been paid to the synthesis and characterization of inorganic–organic hybrid materials, as obtained from sol–gel processes. They indeed offer a unique opportunity of combining the desirable properties of organic polymers (flexibility, modulus) and of inorganic glasses (high thermal stability, good optical properties) in a controlled manner. However, the resulting properties are critically dependent on average phase size, phase continuity, and molecular mixing at phase boundaries.

An optimization of such a sol–gel process was performed on a tetraalkoxysilane/α,ω-hydroxyl end-capped polycaprolactone system (PCL, a polyester known for an attractive set of properties, i.e., permeability, biocompatibility, and biodegradability, improved further by copolymerization with lactides and glycolide).

Careful control of the reaction conditions, followed by annealing, allows a

35% nonreversible incorporation of the PCL (26a). In a material containing 54 wt% SiO_2, TEM shows an extremely fine phase morphology (ca. 30 Å mean size). Again, image analysis techniques revealed an important cocontinuity (26b), which is clearly related to the interesting set of properties of the materials.

They already have been investigated for different types of applications, in particular for glial (neuronal) cell culture and for the production of silica displaying controlled porosity (due to the easy PCL degradation).

C. Continuity of a Carbon Black (CB) Minor Phase in Multiphase Polymer Blends

A conducting polymer composite material usually consists of a randomly distributed conducting filler throughout an insulating polymer matrix. Such materials deserve interest in several application fields: antistatic materials, low-temperature heaters, electromagnetic radiation shielding, and so on. The choice of the components is crucial for efficient results, and most importantly, there must be a good balance between the filler–filler and the filler–polymer interactions. In this respect, CB is probably the most widely used conducting filler. Another major problem raised by such composites is that the filler content must be as low as possible in order to avoid difficult processing, poor mechanical properties, and high cost.

In an amorphous PS matrix, the percolation threshold (i.e., the lowest concentration of conducting particles at which continuous conducting structures are formed) is as high as 8 wt%. The only way to reduce that CB content is to favor inhomogeneities in the material: e.g., the threshold comes down to 5% in a semicrystalline polyethylene (PE) matrix. More recently, two-phase polymer blends have been considered as possible alternatives. But either one of the phases must be continuous and contain the CB particles, or the two phases must be cocontinuous, and the filler preferably located in the minor one, or better still at the interface (double percolation).

A careful control of PE/PS blend morphology in the presence of CB led to very interesting results (27). First, CB stabilizes a cocontinuous polymer morphology against shear and temperature (200°C), either through its own interfacial activity or by its kinetic effect on the coalescence process. With 4 w% CB, this morphology is preserved over a wide composition range of 5 to 70% PE (i.e., more than the values predicted by percolation theory for the formation of an infinite cluster). For an optimum cocontinuity, the threshold further comes down to 2 w% CB (localized in the PE phase). Fine-tuning allows a selective localization of CB at the interface of the cocontinuous polymer blend. When CB particles are dispersed in this two-dimensional space, the interfacial area has to be decreased down to the point where they are touching each other whilst preserving the interface continuity. The best strategy to reach this target was annealing of the cocontinuous two-phase polyblends, further favoring localization of conductive particles at the interface, as a web of very fine conducting features revealed on optical micrographs (27).

The universal percolation theory was applied to the conductivity measurements performed in that situation. It is known that in the general equation $\sigma \approx \sigma_0(p - p_c)^t$, the critical exponent t is only related to the dimensionality of the system, theoretical values being 1.9 in 3D and 1.3 in 2D systems. Rewardingly, calculated t values (from conductivity and weight data) amount to 2.0 (± 0.2) when CB is dispersed within the amorphous phase of pure PE (3D) but to 1.3 (± 0.2) when CB is at the interface of a PS/PE blend with dual phase continuity, a typical 2D space indeed.

Under the best optimization conditions, the percolation threshold is now 0.4%, i.e., a striking 0.002 volume fraction, and plaques of that three-phase blend are practically transparent.

D. Minor Polymer Phase Continuity by In-Situ Polymerization

Another rather straightforward and easy way to establish minor phase continuity in a two (or three)-phase alloy has been found in the bulk polymerization of a monomer containing a few wt% of a preformed network in "solution." A good example of the feasibility of this approach comes from the use of halatotelechelic polymers (HTPs), a typical structure of which can be sketched as —[OCO—PBD —CO—O—M^{++}]$_n$. Easily obtained by neutralization of divalent (or even tetravalent) metal alkoxides with telechelic dicarboxypolybutadiene ($M_n = 4000$), these products are readily soluble in hydrocarbon media. Above a critical concentration of a few wt% (typically 2 to 5), their ion pairs associate intermolecularly into small aggregates yielding thermally (or chemically) reversible networks and thus forming gels stable up to ca. 80 to 150°C and displaying a number of peculiar properties (28). If one dissolves ca. 8 wt% of such a polymer in styrene monomer, together with a small amount of AIBN, and polymerizes the resulting gel at 60°C for an appropriate time, a multiphase alloy is obtained that displays quite intriguing and attractive properties. The first one is a specific morphology, as revealed by TEM and SEM, wherein the HTP is organized in a web-like 3D network of pentagonal and hexagonal thin cell walls (Fig. 5).

It is most likely that this kind of morphology, corresponding to the lowest free energy pattern, imparts a high impact strength and a remarkable rebound capacity to the predominantly PS resin.

In conclusion, although the optimal conditions might not always be easy to determine, a similar type of strategy, giving access to a morphology that might also be described as inverse "snake cage" or semi-IPN, is certainly worthy of further investigation and extension. This is well supported by the technologically significant fabrication of high-impact-strength, high-transparency PMMA cast

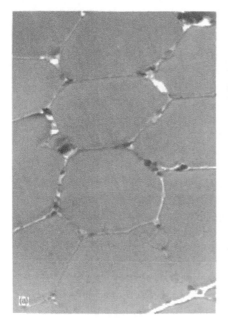 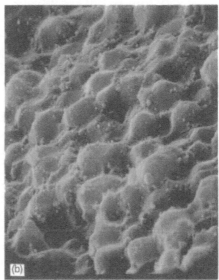

Figure 5 Blend of 10% Mg-dicarboxylato-PBD and 90% styrene, after poly-merization: (a) TEM (×10,000); (b) SEM (×2,500).

sheets, obtained by polymerization of the monomer containing 4 to 5% of a swollen elastomeric polyurethane network (45).

IV. DIRECT DESIGNING OF ORGANIC/INORGANIC HYBRID COMPOSITES BY ON-SITE POLYMERIZATION

In many cases of great potential interest, it is impossible to achieve desired alloys by using standard processing techniques, such as blending or compression molding, for many conjunctional reasons: lack of adhesion, melt viscosity mismatch, etc. Strategies have thus been developed in which direct monomer polymerization from or/and onto the filler was foreseen to circumvent the above-mentioned difficulties, while possibly resulting in better interaction (if not covalent bonding) between the two phases. As a basic illustration of those strategies, two recent developments of great technological promise will now be described.

A. The Polymerization-Filling Technique: A General Approach to Polyolefin-Based Composites

Academic and industrial interest in filled polymers is increasing as the result of an appreciable cost reduction and the opportunity of producing materials with a new set of selected properties. Historically, this idea has been successfully implemented for elastomers blended with a high level of filler (i.e., CB, SiO_2), increasing tensile strength, modulus, and hardness. In contrast, a similar approach involving the addition of mineral fillers into a semicrystalline thermoplastic such as a polyolefin has usually resulted in a detrimental effect on the mechanical performance. This was essentially due to a weak interfacial adhesion and also to a complex interplay of the characteristic features of the filler, the polymer matrix, and the dispersion technique. Since melt blending was quite inefficient, rather complex and expensive answers were proposed, such as filler encapsulation by polymer coating or chemical modification of its surface by functional adhesion promoters. As a more attractive alternative, the polymerization-filling technique was developed both in Russia (29) and in the USA (30). It consists of attaching a Ziegler–Natta type of catalyst onto the filler surface, which upon olefin polyme-rization allows a very high filler loading (up to 95 vol%). Acceptable mechanical properties [due to the ultrahigh-molecular-weight polyethylene (UHMWPE) produced] and the high filling degree (providing for properties such as good abrasion resistance and low flammability, and even making possible use as a ceramic precursor) are the basis of materials with an entirely new composition, that cannot be produced by the standard mixing methods (31). Their key advantage is an unusual combination of high stiffness and high impact resistance, even for filler contents as high as 60–70%.

In spite of these valuable properties, interest in such composites was limited, mainly owing to their very poor processibility, either by compression molding at high temperature or by blending with low-molecular-weight PE (so as to reach an acceptable melt viscosity, as in Russia for "Norplastic" composites). Moreover, elongation at break decreases with increasing filler content, and a number of other important requirements still were not fully met: control of a maximum filler dispersion, optimization of filler–polymer interfacial adhesion, and tailoring of the matrix molecular weight values over at least one decade.

In order better to understand and tailor the key structure–properties relationships at the basis of that unique combination of high impact strength, ductile fracture, and high modulus that should be achieved with the PFCs, a systematic study was recently undertaken that yielded very encouraging results and significantly broadened the scope of the technique (31,32). In summary, it was found that

> The nature and processing of the catalytic system are of prime importance and were optimal with an Al/Ti/Mg system (of 120/0.75/10 composition), the synthesis of which is sketched in Fig. 6. The surface coverage was limited to 25–50% of the OH groups (i.e., on kaolin) in order to avoid undesirable formation of free catalyst.

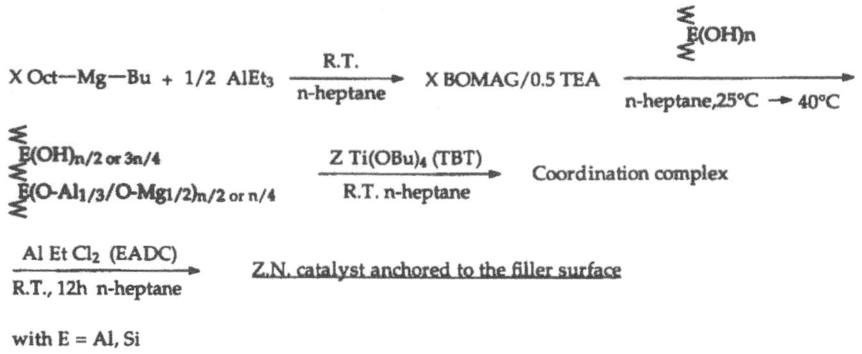

$$X \text{ Oct}-\text{Mg}-\text{Bu} + 1/2 \text{ AlEt}_3 \xrightarrow[\text{n-heptane}]{\text{R.T.}} X \text{ BOMAG}/0.5 \text{ TEA} \xrightarrow[\text{n-heptane},25°C \rightarrow 40°C]{\overset{\displaystyle E(OH)n}{}}$$

$$\underset{E(O\text{-Al}_{1/3}/O\text{-Mg}_{1/2})_{n/2 \text{ or } n/4}}{\overset{E(OH)_{n/2 \text{ or } 3n/4}}{}} \xrightarrow[\text{R.T. n-heptane}]{Z \text{ Ti}(OBu)_4 \text{ (TBT)}} \text{Coordination complex}$$

$$\xrightarrow[\text{R.T., 12h n-heptane}]{\text{Al Et Cl}_2 \text{ (EADC)}} \underline{\text{Z.N. catalyst anchored to the filler surface}}$$

with E = Al, Si

Figure 6 Anchored catalyst synthesis for the production of polyolefin-based PFCs.

Under these conditions, hydrogen is a convenient regulator of molecular weight, which has been found to decrease as $[P_{H2}]^{0.5}$ (Keii's law). Moreover, H_2 simultaneously acts as an efficient activator, leading to a catalyst productivity as high as 300 kg PE/gTi × h (i.e., 14.6 Tons PE/mol·Ti × h). Interestingly, α-olefins such as 1-octene also behave as active transfer agents, although in a rather complex way. Using a suitable combination of H_2 and octene, the melt index (MI) of a 32 wt% kaolin-containing composite was increased from 0.01 g/10 min (6 bar of H_2) up to 2.1 g/10 min ($H_2 + 8.10^{-3}$ mole of C_8). Although this MI is still rather low, those PFCs can now be processed and shaped by standard techniques.

Finally, SEM investigations have fully demonstrated the remarkable improvement of the interfacial adhesion in such materials. Small filler particles are found regularly dispersed within the matrix and interconnected through stretched polymer threads (even after impact testing), a morphology quite different from that of melt-blended composites. Also, the final structure observed is completely different from that observed for PE produced by traditional catalysts giving globular particles.

As a result of these optimized approaches, materials displaying a set of remarkable properties have been prepared that are definitely superior to melt-blended composites and even to classical PFCs, further enjoying a much better processibility, one close to that of HDPE. Although comparison of PFCs with melt-blended composites is approximate because of the difficulty of matching M.W.s and their distribution, some convincing data are presented in Table 1.

The superiority of PCFs clearly emerges from the comparison of impact energy, elongation at break, and σ_B/σ_y tensile strength ratio values. Moreover, in optimized samples, the impact resistance was found to be remarkably high at low

Table 1   Comparison of PFCs and Melt-Blended Composites

Composite type	Filler		E (GPa)	σy (MPa)	εy (%)	σB (MPa)	εB (%)	I.En. (KJ/m
	Nature	Content						
Blend[a]	Kaolin	32	1.7	26	3.0	26	5	4
PFC	(1.4 μm)	"	1.6	30	6.8	22	180	84
Blend[a]	Barite	31	1.3	25	2.2	12	9	8
PFC	(3 μm)	"	1.2	26	3.3	33	431	≫[b]

[a] Blends prepared with HDPE, M_n = 20.000, M_w/M_n = 3, MI = 9.4 g/10 min.
[b] Sample bends rather than breaking.

temperature, even in comparison with polycarbonate and flame-resistant ABS. Another important feature is the very low smoke emission of $Al(OH)_3$-based PFCs, when they burn under forced conditions.

It is of course of prime importance to extend as far as possible this polymer-filling technique. Very encouraging results are already being obtained (33a) by using the latest generation of metallocene catalysts. This new development allows the synthesis of a broader range of matrices, e.g., HDPE, LLDPE, PP (33b), on virtually every type of filler particles such as kaolin, barite, magnesium hydroxyde, graphite, carbon black, and even ferrites and nickel powder (33c). Interestingly, the use of very small particles should also provide transparency for composite plaques. All these new achievements obviously open bright prospects in the field of materials science and technology.

B.   "Bonding" Electropolymerization: A New General Strategy Towards Multilayer Alloys and Films Based on Conducting Substrates

During the past decades, increasing attention has been paid to electrochemistry as an interesting technique for polymerizing different monomers (34), the obvious motivation being to combine the potentialities of electrochemical methods and the attractive features of polymer materials science. This time-honored approach was, however, suffering from a double drawback, i.e., weakness of both interfaces, substrate/polymer and polymer/outside medium (this latter being either air or a polymer matrix, a biological medium, etc.). Those problems were somewhat alleviated by using physical treatments (plasma, etching) although there were other difficulties such as delicate processing and high cost.

More recently, Lécayon in France aroused great excitement by claiming (35) that polyacrylonitrile (PAN) could be grafted onto a number of common metals (e.g., Ni, Fe) by controlled cathodic polymerization of AN monomer. It was realized at that time, thanks to very sensitive equipment, that a voltammogram of the AN reaction on a clean Ni plate shows the existence of two peaks, a very weak one corresponding to a passivation phenomenon (P.I, around -1.8 V) and a much more intense one beyond ca. -2.2 V (P.II), a typical "diffusion-like" peak. Lécayon always used acetonitrile (AcN) as a reaction medium, i.e., a nonsolvent of PAN, which accordingly precipitates onto the Ni plate. Its substitution for dimethylformamide, a good PAN solvent, unambiguously demonstrated that the polymer formed at P.I was strongly bonded to the metal, contrary to the one formed at P.II, which immediately and quantitatively dissolved (36). Besides adhesion (or not), there are other striking differences between the films obtained at these two potentials. A film at P.I is thin (typically from 50 to 750 Å), transparent, and colorless, rich in isotactic microstructure and resistant to abrasion, all characteristics different from those of the film at P.II. Moreover, dual-cantilever DMTA studies demonstrate a spectacular difference between that electrografted film and one of commercial PAN cast from DMF. The polymerization mechanism at P.II is undoubtedly similar to the classical dianionic process (from one-electron transfer to monomer and dimerization of the formed radical anion, as known for AN dimerization and solution polymerization). However, the mechanism at P.I is still subject to controversy. In spite of a number of experiments involving different types of probes, it was not possible up to now to distinguish between an anionic process involving an adsorbed monoactive species and a radical type process induced by an electronic rearrangement between substrate and adsorbed monomer (both processes implying a one-electron transfer). This issue is of great practical importance, since an in-depth knowledge of the active species at P.I might give better clues for a selective termination (or further propagation) of the growing chains, i.e., for functionalization of the film surface.

Up to now, only AN monomer had been cleanly grafted onto metals in AcN. But a careful study of the solvent influence has put in evidence a strong competition for the electrode between that solvent and the monomer. In fact, competition by the solvent decreases when its electron donicity increases, and it was accordingly shown that in solvents such as DMF, but still better pyridine or HMPA, most (meth)acrylates could be efficiently grafted (37). As expected, with a negatively charged electrode, decreasing competition closely followed the Gutmann scale (38) of increasing solvent donicity. This represents a major breakthrough, since that broad class of monomers practically provides for any type of physical mechanical behavior and any desired functionality.

Another breakthrough extension came from a study of the conducting substrate nature. Certainly a very interesting mechanistic feature is that only transition metals (all those with easily mobilized nonbonding electrons from d-orbitals) are efficient in promoting "bonding" polymerization. Remarkably, on "saturated" metals (e.g., Zn), polymerization only takes place at the higher potential (P.II), to yield a nonbonded soluble polymer. A decisive confirmation of this fundamental difference has been obtained by using another type of conducting substrate, i.e., carbon, the implication being that easily extracted d-electrons could be replaced by equally "easy" π-electrons, such as those engaged in large electron clouds of polyconjugated aromatic structures. It was rewarding to observe, for vitreous carbon plaques (39), exactly the same type of voltammogram as for transition metals: P.I is witnessed again (as well as P.II) and displays a typical "passivation" behavior due to surface coverage. Furthermore, the bond formed between substrate and polymer is now still stronger and cannot be broken even under strong hydrolytic conditions.

Due to the good electroreaction control described above, and especially to its extension to other conducting substrates (e.g., carbon) and to a very broad and versatile class of polymers (i.e., those from (meth)acrylic esters), a wealth of exciting applications can be envisioned (some of which are already implemented) in three main areas:

Surface coating for corrosion protection. These reactions are very fast (in the order of the second) and happen in an electrolytic solution wherein the structure to be protected can be immersed as one of the electrodes, as it contains a transition metal (the major type of metal). Those are highly favorable conditions for a convenient and efficient protective coating process, even for objects displaying complicated shapes (art pieces, e.g.).

Tailoring of multiphase composites. Simultaneous reduction of an inorganic derivative (in the range of the polymerization potential) may lead to electroinsertion of a very fine and homogeneous dispersion of the inorganic reduced species, in a rather thin polymer film deposited on the conducting substrate. A particularly interesting application is the reduction of uranium salts, leading to thin α-emitter devices that display an unusually well-resolved α-spectrum (40).

Tailoring of interfaces and interphases. As has already been alluded to above, chain termination by specific additives may be a strategy for surface functionalization. However, the most substantial progress in this prospect comes from the new possibility of polymerizing (meth)acrylic esters. These monomers indeed offer a number of interesting functional groups able to interact strongly with the surrounding medium. After eventual deprotection, one is able, in principle, to promote compatibility and even bonding of the coated substrate with many other types of phases, e.g., polymer matrix and aqueous biological medium, through reactive groups such as epoxide, hydroxyl, and carboxyl. This approach is the more interesting as it can be applied to plaques, wires, fibers, and even powders (when their volume fraction is high enough to ensure current percolation, under good stirring).

In conclusion, we witness here the opening of a new field, promising countless applications for original multiphase materials.

GENERAL CONCLUSIONS

Considering the diversity and the flexibility of the examples discussed in this presentation, it is obvious that what can be rightly called a "macromolecular engineering" of multiphase polymer-based materials is an extremely lively field for present and future research and developments. The capability of answering a list of requirements shaping the profile of a desired material (always improved, and sometimes completely new) is certainly one of the major conquests of polymer science in the last two decades though it is still far from having answered all of the important challenges ahead.

Although the examples presented here were purposedly chosen in different fields and aimed at different goals, they are unified by a basic "molecular" and "supramolecular" way of thinking. It is indeed the precise knowledge and control of polymer microstructure and architecture, resulting in well-defined reproducible morphologies (at the nano scale), that is going to give us the capability of tailoring at will the final set of properties and the behavior of these increasingly complex materials. This approach obviously requires a better understanding of structure/property relationships, and even for the specialist it is amazing to see how a rather small change at the microstructural and/or morphological level can promote extremely significant and valuable modifications at the macroscopic (bulk property) level.

It must be clearly realized that "molecular modeling" is still far from perfect and is often based on rules of thumb. Fortunately, it is possible to circumvent the remaining black boxes by a usually efficient feedback optimization strategy in three steps. It implies, after a careful control of the fine structure of the components (mainly polymers, but also other ones), an in-depth morphological study using all the sophisticated methods offered by modern physics. From there on, one has to evaluate all the bulk physicomechanical properties of the material, particularly those that are going to be significant for the envisioned applications. The strategic key point is to send back the material for synthetic modifications after the second and/or third step, modifications precisely directed by the results of the characterizations. After a few of these "educated" experimental feedback cycles, an optimized situation is obtained, which often is a very narrowly defined one. Of course, the main difficulty here is that, to be time-efficient, these cycles have to be performed in the same laboratory or at least the same research center. But as a whole, the process is remarkably productive.

The successful cases reported here were deliberately selected from the relatively simple ones so as to allow a clear illustration of the underlying molecular engineering principles. The general trend already emerging in the field is however to design more and more sophisticated "smart" materials able to adapt themselves to complex situations and precise external stimuli, in other words high-performance "functional systems." Suffice it to cite here a few rather spectacular ones: structuration of surfaces for electronic devices (41), achievement of an all-polymer transistor (42), production of drug-loaded microspheres for in-situ embolization of vascularized tumors (43), design of macroporous tapes and tubes used, after neuronal cell growth, in restoration of large traumas such as medullar ones (44), and the many studies on polymer-based materials for applications in optoelectronics (frequency multiplication, etc.).

The field is immense and calls for many enthusiastic chemists, physicists, and engineers who are intellectually well-prepared and experimentally well-equipped.

ACKNOWLEDGMENTS

The author is deeply indebted to all his coworkers, too many to be cited here, whose names appear in the references. He wants however to express his special gratitude to Dr. R. Fayt and Professor R. Jérôme, as well as to the industrial companies that kindly financed and discussed much of his research (Agfa-Gevaert, Atochem, Cockerill-Sambre, Dow Chemicals, D.S.M., B. F. Goodrich, Himont, New Carbochim, Labofina, Solvay) and to the Belgian research agencies F.N.R.S., F.R.I.A., and S.S.T.C. for their continuous support.

REFERENCES

1. M. C. Gabriele. Plastics Technology, June 1992, p. 58.
2. C. Pagnoulle, R. Jérôme, C. Koning, M. Van Duin. Progress in Polymer Science, in press.
3. P. Cigana, B. D. Favis, R. Jérôme. J. Polym. Sci. B. Pol. Phys. 34:1691, 1996.
4. J. F. Gohy, R. Jérôme. Macromol. Chem. Phys. 197:2209, 1996.
5. C. E. Koning et al. Makromol. Chem. Macromol. Symp. 75:159, 1993.
6. S. Petitjean, G. Ghitti, R. Jérôme, Ph Teyssié, J. Marien, J. Riga, J. Verbist. Macromolecules 27:4127, 1994.
7. Y. Yamashita, Y. Tsukahara, J. Macromol. Sci. Chem. A21:997, 1984.
8. R. Fayt, Ph Teyssié, F. Zerega, G. Cecchin. Europ Pat Applic 91108892.0, 1991.
9. Z. H. Liu, Ph Maréchal, R. Jérôme. Polymer 37:5317, 1996; Polymer 38:5149, 1997; Polymer 38, 1998, accepted.
10. N. Moussaif, Ph Maréchal, R. Jérôme. Macromolecules 30:658, 1997.
11. T. P. Russell, R. Jérôme, Ph Dubois, M. Foucart. Macromolecules 21:1709, 1988.
12. Schrijnemackers J. PhD. diss., University of Liège, Belgium, 1989; R. Vankan, R. Fayt, R. Jérôme, Ph Teyssié. Polym. Engin. Sci. 36:1675, 1996.

13. Ph Degée, R. Vankan, Ph Teyssié, R. Jérôme. Polymer 38:3861, 1997.
14. C. Maus, R. Fayt, R. Jérôme, Ph Teyssié. Polymer 36:2083, 1995.
15. Presented in part at the Fifth Symposium on Polymer Blends, Maastricht, 1996.
16. L. Leemans, R. Fayt, Ph Teyssié. Polymer 31:106, 1990.
17. J. P. Masy, R. Jérôme, Ph Teyssié, E. Goethals. Polym. Mat. Sci. Eng. 64:149, 1991; US Pat 5.081.180, 1992.
18. J. M. Borsus, P. Merckaert, R. Jérôme, Ph Teyssié. J. Appl Polym Sci 29:1857, 1984.
19. L. Leemans, R. Fayt, Ph Teyssié, N. C. de Jaeger. Macromolecules 24:5922, 1991.
20. L. Leemans, H. J. Uytterhoeven, Ph Teyssié, R. Fayt, N. C. de Jaeger. US Pat 4.908.155, 1990.
21. Ph Teyssié, L. Leemans, W. Verdijck, N. C. de Jaeger. US Pat 5.200.456, 1993; Leemans L. PhD. diss., University of Liège, Belgium, 1995.
22. Creutz S. PhD. diss., University of Liège, in preparation; S. Creutz, R. Jérôme, G. Kaptijn, A. Van der Werf, J. Akkerman. J Coatings Technology, submitted (see also other papers in preparation by S. Creutz and R. Jérôme).
23. L. Leemans, R. Jérôme, Ph Teyssié. Submitted to Macromolecules.
24. J. Lyngaae-Jorgensen, L. A. Utracki. Makromol. Chem. Macromol. Symp. 48/49:189, 1991.
25. (a) S. Blacher, F. Brouers, R. Fayt, Ph Teyssié. J. Polym Sci Polym Phys Ed 31:655, 1993; (b) C. Harrats, S. Blacher, R. Fayt, R. Jérôme, Ph Teyssié. J. Polym. Sci. Polym. Phys. Ed. 33:801, 1995.
26. (a) D. Tian, Ph Dubois, R. Jérôme. Polymer 37:3983, 1996; (b) D. Tian, S. Blacher, Ph Dubois, R. Jérôme. Polymer, in press.
27. F. Gubbels, R. Jérôme, et al. Macromolecules 27:1972, 1994; F. Gubbels, S. Blacher, E. Vanlathem, R. Jérôme, R. Deltour, F. Brouers, Ph Teyssié. Macromolecules 28:1559, 1995.
28. G. Broze, R. Jérôme, Ph Teyssie, et al. Macromolecules 14:224, 1981; Macromolecules 15:1300, 1982; 16:996 and 1771, 1983.
29. N. S. Enikolopian, USSR Pat 763379, 1976.
30. E. J. Howard, J. W. Collette et al. Ind. Eng. Chem. Prod. Res. Dev. 20:421 and 429, 1981; E. J. Howard, US Pat 4.187.210, 1980.
31. For a review, see Ph Dubois, M. Alexandre, F. Hindryckx, R. Jérôme. J. Macromol. Sci. Rev., in press.
32. F. Hindryckx, Ph Dubois, R. Jérôme, Ph Teyssié, M. G. Marti. J. Appl. Polym. Sci. 64:423 and 439, 1997.
33. (a) M. Alexandre, F. Hindryckx, Ph Dubois, R. Jérôme, M. G. Marti, PCT Pat pending; (b) Ph Dubois, M. Alexandre, R. Jérôme, M. G. Marti, Euro Fillers Conference, Manchester, UK, Sept 1997; (c) M Alexandre, E. Martin, Ph Dubois, R. Jérôme, M. G. Marti, IPCM, Eger, Hungary, Sept 1997.
34. G. Mengoli. Adv Polym Sci 33:2, 1979; W. Yuan, J. O. Iroh. Trends Polym. Sci. 1(12):388, 1993.
35. G. Lécayon, et al. Eur Pat 0038244, 1981; Chem Phys Lett 91:506, 1982; La Recherche 19:888, 1988; and following papers.
36. M. Mertens, C. Calberg, L. Martinot, R. Jérôme. Macromolecules 29:4910, 1996.
37. N. Baute, Ph Dubois, L. Martinot, M. Mertens, Ph Teyssié, R. Jérôme. Eur. J. Inorg. Chem., in press.

38. V Gutmann. The Donor-Acceptor Approach to Molecular Interactions. Plenum Press, New York, 1978.
39. M. Mertens, R. Jérôme, L. Martinot. Belg Pat Appl 09700608, 1997.
40. L. Martinot, M. Mertens, R. Jérome, et al. US Pat Appl 08221378, 1994; Radiochimica Acta 75:111, 1996.
41. C. J. Hawker, ACS 213th Meeting, San Francisco, April 1997.
42. F. Garnier, R. Hajlaoui, A. Yassar, P. Srivastava. Science 265:1684, 1994.
43. P. Flandroy, C. Grandfils, B. Danen, F. Snaps, R. F. Dondelinger, R. Jérôme, R. Bassleer, E. Heinen. J. Controlled Release 44:153, 1997.
44. Ch Schugens, Ch Grandfils, R. Jérôme, Ph Teyssié, P. Delrée, D. Martin, B. Malgrange, G. Moonen. J. Biomed. Mat. Res. 29:1349, 1995; C. Schugens, V. Maquet, C. Grandfils, R. Jérôme, Ph Teyssié. Polymer 37:1027, 1996.
45. Ph Heim, C. Wrotecki, M. Avenel, P. Gaillard. Polymer 34:1653, 1993.

22

Developments in Reactive Blending

Yasuhisa Tsukahara
Kyoto Institute of Technology, Kyoto, Japan

Hanafi Ismail
Universiti Sains Malaysia, Penang, Malaysia

Gabriel O. Shonaike
Himeji Institute of Technology, Himeji, Hyogo, Japan

I. INTRODUCTION

Polymer alloys and blends constitute one of the major fields in polymer science and technology. Numerous useful and outstanding properties can be obtained by blending different polymers as we do metal alloys (1–3). But simple mechanical blends of incompatible polymers do not usually show the desired properties because of macroscopic phase separation and weak connectivity at the phase boundary. Therefore many kinds of compatibilizers have been studied and used to modify these drawbacks. The addition of a compatibilizer as a third component in such as preformed block and graft polymers is useful to compatibilize binary incompatible polymer blends. Recently, reactive blending of polymer components has been actively studied in which block or graft polymers are formed in-situ during the mixing process, which act as effective compatibilizers. Adding a vector liquid, a reactive low-molecular-weight compound, as a third component, also enhances efficient compatibilization of different polymers (4). Curing of one polymer with an other reactive polymer as the blend component to form multicomponent polymeric networks might be also another reactive blending technique. In this case, end-functionalized polymers possessing functional groups at two chain ends, as well as polymers possessing pendant functional groups, can be used as reactive polymer components. If we define reactive blending as the preparation pro-

Table 1 Examples of Reactive Blending and Alloys

Type	Reaction	Polymers[a]
Reactive compatibilization (improvement of interface)	Coupling reaction	HIPS, ABS
	Polymerization reaction In-situ formation of block and graft copolymer	PP/Nylon
Self-curing or self-vulcanization	Cross-linking reaction Random site type and telechelic type	CP/PAA, ENR/XPCl
Reactive filler	Polymerization	PMMA/PBA
Dynamic vulcanization	Cross-linking reaction	EPDM/PP
Interpenetrating polymer network (IPN)	Polymerization reaction	PEA/PS, PB/PS

[a] HIPS: high impact polystyrene, ABS: acrylonitrile-butadiene-styrene rubber, PP: polyprc lyrene, CP: chloroprene rubber, PAA: polyacrylic acid, ENR: epoxidized natural rubbei XPCL: carboxylated poly (ε-caprolactone), PBA: polybutyleacrylate, PEA: polyethylacrylate PB: polybutadiene.

cedure of multipolymer component materials by blending or mixing accompanied by a certain kind of chemical reaction, the formation of interpenetrating networks (IPNs) (5), reactive polymeric fillers (6), and dynamic vulcanization techniques (7–8) are also included in this category. Examples of reactive blending are shown in Table 1.

The technology of polymer blends and alloys utilizing chemical reactions appeared in 1960–1970 in the production of high-impact polystyrenes (HIPS) and acrylonitrile-butadiene-styrene (ABS) polymers by grafting reactions (1–3). In the case of HIPS, styrene monomers are polymerized in the presence of polybutadiene, and graft polymers are formed in-situ resulting in the mechanical properties being much improved by the grafting reaction in comparison with the corresponding simple mechanical blends.

Reactive polymer blending has several advantages over simple mechanical blending for the production of polymer blends and alloys. First, the connectivity between the phases of different polymer components can much be improved by covalent bond formation. Second, control of phase morphology over a wide range can be achieved by utilizing chemical reactions such as grafting reactions, cross-linking reactions, and polymerization reactions. Complex and sophisticated morphology is often observed, whereas this cannot be obtained through simple mechanical blend systems. Third, fixing or stabilization of the phase-separated domain morphology is possible by covalent bond formation between the phases. These advantages allow us to find ways to create new polymer blends and alloys of higher performance. In addition, simultaneous progression of chemical reactions with processing, such as compounding and molding or extrusion, is an advantage in the economical production of polymer blends and alloys.

1) Grafting onto:

2) Grafting from:

3) Grafting through:

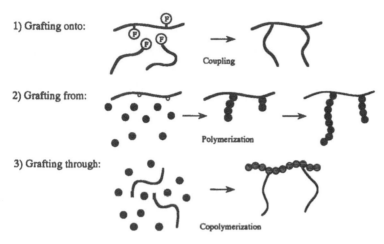

Figure 1 Examples of grafting reactions. (From Ref. 9.)

In-situ grafting reactions are some of the most important reactions in reactive blending of different polymers because the in-situ formed graft polymers act as effective compatibilizers in the blend. There are basically three types of grafting reactions, as shown in Fig. 1 (9). The first is the "grafting onto" method, in which a grafting reaction of an end-reactive polymer onto the second polymer takes place by a coupling reaction. Addition of maleic-anhydride-modified polypropylene in polypropylene/nylon blends to achieve in-situ formation of graft copolymers of PP and nylon is such an example (10). The second is the "grafting from" method. Here the graft polymerization reactions of the second monomer are initiated from the active sites of the coexistent first polymers. Preparations of HIPS and ABS resins are examples of this kind of reaction, and macroinitiators can be used for this purpose (6,11). The third is the "grafting through" method. An example is the chain copolymerization of a conventional monomer with a macromonomer that is a reactive polymer having a polymerizable functional group. In this case, polymer chains are incorporated though its polymerizable end group. Macromonomers have been used in the preparation of a wide range of well-defined graft copolymers (12–15). Main-chain type macroinitiators can be used for in-situ formation of block polymers, which in turn also act as compatibilizers between different polymers (16,17). Therefore the end-reactive polymers or oligomers such as telechelics, macromonomers, and macroinitiators are useful as polymer components for reactive blending.

Reactive blending is performed in reactive polymer processing. One can use reactive compounding/molding processing and reactive extrusion processing, which are widely used in reactive processing technology. It is not an overemphasis to say that the blending or mixing process is of utmost importance in reactive blending, because the extent of reaction as well as the phase morphology strongly depends on the mixing process, including the design of mixing machines. Well designed twin-screw extruders are powerful machines for this purpose (18–20). Since the reaction takes place in the course of mixing in the processing machine, it is often difficult to investigate the reaction process between the different polymer components directly. However, information about the kinds of reaction taking place is important for the quality control of the products (20).

II. REACTIVE COMPATIBILIZATION

Research activities on compatibilization of incompatible polymer blends have led to the development of reactive blending of multipolymer components. This method is becoming more attractive for achieving blend compatibility. In the last couple of years, several investigators both in industry and in academic communities have employed the method successfully. The technique relies on the preparation of one polymer component with a functionalized interface and a second component that is added to form the blend. In reactive blending, no compatibilization is needed, as reaction occurs between the functionalized component and the second (unfunctionalized) component during melt blending.

Among the advantages of this method are adequate temperature and pressure control, the lack of any need for compatibilizer, cost effectiveness (21), and improved compatibility due to a chemical reaction between the components. Research activities on reactive blending are extensive, and a review of such could fill a whole volume. Some of the recent investigations (22–43) indicate that compatibilization via the reactive blending technique relies on the in-situ formation of copolymers or interacting polymers. One of the most successful approaches is to modify one component with maleic anhydride (MA) and blend the modified polymer with an unmodified second component. The MA is a free radical grafted onto a polymer backbone in the presence of an organic peroxide. Thus the MA content in the modified component acts as a chemical "hook" and reacts with an immiscible component, thereby linking itself to the polymer chain (44).

Table 2 highlights some of the MA-functionalized reactive polymers (45). This table a list of some physical and mechanical properties achieved as a result of the blending of various polymers functionalized with maleic anhydride. Reactive agents include peroxides, silanes, organotitates, and phenoxies. Peroxide agents are known to be good initiators in reactive processes, i.e., they initiate the grafting of MA-modified compatibilizers. Apart from MA-functionalized polymers, other categories include carboxylic acid derivatives, primary and secondary amines, carboxylic acid, hydroxyl and epoxide, and other groups capable of ionic interactions (46). However, according to Paul and collaborators (47), the design

Table **2**   Maleic Anhydride–Functionalized Reactive Polymer

Polymer	Reactive group	Blends	Reactive type	Properties	Ref.
	Grafted MA	PA6/EPR	Imidation	Impact	52,54,55
				Crystallization	55
				Morphology	53
		PA6,6/EPR	Imidation	Impact	56
		PBT/EPR	Ester interchange	Impact	57
		PA6,6/EPR	Ester interchange	Morphology	53
EPDM	Grafted MA	PA6/EPDM	Imidation	Impact	24,25,52,58–60
	Grafted MA	PA6/PE	Imidation	Impact	52,59–62
				Morphology	32
		PS/PE	Ring opening	Morphology	63
		PE/EPDM	Ionic	Tensile	64
		PE/NR	Ring opening	Tensile	
	Grafted MA	PA6/PP	Imidation	Impact	23,65,66
				Crystallization	67
				Tensile	68
		PA6/PPE	Imidation	Impact	69
				Morphology	69
PPE	Grafted MA	PA6/PPE	Imidation	Impact	33
ABS	Grafted MA	PA6/ABS	Imidation	Tensile	70
SEBD	Grafted MA	PE/PA6	Impact	Impact	26
		PP/PA6	Imidation	Impact	59,65
				Crystallization	68
SMA	Copolymerized MA	PA6/SAN	Imidation	Impact	71
		PA6/ABS	Imidation	Impact	71
		PA6/SMA	Imidation	HDT	36
		PA6/PS	Imidation	Morphology	69
				Impact	69
AC	Copolymerized	PA6/AC	Imidation	Impact	66,72
EMA	Copolymerized MA	PA6/PE	Imidation	Permeability	73
				Morphology	74

*Source:*Ref. 45.

and implementation of reactive blending involves the following: selection of the chemistry to be used (types of functional groups and reactions to incorporate functional groups), processing rheology, blend analysis (chemical and morphological), interfacial properties, mechanical properties, and fracture characteristics.

A. Mechanism of Reactive Blending

As in the case of the processing of simple mechanical blend systems, the morphological development during reactive processing is governed by thermodynamic criteria for phase separation, the phase separation kinetics or mechanism such as spinodal decomposition and nucleation and growth, and rheological factors. In addition, in reactive processing, the reaction between unlike polymer components creates covalent bonds at the interface that definitely affect the above terms. Furthermore, the reaction itself is not simple under processing at high temperatures, and there may be undesired side reactions.

The thermodynamic criteria for phase separation in binary polymer blends is generally described in terms of the Gibbs free energy of mixing ΔG_m.

$$\Delta G_m = \Delta H_m - T\Delta S_m \tag{1}$$

According to the Flory–Huggins lattice theory, the criteria for the phase separation in the binary blend is given by

$$\chi_{AB} > \chi_{AB,c} = \frac{1}{2(r_A^{-1/2} + r_B^{-1/2})^2} \tag{2}$$

in which r_A and r_B are the degrees of polymerization of the polymers A and B, and χ_{AB} is the interaction parameter between the polymers, which is given by

$$\chi_{AB} = \frac{v_s}{kT}(\delta_A - \delta_B)^2 \tag{3}$$

in which δ_A and δ_B are the solubility parameters of the polymers A and B. Therefore the occurrence of phase separation basically depends on the values of δ_A and δ_B, i.e., the chemical structure and the degree of polymerization of the component polymers, and temperature. The morphological development during the cooling process of the polymer melt in the processing relates to the phase separation mechanism, which depends on whether the polymer component is a crystalline polymer or an amorphous one. During the spinodal decomposition process, the specific percolation network structure is created at the early stage, and then the structure breaks and incorporates into the final domain morphology. The in-situ formation of block and graft copolymers in reactive processing reduces the interfacial free energy and will stabilize the phase morphology. It may be possible to fix the percolation network structure by a chemical reaction. The interfacial thickness l is expressed in terms of χ_{AB} and the segment length b (1,2) as

$$l = \frac{2b}{(6\chi_{AB})^{1/2}} \tag{4}$$

the value of which might influence the chemical reaction between the polymer components at the interface.

The rheological environment in the processing is described in terms of the viscosity ratio λ, the capillarity number κ, and the reduced time t' according to Utracki (2). These are given by

$$\lambda = \frac{\eta_A}{\eta_B} \tag{5}$$

$$\kappa = \frac{\sigma d}{v_{AB}} \tag{6}$$

$$t' = \frac{t\gamma}{\kappa} \tag{7}$$

where η_A and η_B are the viscosity of the polymers A and B, σ is the shear stress, d is the diameter, and γ is the rate of deformation. Therefore morphology formed during melt processing depends on the relative viscosity of the component polymers, the capillary number, the time the blend is in the stress field, and the rate of deformation. For instance, the increase in the shear stress decreases the dispersed domain size. In addition, during the breakup and coalescence of the phase domains, chemical reactions take place at the interfaces that change the stability of the generated small domains. Thus the morphological development during reactive processing is very complicated. In other words, it is therefore possible to control the morphology of a wide variety.

In reactive blending a chemical reaction occurrs between the functionalized and unfunctionalized components during melt blending. The reactive mechanism results in in-situ formation of small interacting polymers resulting in fine dispersion and adhesion between the two phases (48). Bonner and Hope (49) have identified a number of reactive blending mechanisms as follows:

1. An in-situ formation of graft or block copolymers by chemical bonding reactions between reactive groups on component polymers or by addition of a free radical initiator during blending.
2. Formation of a block copolymer by an interchange reaction in the backbone bonds of the components. This is most likely to arise in the condensation reaction.
3. Mechanical scission and recombination of component polymers to form graft or block copolymers, which may be induced by high shear levels during processing.
4. Promotion of the reaction by catalysis.

In an extensive review on reactive blending, an example of in-situ reaction as shown by Chang (50) is depicted in Fig. 2. In this case, a C—X reactive copolymer (where X is a reactive group that could be an end group or could be randomly distributed on the chain backbone) could be a compatibilizer for a polymer blend A/B, provided that C is structurally identical or miscible with A. Thus X could react with the B component (mostly at chain ends) to form an in-situ C—X—B graft or block.

One basic requirement is that for good compatibilization to occur via the reactive blending process, the polymer must possess functional chain end groups,

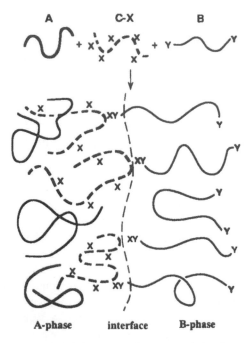

Figure 2 Schematic diagram of the formation of C-X-B grafted copolymers at interface for the ternary A/A-X/B blend. (From Ref. 50.)

such as —COOH, —OH, —NH$_2$, etc. The formation mechanism (morphology) of reactive blending of 2% poly(styrene-acrylonitrile-glycidyl methacrylate) (SAG), polyamide 66 (PA66), and poly(acrylonitrile-butadiene-styrene) (ABS) is shown in Fig. 3 (51). The epoxy functional groups in SAG that are used to compatibilize ABS/nylon 6,6 react with the PA 66 amine end group in the melt to form the SAG-g-PA66 copolymer. The preblend SAG-g-PA66 is mixed with ABS to form the compatibilized ABS/SAG/PA66. The mechanism involved preblending of 20/0 GMA (glycidyl methacrylate) (i.e., 2% GMA in SAG) with PA66 to form lightly grafted SAG2-g-PA66 copolymers that contain a greater fraction of freely exposed, ungrafted SAG segments. The ungrafted SAG segments tend to migrate into the ABS phase, whilst the grafted PA66 chain(s) tends to prevent the SAG-g-PA66 from completely migrating into the ABS phase. Thus the graft copolymers tend to locate at the interface where they function as effective compatibilizers. However, since compatibility via the reactive blending technique is due to surface chemical reactions, a probable mechanism for the interfa-

Preblending (SAG2 + PA66)   ABS

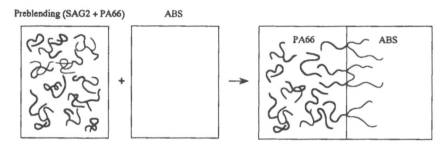

Figure 3 Schematic representation of the formation mechanism of SAG-g-PA66 copolymers and their final distribution by sequential blending. (From Ref. 50.)

cial reaction for functionalized polyethylene with different rubbers is shown in Fig. 4 (64). Both epo-xidized natural rubber and S-EPDM rubbers shown below contain polar groups that can introduce chemical cross-linking at the interface, thereby improving the interfacial adhesion. The authors (64) commented that the mechanism for EPDM is physical, i.e., it has a partial structural similarity with PE and distributes itself between the two phases and adheres better with improved adhesion.

III. SELF-VULCANIZATION

In this section, we describe reactive blending with curing reactions by showing the self-vulcanization behavior and the properties of polymer blends of epoxidized natural rubber with (1) carboxylated nitrile rubber, chlorosulfonated polyethylene, and polychloroprene as a random site cross-linking type, and with (2) telechelic dicarboxylated poly(ε-caprolactone) as a telechelic type.

Generally, vulcanization of rubbers requires many additives in addition to sulfur as a cross-linking reagent, such as an assisting reagent, an accelerator, a retarder, a stabilizer, and so on. In contrast to this, mixing or compounding of raw rubber materials with some sort of polymers possessing reactive functional groups with rubber molecules as polymeric crosslinkers, followed by an appropriate curing by molding, gives the so-called self-vulcanization system. This is a kind of reactive blending of two polymers. In this case, the rubber materials should also possess some sorts of functional groups for the cross-linking reaction.

For this purpose, epoxidized natural rubber (ENR) is a useful and interesting rubber material, since the presence of the oxirane groups make it possible to per-

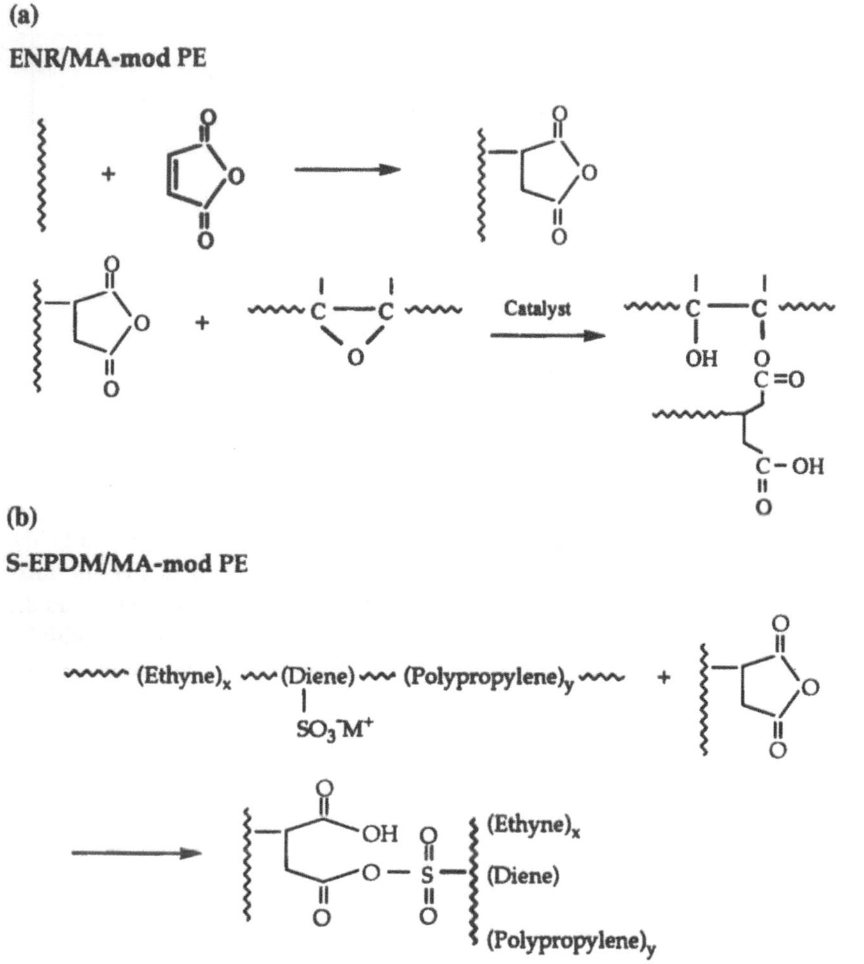

(a)

ENR/MA-mod PE

(b)

S-EPDM/MA-mod PE

Figure 4 Reaction of functionalized polyethylene with (a) ENR and (b) S-EPDM. (From Ref. 64.)

form various types of cross-linking reactions in addition to the normal sulfur vul-canization through the double bonds (75,76). Epoxidation is perhaps the only known chemical modification of natural rubber (NR). The presence of the oxirane group in NR also improves the polarity of NR, and the resulting ENR shows var-ious useful properties depending on the degree of epoxidation, such as oil resis-tance, low gas permeability, and better adhesion (77–80).

$$\text{ENR} \qquad \begin{array}{c} \\ \end{array} \left[\begin{array}{c} \overset{\displaystyle CH_3}{\underset{\displaystyle |}{}} \\ CH_2\!-C\!=\!CH-CH_2 \end{array} \right]_m \left[\begin{array}{c} \overset{\displaystyle CH_3}{\underset{\displaystyle |}{}} \\ CH_2\!-\!C\!\!\underset{\displaystyle O}{\overset{\displaystyle }{\diagdown\!\diagup}}\!\!CH-CH_2 \end{array} \right]_n$$

A. Random Site Type

De and coworkers (81–83) have developed reactive blends or self-vulcanizable blends by blending ENR with various rubbers such as carboxylated nitrile rubber (XNBR), chlorosulfonated polyethylene rubber (CSM), and polychloroprene rubber (CR). These rubbers cross-link during molding in the absence of any vulcanizing agent and can be reinforced with fillers like carbon black and silica. Cross-linking is believed to take place between the epoxy groups of ENR and the functional groups of the other rubbers. Table 3 shows the formulation of the two ENR/XNBR blends and the physical properties of the molded blends. Blend A contains XNBR : ENR in the ratio 100 : 50 by weight. Blend A_c is similar to blend A but contains 45 parts by weight of ISAF type carbon black (average diameter is 20–25 nm).

Table 3 Composition and Properties of ENR-XNBR Blend Molded at 140°C for 45 min

	Blend designation	
Composition (part by weight)	A	A_c
ENR 50	50	50
XNBR	100	100
ISAF carbon black	—	45
Properties		
100% modulus (MPa)	1.02	1.59
Tensile strength (MPa)	3.70	20.56
Elongation at break (%)	396	538
Tear strength (kNm^{-1})	14.40	38.00
Compression set at constant stress (%)	11	14
Compression set at constant strain (%)	20	32
Heat buildup (ΔT) at 50°C (°C)	24	40
Resilience (%)	62	47
Hardness (Shore A)	43	55
Swelling in chloroform (% increase in weight)	872	517

Source: Ref. 9.

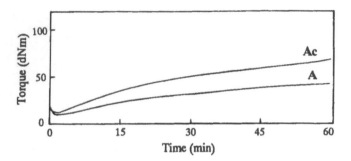

Figure 5 Torque–time curves of ENR and carboxylated nitrile rubber (XNBR) blends A and A$_c$ at 140°C. The blends A$_c$ contains 45 parts of carbon black filler by weight as shown Table 3. (From Ref. 83.)

Figure 5 shows the rheographs of the two blends at 140°C. An increase in rheometer torque with vulcanization time indicates progressive cross-linking of the system. A marching increase in modulus with cure time, as shown in the rheograph, implies that cure reversion is absent and that the vulcanized network is thermally stable. Table 3 shows that gum blend vulcanizate has poor physical properties. However, addition of reinforcing carbon black increases the tensile strength,

Table 4 Effect of Molding/Vulcanization Time on Physical Properties of Black Filled XNBR–ENR Blend (A$_c$)

	Molding/vulcanizing time (min)		
Physical property	30	45	90
Modulus at 100% strain (MPa)	1.15	1.57	3.57
Tensile strength (MPa)	20.17	20.56	19.28
Elongation at break (%)	737	538	374
Tear resistance (kNm^{-1})	41.75	38.00	33.02
Heat buildup at 50°C (ΔT) (°C)	44	40	30
Dynamic set (%)	9.3	3.6	0.76
Resilience (%)	44	47	48
Swelling in chloroform (% increase in weight)	652	517	426

Source: Ref. 9.

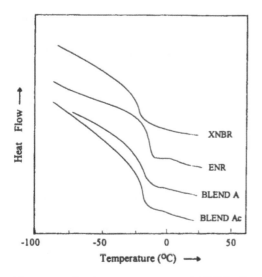

Figure 6 DSC thermograms of XNBR, ENR, and XNBR/ENR blends (A and A_c). (From Ref. 83.)

tear resistance, tensile modulus, heat buildup, compression set, and hardness and causes reduction in resilence. Reduction in percentage of swelling in solvent (chloroform) for the filled system shows increased restriction to solvent swelling due to polymer–filler interaction.

The degree of cross-linking can be changed by varying the curing or molding time. Table 4 shows that with increase of cure time, swelling decreases due to the increase in cross-linking density, which causes formation of a tighter network resulting in low dynamic set, low heat buildup, lower elongation at break, higher modulus, and higher resilience.

Figure 6 shows the DSC thermograms of ENR, XNBR, and blends of ENR and XNBR, whereas Table 5 summarizes their glass transition temperatures. The

Table 5 Glass Transition Temperature (T_gs) of the Systems

Rubber/blend	T_g (°C)
XNBR	−25
ENR	−15
Blend A	−19
Blend A_c	−19

Source: Ref. 9.

occurrence of a single T_g in the blend and the transparent nature of the gum blend indicates complete miscibility of ENR and XNBR.

The dynamic mechanical analysis results also show that the blends of XNBR and ENR form a miscible system. Figure 7 shows that the damping peak of the blend occurred in between that of the individual components, and that the peak width or broadening is similar in the blend as compared to XNBR and ENR. The proposed cross-linking reactions of ENR with XNBR is shown in Fig. 8.

Table 6 shows the formulation and the physical chlorosulfonated-polyethylene (hypalon)/ENR blend. Table 6 indicates that the blend shows a moderately high level of strength, even without any reinforcing filler. The increase in rheometer torque with vulcanization time in Fig. 9 indicates a progressive cross-linking of the system. Both epoxidized natural rubber and hypalon are soluble in chloroform, but the molded blend was insoluble in the same solvent. This shows that during molding each blend constituent is vulcanized by the other component.

The torque values increased as the vulcanization temperature was increased from 150 to 160°C. A marching increase in modulus with cure time indicates that the cure reversion is absent and that the vulcanized network is thermally stable.

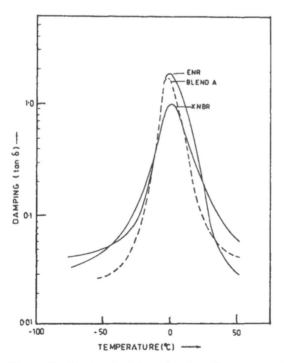

Figure 7 Mechanical damping (tan δ) curves for XNBR, ENR, and XNBR/ENR blends. (From Ref. 83.)

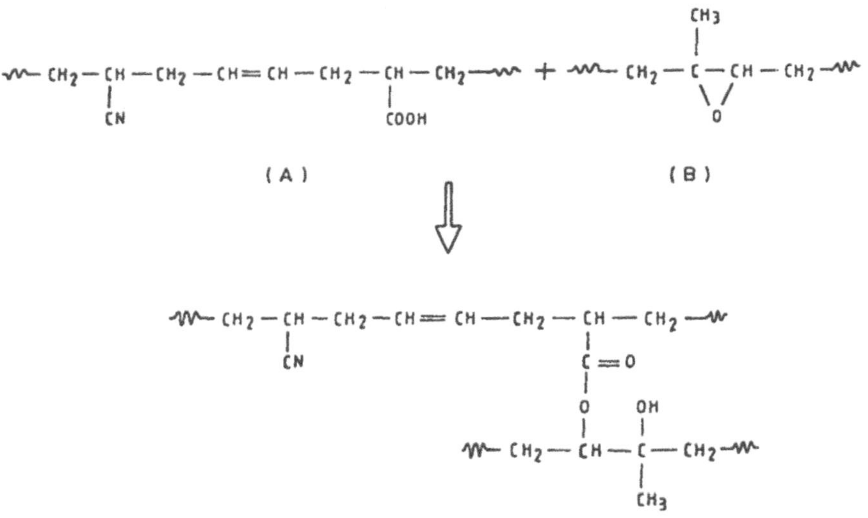

Figure 8 Possible mechanism of cross-linking between carboxylated nitrile rubber (A) and epoxidized natural rubber (B) in XNBR/ENR blends. (From Ref. 84.)

However, at 180°C reversion occurred, and torque values started decreasing beyond 27 min. Figure 10 shows the differential scanning calorimeter thermograms of ENR, hypalon, and their blend. It can be seen that the blend registers no clearcut transition but rather is diffuse, covering a wide temperature range, indicating that the blend is partially miscible.

Table 6 Properties of the Hypalon-ENR Blend

Blend composition (parts by weight), hypalon 100, ENR 100; Mooney viscosity, $ML_{(1+4)}$ at 120°C, 61	
Modulus at 100% extension (MPa)	6.18
Modulus at 200% extension (MPa)	9.33
Tensile strength (MPa)	10.03
Elongation at break (%)	205
Tear strength (kN/m)	12.85
Abrasion loss (c.c./hr)	0.56
Compression set at constant strain (%)	30
Resilience (%)	21
Hardness (shore A)	75
Swelling in chloroform (% increase in weight)	665

Source: Ref. 11.

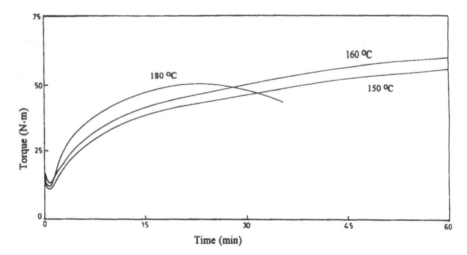

Figure 9 Rheographs of hypalon-epoxidized natural rubber blend at different temperatures. (From Ref. 85.)

The mechanical damping characteristics at different temperatures are shown in Fig. 11. The occurrence of two peaks in the blend indicates that the blend is not completely miscible. A probable network structure of the blend vulcanizate is proposed in Fig. 12 showing how epoxy groups of ENR cross-link with SO_2Cl groups of hypalon. The formulation and processing characteristics of the ENR and poly-

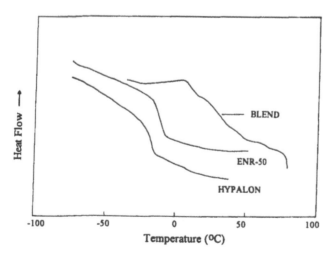

Figure 10 DSC thermograms of hypalon, ENR, and their blend. (From Ref. 85.)

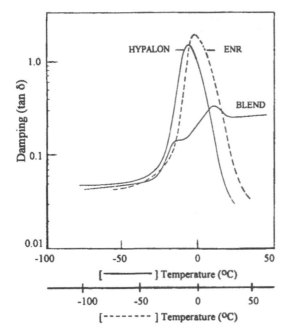

Figure 11 Mechanical damping (tan δ) curves for hypalon, ENR, and their blend. (From Ref. 85.)

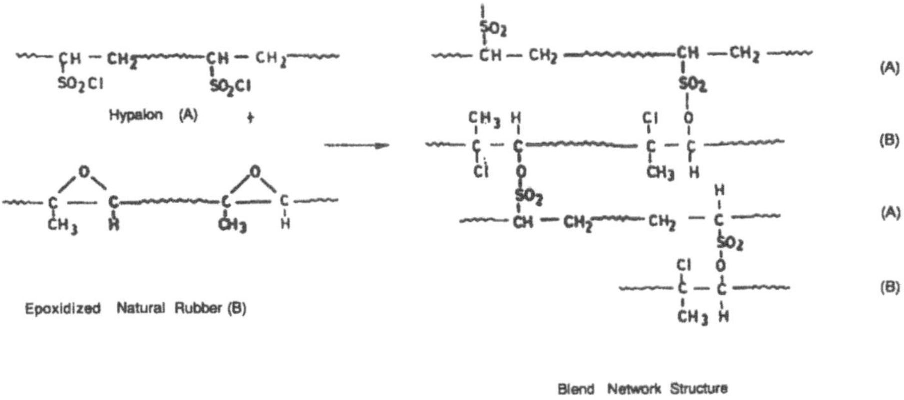

Figure 12 Probable network structure of hypalon cross-linked by ENR. (From Ref. 85.)

Table 7   Formulation and Processing Characteristics of the Blend

	Blend designation	
	N-E 75:25	N-E 50:50
Neoprene AC	75	50
ENR 50	25	50
Minimum Mooney viscosity at 120°C	38	31
Mooney scorch time at 120°C in min	38	132

Source: Ref. 8.

chloroprene (CR) blend are shown in Table 7. Both CR and ENR were masticated on a two-roll mixing mill to the same Mooney viscosity and blended on the mill for about 6 min.

Figure 13 shows the calculated rheographs of the self-vulcanizable Neoprene AC-ENR blends after deducting the effect of thermovulcanization of Neoprene AC. It can be seen that each blend constituent is cross-linked by the other during molding. This is also evident from V (volume fraction of rubber in the swollen vulcanizate) shown in Table 8. The cross-link density, which can be regarded as proportional to V, is high when the Neoprene AC content is high. The physical properties of blends are shown in Table 8. A higher proportion of Neoprene AC in the blend results in improved physical properties except for resilience values. Figure 14 shows the DSC thermograms of the ENR-Neoprene AC blends, which indicate the partial miscibility of the blends. Blends exhibit a single T_g, which is shifted to a higher temperature as ENR content decreases. The T_g zone becomes

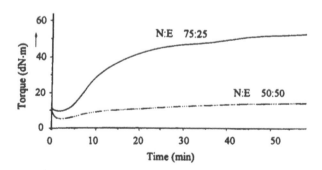

Figure 13   Calculated rheographs of the self-vulcanizable Neoprene AC/ENR blends after deducting the effect due to thermovulcanization of Neoprene AC. (From Ref. 82.)

Table 8 Physical Properties of the Blends Molded at 180°C for 60 min

	Blend designation	
	N-E 75:25	N-E 50:50
Modulus 300% (MPa)	1.8	1.1
Tensile strength (MPa)	5.7	3.8
Elongation at break (%)	460	570
Tear strength (kN/m)	16.8	13.0
Hardness (Shore A)	41.0	30.0
Resilience at 40°C (%)	55.0	55.0
Compression set at constant strain (%)	4.0	11.0
Heat buildup by Goodrich flexometer		
(a) T (°C)	14.0	—[a]
(b) Dynamic set after 25 min (%)	1.0	—[a]
Abrasion loss (cm³/1000 rev)	1.1	4.8
Volume fraction V_r	0.14	0.05

[a] Sample blown out in the 10th min.
Source: Ref. 8.

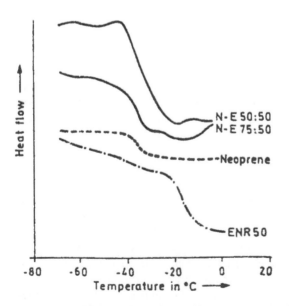

Figure 14 DSC thermograms of Neoprene AC, ENR, and blends of Neoprene AC and ENR. (From Ref. 82.)

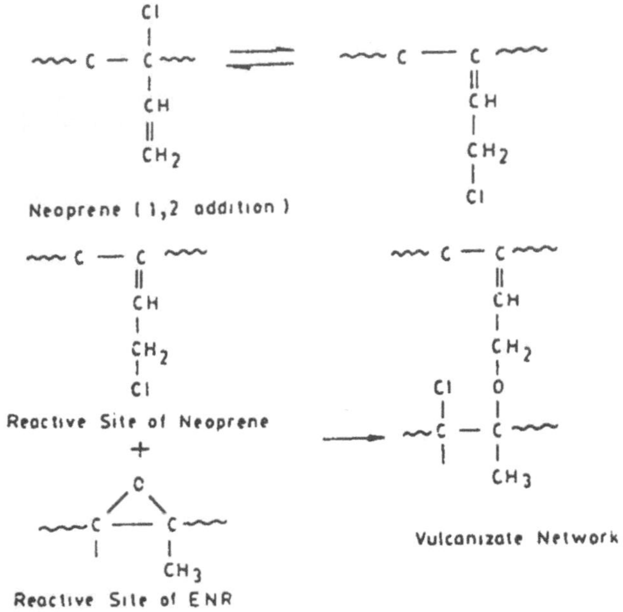

Figure 15 Possible cross-linking reaction between ENR and Neoprene AC in their blend. (From Ref. 82.)

broader in the case of blends. This broadening shows partial miscibility of the components in the blend. A plausible mechanism of self-vulcanization or reactive blend between neoprene and ENR is shown in Fig. 15. Table 9 shows the blend formulations of zinc-sulfonated ethylene propylene diene rubber (SEPDM) and ENR. The blends were prepared in a tight-nipped two-roll mill. First ENR was masticated for 2 min and then the SEPDM was added. Total mixing time was 7 min. The mixing temperature was maintained around 30°C by water circulation. Figure 16 shows the Monsanto rheographs of the blends at different blend ratios and temperatures. The progressive formation of a cross-linked network is manifested by the gradual rise in the torque with curing time. There is no increase in torque with time when either the sulfonate or the epoxy groups are absent in the polymer backbone, as seen in the blends of SEPDM/NR and EPDM/ENR (Table 9). Since the neat polymers (SEPDM or ENR) do not show any torque rise, the possibility of thermovulcanization of the component polymers in the blend is also eliminated. This indicates that the chemical interaction takes place between the functional groups in the two rubbers, namely SEPDM and ENR. The probable reaction pathway for the chemical interaction between SEPDM and ENR is shown in Fig. 17. As the curing temperature increases from 150 to 190°C, the rate and ex-

Table 9 Summary of Monsanto Rheometric Studies

end	Composition	Test temperature (°C)	Minimum torque (dN·m)	Rheometer scorch time[a] (min)	Torque after 120 min (dN·m)
SEPDM25-ENR50	0/100	170	1	—	2
	25/75	170	1	6	24
	50/50	150	5	30	10
		170	4	8	40
		190	4	6	48
	75/25	170	9	4	57
	100/0	170	42	—	39
SEPDM25-ENR25	0/100	170	1	—	1
	25/75	170	1	110	3
	50/50	170	6	11	36
	75/25	170	8	8	47
	100/0	170	42	—	39
SEPDM10-ENR50	0/100	170	1	—	2
	25/75	170	3	—	3
	50/50	170	4	30	18
	75/25	170	8	8	35
	100/0	170	28	—	25
SEPDM10-ENR25	50/50	170	1	—	1
SEPDM25-NR	50/50	170	3	—	3
EPDM-ENR50	50/50	170	1	—	1

[a] Time for a 2-unit torque rise beyond the minimum torque.
Source: Ref. 12.

tent of the cross-linking reaction also increases, as shown in Fig. 16 for the SEPDM25 : ENR50 (50 : 50) blends.

B. Telechelic Type

Utilization of a telechelic polymer as one component in a polymer blend is also effective in achieving the reactive blending of two different polymers, in which desired properties associated with the telechelic polymers and the well-defined cross-linking points between the two polymer components can be introduced simultaneously. However, this subject has been little researched compared with the random site types, so we describe here the details of the self-vulcanization behavior of an epoxidized natural rubber/telechelic poly(ε-caprolactone) (PCL) blend system as an example (88–90). The telechelic poly(ε-caprolactone)s are dicarboxylated telechelic poly(ε-caprolactone)s (XPCL)s of different molecular weights, which were prepared according to Scheme 1. The molecular weight and the polydispersity index (M_w/M_n) of XPCLs determined by GPC are shown in Table 10. Poly(ε-caprolactone), which is a tough crystalline polymer of moderate

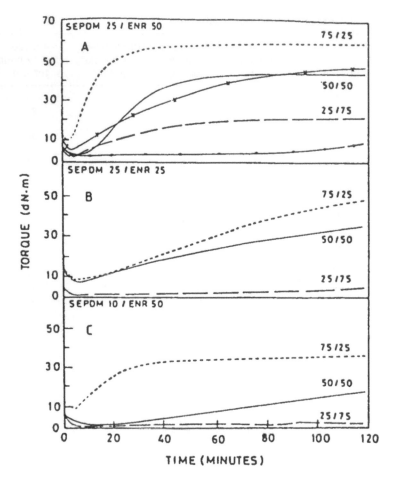

Figure 16 Rheographs for (A) 25/75 SEPDM25/ENR50 (– – –), 75/25 SEPDM25/ENR50 (- - - - -) at 170°C, and 50/50 SEPDM25/ENR50 at 150°C (•—•—•), 170°C (———), and 190°C (*—*—*); (B) SEPDM25/ENR25 blends at 170°C: 25/75 (– – –), 50/50 (———), 75/25 (- - - - -); (C) SEPDM10/ENR50 blends at 170°C: 25/75 (– – –), 50/50 (———), and 75/25 (- - - - -). (From Ref. 86.)

melting point, possesses the unique ability of being miscible with a variety of other polymers over a wide composition range and is thus interesting as a polymer blend component. Even the telechelic polymers with only two functional groups at the chain ends cure ENR well, and the resulting polymer blends show various stress–strain curves without any other additives.

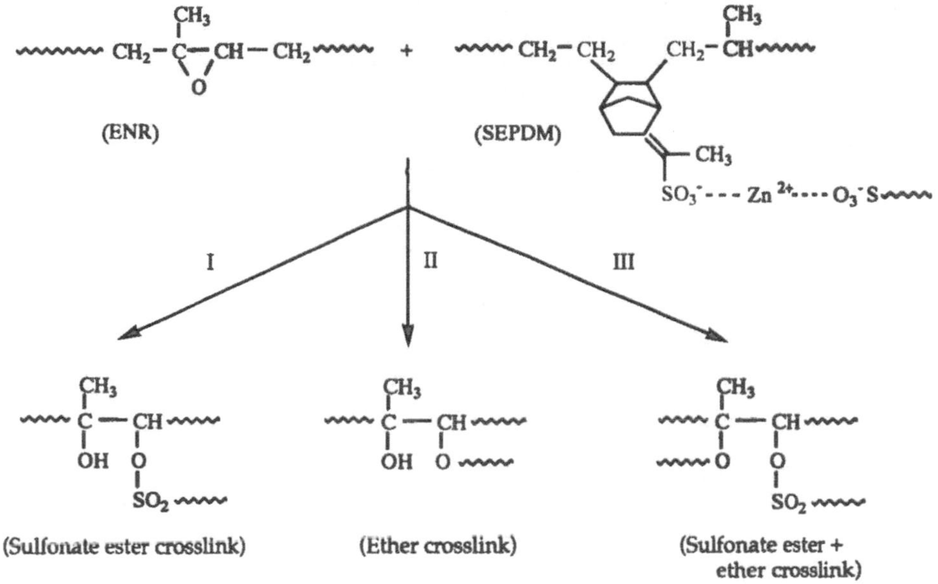

Figure 17 Proposed cross-linking reaction in SEPDM/ENR blends. (From Ref. 86.)

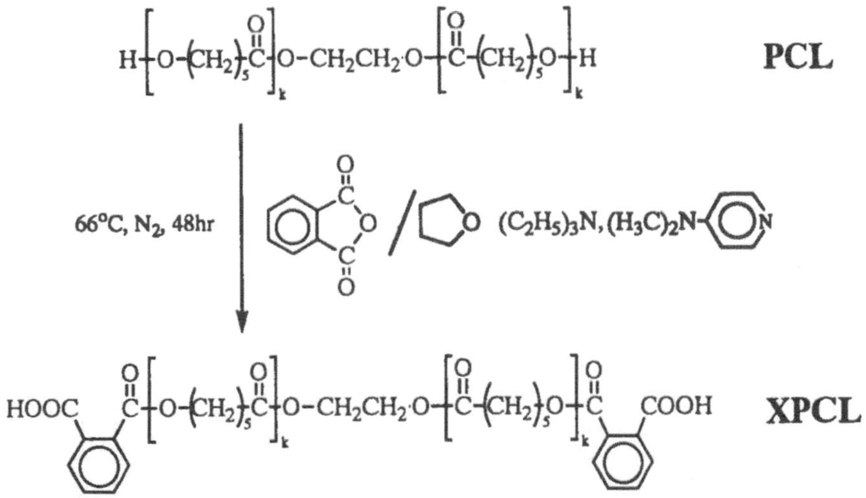

Scheme 1 Preparation of XPCLs.

Table 10 Characteristics of the End Dicarboxylated
Telechelic Poly(ε-caprolactone)s (XPCLs)

Code	GPC		
	M_n	M_w/M_n	f
XPCL-1000	1300	2.04	1.40
XPCL-2000	2500	2.42	2.00
XPCL-4000	6400	1.75	1.52

f: end group functionality estimated by ^{1}H-NMR.

1. Curing Behavior

Blend ratios of the compounds prepared by an open two-roll mill are shown in
Table 11. The blend compounds of ENR with the original PCLs were also prepared for comparison. The time–torque curves of XPCL/ENR binary blend compounds without any other additives at 160–200°C showed a large torque rising af-

Table 11 Blend Ratios in XPCL/ENR and PCL/ENR Binary Blend
Compounds (phr)[a,b]

(XPCL/ENR)						
ENR-50	100	100	100	100	100	100
XPCL-1000	30	—	—	—	—	—
XPCL-2000	—	30	—	—	—	—
XPCL-4000	—	—	20	30	40	50
ENR-25	100	100	100			
XPCL-1000	30	—	—			
XPCL-2000	—	30	—			
XPCL-4000	—	—	30			
(PCL/ENR)						
ENR-50	100	100	100			
PCL-1000	30	—	—			
PCL-2000	—	30	—			
PCL-4000	—	—	30			

[a] Compounds were prepared using an open two-roll mill at r.t.
[b] The numerals in the codes ENR50 and ENR25 represent the mole% of epoxidized isoprene units in the polyisoprene chain of natural rubber.

ter a very short induction period (1–2 min). However, the ENR/PCL blend compounds did not show any increase in torque in the time–torque curve, and the curve was almost the same as those of the green ENRs at the same curing temperature. The hot-pressed sheets of XPCL/ENR blends at 180°C were homogeneous, but the PCL/ENR sheets prepared under the same conditions showed rather heterogeneous appearances and did not cure well. This indicates that ENR can be well cured by the telechelic dicarboxylated PCL, while dihydroxy PCL does not cure ENR.

Figure 18 shows the effect of the molecular weight of the XPCL component on the cure curves measured for XPCL/ENR50 (XPCL : ENR = 30 : 100 by weight) blend compounds. It is seen that the maximum torque (torque at 2 h) rapidly increases with decreasing molecular weight of XPCL. This corresponds to the increase in the cross-linking reaction due to the increase in the concentration of the end functional group with decreasing molecular weight of XPCLs. Figure 19 shows the effect of the cure temperature on the curing behavior of the blend compounds, (a) XPCL/ENR50 and (b) XPCL/ENR25, using the telechelic dicarboxylated poly(ε-caprolactone) of the largest molecular weight (XPCL4000) where the curing rate increases with increasing cure temperature. The compounds with highly epoxidized ENR (ENR50) exhibit higher torque at comparable time than those with a low epoxidized one (ENR25), indicating that the cross-linking formation is influenced by the degree of epoxidation in ENR.

The end carboxylic groups also provide an acidic environment that might cause cationic ring opening of the epoxy groups, like furanization. This may de-

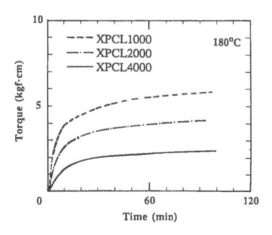

Figure 18   Time–torque curves for XPCLs/ENR50 blend compounds cured at 180°C (XPCLs : ENR50 = 30 : 100 by weight). (From Ref. 89.)

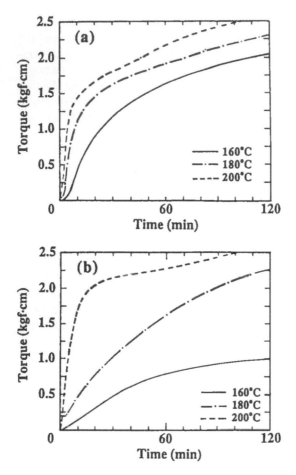

Figure 19 Effect of cure temperature on time–torque curves for (a) XPCLs : ENR50 (30 : 100) blend compounds and (b) XPCLs : ENR25 (30 : 100) blend compounds. (From Ref. 89.)

pend on the molecular weight and the blend ratio of XPCL. The possible cross-linking reactions for ENR/XPCL blends by curing are shown in Scheme 2 (90).

2. Thermal Properties

PCL is a semicrystalline polymer and shows a melting point T_m at around 50–60°C, the temperature of which slightly depends on the molecular weight. Fig-

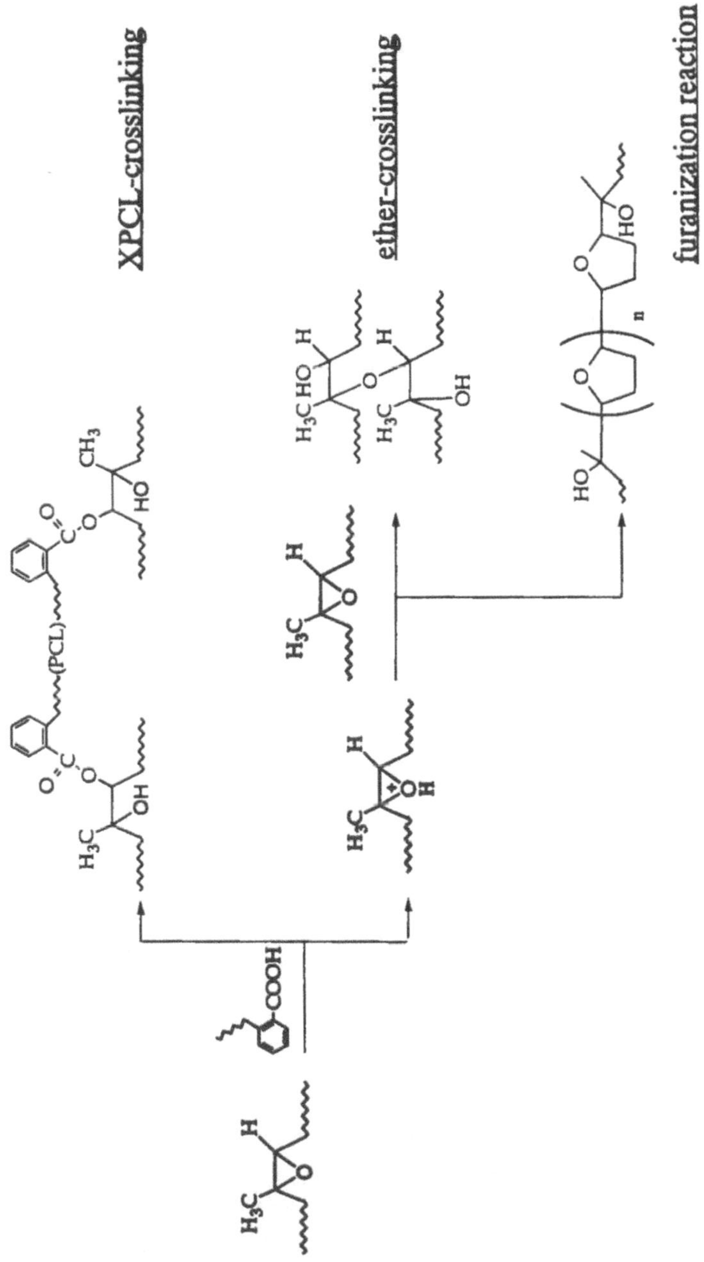

Scheme 2 Cross-linking of reactions.

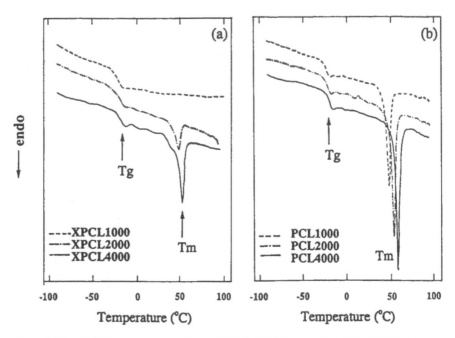

Figure 20 DSC thermograms for (a) XPCL/ENR50 and (b) PCL/ENR50 binary blends. Heating rate was 10°C/min. Cure time was 14 min, and cure temperature was 180°C for all samples. (From Ref. 89.)

ure 20 shows the comparison of DSC curves for (a) XPCL/ENR50 (XPCL : ENR = 30 : 100 by weight) hot-pressed sheets and (b) those for PCL/ENR50 (30 : 100) sheets in the temperature range from −100°C to 100°C, the cure time being 14 min for all samples. The glass transition T_g of the ENR component is observed at the same temperature at around −20°C for all samples in Figs. 20a and 20b. This indicates that the phase-separated structure exists in these blends. However, the endothermic peak corresponding to T_m of the XPCL component at around 50–55°C differs considerably from those of the original dihydroxyl PCL. T_m peaks for PCL/ENR blends are sharp, and the intensity is almost the same, irrespective of the molecular weight of the PCL component. In contrast to this, the peak intensity is much reduced for XPCL/ENR blends, and there is no T_m peak for the blend of XPCL1000 of the lowest molecular weight. The difference in the T_m peak intensity between XPCL/ENR and PCL/ENR might be ascribed to the existence of the cross-linking points, which could be restricting the crystallization of PCL chains.

3. Mechanical Properties

Figures 21a and b show the temperature dependencies of the dynamic mechanical modulus and the loss tangent for the XPCL/ENR50 blend sheets (XPCL : ENR =

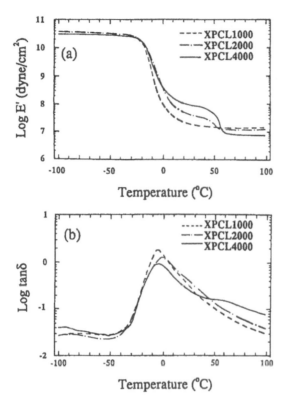

Figure 21 Temperature dependences of (a) dynamic storage modulus and (b) loss tangent for XPCL : ENR50 (30 : 100) blend sheets. Cure time of the compounds was 14 min for all samples. (From Ref. 89.)

30 : 100 by weight). It can be seen that the large lowering of E' corresponding to T_g of the ENR component is observed at around $-20°C$ and that there is a second reduction of E' due to the T_m of the XPCL component at around 50°C. The degree of the second reduction decreases greatly with decreasing molecular weight of XPCL, and there is no second-stage reduction of modulus in the XPCL1000 blend. The temperature dependence of the loss tangent for the XPCL4000/ENR50 blend also shows both maxima corresponding to T_g and T_m, but the second maximum corresponding to T_m at around 60°C disappears in the XPLC1000 blend. These results are consistent with the results of DSC measurements. Accordingly, the chain length of XPCL1000 is too short to form crystalline domains under the restriction by the cross-linking points, while the long poly(ε-caprolactone) chains of XPCL4000 can form a substantial crystalline phase.

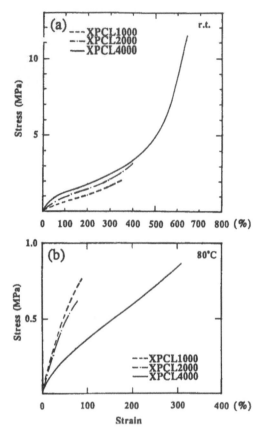

Figure 22 Stress–strain curves for XPCL : ENR50 (30 : 100) blend sheets measured (a) at room temperature (20°C) and (b) in hot water bath (80°C). Elongation rate was 100 mm/min. Cure time of the compounds was 14 min for all samples. (From Ref. 89.)

Figures 22a and b show the effect of the molecular weight of XPCL on the tensile stress–strain curves for the XPCL/ENR50 (XPCL : ENR = 30 : 100 by weight) blend sheets measured at (a) room temperature and (b) above the T_m of the poly(ε-caprolactone) chain (80°C). The cure time was 14 min for all these samples. In Fig. 22a, the modulus and the tensile strength (σ_B) and the elongation at break (ε_B) increase with an increase in the molecular weight of XPCL. The values of T_B and E_B and the stress at 50–600% elongation (M50-M600) are shown in Table 12. The increase in the modulus might be related to both the filler effect and the physical cross-linking of the crystalline phase of the XPCL component.

Table 12 Tensile Properties of XPCL/ENR50 Binary Blends[a]

Sample code	XPCL 1000	XPCL 2000	XPCL 4000
M50(MPa)	0.52	0.59	0.96
M100(MPa)	0.86	0.92	1.27
M200(MPa)	1.44	1.47	1.86
M300(MPa)	2.27	2.15	2.45
M400(MPa)	—	3.17	3.35
M500(MPa)	—	—	4.93
M600(MPa)	—	—	8.77
σB (MPa)	2.57		11.48
εB (%)	325	400	645

[a] Blend ratios were XPCL : ENR=30 : 100 by weight for all samples, which were cured at 180°C for 14 min.

The marked increase in σ_B for the XPCL4000/ENR50 blend in Fig. 22a might be involved with the strain-induced crystallization of the ENR component at large elongation. The degree of the strain-induced crystallization could be expected to be reduced by an increase in the cross-linking density. Therefore the good mechanical properties (an increase in σ_B and ε_B) for XPCL4000/ENR50 in Fig. 22a might be ascribed to the effect of the synergism of the physical cross-linking through the crystalline domains of poly(ε-caprolactone) chains, the strain-induced crystallization of the rubbery component, and the chemical cross-linking between the end carboxyl groups and the epoxy groups. The crystalline domains of XPCL4000 may reinforce the tensile properties of the blend.

Figures 23a, b, and c show the comparison of the hysteresis and the permanent set in the stress–strain curves of XPCL/ENR50 (XPCL : ENR = 30 : 100 by weight) blend sheets during cyclic deformations of up to 300% elongation, where considerable hysteresis is observed in the blend with XPCL4000. The hysteresis as well as the permanent set decreases with decreasing molecular weight of the XPCL component, and there is almost no hysteresis in the XPCL1000 blend. The observed hysteresis and permanent set might be ascribed to the structural change during the elongation, especially the change of the crystalline domains of poly(ε-caprolactone) chains, in which plastic deformation, destruction, and fragmentation of the crystalline domains might occur by the applied large strain, although the fragmented small crystalline domains still may be able to act as physical cross-linking points. It is thus reasonable that the blend with XPCL1000 did not show hysteresis because the crystalline phase was not detected in the DSC curve and the dynamic modulus as already shown in Figs. 20 and 21. It is also seen in Fig. 23

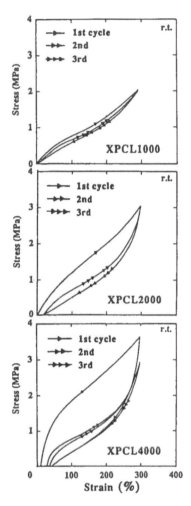

Figure 23 Effect of the molecular weight of XPCL on the elongation and recovery behaviors in the cyclic stress–strain at room temperature (20°C). Elongation and recovery rate was 100 mm/min. (From Ref. 89.)

that the hysteresis of XPCL4000 and 2000 blends in the second and third elongation cycles becomes very small and resembles that of the XPCL1000 blend. This
indicates that the crystalline domains are destroyed and fragmented by the first
elongation, resulting in the suppression of the effect of the crystalline domains,
and the blends show more elastic behavior only through the chemical cross-
linking points.

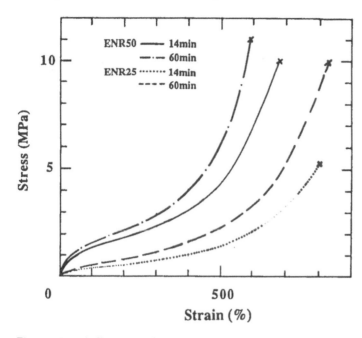

Figure 24 Influence of the degree of epoxidation and cure time on the stress–strain curves at room temperature (20°C). Elongation rate was 100 mm/min. (From Ref. 89.)

Figure 24 shows the effects of the degree of epoxidation of ENR as well as the cure time on the stress–strain curves for ENR50 and ENR25 blends with XPCL4000. These blends were cured for different cure times (14 min and 60 min). The values of σ_B, ε_B, and the stress at 50–800% elongation (M50-M800) of these blend sheets are summarized in Table 13. The results show that the use of less epoxidized ENR produces less rigid and more extensionable rubber materials. It is also seen that increase in the cure time increases both the modulus and σ_B but decreases ε_B.

Figure 25 shows a photograph by transmission electron microscopy of an ultrathin section of XPCL4000/ENR50 (XPCL : RNR50 = 30 : 100 by weight) blend sheet, in which the light islands of the poly(ε-caprolactone) domain dispersed in the dark ENR matrix are seen. It can also be observed in the photograph that many long poly(ε-caprolactone) crystalline lamellae are growing into the rubbery matrix as light strips, the thickness of which is ca. 10 nm and uniform. Presumably, these crystalline lamellae are playing an important role in the reinforcement of the tensile mechanical properties of the XPCL4000/ENR blends.

Table 13   Tensile Properties of XPCL4000/ENR Binary Blends

	ENR50		ENR25	
Sample code	(14 min)	(60 min)	(14 min)	(60 min)
M50 (MPa)	0.99	1.20	0.26	0.33
M100 (MPa)	1.31	1.56	0.35	0.49
M200 (MPa)	1.79	2.19	0.52	0.78
M300 (MPa)	2.33	2.95	0.76	1.12
M400 (MPa)	3.10	4.05	1.04	1.64
M500 (MPa)	4.42	6.29	1.51	2.32
M600 (MPa)	7.15	—	2.18	3.39
M700 (MPa)	—	—	3.31	5.51
M800 (MPa)	—	—	—	8.75
σB (MPa)	10.02	11.08	5.27	9.75
εB (%)	683	591	810	830

Blend ratios were XPCL:ENR=30:100 by weight. Cured at 180°C.

In conclusion, the blends of end carboxylated telechelic XPCLs and ENR are self-vulcanizable binary blends, which can be converted to elastomeric materials by simple curing at 160–200°C in a kind of reactive blending technique. The relative contribution of the physical cross-linking to the chemical one is controlled by the molecular weight of the XPCL component to give various stress–strain curves that reflect the developed morphology in Fig. 25. Therefore, utilization of telechelic polymers might be very effective for reactive rubber processing (91), and for the preparation of a wide variety of self-vulcanizable blend materials.

IV.   REACTIVE POLYMERIC FILLER

A macroinitiator, which is a macromolecular initiator, can be also be useful to achieve reactive blending of two incompatible polymers. Therefore it might be worthwhile to describe macroinitiators briefly for the reactive blending technique. Macroinitiators are classified into two types, the main-chain type and the side-chain type. Nuyken and Voit (6) studied the blending of natural rubber with the terpolymer of acrylonitrile, butadiene, and 3-vinylphenylazo monomer, which is a side-chain type azomacroinitiator. These polymers were coprecipitated from emulsion and then cured at 170°C to produce a rubber blend stabilized by cova-

200 nm

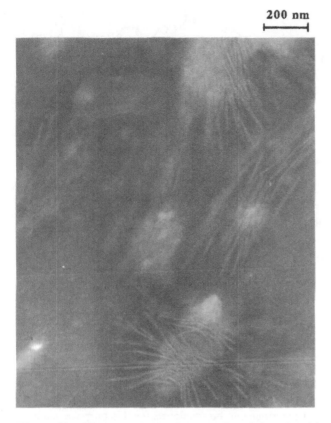

Figure 25 Electron photomicrographs of the ultrathin section of XPCL4000 : ENR50 (30 : 100) blend cured at 180°C for 14 min. Stained with OsO$_4$. Magnification ×5000. (From Ref. 89.)

lent bonds between the different polymers (6). By this method it is possible to obtain optically clear blends of nitrile rubber and natural rubber.

This method can be extended to prepare reactive fillers (6). The macroazo reactive fillers can be prepared by the copolymerizations of MMA, BDDM, and the monomer containing an azo group as shown in Fig. 26. The terpolymer prepared in emulsion gives beads of diameter nearly 50 nm, the azo content of which is varied by the feed composition of the azo monomer. The reinforcing effect of the reactive latex filler on styrene-butadiene rubber (SBR) was observed in the stress–strain curve. The elongation at break increased by a factor of four compared with SBR containing conventional filler particles.

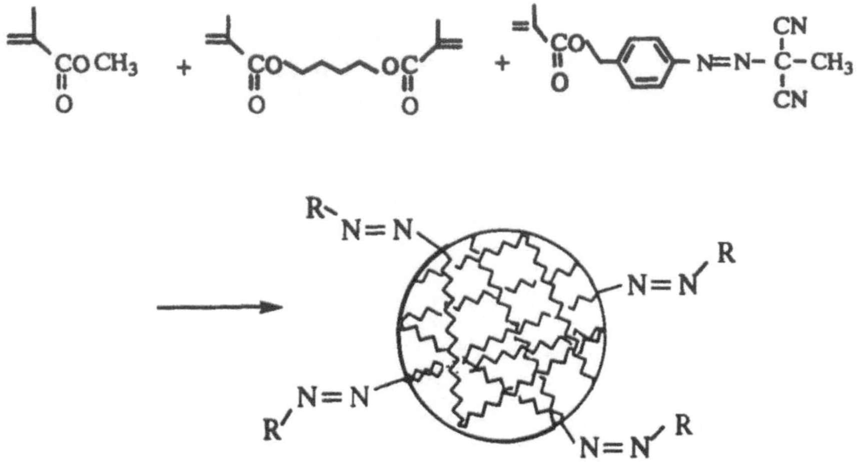

Figure 26 Preparation of polymeric reactive filler using azomonomer. (From Ref. 6.)

V. CONCLUSION

Blending or mixing is itself already a type of processing procedure, and therefore reactive blending is nothing else but a reactive processing that is very efficient and economical in the production of high performance polymeric materials. In addition, a very wide variety of reactions can be employed for this purpose, such as grafting reactions, cross-linking reactions, and polymerization reactions. Self-vulcanization using random copolymers or telechelic polymers possessing suitable functional groups is also a kind of reactive blending technique. Reactive blending may also be the key technology for the recycling of polymeric materials in which polymers collected as polymer blends must have good quality for second use. Therefore its significance in polymer blending and alloying will increase in the future.

REFERENCES

1. D. R. Paul and S. Newman, eds. Polymer Blends. 2 vols. New York: Academic Press, 1978.
2. L. A. Utracki. Polymer Alloys and Blends. New York: Hanser, 1989. Commercial Polymer Blends. London: Chapman and Hall, 1997.
3. Society of Polymer Science, ed. Polymer Alloys. 2d ed. Tokyo: Tokyo Kagaku Dojin, 1993 (in Japanese).

4. W. E. Baker, V. Flaris, Y. J. Sun, T. Pham, and Ph. Lagardere. Novel reactive processing of polymer blends. IPC97 of JSPS in Kusatsu, 1997, p.51.
5. J. A. Manson and L. H. Sperling. Polymer Blends and Composites. New York: Plenum Press, 1976, Chap. 8.
6. O. Nuyken and B. Voit. Polymeric azoinitators. In: M. K. Mishra. Macromolecular Design: Concept and Practice. Polymer Frontier International, 1994, Chap. 8.
7. A. Y. Coran and R. Patel. Rubber–thermoplastic compositions. Part I. EPDM-polypropylene thermoplastic vulcanizates. Rubber Chem. Technol. 53:141, 1980.
8. Y. Kikuchi, T. Fukui, T. Okada, and T. Inoue. Origin of rubber elasticity in thermoplastic elastomers consisting of crosslinked rubber particles and ductile matrix. J. Appl. Polym. Sci., Appl. Polym. Symp. 50:261, 1992.
9. Y. Tsukahara. Macromonomers. In: J. C. Salamone, ed. The Polymeric Materials Encyclopedia. CRC Press, 1996, pp 3918–3927.
10. F. Ide. Design of Polymer Alloys for Practical Use. Tokyo: Kogyochousakai, 1996 (in Japanese).
11. Y. Yagci and M. K. Mishra. Macroinitiators for chain polymerization. In: M. K. Mishra, ed. Macromolecular Design: Concept and Practice. Polymer Frontier International. 1994, Chap. 6.
12. Y. Yamashita, ed. Chemistry and Industry of Macromonomers. Basel: Huthig and Wepf Verlag, 1993; Tokyo: IPC, 1989.
13. P. Rempp and E. Franta. Adv. Polym. Sci. 58, 1, 1984.
14. Y. Yamashita and Y. Tsukahara. In: C. Carraher and J. Moore, eds. Modification of Polymers. New York: Plenum Press, 1983, pp. 131–140.
15 Y Tsukahara. In: M. K. Mishra, ed. Macromolecular Design. Concept and Practice. Polymer Frontier International, 1994, Chap. 5.
16. H. Craubner. Macromolecular N-nitroso-acylamines as initiators for polyreactions. I. Synthesis of block copolymers from polycondensates and olefinic monomers. J. Polym. Sci., Polym. Chem. Ed. 18:2011, 1980.
17. A. Ueda and S. Nagai. Block copolymers derived from azobiscyanopentanoic acid: synthesis of a polyethyleneglycol-polystyrene block copolymer. J. Polym. Sci., Polym. Chem. Ed. 24:405, 1986.
18. H. Kye and J. L. White. Continuous polymerization of caprolactam in a modular intermeshing corotating twin screw extruder integrated with continuous melt spinning of polyamide 6 fiber: influence of screw design and process conditions. J. Appl. Polym. Sci. 52:1249, 1994.
19. G. Samay, T. Nagy, and J. L. White. Grafting maleic anhydride and comonomers onto polyethylene. J. Appl. Polym. Sci. 56:1423, 1995.
20. T. Nishio, Y. Suzuki, K. Kojima, and M. Kakugo. A twin extruder developed for analysis of reactive process and its application for analysis of reactive processing of maleic anhydide grafted polypropyrene/polyamide 6. Kobunshi Ronbunshu 47:331, 1990.
21. R. Vankan, P. Degee, R. Jerome, and P. Teyssie. Design of polymer blend rheology: effect of maleic anhydride containing copolymers on the melt viscosity of polyamides. Polym. Bull. 33:22, 1, 1994.
22. D. R. Paul and J. W. Barlow. Mechanical compatibilization of immiscible blends. Polymer 25:487, 1984.

23. F. Ide and A. Hasegawa. Studies on polymer blend of nylon 6 and polypropylene or nylon 6 and polystyrene using the reaction of polymer. J. Appl. Polym. Sci. 18:963, 1974.

24. R. J. M. Borggreve. R. J. Gaymans, J. Schaijer, and J. F. Ingen Houz. Brittle–tough transition in nylon-rubber blends: effect of rubber concentration and particle size. Polymer 28:1489, 1987.

25. C. Han and H. Chuang. Blends of nylon 6 with an ethylene-based multifunctional polymer. I. Rheology–structure relationships. J. Appl. Polym. Sci. 30:2431, 1985.

26. J. Angola, Y. Fujita, T. Sakai, and T. Inoue. Compatibilizer-aided toughening in polymer blends consisting of brittle polymer particles dispersed in a ductile polymer matrix. J. Polym. Sci., Polym. Phys. 26:807, 1988.

27. W. E. Baker and M. Saleem. Coupling of reactive polystyrene and polyethylene in melt. Polymer. 28:2057, 1987.

28. W. E. Baker and M. Saleem. Polystyrene–polyethylene melt blends obtained through reactive mixture process. Polymer 28:2057, 1987.

29. S. Cimmino, L. D'orazio, R. Greco, G. Maglio, M. Malinconico, C. Mancarella, E. Martuscelli, R. Palumbo, and G. Ragosta. Polym. Eng. Sci. 24:48, 1984.

30. C. S. S. Namboodiri, S. Thomas, S. K. De, and D. Khastgir. Thermoplastic elastomers from epoxidized natural rubber (ENR)-styrene-acrylonitrile copolymer (SAN) blends. Kautsch. Gummi Kunstst. 42:1004, 1989.

31. A. Simmons and W. E. Baker. Compatibility enhancement in polyethylene/styrene-maleic anhydride blends via polar interactions. Polymer Comm. 31:20, 1990.

32. B. K. Kim, S. Y. Park, and S. J. Park. Morphological, thermal and rheological properties of blends: polyethylene/nylon-6, polyethylene/nylon 6/maleic anhydride-g-polyethylene/nylon-6. Eur. Polym. J. 4/5:349, 1991.

33. J. R. Campbell, S. Y. Hobbs, T. J. Shea, and V. H. Watkins. Poly(phenylene o-xide)/polyamide blends via reactive extrusion. Polym. Eng. Sci. 30:1056, 1990.

34. M. Hert, L. Guerdoux, and J. Lebez. Reactive olefinic copolymers: new developments and application. Angrew. Makromol. Chem. 155:111, 1987.

35. M. Xanthios, M. W. Young, and J. Biesenberg. Polypropylene/polyethylene terephthalate blends compatibilized through functionalization. Polym. Eng. Sci. 30:355, 1990.

36. B. K. Kim and S. J. Park. Reactive melt blends of nylon with polystyrene-co-maleic anhydride. J. Appl. Polym. Sci. 43:357, 1991.

37. M. Hara and A. Eisenberg. Miscibility Enhancement via Ion-Dipole Interaction. Macromolecules 17:1335, 1984.

38. C. J. Wu, F. Kuo, and C. Y. Chen. Rubber toughned polyamide 6: the influences of compatibilizer on morphology and impact properties. Polym. Eng. Sci. 33:1329, 1993.

39. C. E. Scott and C. W. Macosko, Morphology development during reactive and non-reactive blending of an ethylene-propylene rubber with two thermoplastic matrices. Polymer 35:5422, 1994.

40. N. K. Kalfoglou, D. S. Skafidas, and J. K. Kallitsis. Blends of poly(ethylene terephthalate) with unmodified and maleic anhydride grafted acrylonitrile-butadiene-styrene terpolymer. Polymer 37:3387, 1996.

41. Pentti Jarvela, L. Shucai, and Pirkko Jarvela. Dynamic mechanical properties and morphology of polypropylene/maleated polypropylene blends. J. Appl. Polym. Sci. 62:813, 1996.

42. S.-W. Lee, C.-S. Ha, and W.-J. Cho. Miscibility of nylon 6 with poly(maleic anhydride-co-vinyl acetate) and hydroxylated poly(maleic ahhydnde-co-vinyl acetate) blends. Polymer 37:3347, 1996.

43. C.-S. Ha, M.-G. Ko, and W.-J. Cho. Miscibility of nylon 46 and ethylene-vinyl alcohol copolymer blends. Polymer 38:1243, 1997.

44. K. Kreisher. Compatibilizers: the secret "glue" binding new alloys. Plast. Technol. 35:67, 1989.

45. N. C. Liu and W. E. Baker. Reactive polymers for blend compatibilization. Adv. Polym. Tech. 11:249, 1992.

46. K. Mai and J. Xu. Toughening thermoplastics. In: O. Olabisi, ed. Handbook of Thermoplastics. New York: Marcel Dekker, 1997, pp. 523–556.

47. V. J. Triacca, S. Ziaee, J. W. Barlow, H. H. Keskkula, and D. R. Paul. Reactive compatibilization of blends of nylon 6 and ABS materials. Polymer 32:1401, 1991.

48. M. Okamoto and T. Inoue. Reactive processing of polymer blends: analysis of the change in morphological and interfacial parameters with processing. Polym. Eng. Sci. 33:175, 1993.

49. J. G. Bonner and P. S. Hope. Compatibilization and reactive blending. In: (P. S. Hope and M. J. Folkes, eds.) Polymer Blends and Alloys. Blackie, 1993, Chap. 3.

50. F. C. Chang. Compatibilized thermoplastic blends. In: (O. Olabisi, ed.) Handbook of Thermoplastics. New York: Marcel Dekker, 1997, pp. 491–521.

51. H. H. Chang, J. S. Wu, and F. C. Chang. Reactive compatibilization of ABSI/Nylon 6,6 blends: effect of reactive group concentration and blend sequence. J. Polym. Res. 1:235, 1994.

52. R. J. M. Borggreve. R. J. Gaymans, and J. Schijer. Impact behavior of nylon–rubber blends. 5. Influence of the mechanical properties of the elastomer. Polymer 30:71, 1979.

53. S. Wu. Formation of dispersed phase in incompatible polymer blends: interfacial and rheological effects. Polym. Eng. Sci. 27:335, 1985.

54. R. Greco, M. Malinconico, E. Martuscelli, G. Ragosta, and G. Scarinzi. Role of degree of grafting of functionalized ethylene–propylene rubber on the properties of rubber modification polyamide 6. Polymer 28:1185, 1987.

55. S. Cimmico, F. Coppola, L. D'Orazio, R. Greco, G. Maglio, M. Malinconico, C. Mancarella, E. Martuscelli, and G. Ragosta. Ternary nylon 6/rubber/modified rubber blends: effect of the mixing procedure on morphology, mechanical and impact properties. Polymer 27:1874, 1986.

56. S. Wu. Phase structure and adhesion in polymer blends: a criterion for rubber toughening. Polymer 26:1855, 1985.

57. A. Cerere, R. Greco, G. Ragosta, G. Scarinzi, and A. Taglialatela. Rubber toughening of polybuthylene terephthalate: influence of processing on morphology and impact properties. Polymer 31:1239, 1990.

58. R. J. M. Borggreve and R. J. Gaymans. Impact behavior of nylon–rubber blends. 4. Effect of the coupling agent: maleic anhydride. Polymer 30:63, 1989.

59. H. Chuang and C. Han. Mechanical properties of blends of nylon 6 with a chemically modified polyolefin. J. Appl. Polym. Sci. 30:165, 1985.

60. H. Chuang and C. Han. Blends of nylon 6 with ethylene-based multifunctional polymer. II. Property–morphology relationship. J. Appl. Polym. Sci. 30:2457, 1985.

61. S. Y. Hobb, R. C. Bopp, and V. H. Watkins. Toughened nylon resins. Polym. Eng. Sci. 23:380, 1983.

62. A. R. Padwa. Compatibilized blends of polyamide 6 and polyethylene. Polym. Eng. Sci. 32:1703, 1992.

63. N. C. Liu, W. E. Baker, and K. E. Russell. Functionalized of polyethylene and their use in reactive blending. J. Appl. Polym. Sci. 41:2285, 1990.

64. N. R. Choudhury and A. K. Bhowmick. Compatibilization of natural rubber–polyolefin thermoplastic elastomeric blends by phase modification. J. Appl. Polym. Sci. 38:1091, 1974.

65. R. Holsti-Mettinen, J. Seppala, and O. T. Ikkala. Effects of compatibilizers on the properties of polyamide/polypropylene blends. Polym. Eng. Sci. 32:868, 1992.

66. F. P. Lamantia. Blends of polypropylene and nylon 6: influence of the compatibilizer molecular weight, and processing conditions. Adv. Polym. Tech. 12:47, 1993.

67. S. J. Park, B. K. Kim, and H. M. Jeong. Morphological, thermal and rheological properties of the blends of polypropylene, nylon 6, polypropylene/nylon 61 (maleic anhydried-g-propylene) and (maleic anhydride-g-polypropylene/nylon 6). Europ. Polym. J. 26:131, 1990.

68. J. Duvall, C. Sellitti, C. Myers, A Hiltner, and E. Baer. Effect of compatibilization on properties of polyethylene/polyamide 66 (75 : 25 wt/wt) blends. J. Appl. Polym. Sci. 52:195, 1994.

69. C. C. Chen, E. Fontan, K. Min, and J. L. White. An investigation of instability of phase morphology of blends of nylons with polyethylene and polystyrene and effects of compatibilizing agents. Polym. Eng. Sci. 30:1056, 1990.

70. C. Carrot, J. Guillet, and J. F. May. Blends of polyamide 6 with ABS: effect of a compatibilizer on adhesion and interfacial tension. Plast. Rubber Proc. Appl. 16:61, 1991.

71. V. J. Triacca, S. Ziaee, J. W. Barlow, and D. R. Paul. Reactive compatibilization of blends of nylon 6 and ABS materials. Polymer 32:1401, 1991.

72. M. Hert, L. Guerdoux, and J. Lebez. Reactive olefinic copolymers: new developments and applications. Angew. Makromol. Chem. 155:111, 1987.

73. Y. J. Kim, C. D. Han, B. K. Song, and E. Kouassi. Mechanical and transport properties of coextruded films. J. Appl. Polym. Sci. 29:2359, 1984.

74. G. Serpe, J. Janin, and F. W. Dawans. Morphology-processing relationships in polyethylene-polyamide blends. Polym. Eng. Sci. 30:553, 1990.

75. I. R. Gelling. Modification of natural rubber latex with peracetic acid. Rubber Chem. Technol. 58:86, 1985.

76. S. Jayawardena, D. Reyx, D. Durand, and C. P. Pinazzi. Synthesis of macromolecular antioxidants by reaction of aromatic amines with epoxidized polyisoprene. 4. Antioxidation efficiency of supported 4-anilinoanilina for the protection of cis-1, 4-polybutadiene. Makromol. Chem. 185:19, 2089, 1984.

77. C. S. L. Baker, I. R. Gelling, and R. Newell. Epoxidized natural rubber. Rubber Chem. Technol. 58:67, 1985.

78. I. R. Gelling and M. Porter. In: A. D. Roberts, ed. Natural Rubber Science and Technology. Oxford: Oxford University Press, 1988, Chap. 10, p. 359.

79. A. S. Hashim and S. Kohjiya. Preparation and properties of epoxidized natural rubber. Kautsch. Gummi Kunstst. 46:203, 1993.
80. Rubber Research Institute of Malaysia. Epoxidized Natural Rubber. Malaysia, 1984.
81. R. Alex, P. P. De, and S. K. De. Self-vulcanizable rubber blend system based on epoxidized natural rubber and carboxylated nitrile rubber. J. Polym. Sci. Part C: Polym. Lett. 27:361, 1989.
82. S. Mukhopadhyay and S. K. De. Self vulcanizable rubber blend system based on epoxidized natural rubber and chlorosulphonated polyethylene. J. Mat. Sci. 25:4027, 1990.
83. R. Alex, P. P. De, and S. K. De. Self-vulcanisable rubber–rubber blends based on epoxidized natural rubber and polychloroprene. Kautsch. Gummi Kunstst. 44:333, 1991.
84. R. Alex, P. P. De, and S. K. De. Epoxidized natural rubber–carboxylated nitrile rubber blend: a self-vulcanizable miscible blend system. Polym. Communication 31:(3)118, 1990.
85. R. Alex, P. P. De, N. M. Mathew, and S. K. De. Effect of fillers and moulding conditions on properties of self-vulcanisable blends of epoxidized natural rubber and carboxylated nitrile rubber. Plast. Rubb. Process. Appl. 14:4, 1990.
86. Sujata Mukhopadnyay, T. K. Chaki, and S. K. De. Self-vulcanizable rubber blend system based on epoxidized natural rubber and hypalon. J. Polym. Sci. Part C: Polym. Letters 28:25, 1990.
87. N. R. Manoj, P. P. De, S. K. De, and D. G. Peiffer. Self-crosslinkable blend of zinc sulfonated EPDM and epoxidized natural rubber. J. Appl. Polym. Sci. 53:361, 1994.
88. Y. Tsukahara, T. Yonemura, A. S. Hashim, and S. Kohjiya. Polym. Prepr. Jpn. 43:1298, 1994, International Rubber Conference (IRC95 Kobe), 1995, p. 74.
89. Y. Tsukahara, T. Yonemura, A. S. Hashim, S. Kohjiya, and K. Kaeriyama. Preparation and properties of epoxidized natural rubber/poly(ε-caprolactone) self-vulcanizable blends. J. Mater. Chem. 6:1865, 1996.
90. Y. Tsukahara, T. Yonemura, H. Ismail, A. S. Hashim, S. Kohjiya, and K. Kaeriyama. Preparation and properties of epoxidized natural rubber/poly(ε-caprolactone) composites. Sixth JSPS-VCC Seminar on Intergrated Engineering, Kyoto, 1996, p. 28.
91. S. Yamashita. Future trends in rubber processing: the chemistry of rubber processing. J. Macromol. Sci., Pure Appl. Chem. A33:1897, 1996.

23

Developments in Poly(vinyl chloride)/Epoxidized Natural Rubber Blends

Umaru Semo Ishiaku and Zainal Arifin Mohd Ishak
Universiti Sains Malaysia, Penang, Malaysia

I. INTRODUCTION

This chapter discusses blends of poly(vinyl chloride) (PVC) and epoxidized natural rubber (ENR). ENR is a relatively new rubber that is a modified form of natural rubber.

Epoxidation may be the only chemical modification of NR, besides cis–trans isomerization, in which the product still retains elasticity above 20 mol% modification (1,2). ENR has been known since Pummerer and Burkard (3) reported the reaction of NR with peroxycarboxylic acids in 1922. However, ENR only became commercially important in 1982 when it was patented by Gelling (4) after success in making clean ENR under controlled conditions by reacting NR latex with peracetic acid. The product possesses characteristics such as retention of the high tensile strength of NR, good oil resistance, low air permeability, high damping properties, and anti-wet skid resistance, although the increased T_g ($\approx$ 1°C/1 mol% epoxidation) with rising levels of epoxidation is not advantageous (1,2,4–8).

Epoxidized natural rubber with 50 mol% epoxidation (ENR-50) is polar and has properties similar to those of synthetic elastomers (1–2,5–8). These qualities generated interest in assessing the compatibility of ENR with other polymers (9–14), particularly poly(vinyl chloride) (PVC) (14–17). ENR-50 was found to be miscible with PVC at all compositions (16,17), as well as at positive deviation from the rule of mixtures in a log–log plot of shear viscosity versus composition at a constant shear stress (18). Rheological studies revealed the pseudoplastic nature of the blends (18). Fracture studies also revealed that ENR-50 is capable of acting as an impact modifier for PVC (17). However, this seemingly attractive blend is beset with aging problems (19,20).

The poor thermo-oxidative aging of PVC/ENR blends was demonstrated in our studies of mechanical properties and Fourier transform infrared (FTIR) spectra (20). In the same study, attempts made to stabilize the system with an antioxidant and a base were reported. The incorporation of a plasticizer was found to be particularly effective in stabilizing the blends.

The details of some of the studies are presented in this chapter. The chapter is broad-based although it is comprehensive. It covers important aspects of blends, which include melt compounding and rheology, the effect of blend composition, thermo-oxidative aging, and effects of plasticizers and dynamic vulcanization.

II. EXPERIMENTAL

Epoxidized natural rubber grade Epoxyprene 50 was purchased from Kumpulan Guthrie Berhad, Seremban, Malaysia; suspension polymerized poly vinyl (chloride) with K value of 65, Mericon HP-65, was supplied by Malayan Electro-chemical Company Sdn Bhd., Penang. Ba/Cd/Zn-based PVC stabilizer Irgastab BC 455s was supplied by Ciba-Geigy (M) Ltd. Di-ethylhexyl phthalate (DOP), etherthioether (ETE), epoxidized soya oil (ESO), 2,2,4-trimethyl-1,2-dihydroquinoline (TMQ), calcium stearate $(Ca(St)_2)$ and vulcanization ingredients such as sulphur, ZnO, stearic acid, MBTS, and TMTD were supplied by Bayer (M) Sdn Bhd.

A Brabender Plasticorder model PLE 331 coupled with a mixer/measuring head (W50H) was used for melt compounding at 150°C with a 50 rpm rotor speed for 6 minutes. This condition varies accordingly in cases involving rheological studies under different conditions. The samples obtained were compression molded at 150°C prior to mechanical/physical testing. Tensile tests were carried out with the Monsanto Tensometer T10, according to ISO 37 Type III at 25 ± 3°C and crosshead speed of 50 cm/min. Tear tests were done according to ISO 34 type III using Y-shaped speciments.

Dynamic mechanical properties were scanned from − 100 to 150°C at the rate of 10°C/min using the resonant mode at 30Hz with the DuPont DMA Model 983 coupled with the 2000 Thermal analyzer.

The rheological properties of the blends were determined with the Shimadzu Rheometer Model CFT-500 in accordance with JIS K 6719–1977. The constant temperature mode was used at 160 or 150°C as required. Samples were extruded with various loads (stress) to determine the flow rate and other rheological information. A circular die of L/D ratio 10 with a die diameter of 1 mm was used except where the effects of L/D ratio were studied.

A Fourier transform infrared spectrometer (FTIR) Perkin-Elmer 1600 series model was used to obtain the IR spectra. The spectrometer was operated with a resolution of 4 cm^{-1}, and the scanning range was 4400–450 cm^{-1}. Thermo-ox-idative aging for all samples was done at 80°C for 168 hours. The spectral changes as a consequence of thermo-oxidative aging can be obtained by subtracting the spectrum of the unoxidized blend from that of the oxidized blend.

Difference spectrum = K_o (oxidized spectrum) − K_u (unoxidized spectrum)

where K_o and K_u are variables that can be used to compensate for variations in film thickness. K_o and K_u were kept constant by scanning an identical spot of the same film before and after aging, so that $K_o/K_u = 1$. Absorptions above the base line in-dicate the formation of new species or an increment in the concentration of exist-ing ones, whereas absorptions below the base line indicate a decrease of existing species.

III. COMPOUNDING VARIABLES

Several studies (13,21,22) have indicated that the mechanical properties of melt compounded polymer blends are greatly influenced by the mixing parameters such as mixing temperature, rotor speed, and/or mixing time. Earlier work on PVC/ENR blends by Nasir and Ratnam (21,22) indicates the need of using suit-able mixing conditions to attain optimum properties. Similar observations have also been reported by George et al. (13) working with PVC/NBR blends.

Tables 1 and 2 present sets of ultimate tensile strength for PVC-dominant and ENR-dominant blends, respectively, obtained by varying either the mixing tem-peratures (120–160°C) or the rotor speeds (20–60 rpm). The optimum tensile properties at respective rotor speeds are indicated. Mixing was carried out 9 min in each case. Judging from optimum properties, there is a correlation between both mixing parameters. For a PVC-dominant blend, optimum properties are obtain-

Table 1 Ultimate Tensile Strength (MPa) of PVC-Dominant Blends (80/20) Prepared with Varying Mixing Temperatures and Rotor Speeds

Rotor speed (rpm)	Mixing temperature (°C)				
	120	130	140	150	160
20	—	—	—	20.4[a]	16.2
30	14.0	14.3	23.1	24.9[a]	20.2
40	18.8	19.8	25.4[a]	25.3	15.5
50	23.4	24.4[a]	20.8	21.2	—
60	22.4	23.6[a]	—	22.2	11.0

[a] Optimum; maximum tensile strength at respective rotor speeds.
Source: Ref. 15.

Table 2 Ultimate Tensile Strength (MPa) of ENR-Dominant Blends (20/80) at Varying Mixing Temperatures and Rotor Speeds

Rotor speed (rpm)	Mixing temperature (°C)				
	120	130	140	150	160
20	—	—	—	10.4[a]	8.9
30	7.8	8.6	10.0	12.8[a]	10.6
40	13.7	13.9[a]	12.3	13.4	11.8
50	15.5	15.7[a]	14.5	13.8	11.2
60	13.2	15.4[a]	15.0	12.7	9.5

[a] Optimum; maximum tensile strength at respective rotor speeds.
Source: Ref. 15.

able by mixing either at high temperature, say 150°C, at low rotor speed, or at higher rotor speeds at a lower temperature. On the contrary, for the ENR-dominant blend, optimum conditions are rather restricted to high temperature and low rotor speed only. The wider choice in ensuring optimum PVC-dominant than ENR-dominant blends can be attributed to the latters' poorer flow properties. As expected, a similar trend is also observed with elongation at break (Fig. 1).

Optimum blending conditions are better illustrated with a composite plot of mixing temperature versus rotor speed for both PVC- and ENR-dominant blends. The higher correlation factor of 0.949 for the PVC-dominant blend as compared to 0.943 for the ENR-dominant blend further indicates the availability of wider choices in blending PVC-dominant blends (Fig. 2). Figure 2 also indicates that the choice of optimum blending conditions is not limited to any particular temperature or rotor speed. Rather, optimum blends can be obtained by selecting any set of conditions within a processing window.

Figure 3 shows the plastograms obtained by varying the rotor speeds for the intermediate PVC/ENR blend (50/50). The mixing temperature was fixed at 150°C. If mixing time is fixed at 9 min, a rotor speed of 30 or 40 rpm ought to be selected. At 50 and 60 rpm, degradation is imminent even after approximately 7 min mixing. Thus at higher rotor speeds, a shorter mixing period should be employed to prevent any degradation of the blend. The effect of mixing temperature for the intermediate blend is illustrated in Fig. 4. It is quite obvious that the mixing time of 9 min is best suited only for lower mixing temperatures, i.e., 130 and 140°C.

Supportive evidences for the effect of mixing parameters on the quality of the intermediate PVC/ENR blends could also be derived from the morphological study using optical microscopy. This is illustrated by the photomicrographs in Figs. 5 and 6, which show the morphological development occurring in the blends.

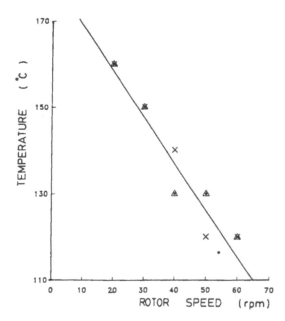

Figure 1 Comparison of mixing temperature versus rotor speed for maximum elongation at break. (■) 80/20; (△) 50/50; (x) 20/80. (From Ref. 15.)

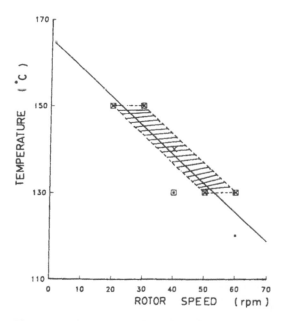

Figure 2 Composite plot of mixing temperature versus rotor speed with ultimate tensile strength as indicated. (□) 80/20; (■) 50/50; (x) 20/80. The shaded area indicates the processing window. (From Ref. 15.)

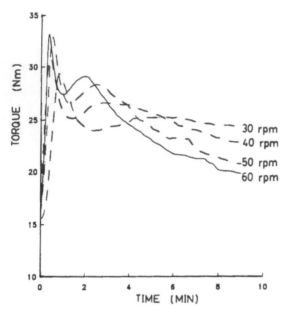

Figure 3 Effect of changing rotor speed on the plastograms of 50/50 PVC/ENR blends, mixed at 150°C. (From Ref. 15.)

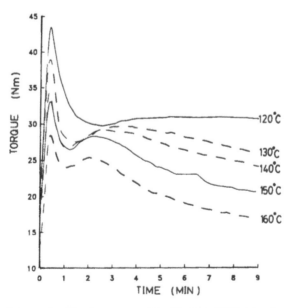

Figure 4 Plastograms of 50/50 PVC/ENR blends obtained between 120 and 160°C at a fixed rotor speed of 50 rpm. (From Ref. 15.)

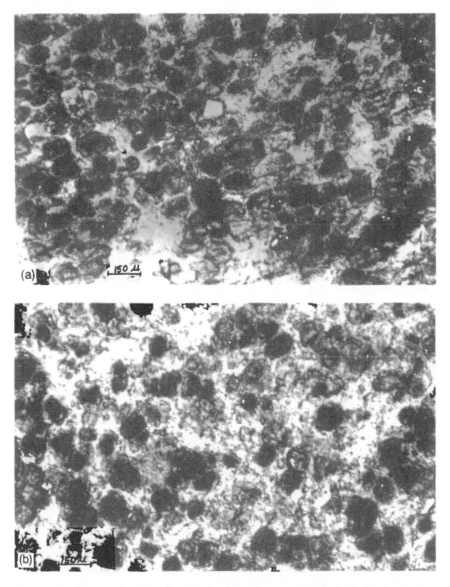

Figure 5 Photomicrographs illustrating the morphological changes occurring with mixing time in the 50/50 blend obtained at 20 rpm and 150°C. (a) 1 min; (b) 2 min; (c) 3 min; (d) 4 min. (From Ref. 15.)

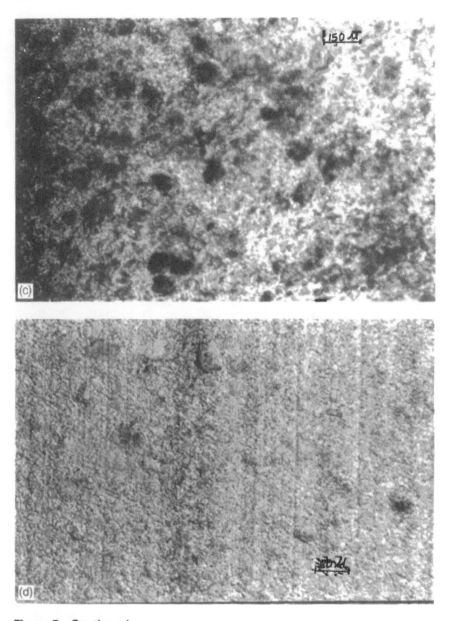

Figure 5 Continued

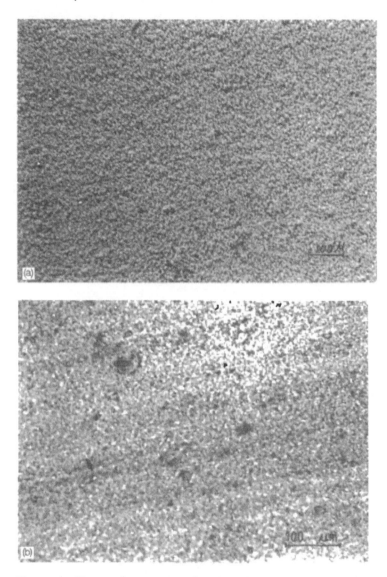

Figure 6 Photomicrographs showing close resemblance in morphologies between those at 20 rpm after 7 min and those at 50 rpm after 2 min mixing. (a) 20 rpm/7 min; (b) 50 rpm/2 min. (From Ref. 15.)

As can be seen from Fig. 6, at low rotor speeds, i.e., 20 rpm, the transformation from the multiphase or heterogenous system to the one-phase system is not fully achieved after 4 min mixing. Traces of unfused PVC particles are still apparent from the photomicrograph. Further, in Fig. 6, the similarity between the morphologies procured at 20 rpm/7 min and 50 rpm/2 min is clearly illustrated, which further supports the above observation. It is thus appropriate to conclude that a close relationship exists between mixing behavior as characterized by the plastogram and the morphologies. Similar attempts have also been reported by several workers (24–26), although they worked on different blends.

In polymer blend studies, the dynamic mechanical analysis technique (DMA) is well accepted as a powerful tool to characterize the compatibility, miscibility, and mechanical properties of blends. In Fig. 7, the temperature dependence of loss tangent, tan δ, for an intermediate blend is compared with that of both the ho-

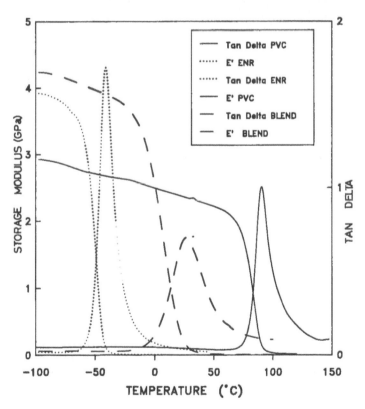

Figure 7   Dynamic mechanical properties of 50/50 PVC/ENR blend, PVC, and ENR homopolymers. (From Ref. 17.)

mopolymers, PVC and ENR. As expected, a compatible blend with a single T_g intermediate between those of constituent materials is obtained. This is in agreement with earlier work reported by other workers (23,24). The variation of tan δ with temperature of 50/50 blends procured using dissimilar rotor speeds after mixing times of 4 and 9 min are depicted in Figs. 8 and 9, respectively. When mixing was stopped after 4 min, only that at 50 rpm rotor speed results in compatible blends, while at 20 rpm traces of incompatibility are still present, shown by the residual peak, which can be attributed to traces of the PVC phases (Fig. 8). However, from Fig. 9 it can be seen that after 9 min, both rotor speeds give rise to compatible blends but slightly different tan δ values.

Apart from torque rheometery and morphological and dynamic mechanical studies, supportive evidence for the effect of mixing parameters on the interaction between PVC and ENR can also be derived from the IR spectroscopy technique.

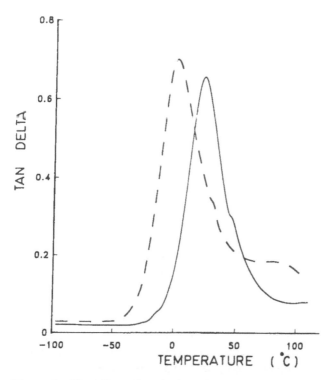

Figure 8 The effect of rotor speeds on the temperature dependence of tan δ of 50/50 PVC/ENR blends after 4 min mixing at 150°C. (·····) 20 rpm; (— —) 50 rpm. (From Ref. 15.)

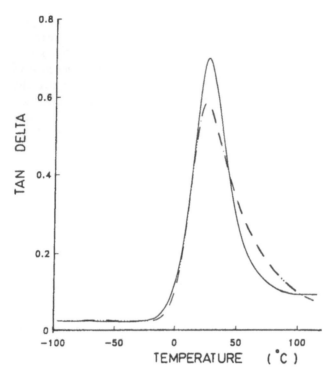

Figure 9 The effect of rotor speeds on the temperature dependence of tan δ of 50/50 PVC/ENR blends after 9 min mixing at 150°C. (·····) 20 rpm; (— —) 50 rpm. (From Ref. 15.)

Figures 10, 11, and 12 show the IR spectra of PVC, ENR-50, and the intermediate blend of PVC/ENR, respectively. Both polymers, i.e., PVC and ENR, are individually capable of hydrogen bonding. The broad peak prominent at 3500 cm^{-1} in the blend absorbs with an intensity four times as great as that of either component. This is very clear evidence of hydrogen bonding. The hydrogen bonding peak is only next in prominence to the C—C and C—H stretching absorptions in the 2900s, thus indicating the important role played by hydrogen bonding in PVC/ENR blends. Several workers also deduced hydrogen bonding from IR spectra of blends (25–27). Other investigators also reported shifts in absorption bands as evidence of specific interactions in polymer blend systems (28).

Figure 12 shows the effect of mixing time on the formation of hydrogen bonding in an intermediate PVC/ENR blend. The intensity of the hydrogen bonding peak was observed to increase with prolonged mixing. It is again worth mentioning that the change in the absorption peak between 4- and 9-min mixing is not sub-

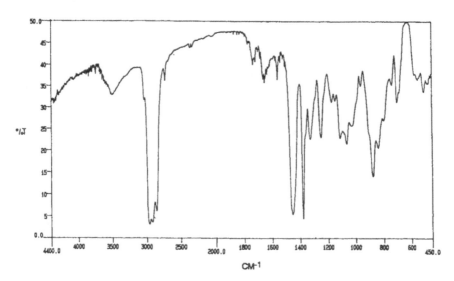

Figure 10 IR spectrum for PVC. (From Ref. 16.)

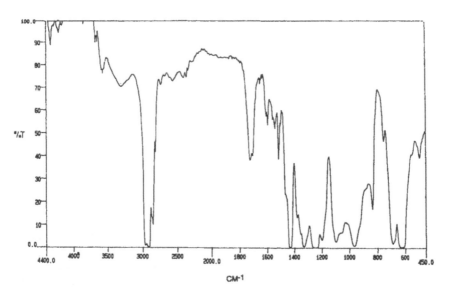

Figure 11 IR spectrum for ENR-50. (From Ref. 16.)

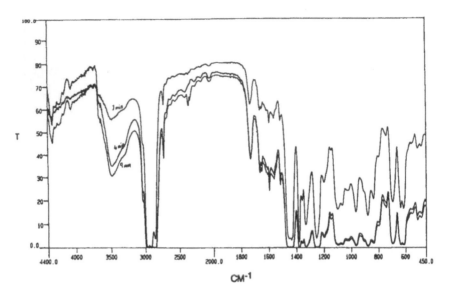

Figure 12 IR spectra for 50/50 PVC/ENR blends prepared at different mixing times. (From Ref. 19.)

stantial. This again supports the earlier observation that under the prevailing conditions (150°C at 50 rpm), 4-min mixing is essentially adequate to impart optimum properties (29). Based on these observations, it could be inferred that the presence of hydrogen bonding could be used to account for the miscibility and hence the enhanced mechanical properties of the blends.

IV. BLEND COMPOSITION

One of the main targets of investigating polymer blend rheology is to optimize the processing conditions of polymer–polymer systems having unique performance characteristics. Rheology is the study of interrelated flow or deformation variables that determine the mechanical and physical properties of polymer blend products. Apart from compounding variables such as shear rate, mixing temperature, and mixing time, the rheological properties of a polymer blend are also influenced by blend composition.

Figure 13 illustrates the typical plot of apparent shear stress versus apparent shear rate for PVC, ENR, and various blends. It is interesting to note that not all the curves fall between those of the constituent materials, as would have been the

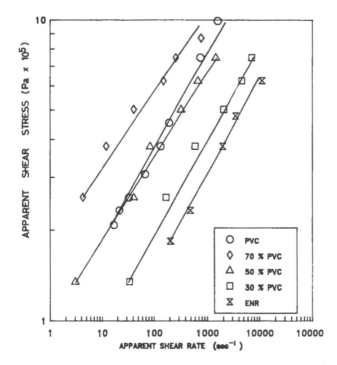

Figure 13 Composition dependence of apparent shear stress–apparent shear rate curves for PVC, ENR, and PVC/ENR blends determined at 160°C. (From Ref. 18.)

case if the rheological properties were merely additive (11,30). This point can be clarified by plotting the composition dependence of viscosity at a constant shear stress, 3.80×10^5 Pa, as illustrated in Fig. 14. Ideally the zero shear viscosity should be used. Since these data are difficult to obtain, the second best criterion, which involves considering the log plot of shear viscosity vs. composition at a constant shear stress, was employed (31). The shear stress of 3.80×10^5 Pa was chosen because all blend compositions and the pure components yield appreciable shear rate values under this condition. A positive deviation from the logarithmic rule of mixtures is observed. This synergistic behavior is typical of thermodynamically miscible systems such as polyphenylenether/polystyrene (32), polymethyl-methacrylate/poly(styrene-co-acrylonitrile) (33), and polyisoprene/polyvinylethylene (34). The composition dependence of rheology, as mentioned above, is in accordance with the earlier reports involving miscibility (17,18). All findings lead to the conclusion that PVC/ENR blends are thermodynamically mis-

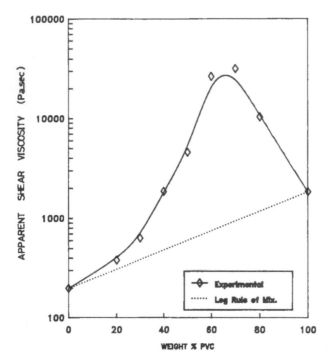

Figure 14 Composition dependence of apparent shear viscosity at a constant apparent shear stress (3.80×10^5 Pa) determined at 160°C. (From Ref. 18.)

cible systems. The clarity of films or sheets made from the blend supports the idea that the miscibility inferred from the melt state is equally true for the solid state. This statement is further supported by the observation of synergism in tensile properties, as will be discussed later (Figs. 15 and 16). Consequently, PVC/ENR blends can be classified as positive-deviation blends (11,35,36).

Figures 15 and 16 show the effect of blend composition on tensile strength and elongation at break, respectively. In the former, significant property enhancement has been achieved. The tensile strength of the PVC-dominant blend, i.e., 40–80% PVC, exceeds the values expected if the properties were merely additive (37). A similar trend is also observed in the case of elongation at break, where the EB exceeds the additivity line at all compositions. Synergism in tensile properties of miscible polymer blend systems has been reported by Fried et al. (38) working with poly(2,6-dimethyl-1,4-phenylene oxide)/polystyrene (PPO/PS), poly(p-chlorostyrene) (Pp-CIS), and random copolymers of styrene with p-chlorostyrene

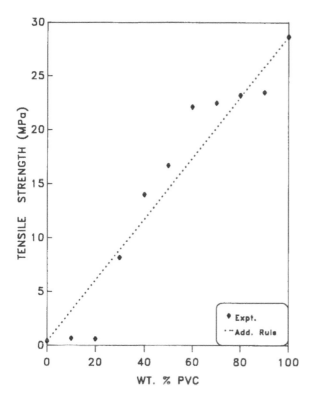

Figure 15 Tensile strength vs. composition for various PVC/ENR blends. (From Ref. 16.)

(Pp-CIS/S). Synergism in PS/PPO blends has also been reported in other instances (39,40). The general improvement in EB at all compositions is a direct contrast to the observation reported by Fried et al. (38) while working with incompatible blends of PP-CIS/PPO. The broad minima in EB shown by PP-CIS/PPO are clear evidence of immiscibility, whereas the indication here is the opposite. Although miscibility between PVC/ENR is well established (23,24,41), the present observation of synergism in tensile properties is new evidence in support of miscibility in PVC/ENR systems.

The relative density vs. composition plot is shown in Fig. 17. Generally, the variation in density with composition closely estimates the values predicted by assuming no volume change on mixing (42). It is worth noting that the relative density at about 60% PVC content is higher than calculated values. Density vs. composition plots have been used by several workers to explore polymer–polymer

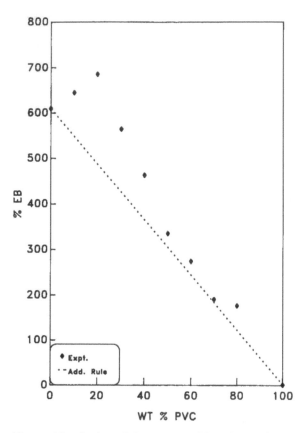

Figure 16 A plot of the composition dependence of percentage elongation at break vs. composition for various PVC/ENR blends. (From Ref. 16.)

miscibility. Increases in the densities of PVC/EVA and PVA/NBR miscible systems have been reported by Shur and Ranby (43,44). Similar observations on other blend systems have been reported by Fried et al. (38) and Jacques and Hopfenberg (42). Shur and Ranby (43) have eliminated the possibility that crystallinity might be responsible for the observed increase in density by demonstrating through x-ray diffraction methods that crystallinity in PVC/EVA is less than 10%. According to these workers, the observed increase in density or decrease in volume could reasonably be attributed to closer packing or better spatial arrangements of polymer chains owing to increased interaction between the two polymers. They further attributed the increasing contraction in the PVC/NBR blend volume (with increasing acrylonitrile content) to increased polymer–polymer interaction causing reduced segmental mobility and thus increasing miscibility (44).

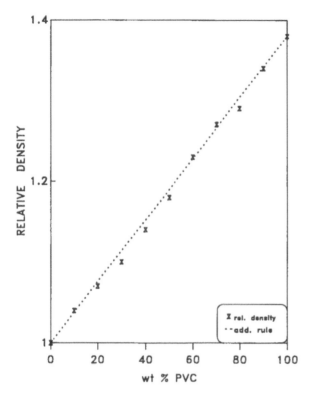

Figure 17 Relative density vs. composition for various PVC/ENR blends. (From Ref. 16.)

These observations are in agreement with those made by Kleiner et al. (39), who postulated that synergism in PS/PPO blends might be due to the observed increase in packing density on blending.

The observations from the present investigation tend to favor these views, as indicated by the peculiar shapes of the property composition curves (Figs. 15, 16, and 18) around the 70% PVC region, which coincide with the observed increase in density. Maximum synergistic effects in tensile strength occurred around this composition region. The sudden deviations from the observed trends of such properties around the 70% PVC composition range could also be a reflection of improved packing.

The composition dependence of the hardness of blends with increasing PVC content is illustrated in Fig. 18. At low PVC content, i.e., up to 40%, the hardness of blends is between 30 and 70 IRHD, which is the hardness range for soft vulcanized rubber (45). At these concentrations, the blends behave like typical filled

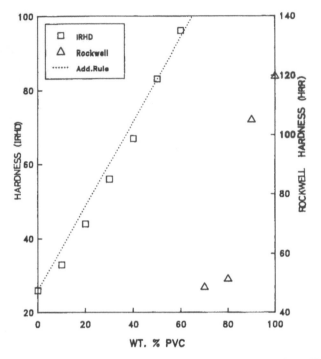

Figure 18 Hardness vs. composition plot for various PVC/ENR blends. (From Ref. 16.)

rubber systems in which ENR is the continuous matrix phase. The properties are therefore rubber-dominant with PVC acting as the filler.

A sharp rise in hardness is observed between 40% and 60% PVC accompanied by a changeover at about 50% PVC from predominantly rubberlike behavior to predominantly plastic behavior. At 60% PVC content, the hardness recorded is almost at the maximum of the IRHD or Shore A durometer hardness, and further measurement along this scale may not be meaningful. For this reason, the Rockwell hardness, more often used for rigid plastics and engineering materials, was employed for the PVC-dominant blends (46). The trend in the values of the Rockwell hardness (Fig. 18) further supports the tendency towards typical PVC behavior. A very sharp rise towards typical rigid PVC values at about 90% to 100% PVC is evident. Typical Rockwell hardness values for PVC reported in the literature are in the range of 115 HRR (46). The incidence of phase inversion also coincides with the changeover from property additivity to synergistic behavior (39). This behavior is again a proof of miscibility, as noted by Ratnam (45).

Figure 7 shows the variation flexural storage moduli and tan δ with temperature for ENR, the intermediate (50/50) blend, and PVC, respectively. Here, for comparison, the tan δ peaks were approximated as corresponding to T_g values. The T_g value of $-48.5°C$ determined for ENR is much lower than the $-20°C$ quoted in the literature (15,41). This rather low value is expected, since the ENR was equally subjected to the same mixing conditions (50 rpm at 150°C for 6 min) as the blends prior to the T_g measurement. Hence this decrease may be attributed to ENR, which, like natural rubber, is known to undergo rapid reduction in molecular weight on mastication (1), more so at high temperatures. On the other hand, the T_g for PVC premixed with stabilizer and subjected to similar mixing conditions is 84.2°C, which is in agreement with most values given in the literature (47). This is an indication that stabilization has been essentially effective (48). A single T_g intermediate between those of PVC and ENR was noted for the intermediate blend. This conforms with what is generally expected of any miscible polymer–polymer system (11,49,50). According to Utracki (11), the width of the transition zone is usually an indication of miscibility or compatibility. The blend shows a sharp transition, as indicated by a sharp drop in the storage modulus (which is almost parallel to those of the pure polymers). The sharp tan δ delta peaks, therefore, are clear evidences of high intermolecular interactions, which minimize interchain chemical hetrogeneity (51). This amply shows that the PVC/ENR blend is a miscible system.

It is worth noting, however, that the base of the tan δ peak of the intermediate blend is slightly wider than the bases of the constituent materials. This could be attributed to the presence of microheterogeneities in PVC/ENR systems arising from the different segmental environments. In addition, the broadening might also be due to the high gel content of ENR-50 (20). Similar observations have been encountered in earlier investigations involving PVC/ENR blends (10,41) and other blends of ENR (52).

Kleiner et al. (39) derived a specific form of the general equation cited by Nielsen (53) for one-phase binary mixtures:

$$E = E_1 X_1 + E_2 X_2 + \beta_{12}^E X_i X_2 \tag{1}$$

E_1, E_2, X_1, and X_2 represent the moduli and compositions of components 1 and 2, respectively, while β_{12}^E expresses the magnitude of deviation from nonlinearity.

A positive value of β_{12}^E represents a nonlinear synergism, while a negative one expresses nonlinear antagonism, which are criteria for miscibility-compatibility or incompatibility, respectively (39). β_{12}^E is readily estimated (54) from

$$\beta_{12}^E = 4E_{12} - 2E_1 - 2E_2 \tag{2}$$

where E_{12} represents the response of the blend (i.e., a 50/50 blend).

The low-temperature storage moduli at $-97°C$ for pure components and blends were then estimated using the Kleiner equation, Eq. (1) (39). At this temperature, both pure components and blends are essentially glassy, and it should therefore be feasible to employ the Kleiner relationship. The β_{12}^E was determined, using Eq. (2), as 2.628 GPa, which was then used to calculate the theoretical values estimated from Eq. (1). Since the value of β_{12}^E is positive, synergism is predicted. The experimental values obtained are compared with those estimated from both the rule of mixtures and the Kleiner relationship, given in Fig. 19. It is apparent that the measured values correspond well to those obtained using the Kleiner equation (39). A similar observation was reported by Fried et al. (38) in their investigations of PS/PPO systems. Similar observations have also been made with regard to PVC/NBR systems (12). This reaffirms that the storage moduli for miscible systems follow the second-order polynomial curve for two-component systems, which is a positive deviation from the composite rules. This observation again supports the quest to predict blend modulus based on the moduli of components for miscible systems (38,39).

The composition dependence of T_g in PVC/ENR blends is illustrated in Fig. 20. The T_g values were those that correspond to the loss modulus (E'') peak, since they coincide with T_g values determined by other methods, for example, free volume studies (55). The experimental values are compared with those obtained us-

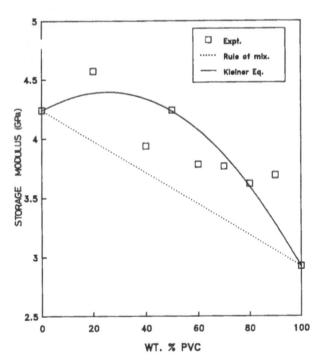

Figure 19 The composition dependence of low-temperature storage modulus ($-97°C$) for PVC/ENR blends. (From Ref. 17.)

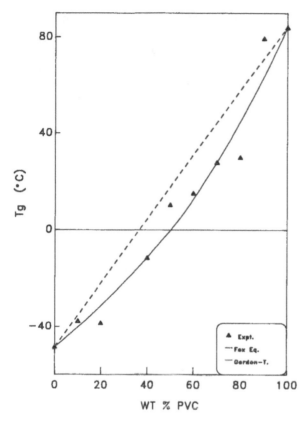

Figure 20 The composition dependence of glass transition temperature (T_g) for PVC/ENR blends. (From Ref. 17.)

ing the Fox equation and the Gordon–Taylor equation given by Eqs. (3) and (4), respectively.

$$\frac{1}{T_{gb}} = \frac{W_1}{T_{g1}} + \frac{W_2}{T_{g2}} \tag{3}$$

$$T_{gb} = \frac{T_{g1} + kW_2}{W_1(T_{g2} - T_{g1})} \tag{4}$$

where T_{gb} is the glass transition temperature of the blend, T_{g1} and T_{g2} are the glass transition temperatures of components 1 and 2, and W_1 and W_2 are the weight fractions of components 1 and 2.

As shown in the figure, a clear deviation from the Fox relationship is evident, while the scatter of experimental points tends to agree with the Gordon–Taylor equation. This finding is in accordance with those reported by Margaritis et al. (10) also working with PVC/ENR blends. Margaritis et al. also established a similar relationship in ENR/CPE systems (10) and ENR/thermoplastic and thermosetting resins (52). Varughese et al. (14) also observed a similar trend for ENR/plasticized PVC systems. A critical assessment of the experimental points also reveals that the property additivity line has been exceeded at very high PVC concentrations, thus describing an S-shaped curve. This type of dependence in miscible systems has always been linked to hydrogen bonding (10), which in PVC/ENR systems has been reported in the earlier part of this chapter.

V. THERMO-OXIDATIVE AGING

In the preceding section, it has been shown that ENR-50 is mutually miscible with PVC at all compositions. The blends showed synergism in mechanical and dynamic mechanical properties, as well as positive deviation from the rule of mixtures. Based on these attributes, ENR may be considered to have the potential to act as a substitute in some of the applications well suited to NBR, particularly those involving blending with PVC. However, it was soon realized that ENR vulcanizates exhibit poor aging properties. The unstable nature of ENR is believed to be due to the residual acidity that was inherited from the natural rubber modification with a peroxide (1,2,4–7). Residual acidity has been claimed to be responsible for the high gel content, high viscosity, and poor storage stability of ENR.

Conventional high-sulphur vulcanizates harden rapidly due to the acidic by-products containing sulphur, which induce epoxide ring-opening reactions via ether cross-links, resulting in an increase in T_g and cross-link density (1,2,56). For this reason, EV and semi-EV vulcanization systems are recommended for ENR. Subsequently, the use of bases, such as sodium carbonate and calcium stearate, has been recommended to provide a basic environment in ENR formulations (1,2,56). Thus it is of paramount importance to characterize the thermo-oxidative stability of any blend associated with ENR if their full potential is to be realized.

Long-term exposure to ambient conditions (27–30°C) in the laboratory conducted by Ishiaku et al. (57) revealed that PVC/ENR blends deteriorate rather rapidly. Flexible blends exposed to laboratory conditions lose their flexibility with time. A hard outer skin develops, which spreads gradually into the inner soft core and ultimately results in a hard and brittle material. This phenomenon is illustrated in Fig. 21. An extrudate of the 50/50 PVC/ENR blend exposed to ambient conditions for eight months was assessed qualitatively by applying a small strain in tension that was then released. Under tension, the hard outer skin cracked open as the strand expanded. After releasing the tension, the strand retracted and the cracks appeared as surface rings. This hardening at ambient conditions is an indication that the blend is unstable to degradation agents such as ultraviolet light and ozone. Recently, attempts to improve the oxidative stability of PVC/ENR with an amine-type antioxidant, 2,2,4-trimethyl-1,2-dihydroquinoline (TMQ), and a base, cal-

1   mm

Figure 21 Bright field image of 50/50 PVC/ENR extrudate strained in tension and released after storage at room temperature for eight months. Darker rings are cracks. (From Ref. 57.)

cium stearate, were reported (20). The effect of both types of additives on the thermal oxidative behavior of PVC/ENR was investigated with studies involving mechanical properties and FTIR. Optimum quantities of the additives were determined by observing the trends obtained from mechanical properties.

Figure 22 shows the effect of TMQ on the tensile strength of PVC/ENR blends before and after thermo-oxidative aging. In the absence of TMQ, an enormous drop in tensile strength was observed after thermo-oxidative aging at 100°C for 168 h. Only about 30% of the original strength was retained. The poor retention in tensile strength may be a result of excessive degradation in the ENR blend, as revealed later by FTIR studies.

From Fig. 22 it can be seen that the addition of TMQ did not have any noticeable influence on the tensile strength of either blend in the unoxidized state. The effect of the antioxidant is clearly shown after oxidative aging by the marked increase in tensile properties with increase in the concentration of TMQ. With addition of 3 phr of TMQ, the retention in tensile strength increased by about 40% from 30 to 70%, as shown in Table 3a. This shows that the antioxidant considerably arrested thermo-oxidative degradation by mopping up free radicals produced during oxidation. Figure 23 shows the variation in elongation at break with TMQ concentration for PVC/ENR blend.

Again, before aging, the EB of the blend was not sensitive to TMQ addition. In the unstabilized PVC/ENR blend, thermo-oxidative aging effected a drastic reduction in elongation at break from about 300% to less than 40%. This could be due to the thermo-oxidation of the rubber chain, which results in extensive chain scission. The addition of TMQ to the ENR blend increased the retention of elon-

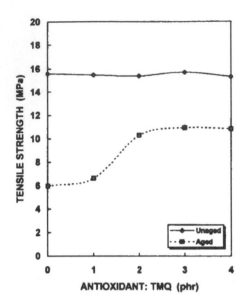

Figure 22 Effect of TMQ on the tensile strength of blends of PVC with ENR before and after aging at 100°C for 168 h. (Redrawn from Ref. 20.)

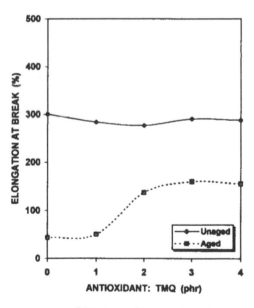

Figure 23 Effect of TMQ on the elongation at break of blends of PVC with ENR before and after aging at 100°C for 168 h. (Redrawn from Ref. 20.)

Table 3 Retention of Mechanical Properties and the Effect of the Additives TMQ and Ca(St)$_2$ after Aging at 100°C for 168 h

(a) Effect of TMQ

TMQ (phr)	Retention (%)		
		ENR Blend	
	Tensile strength	Elongation at break	Tear strength
0	32	15	68
1	43	16	67
2	67	53	67
3	69	56	73
4	71	55	85

(b) Effect of calcium stearate

Ca(St)$_2$ (phr)	Retention (%)		
		ENR blend	
	Tensile strength	Elongation at break	Tear strength
0	39	15	68
1	29	11	81
3	32	12	86
5	35	14	98
7	33	18	89

(c) Effect of Ca(St)$_2$ in the presence of TMQ

TMQ (phr)	Ca(St)$_2$ (phr)	Retention (%)	
		ENR blend	
		Tensile strength	Elongation at break
3	0	70	57
3	1	65	70
3	3	66	69
3	5	64	70
3	7	70	72
3	9	70	69

Source: Ref. 20.

gation at break from 15% to more than 50% with the addition of 2 phr of TMQ. Further addition did not have much effect (Table 3a).

The effect of TMQ concentration on the tear strength of PVC/ENR blend is depicted in Fig. 24. As expected, thermo-oxidative aging has resulted in a substantial reduction in the tear strength. However, from Table 3b it is clear that the addition of TMQ has improved the retention in the tear strength of the blends.

The incorporation of a base, i.e., $Ca(St)_2$, did not have any appreciable effect on the tensile strength of the unaged PVC/ENR blends (Fig. 25). Interestingly, after aging, $Ca(St)_2$ showed contrasting effects. Increasing the concentration of $Ca(St)_2$ did not improve the tensile strength of the blend. This reveals the inability of $Ca(St)_2$ to curb thermo-oxidative degradation in PVC/ENR blends by preventing acid-catalyzed ring-opening reactions. This observation is also supported by FTIR results, which reveal extensive main-chain scission during thermo-oxidation in the presence of $Ca(St)_2$. Thermo-oxidation leads to the lowering of molecular weight, and hence the observed reduction in mechanical properties.

A similar observation was also noted in the case of EB. Increasing the $Ca(St)_2$ concentration had no significant effect on the EB of the unaged PVC/ENR blends (Fig. 26). After thermo-oxidative aging, $Ca(St)_2$ addition did not have any positive influence on the aging properties of either blend. The effect of combining both TMQ and $Ca(St)_2$ on the tensile strength and EB of PVC/ENR blend is illustrated in Figs. 27 and 28, respectively. The addition of $Ca(St)_2$ in the presence of the an-

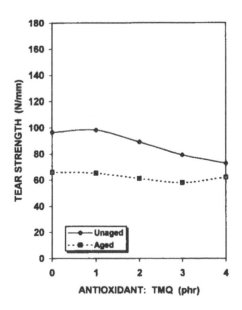

Figure 24   Effect of TMQ on the tear strength of blends of PVC with ENR before and after aging at 100°C for 168 h. (Redrawn from Ref. 20.)

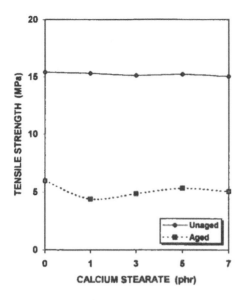

Figure 25 Effect of calcium stearate on the tensile strength of blends of PVC with ENR before and after aging at 100°C for 168 h. (Redrawn from Ref. 20.)

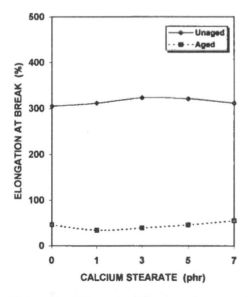

Figure 26 Effect of calcium stearate on elongation at break of blends of PVC with ENR before and after aging at 100°C for 168 h. (Redrawn from Ref. 20.)

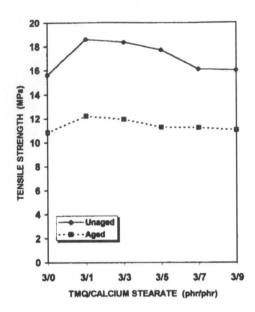

Figure 27 Effect of calcium stearate in the presence of 3 phr of TMQ on the tensile strength of blends of PVC with ENR before and after aging at 100°C for 168 h. (Redrawn from Ref. 20.)

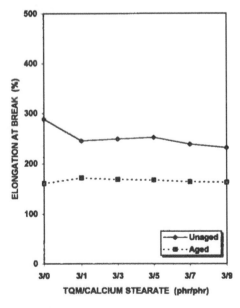

Figure 28 Effect of calcium stearate in the presence of TMQ on the elongation at break of blends of PVC with ENR before and after aging at 100°C for 168 h. (Redrawn from Ref. 20.)

tioxidant improved the tensile properties of the unaged blends, although the tensile strength was effectively unchanged for the 3/9 addition. This is an extreme case of the general observation that further addition, after the optimum amount of $Ca(St)_2$ has been added, leads to a decrease in tensile strength. The ENR blend required just 1 phr of $Ca(St)_2$ for optimum effect. After aging, the tensile strengths were more or less similar to those of blends containing only the antioxidant. In the case of EB, a retention of more than 70% was attained for the ENR blend (Table 3c). This indicates that the combination of TMQ and $Ca(St)_2$ has an advantageous effect as the retention of the PVC/ENR approaches that of PVC/ENR blend under similar conditions (20). The explanation for this observation could be that in the presence of the antioxidant, the $Ca(St)_2$ is available to neutralize HCl, or any other organic acid that may be formed, and hence prevent acid-catalyzed ring-opening reactions in ENR-50, in addition to acting as a filler. It may therefore be appropriate to infer that an antioxidant and a base are required to protect the ENR blend from oxidative aging. A ratio of 3 : 1 for the antioxidant and base, respectively, is required for optimum tensile properties of the unoxidized blend and retention in the aged blends.

Earlier in this chapter, FTIR has been shown to be a powerful tool in assessing the interaction between PVC and ENR. In their investigation of the thermo-oxidative aging of ENR, Poh and Lee (58) demonstrated that FTIR is also capable of providing some supportive evidences to explain the effect of thermo-oxidative aging on ENR. From the difference spectra presented in Fig. 29, it is obvious that excessive thermo-oxidative degradation has occurred in the plain blend, i.e., the blend without stabilizing additive (Fig. 29a). Enormous main-chain scission is indicated by the reduction in the C—H stretching vibrations of the main-chain —CH_2— at 2924 cm^{-1} (asymmetric) and 2860 cm^{-1} (symmetric). The absorption of 2860 cm^{-1} may have shifted from 2857 because of the influence of the strong C—H stretch of the main-chain —CH_3. This may be expected, since scanning was carried out with a resolution of 4 cm^{-1}, which permits an experimental error of $\pm$ 4 cm^{-1}. Chain scission is also indicated by the reduction in —CH_2— scissoring and rocking vibrations at 1460 cm^{-1} and 700 cm^{-1}, respectively (59). The decrease in main-chain —CH_3 is indicated by the immense reduction in peaks at 2975 cm^{-1} and 2860 cm^{-1} for the C—H stretch, coupled with reductions in the C—H bending vibrations at 1447 cm^{-1} and 1377 cm^{-1}. Chain scission may have occurred in both polymers in the blend, as shown by the reduction of the out-of-plane C—H bending vibration of the double bond at 835 cm^{-1}. Double bonds provide the initial sites for thermo-oxidative degradation, acting as weak spots (60). There is reason to believe that both ENR and PVC possess these weak spots, which are either present initially or develop during oxidation.

In ENR-50, half the double bonds originally present in NR are retained. Consequently, ENR-50 has been shown to degrade through reaction mechanisms and to yield products identical with those of natural rubber (NR) (58). Furthermore, degradation of the strained epoxide ring is indicated by the negative peaks at 1254,

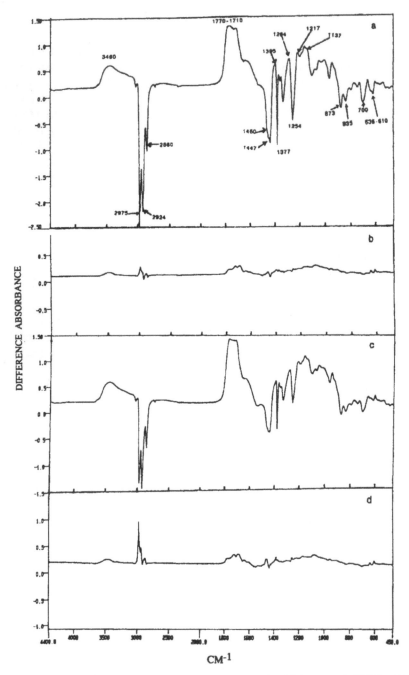

Figure 29 Effect of oxidative aging on PVC/ENR blend at 100°C for 168 h. (a) Plain blend; (b) blend containing TMQ; (c) blend containing Ca(St)$_2$; (d) blend containing combination of TMQ and Ca(St)$_2$, 3 phr each. (From Ref. 20.)

873 and 835 cm^{-1}. This process is believed to proceed via the formation of ether cross-links, particularly in an acidic environment (1,2). Both sources mentioned above lead to the formation of ethers, esters, and alcohols, as indicated by absorptions within the fingerprint region. The increment in the absorption peak at 1770–1710 cm^{-1} is indicative of carbonyl groups from carboxylic acids, aldehydes, and ketones. This is supported by the development of the broad hydrogen bonding peak at 3460 cm^{-1}, the C—H bending of aldehyde at 1395 cm^{-1}, the ketone C—C—C stretching at 1284 cm^{-1}, and the C—CO—C bending of aliphatic ketone at 1230–1100 cm^{-1}. The characteristic aliphatic ether formation is shown at 1137 cm^{-1}. The formation of alcohols is indicated by the increment of the C—O stretching at 1217 cm^{-1} and O—H bending at 1395 cm^{-1}. The band at 1284 cm^{-1} could be associated with C—O stretching.

The degradation of PVC in the blend is shown by the reduction in absorptions at 636–610 cm^{-1} and 1254 cm^{-1}, signifying the destruction of C—Cl and CH$_2$Cl bonds, respectively. The elimination of HCl leads to the production of double bonds, which are subsequently destroyed as thermo-oxidative degradation proceeds. The addition of antioxidant (TMQ) reduced the thermo-oxidative degradation considerably, as shown in Fig. 29b. This observation complements the improvement in mechanical properties noted earlier. There is an indication of cross-link formation (—CH$_2$—) as shown by the increase in absorbance at 2800–2900 cm^{-1}. This may be due to the presence of antioxidant, which reacts with the available oxygen and thus reduces the amount of O$_2$ that reacts with the blend through a free radical process. The reduction in the amount of O$_2$ encourages the formation of cross-links (60).

It is clearly shown in Fig. 29c that Ca(St)$_2$ has not been able to protect the epoxide ring, especially in the presence of PVC. As shown by the IR spectra, PVC has degraded, and this may have proceeded via the production of HCl (a strong acid). HCl could react with Ca(St)$_2$ to give stearic acid and CaCl$_2$ (61). The presence of HCl coupled with an organic acid will accelerate the epoxide ring-opening reaction to yield free radicals. This will lead to autocatalyzed decomposition of the chain via ether cross-links. This reaction may have contributed considerably to the main-chain scission in the ENR blend. The addition of Ca(St)$_2$ in the presence of the antioxidant (Fig. 29d) gave a spectrum similar to that containing the antioxidant alone, except that cross-link formation is more prominent.

VI. PLASTICIZATION

Plasticization is known to increase the workability, flexibility, and stability of plastics, particularly PVC, which is unique in its acceptance of plasticizers (62). Consequently, the PVC industry consumes more than three-quarters of the total plasticizer production. The effect of plasticizer on the PVC/ENR system has been reported by Varughese et al. (14). They observed a remarkable reduction in the width of the glass transition region, which is intermediate between those of the blend components.

Plasticizers such as di-ethylhexylphthalate (DOP), epoxidized soya oil (ESO), and etherthioether (ETE) have recently been evaluated by Ishiaku et al. (63) for potential application in PVC/ENR thermoplastic elastomers (TPE). The effects of these plasticizers on the mechanical and thermo-oxidative aging (TOA) of PVC/ENR was studied.

Figure 30 shows the effect of plasticizer content on the tensile strength of PVC/ENR blend. As expected, irrespective of types of plasticizer, tensile strength drops significantly with an increase in plasticizer concentration. ETE initially gives better strength at lower concentrations (below 15 phr) followed by DOP, then ESO. However, this trend is reversed at higher concentrations. The rate of decrease in the tensile strength for the ETE-plasticized blends is so high that the tensile strength is too low at higher concentrations.

The decrease in tensile strength with increasing plasticizer content may be due to the masking of the centers of force holding the PVC molecules together if viewed from the gel concept, or it may be due to the lubrication effect, which makes it easier to pull chains apart according to the lubricity theory, which mostly applies to crystalline polymers. Both concepts have been applied to PVC (64).

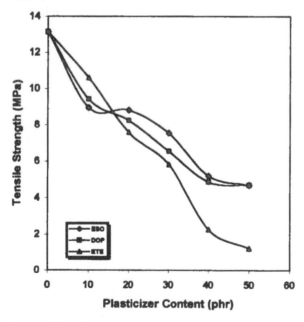

Figure 30 The effect of plasticizer concentration on the tensile strength of PVC/ENR TPE. (From Ref. 63.)

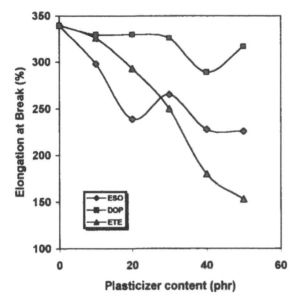

Figure 31 The effect of plasticizer concentration on the percentage EB of PVC/ENR TPEs. (From Ref. 63.)

In the case of EB, it can be seen (Fig. 31) that DOP-plasticized blends show higher EB at all concentrations, while other plasticizers do not perform so well, particularly not ETE, which yields values well below acceptable levels at higher concentrations. The effect of plasticizer content on the tear strength of PVC/ENR blend is illustrated in Fig. 32. Again, as expected, the incorporation of both types of plasticizer, i.e., DOP and ESO, has reduced the tear strength significantly. This could be due to excessive plasticization, which makes it easier to pull chains apart.

The extent to which a plasticizer is capable of lowering T_g is a measure of plasticizer efficiency (64,65). Figure 33 shows the effect of plasticizer concentration on the T_g of PVC/ENR blend. In this respect, ESO is superior to DOP, as indicated by the lower T_gs shown by the ESO-plasticized blends at all concentrations.

In the preceding section, attempts have been made to improve the thermo-oxidative stability of PVC/ENR blends via the incorporation of an antioxidant, i.e., TMQ, and a base, i.e., $Ca(St)_2$. Further investigation to improve the stability of the blend has recently been reported (63). This involves studying the effect of plasticizers, i.e., DOP, ESO, and ETE, on the thermo-oxidative aging of PVC/ENR. Thermo-oxidative aging studies were carried out by evaluating the mechanical properties, viz: tensile strength, tear before aging, and tear after aging in an air oven at 80°C for 168 h.

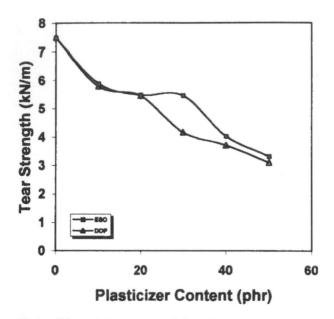

Figure 32 A comparison of the effects of varying DOP and ESO concentrations on the tear strength PVC/ENR TPEs. (From Ref. 63.)

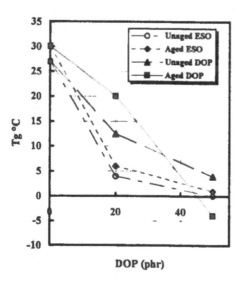

Figure 33 A comparison of the effects of varying concentrations of DOP and ESO on the T_gs of 50/50 PVC/ENR TPEs before and after thermo-oxidative aging. (From Ref. 63.)

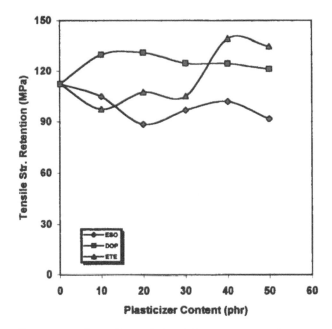

Figure 34 The effect of plasticizer concentration on percentage retention of tensile strength. (From Ref. 63.)

Figure 34 shows the effect of plasticizer concentration on the retention of tensile strength of the blend. The effect of TOA is such that DOP-plasticized blends show greater consistency in tensile strength retention at all concentrations. The retention is higher than 100% at all concentrations, indicating better tensile strength after aging. This synergistic effect predicts good service life (66). ETE-plasticized blends show superior retention at high concentrations (40 to 50 phr). This trend is expected since ETE is specifically recommended for high-temperature applications.

In the case of EB, it can be seen from Fig. 35 that the percentage of retention generally increases with increasing concentration for the DOP and ETE-plasticized blends, whereas a continuous decrease is the case for ESO. The origin of the instability of the ESO-plasticized blends can be traced to the study of the effect of epoxidized plasticizer on PVC by Anderson and McKenzie (67) based on small molecular models. They found that balanced molar quantities of the epoxide and mixed metal carboxylates were required for effective and, to an extent, synergistic stabilization. When the concentration of the epoxide is higher, heavy metal chlorides result, which in turn catalyze degradation. In a previous study, the present authors used this concept to elucidate the instability of PVC/ENR systems, which could be attributed to the high concentration of the epoxide groups (57). DOP curbs degradation by the dilution effect, which effectively masks the effect

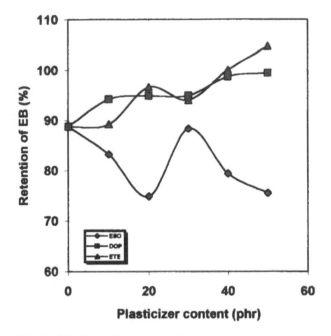

Figure 35   The effect of plasticizer concentration on the percentage retention of EB. (From Ref. 63.)

of the epoxide groups. On the contrary, ESO increases epoxide concentration, which accounts for the instability of ESO-plasticized blends. Consequently this observation confirms the earlier hypothesis advanced by the authors (57).

The tear strength of the blends after aging is better than before aging, as indicated by a very good retention of tear strength shown in Fig. 36. Tear strength retention for ESO is comparable to that of DOP and even better at lower concentrations. However, DOP exhibits superior retention at higher concentrations. The superior retention of tear strength at low ESO concentrations may be accounted for by cross-link formation via ether bonds, which increases tear strength.

Figure 33 also shows the effect of thermo-oxidative aging on the T_g of PVC/ENR blends. It can be seen that aged ESO-plasticized blends show higher T_gs than the unaged ones. This could also signal degradation. DOP-plasticized blends also show increase in T_g with thermo-oxidative aging, albeit at lower concentrations only. At higher loadings of DOP, blend softening accompanies thermo-oxidative aging, as shown by the sharp lowering of T_g.

The increased T_g after thermo-oxidative aging of the ESO blends is consistent with increased modulus (Fig. 37) and hardness (Fig. 38) earlier discussed, which have been associated with PVC/ENR blend degradation (20,57). The instability problem is also evidenced by poor tensile properties, particularly with EB (Fig.

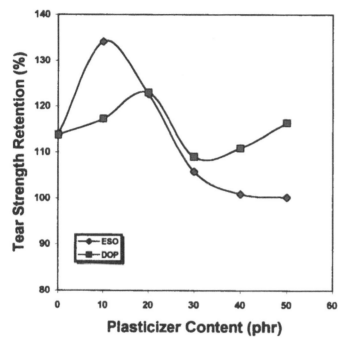

Figure 36 A comparison of the tear strength retention (%) after thermo-oxidative aging for ESO- and DOP-plasticized PVC/ENR TPEs. (From Ref. 63.)

31). Poor tensile properties can be accounted for by main-chain scission, which can be inferred from FTIR studies.

Hardness increase in the presence of ESO plasticizer after thermo-oxidative aging is shown in Fig. 38. Hardening is an indication of degradation in PVC/ENR systems, as demonstrated earlier (20,57). This observation also agrees with that for tensile properties, whereby an increase in modulus is accompanied by loss in tensile strength and EB. ETE-plasticized blend also shows considerable hardening as a result of TOA. On the contrary, the DOP blend does not harden under similar conditions (66).

Based on the above discussion, it can be inferred that each type of plasticizer has its own potential and its own limitations. For instance, the incorporation of DOP generally decreases the mechanical properties of the blends, although it simultaneously stabilizes the blends against thermo-oxidative degradation.

The improvement in the degradation properties of PVC/ENR with DOP incorporation implies that the presence of DOP is effective in altering the epoxide/heavy metal salt balance. According to the gel theory of plasticization, the plasticizer breaks the loose attachments in PVC and masks the centers of force, consequently reducing the number of attachments in the dynamic equilibrium of

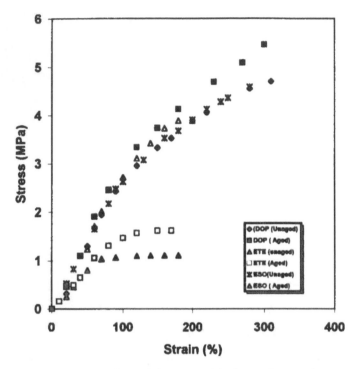

Figure 37   The effect of thermo-oxidative aging on the stress–strain properties at 50 phr concentration of plasticizers. (From Ref. 63.)

solvation and desolvation (65). This action is known to confer stability to PVC
formulations, and thus the incidence of release of HCl is reduced. One of the three
main requirements for plasticization is compatibility and ease of mixing (64).
Based on the narrowing of the glass transition region observed by Varughese et al.
(14) in DOP-plasticized PVC/ENR blend, it can be inferred that DOP is intimately
miscible with ENR, as is the case with PVC. Therefore DOP is also capable of
plasticizing ENR and will break the loose attachments and mask the centers of
force and thus reduce the efficacy of the epoxide groups due to the dilution effect.
Consequently, as the concentration of DOP increases, the effective concentration
of the epoxide decreases, until a stoichiometric balance between epoxide/metal
carboxylate ratio is known to confer synergistic effects and hence improves the
tensile strength of the aged blends in contrast to the unaged ones at 50 phr of DOP.
Although DMA studies reveal that ESO exhibits better plasticizer efficiency than
DOP in terms of lowering the T_gs of the blend, the former suffers a major drawback in that it exhibits poor retention in properties after thermo-oxidative aging,
as indicated by hardening and embrittlement of the blend. ETE-plasticized blends,
on the contrary, show a good retention in properties after thermo-oxidative aging.

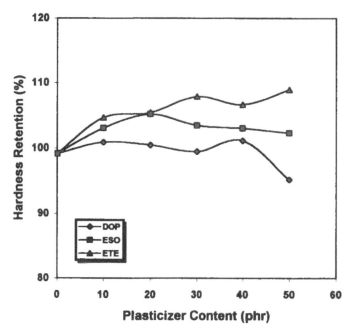

Figure 38 The hardness of PVC/ENR TPEs for different plasticizers at different concentrations. (From Ref. 63.)

However, the mechanical properties of the unaged blends is much poorer than those of ESO- and DOP-plasticized blends. The mechanisms of DOP stabilization in PVC/ENR blends have been investigated by Ishiaku et al. (57) using the FTIR technique.

Figure 39a shows the difference spectrum of the unplasticized PVC/ENR blend, i.e., the spectrum obtained after subtracting the spectrum of the unaged blend from that of the aged blend. The reduction in absorption around 2900 to 2800 cm^{-1} for C—H vibrations of main chain —CH$_2$— indicates main-chain scission. Reduction in main-chain —CH$_2$— is also shown at 1447 and 1377 cm^{-1}. The degradation of the strained epoxide ring is indicated by the reductions in absorption at 873 and 835 cm^{-1}. PVC/ENR degradation leads to the formation of ethers, esters, and alcohols, as indicated by the absorptions within the fingerprint region. The peak prominent at 1770–1710 cm^{-1} is indicative of carbonyl groups from carboxylic acids, aldehydes, and ketones. This is supported by the hydrogen bonding peak (59) at 3435 cm^{-1}.

With the addition of 20 phr of DOP (Fig. 39b) it is obvious that the main-chain scission has been arrested, and network formation is indicated by the increase in absorptions around 2900–2800 cm^{-1}, although some degradation of the epoxide ring still occurs, as indicated in Fig. 39b and the higher modulus of the

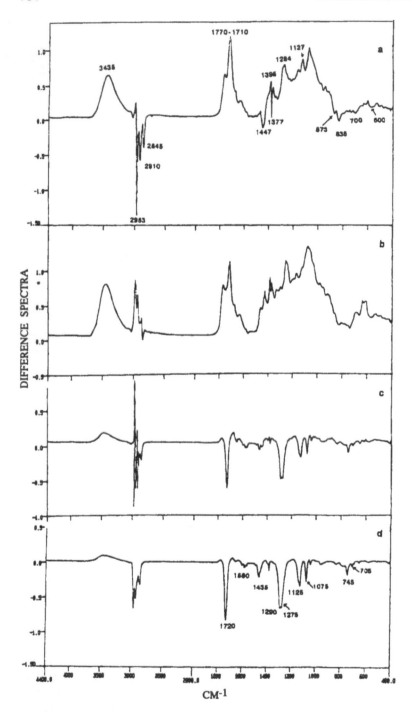

aged blend in Fig. 40. This observation is consistent with the remarkable retention of mechanical properties. Further addition of DOP has resulted in the prevention of epoxide degradation, thus stabilizing the PVC/ENR blend as shown in Figs. 39c and 39d, which are for 30 and 50 phr of DOP, respectively. The negative peaks in Figs. 39c and 39d, which increase in that order, indicate DOP migration. DOP migration is indicated by the absence of absorbance due to degradation products. Instead, the negative peaks exactly match the DOP spectrum as enumerated below. Such negative peaks signify decrease in concentration since no bonds are being broken.

The aromatic C—H and aliphatic —CH_2— and —CH_3 units are indicated at 3027–2920 cm^{-1}. The C=C of the conjugated aromatic ester is prominent at 1720 cm^{-1}, while the peaks at 1580, 1470, and 1435 cm^{-1} indicate the C=C aromatic ring stretch. The most prominent band at 1290–1275 cm^{-1} indicates conjugated aromatic ester COO groups. The bands at 1125 and 107 cm^{-1} signify aromatic aryl —O—CH_2 units, while aromatic C—H bending is indicated at 745 and 705 cm^{-1} (59).

As a control, a set of spectra obtained from a thermally stable system containing the antioxidant TMQ and varying amounts of DOP are presented in Fig. 41. Figure 41a shows the spectrum of PVC/ENR stabilized with TMQ (20). As in the previous case, degradation is effectively controlled by the absence of absorptions due to degradation products. With the addition of 20 phr of DOP (Fig. 41b), degradation is controlled, and there is no plasticizer migration, indicating that 20 phr is the optimum amount of DOP required to forestall DOP migration; it is also consistent with adequate retention of mechanical properties. This indicates that the addition of TMQ in the presence of DOP further stabilizes the blend. In the presence of excess plasticizer, migration occurs in adequately stabilized systems as shown in Figs. 41c and 41d containing and 30 and 50 phr of DOP respectively.

VII. DYNAMIC VULCANIZATION

Dynamic vulcanization plays a significant role in thermoplastic elastomer (TPE) technology by producing thermoplastic vulcanizates (TPVs), which possess a unique combination of properties. TPVs constitute a class of TPEs of rubber/plastic polymer blend in which the rubber is fully vulcanized. Thus products that look,

---

Figure 39 The difference absorbance spectra showing the effect of DOP and thermo-oxidative aging on FTIR spectra of 50/50 PVC/ENR blends. (a) 0 phr DOP; (b) 20 phr DOP; (c) 30 phr DOP; (d) 50 phr DOP. Thermo-oxidative aging was done at 80°C for 168 h, and the spectra of unaged samples were subtracted from those of the aged ones. (From Ref. 57.)

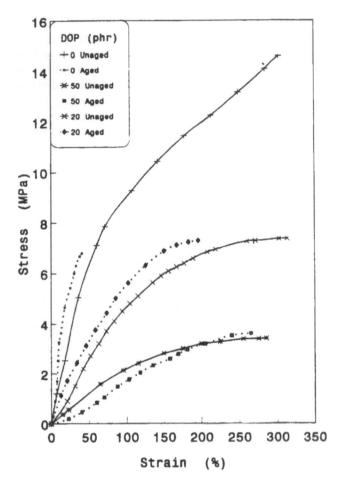

Figure 40 The effects of DOP and thermo-oxidative aging on the stress/strain properties of 50/50 PVC/ENR blends. (From Ref. 57.)

feel, and perform like vulcanized rubber, and yet are processible as thermoplastics, can be manufactured. Dynamic vulcanization confers an increase in tensile strength and modulus, a pronounced decrease in set, and a considerable reduction in swelling by oils (68). Vulcanization of the rubber phase also enhances the retention of the properties at elevated temperatures.

Figure 41 The effect of TMQ and thermo-oxidative aging on the FTIR spectra of DOP-plasticized 50/50 PVC/ENR blends. (a) 0 phr DOP; (b) 20 phr DOP; (c) 30 phr DOP; (d) 50 phr DOP. (From Ref. 57.)

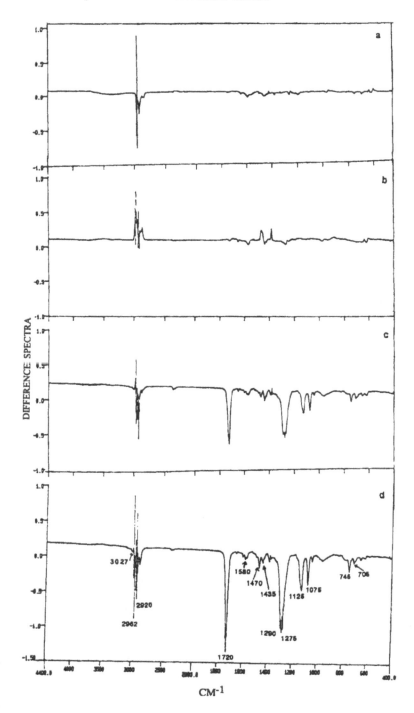

DIFFERENCE SPECTRA

CM⁻¹

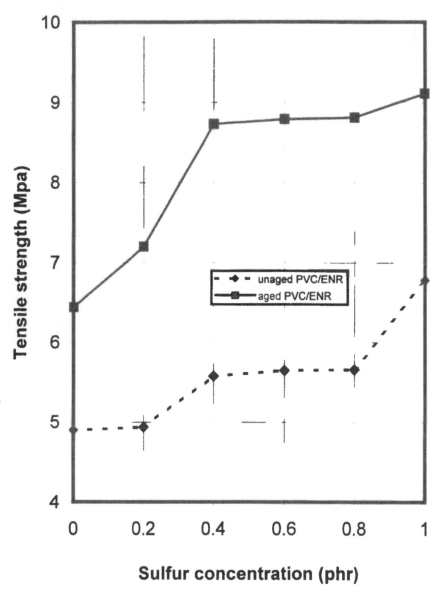

Figure 42 The effect of sulfur loading on tensile strength of unaged and aged PVC/ENR TPEs.

The effect of sulfur loading on the tensile properties of the PVC/ENR TPEs is presented in Figs. 42–44. In Fig. 42, it can be seen that there is a moderate increase in tensile strength with sulfur concentration. The swelling index, on the other hand, decreases continuously with increase in sulfur concentration, as shown in Fig. 45. This indicates that cross-link density increases with an increase in sulfur concentration. Thus higher applied stress is required to cause rupture. ENR like NR is capable of undergoing strain-induced crystallization. Thus the tensile strength is governed by the degree of crystallization attained as the break point is approached. The increase in tensile strength with increase in sulfur loading can be related to the ability of ENR to crystallize when stretched due to the orientation of the intermolecular chains of ENR. The formation of crystals will increase the number of intermolecular network chains per unit volume in the direction of the extension. Thus the crystallized samples become firmer, and this accounts for the enhancement in tensile strength. Coran and Patel (69) also observed a similar increase in the tensile strength of TPEs. The same explanation could also be used to account for the effect of sulfur concentration on the modu-

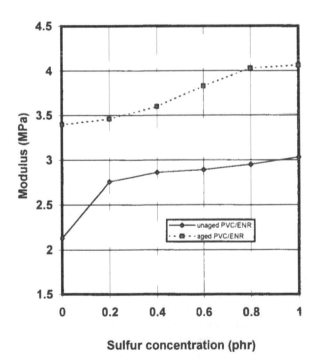

Figure 43   The effect of sulfur concentration on modulus at 100% elongation (M100) of unaged and aged PVC/ENR TPEs.

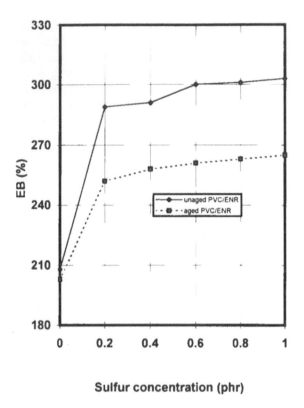

Sulfur concentration (phr)

Figure 44 The effect of sulfur loading on EB of unaged and aged PVC/ENR TPEs.

lus at 100 elongation (M100) as shown in Fig. 43. The incorporation of 0.2 phr sulfur has to a certain extent increased the M100 of PVC/ENR TPEs. Since M100 is directly proportional to the number of cross-links formed (86), the data obtained provide a suitable means for assessing the cross-link density. This notion is strongly supported by the decrease in swelling index in Fig. 45. The percentage of elongation at break (EB) initially increases sharply with 0.2 phr loading of sulfur, as shown in Fig. 44. This is followed by a gradual increase with subsequent addition of sulfur. The trend is expected, since EB is known to increase with increase in cross-link density for lightly vulcanized rubber (70). The continuous rise in EB with increase in cross-link density indicates that cross-link formation in the ENR phase of the blend has not yet peaked off. This suggests that the ENR in the blend is slightly cross-linked. Tear strength is also increased with an increase in sulfur loading, as shown in Fig. 46. Tear strength is a toughness-related property that is

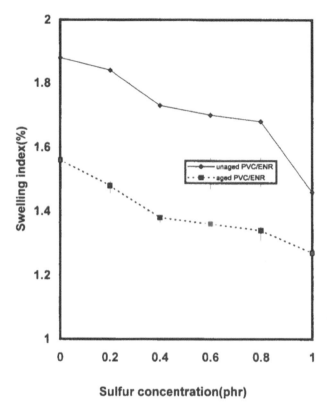

Figure 45 The effect of sulfur concentration on swelling index of unaged and aged PVC/ENR TPEs.

known to increase with an increase in cross-link density for lightly cross-linked elastomers.

The continuous increase in hardness with increasing sulfur concentration shown in Fig. 47 is also expected, since hardness is known to be strongly related to cross-link density (71). The increase could be attributed to the shorter, and therefore more rigid, network chains as the level of cross-linking increases. In the preceding section it was stressed that the uncross-linked PVC/ENR blends suffered a major drawback in that they exhibited poor thermo-oxidative stability. Several attempts have been made to improve the aging behavior of the blends, and these include the incorporation of a plasticizer, di-2-diethylhexylphthalate (DOP) (57), an antioxidant, 2,2,4-trimethyl-1,2-dihydroquinoline (TMQ), and a base such as calcium stearate (20). Plasticizer addition was found to be particularly ef-

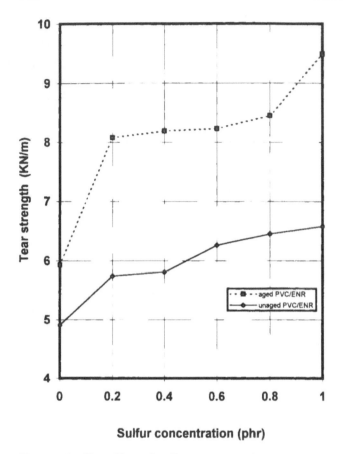

Figure 46   The effect of sulfur concentration on tear strength of unaged and aged PVC/ENR TPEs.

fective in curbing degradation (57). Thus it is interesting to investigate the effect of dynamic vulcanization on the thermo-oxidative stability of PVC/ENR TPEs.

The effect of thermo-oxidative aging on tensile strength is depicted in Figs. 42–44. The aging process has remarkably increased the tensile strength of the dynamically vulcanized PVC/ENR TPEs. This observation is in agreement with a previous work reported by Ishiaku et al. (57). The significant increase in tensile strength is accompanied by a reduction in elongation at break and an increment in hardness, as shown in Figs. 44 and 47, respectively. The formation of new intermolecular bonds will obviously lead to an increase in cross-link density. Thus increase in modulus and reduction in swelling index, as shown in Figs. 43 and 45, respectively, can be expected. The same reasons account for tensile strength en-

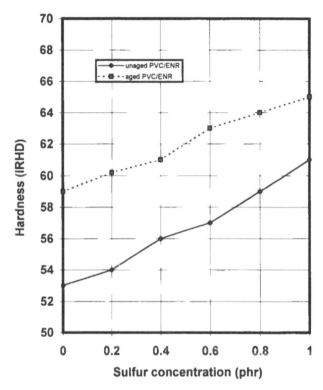

Figure 47 The effect of sulfur concentration on hardness of unaged and aged PVC/ENR TPEs.

hancement. A possible influence by the accelerator MBTS, which is structurally related to the antioxidant 2-mercaptobenzimidazole, is believed to prevent the oxidation of the plasticizer (DOP). The overall improvement in strength and modulus of PVC/ENR TPEs with sulfur loading clearly indicate that some microstructure changes have taken place after aging. As the cross-link density increases, the resistance of a material towards external deformation will increase. The increase in cross-link density, as indicated by swelling index data, is characteristic of postcuring, which arises from the formation of more sulfur cross-links after vulcanization. This is related to the shortening of polythioether bridges with the formation of new cross-links that have taken place as a result of the aging process (71).

Figure 46 shows the effect of sulfur loading on the tear strength of PVC/ENR after aging. It can be seen that the tear strength increases with an increase in sulfur loading. As before, this can be attributed to the increase in cross-link density as a result of thermo-oxidative aging. The effect of postcuring should be taken into consideration to account for the increase in cross-link formation.

VIII. CONCLUDING REMARKS

Some of the important conclusions that can be drawn from this report are as follows.

1.   The properties of PVC/ENR blends are strongly influenced by the compounding variables, viz., mixing time, temperature, and rotor speed. By utilizing the correlation between these mixing parameters, good blends can be easily procured.

2.   Miscibility in PVC/ENR systems has been reaffirmed through studies involving mechanical, physical, rheological, and morphological properties.

3.   PVC/ENR blends are prone to thermo-oxidative aging (TOA). This has been attributed to initial PVC degradation, which releases HCl that catalyzes the ring-opening reaction of the epoxide group. The combination of an optimum amount of antioxidant, 2,2,4-trimethyl-1,2-dihydroquinoline (TMQ), and a base, calcium stearate $[Ca(St)_2]$, has been proven to be effective in curbing the TOA of the blends. Further stabilization of the blends has been achieved by incorporating an optimum amount of diethylhexylphthalate (DOP) in conjunction with TMQ.

4.   Significant enhancement in mechanical properties and improvement in resistance against TOA has also been observed by incorporating sulfur into the plasticized PVC/ENR TPEs. This has been attributed to the increase in cross-link density of the dynamically vulcanized TPEs.

ACKNOWLEDGMENTS

The authors wish to express their sincere appreciation to their colleagues M. Nasir, B. T. Poh, and I. Hanafi, and to the talented students D. Ng, A. Mousa, and A. Shaharum for their contributions, which facilitated the writing of this chapter. Financial assistance in the form of a research grant by the Universiti Sains Malaysia and the Ministry of Science, Technology and the Environment is gratefully acknowledged.

REFERENCES

1.   I. R. Gelling and M. Porter. Epoxidized natural rubber. In: A. D. Roberts, ed. Natural Rubber Science and Technology. Oxford: Oxford Univ. Press, 1988, pp. 359–456.

2. A. S. Hashim and S. Kohjiya. Preparation and properties of epoxidized natural rubber. Kautsch. Gummi Kunstst. 46:208–281, 1993.
3. R. Pummerer and P. A. Burkard. Ber. Dtsch. Chem. Ges. 55:3458, 1922.
4. I. R. Gelling. Epoxidized natural rubber. UK Pat. 2, 133:692, 1982.
5. I. R. Gelling. Sulphur vulcanization and oxidative ageing of ENR. Rubb. Chem. Technol. 58:243–257, 1984.
6. S. T. Wong, L. M. Ong, and I. R. Gelling. Storage stability of epoxyprene. Kautsch. Gummi Kunstst. 45:284–286, 1992.
7. C. S. L. Baker, I. R. Gelling, and R. Newal. Epoxidized natural rubber. Rubb. Chem. Technol. 58:67–78, 1985.
8. Kumpulan Guthrie Sdn. Bhd. Malaysia. Epoxyprene-epoxidized natural rubber. Technical Specification Document ENR-01 to 07, 1990.
9. A. G. Margaritis and N. K. Kalfoglou. Miscibility of chlorinated polymers with epoxidized poly(hydrocarbons): 1. ENR/PVC blends. Polymer 28:497–502, 1987.
10. A. G. Margaritis, J. K. Kallitsis, and N. K. Kalfoglou. Miscibility of chlorinated polymers with ENR: 3. Blends with chlorinated polyethylenes. Polymer 28:2122–2129, 1987.
11. L. A. Utraki. Polymer Alloys and Blends. Munich: Hanser, 1990.
12. S. K. Khanna and W. I. Congdon. Engineering and molding properties of PVC/NBR and polyester blends. Polym. Eng. Sci. 23:627–631, 1983.
13. K. E. George, R. Joseph, and J. Francis. Studies on NBR/PVC blends. J. Appl. Polym. Sci. 32:2867–2873, 1986.
14. K. T. Varughese, P. P. De, S. K. Sanyal, and S. K. De. Miscible blends from PVC and epoxidized natural rubber. J. Appl. Polym. Sci. 37:2537–2548, 1989.
15. Z. A. Nasir, U. S. Ishiaku, and Z. A. Mohd Ishak. Determination of optimum blending conditions of PVC/ENR blends. J. Appl. Polym. Sci. 47:951–959, 1993.
16. U. S. Ishiaku, M. Nasir, and Z. A. Mohd Ishak. Aspects of miscibility in PVC/ENR blends, Part I: Mechanical and morphological properties. J. Vinyl Technol. 16:219–222, 1994.
17. U. S. Ishiaku, M. Nasir, and Z. A. Mohd Ishak. Aspects of miscibility in PVC/ENR blends, Part II: Dynamic mechanical properties. J. Vinyl Technol. 16:226–230, 1994.
18. U. S. Ishiaku, M. Nasir, and Z. A. Mohd Ishak. Rheological properties of PVC/ENR blends. Part I: Composition dependence and the effect of compounding conditions. J. Vinyl Additive Technol. 1:142–147, 1995.
19. U. S. Ishiaku. The rheological and dynamic mechanical properties of PVC/ENR blends. Ph.D. thesis, Universiti Sains Malaysia, 1993.
20. U. S. Ishiaku, B. T. Poh, Z. A. Mohd Ishak, and D. Ng. Thermo-oxidative properties of blends of PVC with ENR and NBR rubbers in the presence of an antioxidant and a base. Polymer International 39:67, 1996.
21. C. T. Ratnam and M. Nasir. In: J. C. Rajoo, ed. Developments in the Plastics and Rubber Product Industries. Malaysia: CAPS Enterprise, 1987, pp. 403–415.
22. Z. A. Nasir and C. T. Ratnam. Internal mixer studies of PVC/ENR blends. J. Appl. Polym. Sci. 38:1219–1227, 1989.
23. A. G. Margaritis and N. K. Kalfoglou. Compatibility of PVC with epoxidized polybutadiene. Eur. Polym. J. 24:1043–1047, 1988.

24. K. T. Varughese and P. P. De. Miscibile blends from plasticized PVC and epoxidized natural rubber. J. Appl. Polym. Sci. 37:2537–2548, 1989.

25. B. Albert, R. Jerome, and P. Teyssié. Investigation of polymer miscibility by spectroscopic methods IV. How far is PVC miscible with S-PMMA and poly(styrene-co-acrylonitrile)? An answer from narrative energy transfer. J. Appl. Polym. Sci. 24:551–558, 1986.

26. P. C. Painter, Y. Park, and M. M. Coleman. Hydrogen bonding in polymer blends 2. Theory. Macromolecules 21:66–72, 1988.

27. A. Garton. Some observations on kinetic and steric limitations to specific interactions in miscible polymer blends. Polym. Eng. Sci. 24:112–116, 1984.

28. K. Matsuo, M. L. Mansfield, and W. H. Stockmayer. Comment on the paper of Fahrenholtz and Kwei concerning the compatibility of polymer mixtures containing Novolac resins. Macromolecules 15:937, 1982.

29. M. Nasir, U.S. Ishiaku, and Z. A. Mohd Ishak. Dependence of mechanical properties in polyvinyl chloride/epoxidized natural rubber blends on mixing conditions. Second ASEAN-JAPAN Symposium on Polymers (Fourth Indonesia-JICA Polymers Symposium-Cum-Workshop). Bandung, 1992.

30. G. Schramm. HAKKE materials testing. Sonderdruck aus Kautsch. Gummi Kunstst. 12:1074, 1990.

31. L. A. Utracki. Melt flow of polymer blends. Polym. Eng. Sci. 23:602, 1983.

32. W. M. Prest, Jr., and R. S. Porter. Rheological properties of poly(2,6-dimethylphenylene oxide)–polystyrene blends. J. Polym. Sci. 10(part A-Z):1639–1655, 1972.

33. S. Wu. Chain entanglement and melt viscosity of compatible polymer blends: poly(methyl methacrylate) and poly(styrene-acrylonitrile). Polymer 28:1144–1148, 1987.

34. C. M. Roland. J. Polym. Sci. Part B Polym. Phys. 26:839, 1988.

35. A. P. Plochocki. Developments of industrial polyolefin blends using melt rheology data. Polym. Eng. Sci. 22:1153–1165, 1982.

36. A. P. Plochocki. Melt rheology of polymer blends the morphology feedback. Polym. Eng. Sci. 23:618–626, 1983.

37. L. A. Utracki. Economics of polymer blends. Polym. Eng. Sci. 22:1166–1175, 1982.

38. J. R. Fried, W. J. MacKnight, and F. E. Karasz. Compatibility and tensile properties of PPO blends. J. Appl. Phys. 50, Part II, 1979.

39. L. W. Kleiner, F. E. Karasz, and W. J. MacKnight. Compatible glassy polyblends based upon poly(2,6-dimethyl-1,4-phenylene oxide): tensile modulus studies. Polym. Eng. Sci. 19:519–524, 1979.

40. L. A. Utracki. Polymer blends and alloys for molding applications. Polym. Plast. Technol. Eng. 22:27–54, 1984.

41. I. R. Gelling and C. Matherell. Epoxidized natural rubber. Technical update. Hertford: Malaysian Rubber Bureau, 1990.

42. C. H. M. Jacques and H. B. Hopfenbert. Vapour and liquid equilibria in glassy polyblends of polystyrene and poly(2,6-dimethyl-1,4-phenylene oxide). Polym. Eng. Sci. 14:441–448, 1974.

43. Y. J. Shur and B. Ranby. Gas permeation of polymer blends I. PVC/ethylene-vinyl acetate copolymer (EVA). J. Appl. Polym. Sci. 19:1337–1344, 1975.

44. Y. J. Shur and B. Ranby. Gas permeation of polymer blends II. PVC/NBR blends. J. Appl. Polym. Sci. 19:2143–2155, 1975.
45. C. T. Ratnam. Poly(vinyl chloride)/epoxidized natural rubber blend. M.Sc. thesis, Universiti Sains Malaysia, 1989.
46. V. Shah. Handbook of Plastics Testing Technology. New York: John Wiley, 1984.
47. Characteristics and Compatibility of Thermoplastics for Ultrasonic Assembly. Technical Information PW-1. Danbury, CT: Branson Ultrasonics Corporation, 1971.
48. I. L. Gomez. In: I. L. Gomez, ed. Engineering with Rigid PVC; Processability and Applications. New York: Marcel Dekker, 1984.
49. O. Olabisi, L. M. Robeson, and M. T. Shaw. Polymer–Polymer Miscibility. New York: Academic Press, 1979.
50. R. E. Wetton, R. de Blok, and P. J. Corish. DMA studies of polymer blends and the interaction between the components of these blends. Intern. Polym. Sci. Technol. 18:63–67, 1991.
51. T. Murayama. Dynamic Mechanical Analysis of Polymeric Materials. Amsterdam: Elsevier, 1978.
52. J. K. Kallistics and M. K. Kalfoglou. Compatibility of ENR with thermoplastic and thermosetting resins. J. Appl. Polym. Sci. 37:453–465, 1989.
53. L. E. Nielsen. Predicting the Properties of Mixtures: Mixture Rules in Science and Engineering. New York: Marcel Dekker, 1978.
54. K. Dimov, E. Dilova, and S. Stoyanov. Effect of comparison on behaviour of tricomponent polymer blends. Study by means of simplex lattice planning of the experiment. J. Appl. Polym. Sci. 19:2087–2098, 1975.
55. L. E. Nielsen. Mechanical Properties of Polymers. New York: Van Nostrand Reinhold, 1962.
56. A. Amu, S. Dulangali, and I. R. Gelling. Latest Developments in Epoxidized Natural Rubber. Singapore: PLAST, 1986.
57. U. S. Ishiaku, Z. A. Mohd Ishak, H. Ismail, and M. Nasir. The effect of di-2-ethylexylphthalate on the thermo-oxidative ageing of PVC/ENR blends. Polymer International 41:327–336, 1996.
58. B. T. Poh and K. S. Lee. FTIR study of the thermal oxidation of ENR. Eur. Polym. J. 30:17–23, 1994.
59. R. M. Silverstein, G. Clayton Bassler, and T. C. Morrill. Spectrometric Identification of Organic Compounds. Singapore: John Wiley, 1991.
60. N. Grassie. Chemistry of High Polymer Degradation Process. London: Butterworths, 1956.
61. W. I. Hawkins. Polymer Stabilization. 5th ed. New York: Wiley-Interscience, 1971.
62. E. Kohlmetz, C. Levy, and P. Walter. Plasticizers. Encyclopedia of Polymer Science and Engineering. New York: John Wiley, 1985, pp. 568–648.
63. U. S. Ishiaku, H. Ismail, A. Shaharum, and Z. A. Mohd Ishak. Evaluation of plasticizers for PVC/ENR thermoplastic elastomers. International Rubber Conference (IRC), Kuala Lumpur, Malaysia, 6–8 October, 1997.
64. W. V. Titow. PVC Technology. 4th ed. London: Elsevier, 1984, p. 117.
65. L. I. Nass. In: L. I. Nass, ed. Encyclopedia of PVC. New York: Marcel Dekker, 1977, vol. 1, p. 273.

66. U. S. Ishiaku, A. Shaharum, Z. A. Mohd Ishak, and H. Ismail. Thermo-oxidative ageing of PVC based thermoplastic elastomers: the effect of "new" epoxidized natural rubber. Kautsch. Gummi Kunstst. 50:292–298, 1997.
67. D. F. Anderson and D. A. McKenzie. Mechanism of the thermal stabilization of poly(vinyl chloride) with metal carboxylates and epoxy plasticizers. J. Polym. Sci. Part A-1, 8:2905–2922, 1970.
68. B. M. Walker. Introduction. In: B. M. Walker, ed. Handbook of Thermoplastic Elastomers. New York: Van Nostrand Reinhold, 1988.
69. A. Y. Coran and R. Patel. Rubber-thermoplastic compositions. Part II, NBR-nylon thermoplastic elastomeric compositions. Rubb. Chem. Technol. 53:781–793, 1980.
70. A. Y. Coran. Thermoplastic Elastomers, A Comprehensive Review. New York: Hanser, 1987.
71. H. W. Hofmann. Vulcanisation and Vulcanising Agents. London: Maclaren, 1967.

24

Developments of Oil Palm–Based Lignocellulose Polymer Blends

H. J. Din Rozman and Wan Rosli Wan Daud
Universiti Sains Malaysia, Penang, Malaysia

I. INTRODUCTION

Oil palm or *Elais guineensis* was first introduced into Malaysia in 1870 from the Botanic Gardens in Singapore. Like the coconut palm, the oil palm is grown mainly for its oil-producing fruit. Owing to its commercial importance, the botanical and cultivation aspects of the oil palm have been extensively studied (1–3). Its two main products are palm oil and palm kernel oil. Traditionally, these products are used mainly in the manufacture of compound fat and soap, but now their usage has widened and varied considerably. Recently, much attention has been channeled towards finding suitable applications for oil palm industry by-products. Coupled with the scarcity of timber and different environmental issues, various types of by-products have been studied to see whether they can serve as replacements for timber or alleviate environmental problems. At the palm oil mills, the by-products consist of shell, empty fruit bunches, pressed fruit fibers (mesocarp fibers), and palm oil mill effluent (POME). Fronds from pruning are constantly generated in the plantations, and these are mainly used in interrow mulching. An even larger quantity of waste material, in the form of oil palm stems as well as fronds, is produced in the plantations during replanting. Finding appropriate applications for these by-products should become increasingly important economically as well as environmentally. Table 1 shows some present uses of oil palm by-products.

Recent trends show that valuable products can be produced from various biofibers of oil palm by-products. Examples of such products are oil palm component plastic composites (4), oil palm component rubber composites (5), sheet

Table 1 Some Uses of Oil Palm By-Products

By-product	Uses
Palm kernel shells	Cheap boiler fuel
Empty fruit bunches	Potassium fertilizer (in the form of potash-rich ash) oil palm mulchings (to recycle nutrient, conserve soil moisture, and control/reduce soil surface erosion) Boiler fuel
Pressed fruit fibers	Boiler fuel Nursery mulching purposes
Palm oil mill effluent	Animal feed (dried and mixed with other supplement) biogas (from anaerobic digestion) for heat and electricity generation
Fronds	mulching

molding compounds (6), composites (7), and pulp and paper (8). Since the chemical composition of oil palm is similar to that of wood, these wastes could be turned into new raw materials with expanding potentials.

Generally, interest has also been growing in finding the appropriate utilization of lignocellulosic materials other than wood. These materials have been subjected to various investigations recently, either for replacing existing wood species in making conventional panel products (7) or for producing plastics composites (4). The increasing trend in using these nonwood-based materials has been induced by the growing demand for lightweight, high-performance materials in an age of diminishing natural fiber resources (wood in particular) and escalating costs of raw materials and energy. Thus the prospect of using oil palm by-products in various products is increasingly bright in the light of the demand for lignocellulosic materials in vast areas of applications. Lignocellulosic materials, especially wood, have stimulated much interest in the manufacture of composites during the past decade, i.e., to be used as filler materials instead of conventional fillers such as mica, talc, clay, and glass fibers. The utilization of lignocellulosic material in the production of plastic composites is becoming more attractive, particularly for low-cost/high-volume applications. There are several factors responsible for the observed trend. Lignocellulosic-derived fillers possess several advantages compared to inorganic fillers, i.e., lower density, greater deformability, less abrasiveness to equipment, and of course lower cost. More importantly, lignocellulosic-based fillers are derived from a renewable resource, available in relative abundance; the potential has not been really tapped.

Lignocellulosic material including wood and oil palm by-products such as empty fruit bunches and fronds have significantly lower density than the common inorganic fillers mentioned earlier. Thus specific mechanical properties (strength-to-weight ratio) of these lignocellulosic–plastic composites often exceed those of other filled plastics owing to this favorable density difference. In addition to this, lignocellulosic materials offer several benefits to composites:

1. Relatively low cost
2. Ease of processing
3. Low equipment abrasion
4. Ease of surface modification
5. Renewability
6. Improved properties and performance

However, there are some drawbacks posed by these materials, especially in plastic composites, such as

1. Poor compatibility with nonpolar thermoplastics
2. Thermal instability at temperatures above 220°C
3. Hygroscopicity contributed by the polar nature of the lignocellulosic surface
4. Low bulk density
5. Difficulty in mixing in ordinary plastic mixing equipment

Nevertheless, through research and development, most of the processing problems have been overcome. These include mixing processes, types of polymers used, types of compatibilizers or coupling agents used, and suitable form of lignocellulosic materials.

Lignocellulosic material is mixed with thermoplastic material in a process known as compounding. Lignocellulosic is usually used in the form of flour. Lignocellulosic materials in long strands of fibers are not suitable, as they tend to tangle and mat together, making conventional mixing procedures difficult. Furthermore, uncontrolled cutting of the long strands or fibers during mixing makes it difficult to monitor the properties of the composites. Lignocellulosic flour or powder (in certain mesh sizes) is easily compounded in conventional plastic mixing equipment such as the twin-screw extruder. For fibrous material, an internal mixer can be used to disperse the fibers effectively in molten plastic in a relatively short time.

Usually, additives are added to aid fiber dispersion, promote adhesion between component materials, and increase processibility. Such additives include coupling agents and compatibilizers. The additives are usually added in small amounts, about 1–5% of the fiber weight (9).

Looking at the trend in the plastic-related industries, composites are expected to be one of the main areas. Wood-based composites will constitute a bigger portion of the industry, since wood is a largely price-driven commodity and has adequate properties, e.g., low density, favorable strength-to-weight ratio, abundance, and desired performance at relatively low cost.

Market potential for lignocellulosic–thermoplastic composites can be categorized into three main areas:

1. Automotive industries—door trim panels, trunk liners, seat backs, package trays, speaker covers
2. Recreational—lumber-size profiles for playground structures
3. Nonstructural applications—furniture, packaging, building components, fencing, highway guardrail components

As more interest has been developing in wood-filled plastics, and as a better understanding of the wood–plastic interaction has been gained, the move to find substitutes for wood has been stimulated especially by the abundance of similar resources. With increasing pressures on forest industries, coupled with the scarcity of natural resources, the need for efficient use of resources is vital in the solid wood and wood composites industries. This should mean better conversion, better methods that are environmentally friendly, new technologies, or even looking for substitutes for wood.

The move to study and produce oil palm–based lignocellulose polymer blends is based on two reasons; first, the incorporation of high-loading, low-cost lignocellulosic fillers into relatively expensive thermoplastics will be an effective strategy to reduce the cost of composite products. The oil palm materials used in making the composites are derived from oil palm frond (OPF) and empty fruit bunch (EFB), components of an oil palm tree. OPF is a by-product of the palm oil industry and consists of about 80.5% of holocellulose and 18.3% of lignin (10). It is readily available at a typical token price of US$30.00 per tonne. In spite of the cost of grinding the material to the required size, the overall cost is cheap. EFB also is a by-product of the palm oil industry and consists of a bunch of fibers in which palm fruit are embedded; it is about 65% holocellulose and 25% lignin (11). EFB is readily available at a lower cost than OPF at a typical token price of US$10.00 per tonne. Second, the OPF and EFB waste generated by the oil palm industry in Malaysia is estimated to be about 13 and 8 million tonnes per year, respectively. Thus finding useful applications for these materials will surely alleviate environmental problems related to the disposal of oil palm wastes and produce materials that could offer a favorable balance of quality, performance, and cost.

II. OIL PALM COMPONENTS—THERMOPLASTIC COMPOSITES

A. OPF–HDPE Composites

Rozman et al. (12) attempted to produce high-density polyethylene (HDPE) composites with OPF as filler. The OPF fibers were ground to different mesh sizes before use. The materials were compounded with single-screw extruder. Addition of OPF flour for all sizes resulted in a decrease in the modulus of rupture (MOR) of the composites as compared to pure HDPE (Table 2). Samples with smaller sizes displayed greater MOR than the larger ones. This implies that the samples are capable of withstanding higher stress before failure than the ones with larger particle size. In addition, samples with fillers of size 80 mesh showed the least reduction in MOR as compared to other samples with larger size. As shown by studies on other lignocellulosic fillers (13), this can be attributed to the greater interaction and/or dispersion of the finer OPF particles in the PE matrix.

In OPF–HDPE composites, poor interfacial interaction is expected owing to poor compatibility between polar OPF and nonpolar polyethylene matrix, which form weak interfacial regions. As a lignocellulosic material, OPF surfaces are covered by polar hydroxyl groups contributed by cellulose, hemicellulose, and lignin. This wetting is further decreased as more filler is added. As more filler is incorporated in the composite, more incompatible interfacial regions between polar lignocellulosic and nonpolar polyethylene are created. The weak interfacial re-

Table 2   Mechanical Properties of OPF–HDPE Composites

Sample	Bending MOE (MPa)	MOR (MPa)	Toughness (kPa)	Tensile modulus (MPa)	Strength (MPa)	Impact strength (J/m)
Mesh 80						
30%	306.23	15.34	24.11	427.59	11.93	148.63
40%	370.98	15.79	38.26	559.79	10.66	102.28
50%	397.94	14.98	22.33	647.22	8.42	82.96
60%	391.84	14.63	20.27	613.85	8.34	60.98
Mesh 60						
30%	470.57	17.02	30.87	488.87	8.43	80.16
40%	580.08	13.71	28.92	474.85	8.99	71.71
50%	645.84	12.77	27.81	478.63	7.94	61.41
60%	643.94	8.69	10.2	360.74	6.26	44.26
Mesh 35						
30%	248.06	5.86	8.79	393.60	12.48	51.41
40%	444.06	8.69	12.54	574.50	10.24	53.70
50%	474.97	7.25	14.06	497.28	8.85	44.76
60%	462.46	5.87	11.90	487.60	7.35	32.91

gions result in the reduction in the efficiency of stress transfer from the matrix to the reinforcement component. As highlighted by several workers (14–16), the quality of interfacial bonding is determined by several factors, such as the nature of lignocellulosic and thermoplastic materials as well as their compositions, the fiber aspect ratio, the types of incorporation procedures, processing conditions employed, and the treatment of the polymer or fiber with various chemicals, compatibilizers, coupling agents, etc.

In the same study Rozman et al. (12) also found that the stiffness of the sample (modulus of elasticity, MOE results) increased as the filler loading was increased. As shown by various studies, incorporation of fillers is able to impart greater stiffness in the composites (17). Generally, samples with smaller particle size filler showed higher modulus, especially at 30% loading. However, there was no significant difference towards higher loadings. The toughness of the samples decreased as the filler loading was increased. This is very much expected, since the lignocellulosic material was added without any pretreatment with any coupling agents or compatibilizers to improve the compatibility of the polar nature of the lignocellulosic and the nonpolar nature of the hydrocarbon chain of PE. The results showed that samples with smaller particle size displayed higher toughness than those with bigger particle size. As toughness is a measure of energy needed for failure, the results generally demonstrate that more energy is needed to break samples with smaller size fillers. Tensile strength of the OPF composites decreased gradually with filler loading. Filler sizes did not exact any significant changes on the strength of the composites. This is true because incorporation of filler into a thermoplastic matrix does not necessarily increase the tensile strength of a composite. Fibers with uniform circular cross section and a certain aspect ratio normally improve the strength. However, the capability of irregularly shaped fillers with low aspect ratio, as mentioned in the study, to support stresses transferred from the polymer matrix is significantly reduced.

Significant improvement in tensile modulus was observed with increasing filler loading. Composites with smaller filler size displayed better modulus than the larger filler size samples, especially at higher filler loading, e.g., 40–60% filler loading. Similar observations have been reported by Rozman et al. (18) for other lignocellulosic–thermoplastic composites. Smaller or finer particles with larger specific area may impart greater interaction with the polymer matrix and can result in uniform filler dispersion in the composite.

The impact strength of composites decreased as the filler loading was increased. This reflects the reduction in energy absorption at the crack tip. The poor bonding quality between the fibers and the polymer matrix creates weak interfacial regions that result in debonding and frictional pullout of fiber bundles. These failure mechanisms, which inhibit the ductile deformation and mobility of the matrix, will obviously lower the ability of the composite system to absorb energy during fracture propagation (18).

B. EFB–HDPE Composites

Rozman et al. (18) in their study of EFB–HDPE composites found that the MOE of the composites increased as the filler content was increased (Table 3). As shown by OPF composites, incorporation of fillers resulted in the improvement of the stiffness of the composites.

For MOR results, all samples with different sizes of filler showed a decreasing trend as the percentage of filler loading was increased, but the difference was too small to be significant. The samples with smaller filler size displayed slightly higher MOR. Samples with fillers of size 80 mesh showed the least reduction in MOR (with reference to PE alone) compared to other samples with larger particle size.

As also shown by OPF composites, the toughness of the samples was reduced as more filler was added. As highlighted by several workers (20–22), the degree of interfacial bonding is dependent on a number of factors, such as the types of lignocellulosic and thermoplastic, the aspect ratio of the fiber, the method of incorporating the lignocellulosic into the resin, the processing conditions, and the treatment of the polymer or fiber with various chemicals or additives.

Table 3 Mechanical Properties of EFB–HDPE Composites

Sample	Bending MOE (MPa)	MOR (MPa)	Toughness (kPa)	Tensile modulus (MPa)	Strength (MPa)	Impact strength (J/m)
Mesh 80						
30%	370.29	15.48	25.98	322.38	13.02	70.91
40%	383.16	15.33	25.46	312.49	11.85	52.41
50%	431.32	14.91	21.53	341.92	9.18	48.14
60%	438.96	14.04	19.00	347.83	7.26	57.30
Mesh 60						
30%	318.49	11.24	28.87	226.33	9.74	59.72
40%	409.20	10.86	24.63	283.67	9.69	63.16
50%	470.20	9.92	20.77	338.33	7.51	55.67
60%	450.96	7.70	12.93	358.17	7.31	48.67
Mesh 35						
30%	299.80	9.89	13.68	328.50	13.17	73.65
40%	360.06	9.87	14.11	364.60	11.67	65.52
50%	412.73	9.75	18.18	438.17	11.08	65.15
60%	452.99	7.03	14.60	375.40	9.68	52.44

Incorporation of filler into a thermoplastic matrix may increase or decrease the tensile strength of the resulting composite. The irregularly shaped short fibers used in the study could not support stresses transferred from the polymer matrix; thus there was a reduction of tensile strength as the filler loading was increased. Tensile modulus increased with the increase of filler loading. This indicates the ability of EFB fillers to impart greater stiffness to the HDPE composites. These results are in agreement with the trend observed in other lignocellulosic-filled thermoplastics (23–26).

The impact strength of the composites decreased as the filler loading was increased. This clearly indicates that the presence of EFB has reduced the energy-absorbing capabilities of the composites. The poor adhesion or bonding between the fibers and polymer matrix creates weak interfacial regions, which will result in debonding and frictional pullout of fiber bundles and inhibit the ductile deformation and mobility of the matrix (19). This in turn lowers the ability of the composite system to absorb energy during fracture propagation. A similar trend has been reported by Myers et al. (27) in the case of wood-flour–filled HDPE.

C. Comparison of Properties Between OPF–HDPE and EFB–HDPE Composites

From both results (Tables 2 and 3), it can be seen that both types of composites show comparable properties, i.e., in flexural and tensile strength and flexural toughness. However, OPF–HDPE composites show superiority on both flexural and tensile modulus as compared to the EFB–HDPE composites. This may be due to the greater compressibility of the former (which is lower in density and hollow in the middle) than of the latter.

D. The Effects of Coupling Agents on the EFB–HDPE Composites

Mohd Ishak et al. (28) attempted to produce EFB–HDPE composites by using various chemicals as coupling agents. They found that with 3-aminopropyltrimethoxysilane (APM) $[H_2N(CH_2)_3Si(OCH_3)_3]$ as coupling agent the MOEs of the samples were significantly higher than the ones with 3-aminopropyltriethoxysilane (APE) $[H_2N(CH_2)_3Si(OC_2H_5)_3]$, especially at 1 and 3% (Table 4). The difference shown may be attributed to a stiffer molecular structure of APM with one methyl group as compared to APE with an ethoxy group. Composites with APM showed a decreasing trend as the loading was increased, whilst APE composites do not show significant changes.

Composites with APM also displayed a significantly higher MOR than the ones with APE also at 1 and 3% filler loadings. As shown by MOE results, higher loading did not exact enhancement in the strength of APM composites. Tensile modulus was increased with the addition of 1% of coupling agents (for 3-APM

Table 4 Effects of Coupling Agents on EFB-HDPE Composites

Sample	Bending MOE (MPa)	MOR (MPa)	Toughness (kPa)	Tensile modulus (MPa)	Strength (MPa)	Impact strength (J/m)
APM						
1%	629.60	24.43	61.99	767	12.62	116.53
3%	508.97	22.78	55.42	762	13.20	123.07
5%	288.83	14.78	40.79	784	11.76	123.20
APE						
1%	405.79	20.00	52.69	737	12.02	123.90
3%	389.33	18.99	30.72	651	11.95	117.04
5%	438.24	18.66	43.36	547	12.04	137.28

and 3-APE). As noted by Riley et al. (29), there are three main factors affecting the modulus; (1) filler modulus, (2) filler loading and, (3) filler aspect ratio. Since these three factors have been kept relatively constant, it can be inferred that the presence of coupling agents has led to a significant improvement in the filler–matrix interfacial bonding. The improvement obviously results in an increase in the efficiency of stress transfer from the matrix to the filler, which subsequently gives rise to higher modulus. Unlike tensile modulus, the incorporation of both types of coupling agents did not produce any significant effect on tensile strength.

A significant improvement in the impact strength was observed with 1% loading of coupling agent followed by a stable impact strength at higher coupling agent loading. The improvement in the filler–matrix adhesion, which has produced a pronounced effect on the tensile modulus, has resulted a similar effect on the impact strength of the composites.

E.   The Effects of Compatibilizers on EFB–HDPE Composites

Addition of poly(propylene-acrylic acid) (PPAA) and poly(propylene-ethylene-acrylic acid) (PPEAA) as compatibilizers was found to improve the flexural stiffness (MOE) of the composites (see Table 5) (28). However, it was clearly seen that incorporation of both compatibilizers did not result in flexural strength improvements as shown by MOR results.

As shown by MOE results, modification of the filler–matrix interfacial bonding through the presence of both compatibilizers produced a significant effect on the tensile modulus of the composites. The slightly higher modulus of the PPEAA-treated EFB–HDPE composites as compared to the ones with PPAA in-

Table 5 Effects of Compatibilizers on EFB-HDPE Composites

Sample	Bending MOE (MPa)	MOR (MPa)	Toughness (kPa)	Tensile modulus (MPa)	Strength (MPa)	Impact strength (J/m)
PPAA						
1%	427.22	10.10	24.18	362	9.92	59.92
3%	397.83	9.85	24.96	328	9.31	56.82
5%	575.00	11.44	20.96	292	9.15	59.29
PPEAA						
1%	403.25	10.09	26.70	351	8.77	64.48
3%	435.86	10.51	27.29	396	9.93	69.35
5%	594.66	11.22	24.35	329	10.12	63.23

dicated the significance of more hydrophobicity incurred by the ethylene groups in PPEAA.

Both PPAA and PPEAA failed to impart any positive effect to the tensile and impact strength. This again supports earlier claims that the shape factors outweigh any improvement in the filler–matrix interaction even in the presence of coupling agent or compatibilizer.

F. Comparison of Various EFB–Thermoplastic Composites

Ping (30) attempted to produce EFB composites with various commonly known thermoplastics. MOE results showed that EFB–polystyrene (PS) composites displayed higher stiffness (MOE) than EFB–polyvinyl chloride (PVC) and EFB–polypropylene (PP) composites, respectively (Fig. 1). Incorporation of fillers results in the reduction of MOE in EFB–PS composites. Both EFB–PVC and EFB–PP composites showed improvement, especially at higher filler loading, i.e., 45 and 60%.

All types of composites displayed a decreasing trend in MOR as the filler loading was increased. As explained earlier, incorporation of fillers may disrupt the continuity of polymer matrix, which may result in the creation of more stress points (Fig. 2).

Tensile strength results showed that the strength of the EFB composites with different thermoplastic matrix depended on the type of matrix used (Fig. 3). EFB composites with PVC matrix displayed the highest tensile strength and were followed by PP, PE, and PS, respectively. Tensile modulus of the EFB–PS and

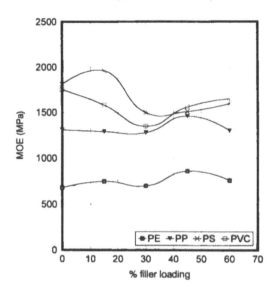

Figure 1   MOE vs. percentage filler loading—various EFB–thermoplastic composites.

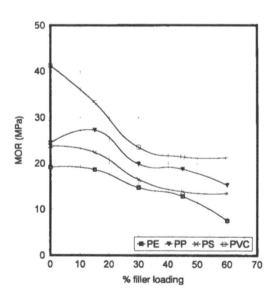

Figure 2   MOR vs. percentage filler loading—various EFB–thermoplastic composites.

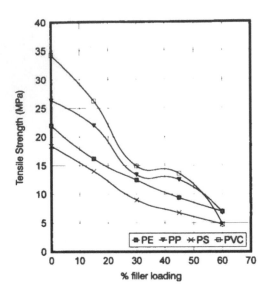

Figure 3 Tensile strength results for EFB composites made from various thermoplastics.

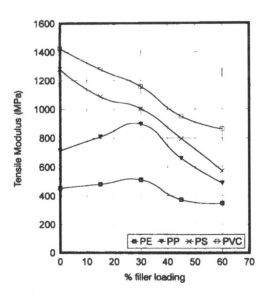

Figure 4 Tensile modulus results for EFB composites made from various thermoplastics.

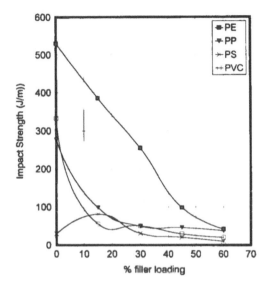

Figure 5 Impact strength results for EFB composites made from various thermoplastics.

EFB–PVC composites decreased as the filler loading was increased (see Fig. 4). However, the ones with PP and PE showed some enhancement especially at 30% filler.

Impact strengths for various composites showed a decreasing trend as the filler loading was increased (Fig. 5). The impact strength of EFB–PE composite was significantly higher than the ones with PP, PS, and PVC, which fall within the same range.

G. The Effect of Compounding Techniques

Two compounding methods have been attempted to produce oil palm empty fruit bunch (EFB)–thermoplastic composites, use of the internal mixer (IM) and use of the single-screw extruder (EX). Composites produced by both techniques show a decreasing trend in tensile strength as the filler loading increases (Fig. 6). This is in agreement with the trend observed in other lignocellulosic filled composites (31,32). As mentioned earlier, this may be caused by the irregular shape of EFB fillers, which affects their capability of supporting stress transmitted from the thermoplastic matrix. Thus the strength enhancement in the filled composite is in general much lower than that of fiber-reinforced systems. The results reveal that IM composites yield a relatively higher strength than EX composites. This indicates that the mode of mixing plays an important role in determining the tensile strength of a composite. As the degree of compounding and distribution of fillers

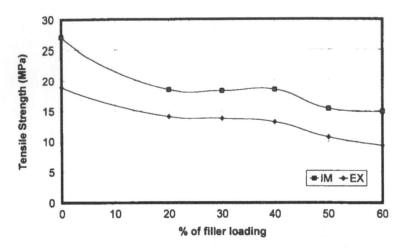

Figure 6 The effect of mixing techniques on the tensile strength of EFB–PP composites. (Redrawn from Ref. 28.)

in the polymer matrix determines the extent of formation of the interfacial region between these two components, it is believed that there is a better transfer of stress in the IM composites than in the EX composites. Tensile modulus for both IM and EX composites increases with increasing filler loading (Fig. 7). This behavior is consistent with the earlier study on PP–wood flour composites (32). The results

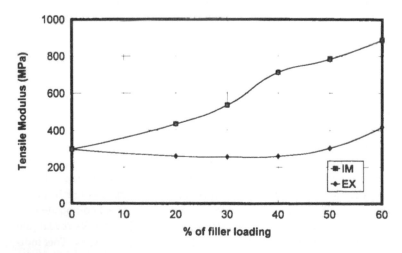

Figure 7 The effect of mixing techniques on the tensile modulus of EFB–PP composites. (Redrawn from Ref. 28.)

suggest that irrespective of compounding techniques, EFB filler is able to impart greater stiffness to the PP composites.

It is interesting to note that IM composites display a markedly higher modulus than EX composites. For composites with 60% filler loading, for instance, the modulus for IM composites reaches as high as approximately 900 MPa as compared to about 400 MPa for EX composites. Thus it can be inferred that the inherent stiffness of the EFB fillers can be greatly exploited by using the internal mixer. Apparently the internal mixer can distribute the filler more evenly throughout the polymer matrix than the extruder.

It is common to observe that the improvement in the tensile modulus is at the expense of the elongation at break (EB) (32). The EB for both composites decreases as the filler loading is increased (Fig. 8). This may be attributed to the reduction in deformability of a rigid interface between filler and the matrix component, which is also reflected in the increase of stiffness as shown earlier in the tensile modulus result. Similar observations have been reported by several workers for other lignocellulosic composites (13,16,23,24,33).

It should be noted that the degree of reduction in EB for EX composites is substantially higher than for IM composites. It can also be seen that IM composites display lower EB than EX composites. As shown by tensile modulus results (Fig. 7), IM composites are stiffer than EX composites. This obviously influences the EB properties of the composites. As better filler distribution is achieved in the internal mixer, more interfacial regions are expected to be formed between the filler and the polymer matrix. This would render the region stiffer and subsequently reduce the deformability of the composite.

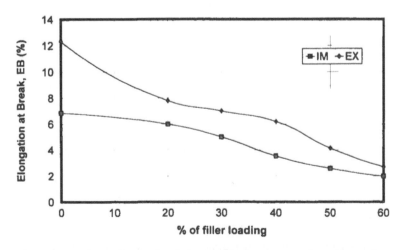

Figure 8 The effect of mixing techniques on the elongation at break strength of EFB–PP composites. (Redrawn from Ref. 28.)

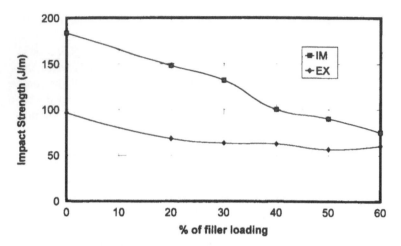

Figure 9 The effect of mixing techniques on the impact strength of EFB–PP composites. (Redrawn from Ref. 28.)

The impact strength results clearly indicate that the presence of EFB has reduced the energy-absorbing capabilities of the composites (Fig. 9). A similar trend has also been reported by Mohd Ishak et al. (28) in the case of EFB-filled HDPE. As mentioned earlier, the filler system used in this study consists of irregularly shaped filler with low aspect ratio. This influences the energy-absorption capability of the composite as well as the capability of supporting stress transferred from the polymer matrix. IM composites exhibit higher impact strength than EX composites. This suggests that composites made by mixing in the internal mixer are of higher strength than those prepared through extrusion. This may be attributed to the mixing characteristics of the former. The mixing process in an internal mixer is done by two counter-rotating rotors that give a high shear rate; the mixture is enclosed and confined in a mixing chamber. The process is more effective in improving the distribution of filler in the thermoplastic matrix. Consequently, the wetting of the filler surface is enhanced significantly. Composites with higher degrees of homogeneity in wettability and filler dispersion can be expected to possess superior mechanical properties.

III. CONCLUSIONS

With their low cost and consistency in availability, oil palm by-products provide a good foundation for their role as components in plastic composites. As shown by the mechanical properties results, it is obvious that the properties of the composites are comparable with those of wood-filled plastic composites. Thus oil palm by-products with reference to EFB and OPF demonstrate a promising future in the lignocellulosic–plastic composite industry. Nevertheless, further studies are required on fundamental issues such as the effect of various chemicals and substances that may affect the properties as well as the parameters involved during the up-scaling stages.

APPENDIX

Modulus of elasticity (MOE):

$$MOE = \frac{L^3 \Delta W}{4bd^3 \Delta S}$$

where
L = the span between the centers of supports (m)
ΔW = the increment in load (N)
b = the mean width of the sample (m)
d = the mean thickness of the sample (m)
ΔS = the increment in deflection (m)

Modulus of rupture (MOR):

$$MOR = \frac{3WL}{2bd^2}$$

where
W = the ultimate failure load (N)
L = the span between centers of support (m)
b = the mean width of the sample (m)
d = the mean thickness of the sample (m)

REFERENCES

1. C. W. S. Hartley. The Expansion of Oil Palm Planting. Advances in Oil Palm Cultivation. Kuala Lumpur: Incorp. Soc. Planters, 1972.
2. C. W. S. Hartley. The Oil Palm. 3d ed. London: Longmans, 1977.
3. R. H. V. Corley, J. J. Hardon, and B. J. Wood. Development in Crop Sciences 1: Oil Palm Research. Netherlands: Elsevier Scientific, 1976.
4. M. J. Zaini, M. Y. A. Fuad, Z. Ismail, M. S. Mansor, and J. Mustafah. The effect of filler content and size on the mechanical properties of propylene/oil palm wood flour composites. Polymer International 40(1):51–55, 1996.

5. H. Ismail, H. D. Rozman, R. M. Jaffri, and Z. A. Mohd Ishak. Oil palm wood flour reinforced epoxidized natural rubber composites: the effect of filler content and size. Eur. Polym. J. 33(10–12):1627–1632, 1997.

6. R. N. Kumar, H. D. Rozman, A. Abusamah, and T. H. Chin. Sheet Moulding Compounds Based on Palm Fruit Pressed Fibre. National Seminar on Utilization of Palm Tree and Other Palms, Forest Research Institute of Malaysia (FRIM), 1994.

7. L. T. Chew and C. L. Ong. Particleboard from oil palm trunk. Proceedings of the National Symposium on Oil Palm By-Products from Agro-based Industries, Kuala Lumpur. PORIM Bulletin 11:99–108, 1985.

8. W. R. Wandaud, K. N. Law, and J. L. Valade. Chemical pulping of oil palm empty fruit bunches. Cellulose Chem. Technol. 32, 1998.

9. B. V. Kokta, R. G. Raj, and C. Daneault. Use of wood flour as filler in polypropylene: studies on mechanical properties. Polym.-Plast. Technol. Eng. 28(3):247–259, 1989.

10. M. Husin, Z. Z. Zakaria, and A. H. H. Hassan. Potentials of oil palm by-products as raw materials for agro-based industries. Workshop Proc. Palm Oil Res. Inst. Malaysia 11:7–15, 1987.

11. Y. Kobayashi, H. Kamishima, I. Akamatsu, A. H. H. Hassan, M. Hussin, K. Hassan, and M. N. N. Yusoff. Workshop Proc. Palm Oil Res. Inst. Malaysia 11:67–78, 1987.

12. H. D. Rozman, H. Ismail, R. M. Jaffri, A. Aminullah, and Z. A. Mohd Ishak. Polyethylene-oil palm frond composites—a preliminary study on mechanical properties. Intern. J. Polymeric Mater. 39:161–172, 1998.

13. R. G. Raj, B. V. Kokta, and C. Daneault. Polypropylene-wood fiber composites: effect of fiber treatment on mechanical properties. Intern. J. Polymeric Material 12:239–250, 1989.

14. R. G. Raj and B. V. Kokta. Compounding of cellulose fibers with polypropylene: effect of fiber treatment on dispersion in the polymer matrix. J. Applied Polym. Sci. 38:1987–1996, 1989.

15. G. E. Myers, C. M. Clemons, J. J. Balatinecz, and R. T. Woodhams. Effects of composition and polypropylene melt flow on polypropylene–waste newspaper composites. Proc. of the 1992 Annual Conference Soc. Plastics Engineers (SPE/ANTEC) 1:602–604, 1992.

16. G. E. Myers, I. S. Chahyadi, C. A. Coberly, and D. S. Ermer. Wood flour/polypropylene composites: influence of maleated polypropylene and process and composition variables on mechanical properties. Intern. J. Polymeric Mater. 15:21–44, 1991.

17. D. Maldas and B. V. Kokta. Effect of extreme conditions on the mechanical properties of wood fiber–polystyrene composites. II. Sawdust as a reinforcing filler. Polym.-Plast. Technol. Eng. 29(1&2):119–165, 1990.

18. H. D. Rozman, H. Ismail, R. M. Jaffri, A. Aminullah, and Z. A. Mohd Ishak. Mechanical properties of polyethylene-oil palm empty fruit bunch composites. Polym. Plast. Technol. Eng. 37(4):493–505, 1998.

19. R. T. Woodhams, G. Thomas, and D. K. Rodgers. Wood fibers as reinforcing fillers for polyolefins. Poly. Eng. Sci. 24(15):1166–1171, 1984.

20. J. M. Felix and P. Gatenholm. The nature of adhesion in composites of modified cellulose fibers and polypropylene. J. App. Polym. Sci. 42:609–620, 1991.

21. M. Xanthos. Processing conditions and coupling agent effects in polypropylene/wood flour composites. Plastics Rubber Processing Appl. 3:223–228, 1983.

22. D. Maldas, B. V. Kokta, R. G. Raj, and C. Daneault. Improvement of the mechanical properties of sawdust wood fibre–polystyrene composites by chemical treatment. Polymer 29:1255–1265, 1988.
23. H. D. Rozman, R. N. Kumar, M. R. M. Adali, A. Abusamah, and Z. A. Mohd Ishak. The effect of lignin and surface activation on the mechanical properties of rubber-wood–polypropylene composites. J. Wood Chem. Tech. 18:471, 1998.
24. L. Yam, B. K. Gogoi, C. C. Lai, and S. E. Elke. Composites from compounding wood fibers with recycled high density polyethylene. Polym. Eng. Sci. 30:693, 1990.
25. J. X. Rietveld and M. J. Simon. Processability and properties of a wood flour filled polypropylene. Intern. J. Polymeric Mater. 18:213, 1992.
26. K. Joseph, S. Thomas, and C. Pavithran. Effect of chemical treatment on the tensile properties of short sisal fibre-reinforced polyethylene composites. Polymer 37:5139, 1996.
27. G. E. Myers, I. S. Chahyadi, C. Gonzales, C. A. Coberly, and D. S. Ermer. Wood flour and polypropylene or high density polyethylene composites: influence of maleated polypropylene concentration and extrusion temperature on properties. Intern. J. Polymeric Mater. 15:171, 1991.
28. Z. A. Mohd Ishak, A. Aminullah, H. Ismail, and H. D. Rozman. The effect of silane based coupling agents and acrylic acid based compatibilisers in mechanical properties of oil palm empty fruit bunch filled high density polyethylene composites. J. App. Poly. Sci. 68:2189, 1998.
29. A. M. Riley, C. D. Paynter, P. M. McGenity, and J. M. Adams. Plast. Rubber Process Appl. 14:85, 1990.
30. L. P. Ping. Lignocellulosic-plastic composites based on empty fruit bunch of oil palm. First degree thesis, Universiti Sains Malaysia, 1997.
31. R. G. Raj, B. V. Kokta, G. Groleau, and C. Daneault. The influence of coupling agents on mechanical properties of composites containing cellulosic fillers. Polym. Plast. Tech. Eng. 29:339, 1990.
32. R. G. Raj, B. V. Kokta, and C. Daneault. Intern. J. Polymeric Mater. 14:223, 1990.
33. R. G. Raj, B. V. Kokta, G. Groleau, and C. Daneault. Use of wood fiber as a filler in polyethylene: studies on mechanical properties. Plast. Rubber Process Appl. 11:215, 1989.

Index